Quantitative
Techniques
in
Participatory
Forest Management

Quantitative Techniques in Participatory Forest Management

Edited by
Eugenio Martínez-Falero
Susana Martín-Fernández
Antonio García-Abril

CRC Press
Taylor & Francis Group
Boca Raton London New York

CRC Press is an imprint of the
Taylor & Francis Group, an **informa** business

CRC Press
Taylor & Francis Group
6000 Broken Sound Parkway NW, Suite 300
Boca Raton, FL 33487-2742

Printed on acid-free paper
Version Date: 20131001

International Standard Book Number-13: 978-1-4665-6924-9 (Hardback)
International Standard Book Number-13: 978-1-138-07514-6 (Paperback)

Library of Congress Cataloging-in-Publication Data

Quantitative techniques in participatory forest management / editors, Eugenio
  Mart?nez-Falero, Susana Mart?n-Fern?ndez, Antonio Garc?a-Abril.
    pages cm
  Includes bibliographical references and index.
  ISBN 978-1-4665-6924-9 (hardcover : alk. paper)
    1. Sustainable forestry--Mathematical models. 2. Sustainable forestry--Citizen
participation. 3. Forest management--Citizen participation. I. Mart?nez-Falero,
Eugenio.

SD387.S87Q33 2014
634.9'2--dc23   2013025843

Visit the Taylor & Francis Web site at
http://www.taylorandfrancis.com

and the CRC Press Web site at
http://www.crcpress.com

To our parents

Contents

Preface

The human view of the forest and the decisions taken with regard to it have changed throughout history. Today, our knowledge is far more comprehensive, enabling us to take a more holistic view and to be more aware of the facets of the forest as an entity, and of its resources and functions. Our decisions are based on a greater understanding of its structure, functioning, and the results of the actions applied.

However, forestry activity continues to have an inherent need for prudence and the long-term view, as decisions taken in a particular situation can cause enduring consequences. Forestry management must therefore of necessity be sustainable, otherwise forestry resources will be condemned to degradation and impoverishment.

Since 1713, when the concept of sustainability first appeared in a forestry publication—specifically in *Sylvicultura oeconomica: Anweisung zur wilden Baumzucht* (Hannβ Carl von Carlowitz, 1713)—this concept, together with forest management principles, has undergone far-reaching changes. In these 300 years, there has been an evolution from a mercantilist and productive view of the forest to a multifunctional management that integrates economic, social, and ecological aspects.

However, the problem of sustainability is not yet resolved. There are currently many viable technical solutions available for addressing this issue, but only a few of them have actually been applied. The starting hypothesis of this work is that universal participation in the transparent and real assessment of sustainability—identifying its social, economic, and natural consequences—push on people's general ability to reduce systemic resistance to adopting new sustainable policies.

The authors, after many years of applied research in sustainable forest management (SFM) and decision making, have pooled their knowledge and experience to produce this work. The main objective of this book is to present the different components involved in a public participatory process to assess sustainability in forest management.

The book is aimed at policy makers, environmental and forest professionals, researchers, university faculty, and postgraduate students, especially in fields related to forest management, landscape planning, recreation, and conservation. It will also be useful for the general public and social groups and associations interested in participating in activities related to forest systems.

The main achievements of the book are as follows:

- It adopts a new approach to the management of sustainability that links human and natural systems.
- It reconsiders our interdependence with the diversity of life and assumes a posture that recognizes our role in a unique and complex system. This approach endorses the design of complex, mature, and highly diverse forests that can provide us with a panoply of services and productions.

Another achievement is the identification of quantitative indices in forest management, which provide a vast amount of information on soil, landscape, and ecological functioning. It also highlights the importance of these indices for public information programs on participatory processes. The application of these indices has served to confirm new trends and paradigms in forest management, for example, the extended coincidence between the visual and ecological landscape (personal perception versus ecological functioning).

From the methodological point of view, other results worth noting refer to inventory, representation of personal and collective preferences, and design of forest plans.

The quantitative techniques for inventory explained in this book allow the identification and geographical location of habitats, structures, and single trees. The inventory is made using two types of input: (1) information collected directly in the field and (2) the more widely used method of consulting existing sources of digital information. As occurs throughout the book, the examples refer to template areas, but the methodology can be applied to any type of ecosystem.

A major achievement deals with the representation of the preferences of an individual from direct comparison between pairs of alternatives. The homogeneous representation of individual preferences makes it possible to compare forest management plans, to contrast a person's preferences on forest management with other evaluators, and to understand how individual preferences change as the majority of the participants modify their overall opinion on forest management. It also facilitates the design and evaluation of forest management alternatives and the transfer of information between the evaluators.

Another central characteristic of the proposed methodology is that it encourages collective decision making. Working with multiple evaluators requires aggregating individuals' information to generate a global solution, and this is done by taking into account individual actions and social interactions. This is a complicated process, but we propose a satisfactory solution that brings additional benefits such as increasing outreach (in order to access people who have not traditionally been included in participatory processes) and also facilitates self-organization, thereby enabling interactions among evaluators (interactions produce aggregated assessments more effectively than the mere sum of individual utilities).

The aggregation of preferences requires the incorporation of additional assumptions into the von Neumann–Morgenstern utility theory. The new hypotheses are based on the notion of empathetic preferences applied to both our own ethical concerns and to those of others. From an operative point of view, there are no particular obstacles to incorporating altruistic preferences into a utility function. Furthermore, in much social decision making, it appears that social evolution tends to favor the survival of the most empathetic. In this context, Binmore argues that in the medium run, equilibrium in empathetic preferences will be achieved. Thus, all evaluators belonging to a same society will tend to share a common standard for making interpersonal comparisons of utility.

The book provides an operating procedure to identify the degree of convergence in the utility of multiple evaluators. The evolution in the degree of convergence of individuals' preferences allows decisions to be made in order to promote or conclude the participatory process. Once the participatory process is halted, the aggregated value is obtained for all evaluators (through the application of procedures of aggregation of individual preferences based on voting systems, other procedures existing in the Web 2.0 and interpersonal comparisons of utility). The methodology allows global participation through the Internet and interaction among the evaluators.

Finally, we highlight the methodology for the design of a forest management plan that best suits a specific preference system (whether this is an individual or a collective system). Given the high number of potential solutions, it is not operative to generate all the feasible alternatives and evaluate each one in order to choose the best (e.g., in a forest of 500,000 trees, in which up to ten different actions [pruning, spraying, soil tillage, etc.] are considered in each tree over a period of time of one year—and over a period of 100 years, the number of different actions would be $500,000^{100}$). In consequence, we have adopted procedures based on combinatorial optimization techniques, which dictate that the best solution will be the one that most likely conforms to the preferences of a given observer. Usually, the algorithms used in risk optimization are a mixture of recursive, neural, and adaptive algorithms. Specifically, we have used a modification of the Metropolis algorithm applied in simulated cooling processes.

The book is accompanied by a computer application that—for a given system of preferences—allows two main issues to be addressed: (1) the assessment of the adaptation of any forest type to the given system of preferences (2) and the identification of the best management plan for such a system of preferences.

The book is linearly organized into ten chapters. The first few chapters focus on sustainable indicators and describe their importance, trends, and application. The subsequent chapters aim to explain the techniques related to the identification and integration of individual and group preferences and to find the best management plan according to these preferences. To conclude the book, the last chapter describes a computer application that integrates the techniques explained in the previous chapters and that can be downloaded from the Internet.

Chapter 1 introduces the reader to the concepts of *forest sustainability and public participation* and discusses the use of both sustainability indicators and quantitative techniques to incorporate public participation into forest management. In social systems involving the human dimension, each individual component of the systems may gain awareness of the emerging phenomenon of which they are a partial cause and therefore react by modifying their behavior. This is the case for sustainable development, where the public is expected to modify their behavior toward becoming aware of the impact they produce.

Chapter 2 shows the different sources of information, from classic sampling in the field to the use of information technologies such as GIS and remote sensing. In particular, the chapter pays special attention to the use of LIDAR data in forest management.

Chapter 3 focuses on two aspects: (1) the different approaches used to assess criteria and indicators for SFM and (2) a case study of computing indicators at the local level based on LIDAR data and yield tables for *Pinus sylvestris*.

Chapter 4 describes soil indices, how to measure them, their relationship with other environmental variables, their role in the study of the impact of land uses, and the conservation of ecosystems.

Chapter 5 describes the indicators of the ecophysiological state of vegetation. These indicators are measures of various important plant functions such as photosynthesis, water balance, and nutrient status, which can reveal what is occurring in a particular ecosystem. Specifically, the chapter emphasizes vegetation indices and models of system functionality and includes a case study on the indicators for the ecophysiological competence of woody species for riparian ecosystem restoration.

Chapter 6 presents a number of landscape indicators that can be currently used in SFM. The chapter starts with a review of the state of the art on landscape indicators and their integration within SFM. The next section focuses on the visual and ecological landscape and discusses examples of man-made landscapes that successfully integrate high biodiversity, production, and landscape beauty. Heterogeneity and diversity are the landscape elements that are required to conserve biodiversity at all scales. The importance of mature forest stages are shown in this context.

The chapter describes the new trends and objectives in forest management and their ecological and visual consequences. In the last 20 years, new management approaches have emerged, such as ecosystem management in the United States, which is an adaptive management in time and space across all scales. Another objective for improving the conservation of biodiversity is to achieve more complex forests and landscapes, including mature stages of the forest succession. Close-to-nature forestry is a European approach that emerged in response to economic objectives but is based on the use of natural processes that integrate economic benefits and complex structures. Its practical experiences, after more than 100 years, are a starting point for the design of complex, mature, and highly diverse forests that can provide us with multiple services and productions.

The spatiotemporal changes in a managed forest are explored against the background of the trends in forestry in the twenty-first century, using the three principal forest-structure models common in traditional silviculture.

Diverse forests and landscapes are also appreciated visually by people; in fact the visual and ecological landscape can coincide. This analysis serves to identify principles and common visual and ecological design criteria where the landscape indicators will be assigned, thereby aiding in ordering the set of indicators as a whole.

The chapter ends with a technical description of the visual and ecological landscape indicators, for which a broad common ground of visual and ecological landscape indicators is identified.

Chapter 7 presents the procedures for preference identification. It describes the procedures for evaluating alternatives based on pair-wise comparison and aggregation of criteria and proposes an alternative valuation method that transforms opinions into a sustainability assessment. It also describes the methods used to characterize the type of rationality and coherence in the opinions of each individual, in addition to the depth of the individual's knowledge of the system to be evaluated. Finally, the previous methodology is applied to the assessment of forest sustainability in a case study.

Chapter 8 describes the methodologies most commonly applied to optimize the sustainable use of forest resources, including an explanatory application of each one to certain stages of forest management. It starts with an introduction to linear programming applied to forest management and then provides a detailed description of heuristic methods such as simulated annealing, genetic algorithms, and tabu search, including forest examples. Artificial neural networks applied to optimization problems are also included. The chapter ends with a case of application that incorporates personal preferences to identify the best forest plan.

Chapter 9 explains the aggregation methods of individual preferences, both with regard to the state of the art and as useful examples. It presents a methodology to describe how sharing opinions with other evaluators allows individual opinions—that is, personal preferences for sustainability assessment—to be modified. To do so, a successful web-based application is described; the model is then adopted to simulate the interactions between evaluators. The last section presents the application of this model to the collective assessment of forest sustainability.

Finally, in Chapter 10, the aforementioned methodologies have been integrated into a computer application. Readers who download this application will find that there are two types of inputs required from the users: one refers to the personal characteristics to be included in a social network; the other consists of individual answers to a set of comparisons of sustainability. Users accessing the application will be offered a map representing their preferred forest management plan in the study zone. They will also be given a map with the results of their corresponding community of evaluators, along with the numerical and qualitative data for both. The system stores a record of the visit, the visitor's profile, and his or her responses in order to progress toward the joint forest management plan.

Acknowledgment

We would like to express our gratitude to our English editor, Prudence Brooke-Turner, for her invaluable help and support in the publication of this book.

Editors

Eugenio Martínez-Falero, PhD, is a full professor at the UPM (Technical University of Madrid), chairman of the Accreditation Board of the Universities of the Madrid Region, and managing director of "Madrid: Fundación para el Conocimiento." In his academic life, Dr. Martínez-Falero is a professor of applied statistics and operational research. He has been director of the Department of Forestry Management at the UPM and a fellow of the College of Environmental Science and Forestry at the State University of New York (SUNY). He has been consultant for the FAO and an external examiner for the University of Wales. His research focuses on developing quantitative techniques for analysis, simulation, and management of natural systems and on methodologies for analyzing systems of preferences in decision making.

He has been director of the Department of Industry, Energy and Mining of the Madrid Regional Government and has also served on the governing boards of various institutions such as the Madrid Institute for Development (IMADE), the Madrid Center for Technology, Gestión y Desarrollo del Medio Ambientede Madrid (GEDESMA), and the Madrid Institute of Food and Agricultural Research. His other positions include being chairman of the Steering Committee of Scientific-Technological Parks in Madrid and a member of the Interdepartmental Commission on Science and Technology.

Susana Martín-Fernández, PhD, has led a multifaceted career in which she has worked as a consultant for leading ICT companies, conducted research into a more effective application of heuristic and combinatorial methods in forest management, and taught statistics and operational research at the Escuela Técnica Superior de Ingenieros de Montes at the Technical University of Madrid. Her research over the past 13 years has centered on improving the decision-making process in forest management and territory planning, specifically in forestry activities and optimal land use assignments. The results of her research have been relayed to forest companies and institutions.

Dr. Martín-Fernández currently serves as associate professor in the Department of Economy and Forest Management and as deputy director of academic planning and faculty at the Escuela Técnica Superior de Ingenieros de Montes. She has previously served as director of the Research Laboratory for Remote Sensing Applied to Natural Resources and Landscape Management.

Antonio D. García-Abril, PhD, currently serves as head of the research group Methods and Techniques for Sustainable Management (Technical University of Madrid). He has been associate professor of landscape planning, landscape ecology, and project engineering at the Escuela Técnica Superior de Ingenieros de Montes in

Madrid (Forestry School, UPM) since 1991. He has worked in collaboration with the European Commission and also has professional experience as a consultant.

Antonio García-Abril has been responsible for many research projects and has worked on a number of projects for companies and government bodies. His research interests include areas such as landscape planning, biodiversity conservation, remote sensing applied to forest structure and forest management, environmental impact studies, and close-to-nature management. He has 68 publications, 27 of which are in international publications.

Contributors

Fernando Arredondo-Ruiz
Department of Forest Engineering
Forestry School
Technical University of Madrid
Madrid, Spain

Esperanza Ayuga-Téllez
Economics and Forest Management
    Department
Forestry School
Technical University of Madrid
Madrid, Spain

Antonio García-Abril
Silvanet Research Group
Forestry School
Technical University of Madrid
Madrid, Spain

Carlos García-Angulo
Silvanet Research Group
Forestry School
Technical University of Madrid
Madrid, Spain

Luis García-Montero
Department of Forest Engineering
Forestry School
Technical University of Madrid
Madrid, Spain

Fernando García-Robredo
Economics and Forest Management
    Department
Forestry School
Technical University of Madrid
Madrid, Spain

Concepción González-García
Silvanet Research Group
Forestry School
Technical University of Madrid
Madrid, Spain

M. Angeles Grande
Silvanet Research Group
Forestry School
Technical University of Madrid
Madrid, Spain

Ana Hernando
Silvanet Research Group
Forestry School
Technical University of Madrid
Madrid, Spain

José Antonio Manzanera
Forestlab Laboratory
Forestry School
Technical University of Madrid
Madrid, Spain

Susana Martín-Fernández
Forestlab Laboratory
Forestry School
Technical University of Madrid
Madrid, Spain

Eugenio Martínez-Falero
Forestlab Laboratory
Forestry School
Technical University of Madrid
Madrid, Spain

Particia Martínez-Obispo
Forestry School
Technical University of Madrid
Madrid, Spain

Francisco Mauro
Forestlab Laboratory
Forestry School
Technical University of Madrid
Madrid, Spain

Cristina Menta
Department of Evolutionary and
    Functional Biology
Section Natural History Museum
University of Parma
Parma, Italy

M. Victoria Núñez
Forestlab Laboratory
Forestry School
Technical University of Madrid
Madrid, Spain

Antonio Orol
Forestry School
Technical University of Madrid
Madrid, Spain

Sigfredo Ortuño
Silvanet Research Group
Forestry School
Technical University of Madrid
Madrid, Spain

Cristina Pascual
Forestlab Laboratory
Forestry School
Technical University of Madrid
Madrid, Spain

Roberto Rodríguez-Solano
Silvanet Research Group
EUIT Forestal
Technical University of Madrid
Madrid, Spain

Rosario Tejera
Silvanet Research Group
Forestry School
Technical University of Madrid
Madrid, Spain

Miguel Valentín-Gamazo
Forestlab Laboratory
Forestry School
Technical University of Madrid
Madrid, Spain

Inmaculada Valverde-Asenjo
Facultad de Farmacia
Department of Soil Science
Universidad Complutense de Madrid
Madrid, Spain

M. Dolores Velarde
Silvanet Research Group
Forestry School
Technical University of Madrid
Madrid, Spain

1 Forest Sustainability and Public Participation

Fernando García-Robredo, Eugenio Martínez-Falero, Sigfredo Ortuño, and Rosario Tejera

CONTENTS

1.1   INTRODUCTION

So far, most of the efforts for a joint consideration of environmental, economic, and social factors have failed (Moore, 2011). However, this trend might change if the static approach to the problem is abandoned, and a decision framework nearest to the complex reality of the integration of man and nature is adopted. From this point of view, the combination of resistance and adaptation mechanisms (such as those triggered in complex systems) will facilitate the evolution toward a more sustainable society (Smith et al., 2011).

For the successful implementation of this new approach, the systemic change resistance must be overcome by enhancing people's general ability to detect manipulative deception (Harich, 2010). Thus, the universal extension of an education that considers *sustainability indices as the key for development* (instead of simply maximizing net profit), combined with *universal and transparent public participation*, can reduce systemic resistance to change. These two concepts are the core of *participatory sustainable forest management* (SFM).

This chapter focuses on explaining the preceding statements: first, we justify the fact that society's acceptance of the complex reality of humans/nature enables measures to be adopted that are conducive to sustainable development. We then describe the current reality of public participation. Finally, we discuss the use of both sustainability indicators and quantitative techniques as a means of incorporating public participation in forest management.

1.1.1 SUSTAINABLE DEVELOPMENT

From an ecological perspective, sustainability is the capacity of biological systems to remain diverse and productive over time. This is an essential precondition for the long-term maintenance of human well-being (global sustainability).

Humanity has been working for decades to solve the problem of sustainability, and we already have a number of viable technical solutions. However, very few of these solutions have been applied. Why have we been unable to solve the sustainability problem? To simplify, two main reasons can be highlighted: the ambiguity inherent in the concept of sustainability and the failure to consider the social aspect of the problem.

To analyze the consequences of ambiguity (see, e.g., Moore, 2011), let us consider the concept of sustainable development. The most widely accepted definition refers to "meeting the needs of the present without compromising the ability of future generations to meet their own needs" (Brundtland Commission*). However, beyond the most basic biological needs, what we need cannot be separated from what we want. As a result, different individuals may have contradictory interpretations of sustainable development, and this may even evolve over time for the same individual. This ambiguity must be removed before a consensus can be reached as to which practices should be adopted in order to improve sustainable development.

Nowhere is this contradiction clearer than when addressing the lack of sociological contributions to the problem of sustainability, and it is for this reason that economic analysis has been used to examine social criteria. Generally speaking, two competing economic theories are used to analyze possible sustainable solutions (Neumayer, 2003): weak versus strong sustainability analysis.

In weak sustainability analysis, all forms of capital (*economic*, the traditional understanding of capital as goods or assets; *social*, covering human resources and their networks of relationships; and *natural*, such as forests, clean water, minerals, and biodiversity) are interchangeable, and an economy is sustainable if the total stock of capital remains constant. This means that an economy is sustainable even if it uses up all of its natural capital, as long as a portion of that income is invested to ensure equally high income for future generations. However, for the requirements of strong sustainability to be met, the stock of natural capital must remain constant. This formulation ensures the environmental protection of economic development. Obviously, both approaches cannot be adopted simultaneously: the formalism of economics leaves no room for ambiguity here.

---

* Formally known as the World Commission on Environment and Development (WCED). Its president (Gro Harlem Brundtland) was appointed in 1983 by the secretary-general of the United Nations, and the commission was officially dissolved in December 1987. The underlying aim was to further the understanding of the terms "environment" and "development" in relation to each other. In the paper "Our Common Future," the Brundtland Commission coined the term "sustainable development" to mean the kind of development that meets the needs of the present without compromising the ability of future generations to meet their own needs.

Two approaches that link sustainability with sociology have emerged in connection with the earlier economic theories: the environmental reform* and the unsustainable economic system.[†] Although they both differ widely, both are rooted in materialism (Clark and York, 2005) and embrace a markedly political economic analysis (Jorgenson and Clark, 2009). Either for this reason or because they are based on the separation between human and natural systems, both approaches miss when addressing important foci for the future (Friedman, 2008). It appears to be necessary (see Catton and Dunlap, 1978) to reconsider our relationship with nature and to abandon our anthropocentric views of nature by taking a position that recognizes our role in a complex system. Two consequences are implicit in this change: the rise of an ecological ethos[‡] that recognizes our interdependence with the diversity of life (Mackey, 2004; Miller and Westra, 2002) in order to drive individual behavior and the adoption of an interdisciplinary approach based on the simulation of complex systems (Holling, 1973; Gunderson and Holling, 2002) in order to conduct our social decision-making.

1.1.2 PUBLIC PARTICIPATION

In knowledge management, the participatory process facilitates the collective intelligence and inclusiveness of the whole community or society in decision-making, on a par with other agents such as the administration and the private sectors.

In environmental resource management, the administration has traditionally been responsible for supervising natural resource management and implementing environmental protection legislation (Selin and Chavez, 1995). In democracy, it has also played a central role in providing professional judgment on behalf of the people through skilled technicians.

The role of the private sector is to manage environmental resources to generate added value. In spite of their historical behavior and whether for ethical reasons or simply to gain a strategic advantage, many private-sector organizations are currently recognizing the considerable benefits—as well as the viability—of sustainable practice.[§]

At the same time, civil society is pressing to be included in environmental decision-making through public participation. We here use the term public participation

---

* The environmental reform argues that change can occur within the current structures of society and that reforms in industrialization and advances in technology will lead to the conditions necessary for ecological sustainability (Mol and Spaargaren, 2000; Mol et al., 2009).

[†] The unsustainable economic system theory considers that economic criteria still remain at the heart of decisions on the design, performance, and evaluation of production and consumption. This primacy tends to overshadow most ecological concerns. In addition, development based only on economic factors is unsustainable (Dietz et al., 2007; Foster, 2005; Jorgenson and Burns, 2007; Schnaiberg and Gould, 2000; York et al., 2009).

[‡] This ecological ethos derives from a realization that *life is fundamentally one*, and its main implication is a sense of universal responsibility that can only be cultivated when we live "with reverence for the mystery of being, gratitude for the gift of life and humility regarding the human place in nature" (Brenes, 2002).

[§] The joint report by the *Boston Consulting Group* and the *MIT Sloan Management Review*: The business of sustainability (2009) shows both the challenges and opportunity that sustainability offers to companies and the number of companies that is recognizing these opportunities.

to refer interchangeably to both the concepts of stakeholder engagement and popular participation. In any case, public participation advances the alternative concept of "more heads are better than one" and argues that public participation can sustain productive and durable change. This has been a global movement since the Rio Declaration of 1992 enshrined public participation in its 27 principles. Particularly, Principle 10 states that "environmental issues are best handled with participation of all concerned citizens, at the relevant level."

1.1.3 Sustainability Criteria and Indicators

The history of the sustainability problem reveals that most of environmental protection legislation appears as a reaction to a catastrophe. Thus, for example, the discovery of the Antarctic ozone hole led to the Montreal Process and the subsequent development of the first set of sustainability indicators. In 1991, the "Montreal Process Working Group" agreed on a framework of criteria and indicators that provides the member countries with a common definition of what characterizes the sustainable management of temperate and boreal forests. Other notable initiatives in parallel to the Montreal Process include the Ministerial Conference on the Protection of Forests in Europe (MCPFE) and the International Tropical Timber Organization. In general, the emergence of a consensus based on regional and international criteria and indicators with regard to seven common thematic areas can be seen. This consensus was acknowledged by the international forest community at the fourth session of the United Nations Forum on Forests and at the 16th session of the Committee on Forestry.

An indicator is a quantitative or qualitative variable that can be measured or described and, when observed over a period of time, can highlight trends. A criterion is a category (characterized by a set of related indicators) whereby SFM can be assessed. Sustainability criteria and indicators are tools used to conceptualize, evaluate, and implement SFM (Prabhu et al., 1999). They are widely used, and many countries produce national reports that assess their progress toward SFM.

The final stage in the development of indicators is to define acceptable standards or measures of indicators that identify a forest as sustainable. It is generally accepted that the first step in this direction was the certification of forest sustainability by independent bodies. This ensures that the comparison of standards for the existing certification systems (Clark and Kozar, 2011) provides relevant information.

1.1.4 Techniques for Participatory Forest Management

As stated by Lawrence and Stewart (2011), the involvement of stakeholders in forest decision-making is not so much a technical challenge as a cultural one. However, not all the technical issues have been resolved. Current developments in public participation show that

- The self-selection of representatives on decision-making panels displays a tendency for committees to include those with real contacts and power (Parkins, 2006, 2010; Reed and Varghese, 2007).
- There is clear evidence that different social and environmental contexts affect stakeholders' preferences for participation (Tuler and Webler, 2010).

- The envisioning of various forest future scenarios reveals that participants differ in their ability to "suspend disbelief" and are partly affected by their past experiences and expertise (Frittaion et al., 2010).
- The prevailing approach to managing nature is described as failing to set store by reflection, learning, and complexity (Allan and Curtis, 2005), and natural resource management organizations point to an established mind-set, which seeks to achieve optimization (e.g., of timber production) rather than adaptation (Linkov et al., 2006).

1.2 UNDERSTANDING THE SUSTAINABILITY PROBLEM THROUGH ANALYZING COMPLEX SYSTEMS

1.2.1 COMPLEX SYSTEMS

Basically, a complex system is defined as "a set of interacting elements" (Bertalanffy, 1968) whenever the interaction makes the whole to be more than the mere sum of its parts. This is called emergent behavior of the system and comes from self-organization of its components.

Complex systems require that

- The system is defined as a set of components that interact.
- Each component has its own rules and responsibilities.
- Some components may have more influence than others, but none completely controls the behavior of the system.
- All components contribute to a greater or lesser extent to the final result.

Additionally, complex systems are used to being adaptive: the system's behavior evolves over time, leading to a certain capacity to respond to changes in the environment. This means that systems react, learn from the environment, and modify their behavior to achieve some goal.

1.2.2 SIMULATION TECHNIQUES FOR COMPLEX SYSTEMS

Until recently, not to solve mathematically, a formal model was a major disadvantage, since there were no other tools to deduce the logical implications resulting from the model. This has changed with the development of the computer. Today, it is possible to explore and analyze formal models that cannot be solved mathematically. Thus, using the new technologies, we can implement and rigorously analyze the behavior of formal models of complex systems, something not feasible until recently.

Based in North and Macal (2007), the main features of the most widely used computer simulation techniques are described in the following:

- *Discrete event modeling.* It models, with great detail, the inner workings of a dynamic process, through programming the occurrence of discrete event in specific instants of time. In turn, each event causes a chain of future events that must also be programmed as the simulation evolves in time.

- *Participatory simulation.* It is a version of crowdsourcing to simulate the interactions between system components and understand their behavior. Each component is replaced by a person or a group of people who make the decisions of the system.
- *Risk analysis.* It is a modeling technique that evaluates the exposure of companies to events that affect their value. It requires the following:
  - Identify negative events.
  - Transform these events into measurable impacts.
- *Statistical methods.* It is a modeling approach that treats the systems or their components as a black box and seeks to determine the system outputs from the inputs to it, without considering the internal structure of the system or causal processes.
- *Optimization methods.* They are methods aimed at finding the best solution for a well-defined problem in a very large set of possible solutions. Linear and nonlinear programming (with or without constraints) can be applied when the problem can be formulated mathematically by continuous functions. Heuristic methods like combinatorial optimization techniques and genetic algorithms are used for the optimization of discrete events.
- *Artificial intelligence methods.* It is the branch of computer science devoted to the development of nonliving rational agents. It includes logic programming, artificial neural networks, and swarm intelligence.

However, the most used techniques are *agent-based simulation* (ABS) and *system dynamics* (SD).

1.2.2.1 ABS

Agent-based methods facilitate the study and modeling of complex systems from the attributes and behavior of their component units. The basic components of the real system are explicitly and individually represented by agents, and the interactions that occur between the basic components of the real system are represented by the interactions that occur between agents (Edmonds et al., 2001). A proper application of ABS requires the following:

- Systems with heterogeneous individual components, when the hypothesis of "representative agent" cannot be successfully applied. As Ostrom et al. (1994) have established, this happens in systems with strong externalities (e.g., the exploitation of the environment, management of common resources).
- Adaptive systems, that is, systems whose individual components are capable of learning, modifying, and redirecting their behavior to achieve specific goals (adaptation at the individual level) where it seems clear that it is convenient to represent explicitly and individually each system component.

- Systems in which geographic space has a major influence (to more easily represent the physical space in which agents move) and/or systems where the social networks are remarkable in order to represent the interaction between agents.
- Systems in which the analysis of the relationship between the attributes and behaviors of individuals (the "microscale") is more important than the global properties of the group (the "macroscale") (Gilbert and Troitzsch, 1999; Squazzoni, 2008).

ABS fits very well when emergence arises from decentralized interactions of simpler individual components (Holland, 1998). What characterizes these emergent phenomena is that their presence or appearance is not apparent from a description of the system consisting of the specification of the behavior of its individual components and the rules of interaction between them (Gilbert, 2002; Gilbert and Terna, 2000; Squazzoni, 2008).

From an operative point of view, ABS incorporates

- Adaptive capacity to learn from the experience
- Perceptual abilities to understand the environment
- Internal models to project consequences of decisions
- Decision rules for selecting measures (both basic levels, as rules that modify the rules of basic level)

For our purpose, the more relevant applications belong to the fields of management of natural resources and ecology (Bousquet and Le Page, 2004; López and Hernandez, 2008), sociology (Conte et al., 1997; Gilbert, 2008; Gilbert and Troitzsch, 1999), and biology (Paton et al., 2004; Walker et al., 2004a,b).

1.2.2.2　SD

SD were initially developed to solve industrial dynamic problems (Forrester, 1961), but their applications have grown to simulate all types of dynamic problems arising in systems characterized by interdependence, mutual interaction, information feedback, and circular causality, so it has become generalized to *SD* (Richardson 1991–1999).

It is a technique for the modeling of complex systems that simulates, on a general level of detail, the inner workings of a process,* through equations that reflect the state of its variables at any moment.

SD conceptualizes the structure of a complex system with diagrams of loops of information feedback and circular causality (*feedback loops*). A feedback loop is a diagram that enables the visualization of causal relationships among variables, showing how the interrelated variables affect each other.† These systems use both stock and flow variables. A *stock variable* is measured at one specific time and

---

* Exogenous disturbances are seen at most as triggers of system behavior.
† It is not enough however: the explanatory power and insightfulness of feedback understandings also rest on the notions of active structure and loop dominance.

represents a quantity existing at that point in time. A *flow variable* is measured over an interval of time.

The emphasis is on the causal relationships that link critical system variables and on the identification of complex causal links between them. The abstraction process that identifies relationships is run by an expert in the system and is prior to the creation of the formal model. The relationships are expressed with a system of coupled, nonlinear, first-order differential (or integral) equations:

$$\frac{d}{dt}\mathbf{x}(t) = f(\mathbf{x}, \mathbf{p}) \tag{1.1}$$

where
  $\mathbf{x}$ is a vector of variables (stocks or flow)
  $\mathbf{p}$ is a set of parameters
  f is a nonlinear vector-valued function

Each state variable is computed from its previous value and its net rate of change $\mathbf{x}'(t)$: $\mathbf{x}(t) = \mathbf{x}(t - dt) + dt \cdot \mathbf{x}'(t - dt)$. Although original work stressed a continuous approach, modern applications contain a mix of discrete difference equations and continuous differential or integral equations.

Simulation of such systems is easily accomplished by partitioning simulated time into discrete intervals and then by computing the value for the variables on each time span (or by applying more sophisticated integration schemes). However, the main work focuses on understanding the dynamics of complex systems—including feedback thinking, stocks and flows, the concept of feedback loop dominance, and an endogenous point of view. These tasks are as important for the purpose of policy analysis and design as the simulation methods used.

As *SD* applies on complex systems, there are solutions for social, managerial, economic, and ecological systems. However, from our perspective, we can highlight the use of model-based insights for organizational learning, specifically to build models with relatively large groups of experts and stakeholders, known as group model building (described in Richardson and Andersen (2010) and Vennix (1996)).

1.2.2.3 Joint Use of ABS and SD

The use of dynamic systems will be more convenient when the prior knowledge of the system and of the objectives to be achieved allows us to carry out the abstraction of the process of emergence in a solid and well-founded way. In general, the SD, by providing a higher abstraction level of the developed agent-based models, will result in lower-complexity models, which will facilitate its implementation, analysis, and interpretation.

However, if the abstraction of the process of emergence cannot be carried out in a scientifically valid priority given our objectives, then it is more appropriate to model the process of emergence explicitly (using the ABS) to study it in detail. The model thus constructed will be scientifically rigorous but significantly more complex, with the drawbacks that this entails.

1.2.3 CONCLUSIONS ON UNDERSTANDING SUSTAINABILITY THROUGH SIMULATION OF COMPLEX SYSTEMS

There are emergent phenomena in many different disciplines (see, e.g., Holland, 1998; Johnson, 2001; Reynolds, 1987), but it is in the social sciences where the idea of emergence takes an additional dimension of complexity and importance. In social systems involving human agents, it is possible that each individual component of the system takes awareness of the emerging phenomenon of which he or she is a partial cause and, therefore, reacts by modifying his or her behavior. This phenomenon, known as second-order emergence (Gilbert, 2002; Squazzoni, 2008), underlies the complexity of many social systems. This is the case for sustainable development, where the public is expected to modify their behavior to be aware of the environmental impact they produce.

In the complex systems approach, the social–ecological systems are interlinked in never-ending adaptive cycles of growth or exploitation, conservation or accumulation, collapse or release, and renewal or reorganization. As Holling (1973) has confirmed, the longer a system is "locked in" to its growth phase, "the greater its vulnerability and the bigger and more dramatic its collapse will be." We can conclude that "for public policy to be grounded in the hard-won results of climate (natural) science, we must now turn our attention to the dynamics of social and political change" (Sterman, 2000). Ultimately (Smith et al., 2011), the mechanisms for building resilience and adaptation and reducing vulnerability rely upon the capacity of understanding "true" sustainability (Freese, 1997). These mechanisms enable the global social–ecological crisis to be explained as part of a long-term process of change (resilience) and adaptation. It is evident that, in terms of adaptation, public participation in designing forest management can push to improve sustainable development. The same can be applied in terms of change (and otherwise with respect to the resistance to change): public participation can be an accelerator of change.

It is worth trying to promote sustainable development through public participation because, so far, the evolution of the society is not toward this type of development: social forces favoring the resistance have managed to counter those favoring change. Haric (2010) has shown this fact by developing a quantitative dynamic simulation from the critical actions taken by the agents involved in the failure to adopt sustainability. His analysis has revealed that unless deception effectiveness is absurdly low, change resistance is high enough to dramatically slow down the rate of adoption of proper practices. Deception appears to be high enough to thwart, weaken, or delay changes that run counter to the goal of maximizing net profit. It creates mistaken or false beliefs/values that become premises for further beliefs and/or actions. The more impact a belief causes and the more people who believe it, the greater the total impact. Society is aware of the proper practices required to live sustainably and the need to do so. But society has a strong aversion to adopting these practices.

Indeed, the main objective of the reforms carried out in relation to the environment over the last decades has been "to buy time" regarding broad-range decision-making facing the challenge of serious problems such as climate change or ecosystem degradation. These reforms do not lead to a sustainable model for our society. It is necessary, therefore, to address these problems from a new ecological theory, whose foundations may be very close to the following (Smith et al., 2011):

- A new evolution of the capitalist system into a production model based on multiple decision criteria (Mol and Spaargaren, 2000)
- The establishment of a new social paradigm that involves greater interdependence between man and the natural environment, using an egocentric view, with greater knowledge of the interactions between human activities and a world of finite nature (Catton and Dunlap, 1978)
- A sense of global responsibility, based on greater humility before nature and gratitude for our existence on Earth (Brenes, 2002)
- The acceptance that crises are an opportunity for the difficult process of change, through better learning and incorporating social changes (Gunderson and Holling, 2002), "panarchy theory"
- Finally, the understanding that small-scale experiments can have large-scale consequences for achieving a more just and ecologically sustainable future

Participatory processes favor the transmission of the previous foundations to the whole society, and in this context, it seems evident that universal participation on transparent and real assessments of sustainability—with identification of its social, economic, and natural consequences—pushes on the general ability of people to detect manipulative deception.

1.3 PUBLIC PARTICIPATION

Public participation is a process dealing with the incorporation of society's views into public decision-making concerning the common good. It includes participation in decisions on public projects or initiatives as well as in decisions by private corporations requiring some kind of permit, concession, or authorization by a public administration.

Environmental protection is one of the fields in which public participation is crucial. The historical evolution of public participation related to the environment is discussed in the following text. There are different levels of public participation and stakeholders' involvement, which are also described under this heading. Finally, the core values of public participation are presented.

1.3.1 HISTORICAL EVOLUTION OF PUBLIC PARTICIPATION*

From the beginning of the conservation movement in the late 1800s, when the first wildland areas were set aside for nature protection, the main goal of these actions was to preserve the natural environment for the benefit and enjoyment of the people, the present and the future generations. This was the objective of the law that established Yellowstone National Park on March 1, 1872.

---

* The authors have identified the four stages of public participation described in this section.

Therefore, public interest led the protected areas declaration policy. However, even though the people were the main beneficiaries of the conservation measures, they were not asked about the decisions involved. This stage could be described as *"Phase 1: For the people without the people,"* and it is characterized by the idea that the government knows what people need.

Later on, the government started informing the people about the decisions made on environmental conservation and natural resource management. This public information stage can be called *"Phase 2: People have the right to know,"* and it is an extension of Phase 1. No public involvement mechanism is implemented at this stage.

The next logical step was *"Phase 3: Maybe the public has something to say about it,"* a stage in which the people with an interest in the issue at hand were given the opportunity to express their opinion and influence the final decision. Once the alternatives were analyzed and studied, they were presented in front of the public and their opinion was considered in the final decision. Although this stage is a step forward with regard to the former one, the limited number of considered alternatives constraints the choice possibility of stakeholders.

To overcome this problem, a possible solution was that the alternatives considered were designed taking into account input from the stakeholders. At this *"Phase 4: People know better,"* the role of society along the whole decision-making process is finally recognized. It means that people with an interest in a particular decision must have the right to participate from the beginning of the process.

Of course, the key question on this matter is how to put in place the right mechanisms and procedures to allow this participatory process to be carried out efficiently. This book is an attempt to provide the methodology to carry this task out.

In correspondence with this chronological evolution, different levels of public participation can be identified. Following the International Association for Public Participation (IAP2, 2000), these levels can be classified into the following categories:

- Information: People must be informed about the issue at hand, problems, options, and solutions.
- Consultation: Feedback from the stakeholders must be obtained.
- Involvement: The opinion and concerns of the people must be considered in the final decision.
- Collaboration: The people must play a role along the whole process, including the development of alternatives and the election of the best one.
- Empowerment: The public is given the power to make the final decision.

Not all these levels must always be present in a stakeholder engagement process. Most times, the process will stop at the involvement or collaboration level, and very few times, full empowerment will be given to the public.

When addressing the historical evolution of public participation, it is important to analyze the legal framework in which this evolution has taken place. The importance of public participation in environmental decision-making has been recognized in the 1992 Rio Earth Summit (Principle 10 of the Rio Declaration on Environment and Development, Agenda 21, and Principle 2 of the Forest Principles) and has been regulated for the first time in the Aarhus Convention* that was ratified by the European Union (EU) in 2005 and established the right to access environmental information, participate in environmental decision-making, and achieve justice on environmental matters. Before the Aarhus Convention was ratified, it was implemented in the EU through the Directive 2003/35/EC of the European Parliament and of the Council, known as the Public Participation Directive, which sets minimum standards for public participation in decisions related to specific projects, programs, plans, and policies.

Besides the Public Participation Directive, there are other European directives where public participation is emphasized, such as the Water Framework Directive (WFD), which requires consultation and stakeholders' involvement, or the Directive on Strategic Environmental Assessment.

1.3.1.1 EIA Participation

Environmental assessment embraces two types of operating tools: environmental impact assessment (EIA) of projects and strategic environmental assessment (SEA) of plans and programs. For a description of EIA and SEA, see Sadler (1996).

Environmental protection requires not only "a posteriori" actions to restore the damage produced but precautionary measures to prevent damage from happening. This is called "precautionary principle." With that purpose in mind, environmental assessment has the objective of incorporating environmental constraints into the decision-making process. According to the International Association for Impact Assessment, one of the principles of environmental assessment is its participatory character (André et al., 2006).

Public participation must be included in all decision-making levels, including policy definition, plan and program elaboration, and project design and implementation. The role of public participation in the last implementation stage is limited to controlling that the implemented actions correspond to the decisions made and the techniques used are consistent with them and do not generate any conflict.

The EU Directive on Strategic Environmental Assessment mentioned earlier addresses this public participation subject.

1.3.1.2 Participation in Forest and Natural Resource Management

Foresters are among the first professionals that developed the foundation principles for the concept of sustainability. The need for sustainable forests was first expressed in Germany by H.K. von Carlowitz in 1713 (Grober, 2007). The origin of forest management in Europe seems to lie in the need to address localized wood shortage since the end of the seventeenth century. However, the goal of attaining a sustained timber production soon led to the consideration of other forest management objectives, particularly those related to the protection of forests.

The appearance of new demands from society in the mid-twentieth century, and the emergence of the concept of sustainable development in the decade of the 1980s,

---

* "UNECE Convention on Access to Information, Public Participation in Decision-Making and Access to Justice in Environmental Matters" (UNECE, 1998).

brought about the evolution of the concept of forest management toward a model that has been called "SFM." This concept incorporates the general idea of sustainability based on meeting the present needs of society without compromising the rights of future generations to use and exploit natural resources.

The need to incorporate the demands of society into forest management involves the development of mechanisms based on public participation, which allow the identification of such claims and the implementation of actions to meet them. Participatory approaches have been introduced in the 1970s. Before this decade, management decisions related to forestry and natural resources in general were top-down oriented.

At present, in most cases, public participation is restricted to the ability to file public comments on the plans made by the local, regional, or national governments before they are considered for approval. According to the levels of participation seen earlier, this situation corresponds to the level of involvement in which the people express their opinions, and these opinions are considered in the final decision. Increased public influence would require the incorporation of the public's preferences along the whole decision-making process, including the development of alternatives.

In the field of public participation in forestry and natural resource management, Buchy and Hoverman (2000) conducted a thorough review of the methodologies used and proposed a set of principles of good practice. Germain et al. (2001) describe the types of public participation and the experience of the U.S. Forest Service over the last 50 years.

Public participation in natural resource management is a growing concern. Besides forests, other basic resources such as soil or water need to be managed in a sustainable and participatory way. Maestu et al. (2003) carry on an analysis of past and present public participation in river basin management in Spain to find out that nowadays consultation and participation are explicitly considered, but public participation is seen as an instrument and not yet as a driving force for change. Moreover, the inception of the WFD requires the adoption of a sound participatory approach not only in water resource planning but in water resource management as well. The implementation of the new policies of the WFD will foster new cooperative agreements among users, environmentalists, consumers' associations, and other stakeholders.

1.3.2 STAKEHOLDER ENGAGEMENT, INVOLVING LOCAL POPULATION AND POPULAR PARTICIPATION

The term stakeholder refers to any person, or group of people, who has an interest in a particular project or could be affected by its outcomes. They can be classified into interest groups such as governments (local, regional, national), institutions (research, academic, religious, etc.), civil society organizations (NGOs, labor unions, other associations), or companies (industrial, commercial, etc.).

The expression "stakeholder engagement" embraces a range of policies, principles, and techniques, which ensure that the stakeholders have the opportunity to participate in the decision-making process regarding a particular project or plan.

Sometimes, stakeholder engagement has been used interchangeably with public participation.

Stakeholder identification is a critical component of the public participation process, and it must be carried out at the very beginning of the process. The key question at this point is: Who should take part in a public participation process? In principle, we must assume that if somebody shows an interest in a particular project, there must be a legitimate reason for it. On the other hand, some social groups are more active than others and they may have a better organization and a stronger involvement in the process. As an example, environmentalists play a very active role in the governing bodies of protected areas, whereas local groups that have a stronger relationship with the land itself, like farmers or livestock breeders, are not so much involved in the process even though they are going to undergo the consequences of the decisions made.

Of course, local actions can have a global dimension and the right to express an opinion on the convenience of those actions is not restricted to the local population. But still the degree of involvement of the different stakeholders in the public participation process should be proportionate to their interest and not to their size or political influence. Once the stakeholders have been identified, their role in the different stages of the process must be defined.

A second element to be defined is the stakeholders' participation level. Depending on the objectives of the process, we could be interested in just informing them, in seeking their opinion to get some input from them, or in working with them along the whole decision-making process (Reed, 2008). For some projects or plans affecting the common good, the participation level is legally established, but some private initiatives that are not legally bound by these regulations may be more flexible in this respect. Different levels of engagement may be appropriate in different contexts depending on the project goals. The engagement levels have been defined in Section 1.3.1.

The reasons for (and the subsequent benefits of) involving the affected communities in the decision-making process can be classified into two types:

1. Ethical and legal reasons
   a. Protection of the right to participate and the right to environmental conservation
   b. Satisfaction of the demand for public participation
   c. Promotion of active citizenship
   d. Meeting policy requirements and regulations
2. Practical reasons
   a. Better knowledge and information on which to find the decisions
   b. Gaining new insight and better understanding from a broader range of perspectives and opinions
   c. Better-quality decisions and enhanced effectiveness
   d. Acceptance by the public of the decisions made
   e. Improvement of the relationships with the local communities and other stakeholders
   f. Improvement of the perception and reputation of public decision-makers
   g. Cost and time savings

The method of stakeholder engagement is also very important. There is a wide range of methodologies/techniques that can be employed in stakeholder engagement. They are not the objective of this chapter, but some of them include fact sheets, websites, open house, public comments, focus groups, surveys, public meetings, workshops, deliberative polling, citizen advisory committees, consensus building, participatory decision-making, citizen juries, ballots, or delegated decisions, among others.

1.3.3 PUBLIC PARTICIPATION CORE VALUES

The general ideas guiding the implementation of a public participation process can be described along the following lines:

1. *Willingness of improvement*: As seen in the previous section, public participation leads to better decisions and thus better management. Therefore, all the process must be guided by the search for improvement.
2. *Democracy*: In developed countries, public participation has been accepted as a right. People have the right to get involved in the decision-making process. Therefore, they must be guaranteed the opportunity to be informed and express their opinion, and no discrimination must be allowed.
3. *Transparency*: Public participation must be a clear and open process, which provides the relevant information and the opportunity to debate in an open space characterized by the receptivity to ideas and initiatives. Once the decision is made, the public must be informed about the outcomes and how their input influenced them.
4. *Involvement–engagement*: Public participation is not possible if the local population and the people who may have an interest in a particular issue do not get involved in the process. It is crucial to put in place a communication strategy that informs about the needs and interests of all the parties involved in order to facilitate stakeholders' engagement. The design and communication of the participation procedure is a key element at this point.
5. *Commitment*: The improvement of education and the development of a public awareness on the search for sustainable development (instead of plain economic growth) will contribute to the emergence of a collective need for public involvement, which eventually develops into a commitment to participate in environmental management.
6. *Credibility*: Public participation must be carried out in a way that all the stakeholders can trust it. In particular, the process must ensure that the public's contribution will influence the final decision.
7. *Effectiveness*: The goal of public participation is to improve decisions, and thus, the whole process must be outcome oriented. The results obtained must be useful and effective.

All the preceding lines must be guaranteed for the process to be successful. The value and effectiveness of public participation lie in the process itself, and it is very important to design guidelines and procedures to request, receive, process, and disseminate the relevant information.

These values are consistent with the "IAP2 Core Values of Public Participation" developed by the IAP2 (2007).

If these values hold, systemic resistance to change can be reduced significantly (Harich, 2010).

1.4 SUSTAINABLE FOREST MANAGEMENT STANDARDS AND INDICATORS

1.4.1 SUSTAINABILITY INDICATORS

Since the 1970s, the need to infer a change in the economic model toward what was called, in the late 1980s (WCED, 1987), "sustainable development" triggered a process of developing indicators of sustainability that has been maintained to date, despite the multiple problems inherent in the very concept of sustainable development:

- The complexity of the concept of sustainable development itself (a meta-concept), which includes not only a change in the production model but also very profound changes in the consumption and educational models, which are not always assimilated by society (Durán, 2000).
- The subjectivity associated with the concepts of "social welfare" and "quality of life" and the different conceptions between regions, countries, and individuals, as well as the "refusal" to eliminate the idea of "living standard" as the axis around which social and economic development revolves (Falconí, 2001).
- The absence of an objective framework that determines when sustainability has been attained or to what degree sustainability achievements are being met. In short, there are no indicators to establish the ultimate goal, which adds more indeterminacy to the very concept.

Regarding the lack of sustainability indicators, the work done in recent decades has been very important, mainly in two areas: on the ecological sustainability indicators (with corresponding plots concerning forests, water, soil, biodiversity, etc.) and the economic sustainability indicators, where different methods have been proposed. None of those methods has reached a global consensus so far, except the need to abandon the traditional system of indicators based on GDP or GNP,* which has proven to be totally inadequate (Daly, 1989).

---

* GDP: gross domestic product, GNP: gross national product.

Specifically indicators need to be established to ensure three key aspects of the new model: the economic, ecological, and social sustainability:

1. Economic sustainability indicators: A new accounting model that includes the value of externalities for greater reliability in the macroeconomic accounts is needed. The most advanced lines of work are as follows:
   • Replacement of GDP by net national product (NNP), where environmental damage is discounted to determine the real economic growth. The idea of fixed NNP has been developed from the work of Solow (1986).
   • The incorporation of satellite accounts, which take into account the evolution of natural resources.
   • Update of the Hicksian income concept, used as an indicator of weak sustainability (Pearce and Atkinson, 1995), allowing the exchange between different types of capital in an economy, as long as the end result of their sum is positive (genuine savings or Hartwick's rule, method of El Serafy, etc.). The objective in all cases would be to calculate a sustainable national income (SNI), defining an optimal consumption level.
   • The development of strong sustainability indicators, which reject the idea of replaceability of natural capital and which include among them the Index of Sustainable Economic Welfare (ISEW) and the Genuine Progress Indicator (GPI), both closer to the idea of social indicators than to economic indicators.
        The most commonly used economic indicators of sustainability are described in Section 1.5.2.
2. Ecological sustainability indicators: These indicators have been widely developed in recent decades and the following can be highlighted:
   • Physical indicators of sustainability: consolidated natural heritage
   • Critical natural capital
   • Ecological footprint (EF) or Biocapacity
   • Energy indicators of sustainability
   • Dematerialization indicators
   • Ecosystem indicators
   Chapters 3 through 6 focus on describing the most recent advances in ecological indicators.
3. Social sustainability indicators: The dissociation between the concepts of growth and development has been the cause of the creation of the social indicators of sustainability. The basis for this type of indicators lies on the idea of measuring the "quality of life," which, in turn, depends on the ability of the individual to freely elect one quality or another (Sugden, 1993). These indicators include the following:
   • Human Development Index (HDI) developed by the United Nations Development Programme (UNDP, 1992). The United Nations has developed some complementary indices such as the index of physical quality of life, the Human Poverty Index (HPI), or the Gender Inequality Index.

- Environmental Sustainability Index (ESI) developed by the World Economic Forum (WEF, 2001).

The most commonly used social indicators of sustainability are described in Section 1.5.3.

1.4.2 International Sustainability Criteria

Sustainability was first used in a United Nations document in 1978. Normative concepts, encapsulated in the term ecodevelopment, were prominent in the United Nations publications. The roots of the term sustainability are so deeply embedded in fundamentally different concepts, each of which has valid claims to validity, that a search for a single definition seems futile. The existence of multiple meanings is tolerable if each analyst describes clearly what he means by sustainability (Kidd, 1992).

Since the Earth Summit in Rio de Janeiro in 1992, countries are urged to ensure sustainable management of forests. In fact, general guidelines for the proper management of forests are set in the Declaration of Principles on Forests, as well as in the establishment of the conventions on biodiversity, climate change, and desertification.

So, in 1993, the "Montreal Process Working Group" agreed in a framework of criteria and indicators that provided the member countries with a common definition of what characterizes sustainable management of temperate and boreal forests.

At the same time, the European countries decided to work as a single geographic region to set its criteria and indicators for SFM, giving rise to the Pan-European Forest Process or Helsinki Process. The process is supervised by the MCPFE, which have held periodic meetings since 1990, prior to the Montreal Process. At the Third Ministerial Conference, held in Lisbon in 1998, the six national-level criteria identified within this process were officially adopted (Pan-European SFM criteria), and the corresponding 27 indicators were endorsed.

Other regional groupings sharing the same goal of setting criteria and indicators for SFM were formed in other regions of the world, giving rise to several processes, which include the following:

- Tarapoto Proposal for the SFM of Amazonian forest, 1995
- African Timber Organization (ATO) Process
- Africa Arid Zone Process, 1995
- Near East Process, 1996
- Lepaterique Process in Central America, 1997
- Initiative of the Arid Zone of Asia, 1999
- International Tropical Timber Organization, 1999

The criteria and indicators developed within these processes are thoroughly addressed in Chapter 3.

Several international meetings for the harmonization of the different criteria have been held, and indeed there is a growing consensus on seven common thematic areas based on the international initiatives for the development of criteria and indicators:

a. Extent of forest resources
b. Biological diversity
c. Forest health and vitality
d. Protective functions of forests
e. Productive functions of forests
f. Socioeconomic functions
g. Legal, policy, and institutional framework

This consensus has been acknowledged by the international forest community at the fourth session of the United Nations Forum on Forests and the 16th session of the Committee on Forestry held in Rome in 2003.

These seven major areas become the seven reference criteria to which the corresponding sustainability indicators are associated in each geographic area. It is important to distinguish between the concepts of criteria and indicators (FAO, 2001):

Criteria define the essential elements against which sustainability is assessed, with due consideration paid to the productive, protective, and social roles of forests and forest ecosystems. Each criterion relates to a key element of sustainability and may be described by one or more indicators.

Indicators are parameters that can be measured and correspond to a particular criterion. They measure and help monitor the status and changes of forests in quantitative, qualitative, and descriptive terms that reflect forest values as seen by those who defined each criterion.

Criteria and indicators are applied at three different levels:

• Regional (international)
• National
• Forest management unit level

1.4.3 SUSTAINABLE FOREST MANAGEMENT STANDARDS

SFM deals with the environmentally appropriate, socially beneficial, and economically viable management of forests for present and future generations.

To assess and monitor SFM, a valid global approach is needed, which means that the same principles must be applied worldwide. SFM is based on a set of principles and criteria (P&C), which have been defined in different international forums described in Section 1.4.2. Pan-European indicators for SFM (MCPFE, 2003) are a key element to assess forest sustainability.

One of the ways to ensure that the criteria of SFM are applied is forest certification. Forest certification emerged at the end of the 1980s to slow down deforestation in the tropics through the implementation of a system that encouraged consumption of products from forests managed in a sustainable way. This proposal was subsequently exported to the rest of the world's forests.

According to Bass (2004), forest certification is a voluntary process by which an independent third party issues a written certificate guaranteeing that forest management in a particular management unit is done according to standards considering ecological, economic, and social aspects.

The objectives of forest certification are to improve forest management and to facilitate market access for products from certified forests. It tries to incorporate sustainability criteria into economic decision-making by changing consumer preferences through information and awareness (Gafo et al., 2011).

There are over 50 forest certification programs worldwide, but the two largest international forest certification standards are the Forest Stewardship Council (FSC) and the Programme for the Endorsement of Forest Certification (PEFC).

The two standards are based on a set of criteria that are similar to the SFM criteria developed by the main international initiatives described in Section 1.4.2. The FSC standard has developed its own P&C, which are very general and applicable worldwide. The PEFC initiative, on the other hand, has adopted the Pan-European criteria from the Lisbon Conference. Therefore, the criteria used by both standards are very similar.

The P&C must be translated into regional and national indicators, which must guide forest policy and management. Criteria at the national level help to define the concept of SFM and the aspects that must be addressed to assess it. Each criterion is related to an important element of sustainability, described by one or more indicators.

Indicators are instruments for assessing and monitoring status, changes, and trends over time. They are used to study the evolution of quantitative and qualitative attributes that show the values encompassed in each criterion. Changes along time will indicate if a country is moving forward toward sustainability of forest management or moving away from it according to the established criteria.

As mentioned earlier, the two main international forest certification standards are the FSC and the PEFC. How these programs differ is a highly contested issue. They have common features, but they emphasize their differences.

The FSC standard is based on 10 principles and 56 criteria applicable worldwide. The indicators are defined on a national or regional scale and are quite specific. There is a group certification schema so that small forest landowners can have access to certification. Certification is carried out by an independent third party, but the FSC is the accreditation authority. Forestry professionals can be members of the certification team, but the team leader does not have to be a forester. The evaluations include a field inspection, as well as a thorough review of the management plan, harvest information, maps, and other data.

The PEFC is based on the Pan-European criteria and a set of common rules on the certification procedure. There is a regional approach to meet the needs of small forest landowners. Both accreditation and certification are carried out by independent third parties. Forest auditors must meet specific requirements and the leader of the certification team must be a professional forester. National certification systems that have developed standards in line with PEFC requirements can apply for endorsement to gain access to global recognition and market access through PEFC International.

In both systems, there are a logo and a trademark and the certificate is valid for a period of 5 years, with annual inspections in the case of FSC and biannual inspections in the case of PEFC.

Clark and Kozar (2011) carried out a comparison of three certification standards: FSC and two PEFC-endorsed certification systems in Canada, the Canadian Standards Association–Sustainable Forest Management Standard (CSA-SFM), and the Sustainable Forestry Initiative (SFI). A selection of 35 literature sources that met certain search criteria were analyzed to determine the system that most effectively meets SFM goals.

The information in the 35 studies was analyzed according to 12 criteria: labeling systems, certification, stakeholder participation, public input, repeatability and consistency, adaptability, applicability, transparency, credibility, monitoring and research, ecological issues, and socioeconomic issues.

Since the methodology and the data of the studies were different, the qualitative and quantitative data for each indicator had to be converted to a binary value (1 if the system met the goal established by the indicator and 0 otherwise). A score for each criterion was calculated as the proportion of indicators that met SFM goals.

FSC seemed to meet the SFM indicators better than the other 2 systems for 8 out of the 12 criteria. It outscored its opponents especially in transparency, credibility, and ecological and socioeconomic issues. On the other hand, CSA-SFM performed better than FSC in repeatability and consistency, as well as in monitoring and research.

Regarding public participation, both FSC and CSA-SFM had a balanced representation of all types of stakeholders, while SFI failed to have social participation and it was biased toward economic stakeholders. All three systems encourage public participation in the development of the standard and in judging conformance to it.

The systems have been compared on the basis of the wording of certifiers' P&C and on user survey analyses, but there is not empirical evidence on the performance of the certification systems so far.

The impact of forest certification on the EU forestry sector and its contribution to SFM in Europe have been addressed in a study by Gafo et al. (2011). The authors carried out a two-round survey for different stakeholders by means of the Delphi method and used the contingent valuation method in some of the questions. Some of the results obtained are summarized along the following lines:

In general, FSC is more present in countries with a larger forest area under public ownership, while PEFC is more important in countries where private forest property is predominant.

Most experts estimated that certification improves the image of forest products and a large majority of respondents considered that the changes required in forest management to obtain certification were either very little or none at all.

From the ecological point of view, a consensus was achieved on the positive impact of certification on biodiversity, as well as on forest area, structure, and functioning.

Regarding the economic aspects, in most cases, certified wood is sold at the same price than noncertified wood, and forest owners would see a 7% increase in price as a reasonable incentive to certify their forests. The same situation holds for certified and noncertified wood products, which are normally sold at the same price, with industry experts considering that they would require a 3%–5% price increase in order to buy certified wood.

According to NGOs, certifiers, and certification bodies, forest certification leads to an improvement of the conditions of workers, but there is not a consensus on this, and in fact other groups such as owners, industry, research, and public service give a neutral or negative response. Other social positive impact of certification is an improvement in the information provided to society and in consumers' education. Despite this positive effect, the authors conclude that an improvement in the information to both society and local people by the actors involved in forest certification could increase the positive impact on the sector.

1.4.4 Information and Models to Build Sustainability Indicators

As seen in previous sections of this chapter, the main objective of public participation is to improve the quality of the decision-making process. Decisions are based on available information, and since a participatory process is carried out, the information is addressed not just to high-ranking officials but to all the stakeholders. It means that the data collected must be processed to generate information, which is easily comprehensible, and this information must be presented in a clear and accessible way.

The development of the sustainability indicators referred to in Section 1.4.1 is a key element in this process. In particular, the assessment of ecological sustainability indicators such as those related to diversity, dead wood, endangered species, forest regeneration, wildlife habitat, and forest health requires a significant effort in field data collection.

Field observations to characterize wildlife populations, forest stands, or rangeland are obtained through the use of surveying and sampling techniques. These techniques are also used to collect data to assess pollution or environmental quality in general. A detailed description of the sampling methods is included in Chapter 2.

Most of the data collected have been obtained at a particular location (with its geographic coordinates) and can be incorporated into a geographic information system (GIS), together with information coming from other sources. GISs are particularly powerful when it comes to putting together different types of information and performing complex analyses with them. The assessment of forest sustainability indicators is carried out through the development and use of computer models, which include a GIS component.

Remote sensing is another technology to capture, analyze, and generate spatial information. In particular, laser imaging detection and ranging (LIDAR) data are becoming more and more important in forestry applications. The graphic capabilities of GIS and remote sensing are a powerful tool to design the communication strategy and the way the information is going to be presented in front of the stakeholder.

These information technologies, as well as the role they play in the decision support system for participatory forest management, are described in Chapter 2.

1.5 SOCIOECONOMIC INDICATORS OF SUSTAINABILITY

The intent to separate environment and society stems from the obsolete man's dream to control and dominate nature (Aledo et al. 2001). Ecology has contributed to other sciences the idea of belonging to an interrelated system, called ecosystem, in such a way that the isolated study of the component parts does not make sense.

Therefore, the solutions proposed to solve the serious environmental problems we are facing will be wrong if done from a reductionist perspective, that is, without taking into account the relationship between society and nature.

The joint analysis of society and environment responds to the following causes:

a. The environment is only understandable if we include the history of the people who live there and their environmental impacts over time.
b. Human society also depends on environmental factors that have influenced its social dynamics.
c. Human action modifies ecosystems, but also environmental factors determine human development in an interdependent relationship.

The subject that deals with this issue is ecological economics, which could be defined as the science and management of sustainability (Costanza, 1991; Kates et al., 2001). It maintains that sustainability is that relationship between economic and ecological systems in which human life can continue indefinitely, with human activities remaining within limits that do not destroy the ecological systems.

There are two approaches to the economic analysis of sustainability:

• The weak sustainability approach, which allows flexibility in the conservation of natural capital, accepting its decrease in exchange of an increase in other types of capital
• The strong sustainability approach, which does not allow any reduction of natural capital within its concept of sustainability

However, the essential point is not to properly define the concept of sustainable development or sustainability but to establish the conditions necessary to achieve it. Under what assumptions a country or an economic sector can be considered sustainable? To answer this question, it is necessary to create indicators to analyze the evolution of the development model and to evaluate its path toward sustainability as the ultimate goal (Lavandeira et al., 2007).

1.5.1 WEAK SUSTAINABILITY VERSUS STRONG SUSTAINABILITY

Weak sustainability: The fundamental principle of this type of sustainability is that natural capital is simply another form of capital and therefore can be exchanged with others (Pearce and Turner, 1990). It has its conceptual basis in the work of Hotelling (1931). Hotelling's rule states that the optimal extraction path of a nonrenewable natural resource is obtained by maximizing its net present value, which leads to the following equations, excluding and including operating costs, respectively:

$$i = \frac{P'(t)}{P(t)} \quad (1.2)$$

$$i = \frac{P'(t)}{P(t) - C} \dagger \qquad (1.3)$$

where
  $P(t)$ is the price of the resource at time t
  $P'(t)$ is the derivative of price with respect to time
  C is the operating costs
  I is the discount rate

In the long run, the relative price growth of the natural resource should be higher than the interest rate in order to ensure its conservation; otherwise, there is overexploitation of the resource. That is, Hotelling establishes an economic relationship between the benefits of conserving or saving the use of a renewable natural resource and the costs associated to this conservation. Hotelling's rule leads to the conclusion that the growth of natural resource value must exceed the rate of interest to ensure preservation. This basic idea will be further developed in his works on sustainability and intergenerational equity.

Later on, Solow's work (1974) incorporates the natural capital in economic growth models, showing how an economy can grow indefinitely in the presence of limited natural resources.

Hartwick in 1977, building on Solow's work, established the so-called rule of constant capital considering consumption as the interests generated by a capital sum in each time period. From this premise, the rules of weak sustainability are established.

In 1992, Pearce and Atkinson stated that an economy is sustainable if savings are greater than capital depreciation (both man-made and natural capital), that is, $K_T > 0$.

The problem with this approach is that, according to it, the global economy as a whole has been in recent decades in a state of sustainability in the weak sense, as it has fulfilled the preceding condition.

1.5.1.1 Intergenerational Equity

One of the most important implications of sustainable development is the concern for the legacy to future generations. The problem is how to incorporate into a welfare function the value of natural resources for coming generations. The subject has not been solved at all and intergenerational equity is only considered taking into account the effects of the discount rate.

According to Jevons' equimarginality principle, the higher the interest or discount rate, the greater the preference for the present and the lower the willingness to forgo current for future well-being. The equation which determines the equilibrium arising from Hotelling's principle is

$$i = \frac{C'(t)}{C(t)} \dagger \qquad (1.4)$$

C being the present or future consumption level and i the discount rate.

The function determining well-being is a function of present and future consumption:

$$W = f\ (C_0, C_1, \ldots, C_t)\tag{1.5}$$

where
  W is the welfare function
  C_i is the consumption at time i

In order to maximize welfare, the following function should be maximized subject to the constraints imposed by the available technology (Lavandeira et al., 2007):

$$\int_0^\infty U(C)e^{-it}dt\tag{1.6}$$

where $U(C)$ is the utility of consumption and t is time.

The result of this expression leads to consumption patterns, which may or may not be sustainable:

$$\text{If }\quad i > \frac{C'(t)}{C(t)}\quad \text{the situation will not be sustainable}$$

$$\text{If }\quad i \le \frac{C'(t)}{C(t)}\quad \text{the situation will be sustainable}$$

Some authors argue that the solution to this problem lies in the demonstration of the existence of the environmental Kuznets' curve (Figure 1.1), which has an inverted U shape. According to this curve, when a high income level is reached, the demand for natural resources decreases.

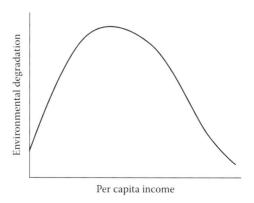

FIGURE 1.1 Environmental Kuznets' curve.

Weak sustainability models are based on perfect substitutability between natural and man-made capital in such a way that the elasticity of substitution is always greater than one. But there are many cases in which this elasticity of substitution is less than unity, namely,

    a. Life-support functions: atmosphere, water, carbon, etc.
    b. Ignorance of the interrelationships among the components of various ecosystems
    c. Irreversible loss of environmental assets

Strong sustainability considers that natural capital is not fully replaceable by artificial capital, forcing the maintenance of the natural capital stock. The constant natural capital rule implies that

$$\frac{\partial K_N}{\partial t} \geq 0 \tag{1.7}$$

where
    K_N is the natural capital
    Which in turn can be expressed as $K_N = K_{NNC} + K_{NC}$
    K_{NNC} is the noncritical capital, with limited substitution capacity
    K_{NC} is the critical capital, atmosphere, ozone, climate, biodiversity, etc.

The critical natural capital represents the minimum level of security. Its objective is to set the maximum limit to which an ecosystem can deteriorate.

The conditions of strong sustainability are the following:

    1. Decreased use of nonrenewable natural resources
    2. Replacement of fossil fuels by renewable energy sources
    3. Biodiversity conservation
    4. Sustainable use of renewable natural resources

1.5.2 ECONOMIC INDICATORS

These indices are included in the concept of weak sustainability and their objective is to achieve a sustainable product or SNI.

1.5.2.1 Indices Based on National Accounting Systems

The basis of these indicators is the *NNP*, which is considered a sustainable income over time, as it is calculated as the sum of net added value plus capital gains or losses, that is,

$$NNP = GDP - \delta_M K_M \tag{1.8}$$

where
    GDP is the gross domestic product
    $\delta_M K_M$ is the depreciation of man-made (artificial) capital

On the other hand, another indicator that has been developed is *green NNP* (*gNNP*). It allows a more rigorous approach to the concept of sustainable development and is defined as *NNP* minus depreciation of natural capital, that is,

$$gNNP = NNP - \delta_N K_N \tag{1.9}$$

Weitzman (1976) showed that *NNP* equals the linearized Hamiltonian, namely,

$$H = pC + \lambda K \tag{1.10}$$

where
  p is the price
  C is the consumption
  λ is the multiplier and shadow price
  K is the capital

In this optimal control problem, the optimal path is obtained by maximizing the Hamiltonian; thus, maximizing *NNP* is the best strategy to follow an economy's ideal path.

Later on, Weitzman (2001) starts working on consumer surplus:

$$H = U(C, E(R)) + \lambda K + \mu E(R) \tag{1.11}$$

where
  p is the price
  C is the consumption
  $U(C, E(R))$ is the utility
  λ, μ are the Hamiltonian multipliers and shadow prices
  K is the capital
  $E(R)$ is the income surplus

Therefore, the *NNP* would be obtained by solving the first-order conditions of the Hamiltonian, with the solution being given by the following expression:

$$NNP = pC + \left(\frac{dU}{dE}\right) \cdot E(R) + P_i K + P_w E(R) \tag{1.12}$$

where
  p is the price
  C is the consumption
  dU/dE is the derivative of utility with respect to income surplus
  P_i is the capital cost
  K is the capital
  P_w is the surplus cost
  $E(R)$ is the income surplus

Other prominent indicators based on the measurement of *NNP* are the following:
Adjusted net national product (aNNP): It is calculated as

$$aNNP = NNP - \text{Defense expenses} - \text{Depletion of natural capital}$$

This indicator is very similar to *gNNP*, but it presents a disadvantage when it comes to measuring the depletion of natural capital. For this reason, which is common to other indicators, many variations trying to address this problem have been developed.

Net savings: It equals national savings minus depreciation of capital (consumption of physical and human capital).

Moreover, according to a broader consideration, it responds to the concept of adjusted savings, which can be corrected by means of an economic or social adjustment, for example, using the *HDI*, or through an ecological adjustment, for example, using the EF (Bolt et al., 2002).

Hartwick's rule (1977) shows that in order to keep utility constant over time in countries with economies highly dependent on their use of natural resources (especially nonrenewable, e.g., oil), an amount equal to the income generated by natural resources extracted in each moment of time must be invested.

However, in practice, there is a low capital accumulation rate in oil-producing countries, while the capital investment rate in countries without such natural resources is very high.

Therefore, Hartwick's rule states that total capital value must be maintained in order to achieve sustainable consumption:

$$\frac{\partial K}{\partial t} = 0 \qquad (1.13)$$

Genuine savings is obtained as total savings minus depreciation of artificial (manmade) capital and natural capital.* The terms of the formula are defined as follows:

$$\text{Genuine savings} = \text{Total savings} - \delta_M K_M - \delta_N K_N \qquad (1.14)$$

where
$\delta_M K_M$ is the depreciation of man-made (artificial) capital
$\delta_N K_N$ is the depreciation of natural capital

The World Bank (2001) calculates genuine savings as follows:

Genuine savings = Total savings − Consumption of fixed capital + Education Expenditure − Resource Exploitation − Exploitation of forests − Damage due to CO_2 emissions

---

* Genuine savings is an economic concept that measures the true savings rate of a country taking into account natural resource degradation and pollution problems, which are quantified and deducted by means of the term δNKN.

Sustainability implies that genuine savings should be greater than zero, meaning that total savings would offset the depreciation of the physical and natural capital.

Genuine savings rule: An application of the Hartwick's rule developed by Hamilton (2000). The general idea is that, according to Hartwick's rule, consumption can only be understood as the interest earned on the investment of available capital. Therefore, for consumption to remain constant over time, capital stock should not vary. This way, consumption becomes a Hicksian income as it would be permanent in time.

Therefore, for each generation to pass the next generation a capital at least equal to that received, the following condition must be met:

$$\frac{dK}{dt} = \frac{dK_M}{dt} + \frac{dK_N}{dt} + \frac{dK_H}{dt} \geq 0 \tag{1.15}$$

where

   K is the total capital
   t is the time
   K_M is the man-made capital
   K_N is the natural capital
   K_H is the human capital

$$\text{On the other hand,} \quad \frac{dK}{dt} = S(t) - \delta K(t) \geq 0 \tag{1.16}$$

$$S(t) - \delta_M K_M(t) - \delta_N K_N(t) - \delta_H K_H(t) \geq 0 \tag{1.17}$$

If the depreciation of human capital is not considered and all the terms of the whole expression are divided by Y, the genuine savings rule is obtained:

$$\frac{S}{Y} - \frac{\delta_M K_M}{Y} - \frac{\delta_N K_N}{Y} \geq 0 \tag{1.18}$$

where

   S is the national savings
   δ is the capital depreciation rate
   Y is the national income (*GDP*)

An economy is sustainable if its savings rate is greater than the sum of the depreciation rates of its natural and artificial capital.

This sustainability indicator does not reveal whether the economy would be sustainable with a growing population. Hamilton (2000) introduced another indicator to solve this problem:

Per capita wealth: It takes into account the growth of population and resources and can be calculated as

$$\text{Per capita wealth} = \frac{W}{P} \tag{1.19}$$

where
   W is resource stock (wealth)
   P is population

Writing this expression in terms of variation rates, we would get

$$\frac{\partial W}{\partial P} = \frac{W}{P}\left(\frac{\partial W}{W} - \frac{\partial P}{P}\right) \tag{1.20}$$

Moreover, wealth (W) is defined as the present value of current and future consumption of goods, that is,

$$W = \sum \frac{c(1+r)^t}{(1+i)^t} \ddot{\text{ii}} \tag{1.21}$$

where
   c is consumption
   r is consumption growth rate
   i is discount or interest rate

Environmental golden rule: It is obtained by reformulating Hicks' income to maximize human welfare through obtaining the highest consumption level that can be maintained indefinitely subject to given environmental constraints and assuming that population level remains constant.

The highest consumption level that can be maintained indefinitely, under environmental constraints, is determined by using the following dynamic optimization procedures:

$$\frac{\partial K/\partial t}{K} = i \tag{1.22}$$

where
   K is the total capital
   i is the discount rate

The earlier expression is obtained by maximizing capital consumption; hence, the optimal savings rate is one that maximizes consumption level.

1.5.2.2 Indices Based on Savings Incentives: Concept of Hicksian Income

Progress in the development of environmental accounting systems or indicators has occurred along two axes: on the one hand, the concept of Hicksian income and welfare economics and, on the other hand, the efforts carried out by supranational institutions.

Hicks (1939) established the concept of Hicksian income as the income that can be maintained indefinitely through time.

Probably the weak sustainability indicator that best fits the concept of Hicksian income and has had further development, applied in all countries with strong economic dependence on the extraction of nonrenewable natural resources (e.g., oil), has been the so-called El Serafy method and its calculation of user cost or opportunity cost for a sustainable consumption level.

El Serafy method (1989) or user cost approach: It is necessary to invest a portion of the profits R generated by leveraging a natural resource to maintain a steady income flow X. The portion of revenue to be invested is called user cost and is calculated as $R - X$:

$$\frac{R}{(1+i)^1} + \frac{R}{(1+i)^2} + \cdots + \frac{R}{(1+i)^n} = \frac{X}{(1+i)^1} + \frac{X}{(1+i)^2} + \cdots + \frac{X}{(1+i)^\infty} \quad (1.23)$$

Perfect substitution between different forms of capital is assumed.

1.5.2.3   New Methods of National Accounting

The idea of substitutability between natural capital and physical or material capital has led to the development of environmentally adjusted national accounting macroeconomic indicators, such as the following:

SNI developed by Pearce and Warford (1993) states that national income is sustainable when total capital (natural and material) remains constant through time.

Sustainable net national product (*SNNP*) developed by Daly (1989) is defined as

$$SNNP = NNP - GD - \delta_N K_N \quad (1.24)$$

where

NNP is the net national product

GD is the environmental protection expenditure

$\delta_N K_N$ is the natural capital depreciation

From the decade of the 1990s, a new approach to establishing sustainability indicators began to spread. This approach was based on the idea of conceiving the problem in a sequential way, using vector-type information and not based exclusively on statistical data as other previously mentioned indicators had been established (Caparro's Gass, 2009). The idea is to take a systemic approach with three dimensions: economic–social–environmental.

The ecosystem approach will prevail over purely ecological criteria for its holistic nature that is able to incorporate the three components that underpin sustainable development and with the main objective of obtaining models for the sustainable management of natural resources on which human beings depend upon.

This way, other indicators developed from different institutional settings are as follows:

Pressure–state–response model (PSR): It measures the pressure of human activity on the environment, state defines measurable characteristics of the environment under pressure, and response measures the environmental changes generated to solve the environmental problems created by human pressure (Mortensen, 1997).

Driving force–state–response model (DSR) developed by the Organisation for Economic Co-operation and Development (OECD): It replaces the idea of "pressure" by the idea of "driving force" and implies the existence of social and economic pressures (e.g., population growth, consumption level increase). The components of the model are different human activities that have an impact on the natural environment.

It allows the comparison between countries according to their degree of environmental impact but leaves the developing countries out of the model. This is a shortcoming of the model since most of the raw materials used by OECD countries are produced in developing countries.

Pressure–state–impact–response (PSIR) model: It adds to the previous models the concept of impact to measure the effect of pressure on the system.

Driving force–pressure–state–impact–response (DPSIR) model of the European Environment Agency, including underlying forces, pressures, state trends, impacts, and responses from society (Figure 1.2).

The World Bank genuine savings model includes not only environmental aspects but also aspects related to human capital acquired through education.

It is a systemic indicator designed to generate a single value, which shows if a system is experiencing difficulties. It measures the balance between an increase in physical and human capital and a decrease in natural capital.

Classification of environmental protection activities (CEPA) model stems from the Convention on Biological Diversity (CBD) and has been developed by Eurostat in collaboration with the OECD and the United Nations. The activities considered in the model are limited to those that cause environmental degradation including not only the private sector but also the public sector and households.

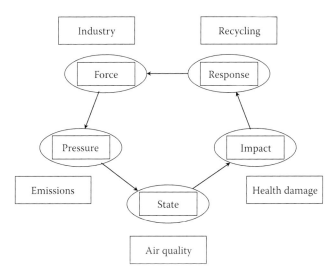

FIGURE 1.2   Diagram of the DPSIR model developed by the European Environment Agency, 1999.

The model classifies environmental protection activities in nine groups and presents the evolution over time of economic investment in each group so that, together with the results obtained in each one of them, the adequacy of such protection investments can be assessed.

The global environmental change currently going on is the result of human activity on the ecosystem, and the answer must be a change in the development model to find solutions to the problem created. This paradigm shift, following the institutional approach previously developed, is based on the participation of various forces of change. It requires a global response in which all stakeholders are involved to rethink the relationship between humans and the ecosystem.

The combination of communication, education, participation, and environmental action causes change. The objective is to achieve a greater public support in environmental management spreading the socioeconomic impact of conservation through the use of opportunities such as tourism or environmental education and awareness (Figure 1.3).

The goal must be to attain a change in cultural, social, political, and economic values and their relationship with the environment, which will lead to an improved quality of life. The implementation of the agreements embodied in Agenda 21 at Rio de Janeiro in 1992 and the subsequent agreements of the 2002 World Summit on Sustainable Development in Johannesburg can be the first step towards this final objective.

Apart from these environmental accounting models, there is an increasing use of additional accounting instruments such as the following:

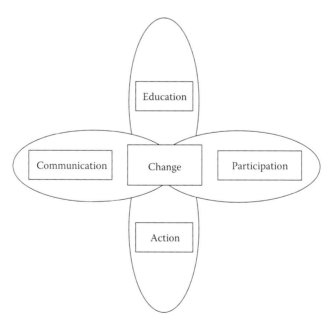

FIGURE 1.3 Diagram showing the relationship between forces and change produced by them. (Own elaboration.)

Natural resource accounts: They measure inflows, initial stock, and resource use, telling apart natural resources (renewable and nonrenewable) from environmental resources, that is, nontradable environmental services.

Satellite accounts: The National Accounting Matrix including Environmental Accounts (NAMEA) model provides financial and economic information.

Integrated accounting systems: Basically two types have been developed: System of Integrated Environmental and Economic Accounting (SEEA; United Nations, 1993, revised in 2002) and European System of Accounts 1995 (ESA 95; European Union).

As for the environmental accounting models relating to forestry, the Manual on Economic Accounts for Agriculture and Forestry called EAA/EAF 97 (within the ESA 95) must be mentioned. The value of forest production is calculated as

$$TP = FPS + FPE + \text{Infrastructure} \qquad (1.25)$$

where
   TP is the total forest production
   FPS is the final production of silviculture (wood and nonwood products)
   FPE is the final production of environmental services

1.5.3   SOCIAL INDICATORS

Social indicators alone do not allow proper measurement of the sustainable development level, but they need to rely on other indicators, economic and ecological, to establish results in this direction. The main purpose of these is to measure the concept of quality of life, based on four pillars: health, education, equity (poverty and gender), and human settlements (population, security, and living conditions).

1.5.3.1   Indicators Based on Life Quality versus Income Level

These indicators are included within the social indicators and have experienced a considerable boom in recent years. However, there are some difficulties for its practical application since they incorporate both objective and subjective elements. These indicators include the following:

   HDI: It is a social indicator developed by the UNDP in 1992 and allows a broader measure of development when compared to economic growth measured by *GDP* or *NNP*. It consists of a combination of indicators of income (*GDP*), health and education, which can detect inequalities, especially in developing countries.

   Other indicators developed similar to the *HDI* are

- Index of physical quality of life, built on indicators such as life expectancy, calorie supply, education, and adult literacy
- HPI (Sen, 1987), in which longevity, knowledge (literacy), quality of life (poverty), and social exclusion (long-time unemployment) are used as indicators
- Gender Inequality Index that measures the disparity between sexes

However, these indicators are criticized because they do not measure the internal distribution of development within a country.

ESI, developed by WEF (2001): It measures pressures and environmental risks, social and institutional capacity, and international cooperation on global issues.

The *ESI* combines 22 environmental indicators for each country, broken down into 67 specific subjects. It measures five essential characteristics:

- State of ecosystems
- Success in the decrease of environmental problems
- Progress related to protection of citizens in environmental matters
- Capacity for action against environmental problems
- Qualification of the public administration in each country

There is a strong correlation between *ESI* and *GCI* (Global Competitiveness Index).

Physical sustainability indices: They use natural heritage accounts associated to GDP, such as

- Critical natural capital: It sets thresholds for the use of natural resources.
- Minimum safety standard: Capital stocks must be maintained as long as the social costs associated to their use are too high.

Pearce and Atkinson index: Natural capital stock cannot decrease through time, that is,

$$\frac{\delta_N K_N}{Y} \geq 0 \tag{1.26}$$

The *ISEW*, based on *aNNP*, is a correction of *NNP*, which takes into account environmental damage. It is calculated as

$$ISEW = C + P + G + W - D - E - N \tag{1.27}$$

with $GDP = C + P + G$

where
    C is the consumption
    P is the nondefense-related public expenditure
    G is the capital growth
    W is the contribution of externalities
    D is the defense expenditure
    E is the environmental degradation costs
    N is the natural capital depreciation

ISEW (Daly and Cobb, 1989) is an economic indicator aiming at replacing *GDP* in the long term.

GPI: It is based on *ISEW* but excludes expenditure on education and health and includes the loss of leisure time and the loss of forest cover.

Moreover, it presents some changes with respect to *GDP*, which are the following:

- Unpaid activities such as domestic work or volunteer and family care are considered.
- Natural resource environmental degradation is included as a cost.
- Income inequality is considered as a cost that increases when the poor lose income.
- External debt (estimated by increasing or decreasing reserves) and crime (measured by prison costs) are also included as costs.

While *GDP* shows growth from 1950, *GPI* shows stagnation since 1970.

1.5.3.2   Indicators of Human Impact on Biosphere

These indicators are based on a combination of economic, ecological, and social aspects with the purpose of establishing an ecosystem approach to the measurement of sustainability but keeping statistical data rather than institutional indicators as the information basis.

1.5.3.3   Biodiversity

Biodiversity indicators include the following:

- Extent and type of forests (FAO). Certification and SFM (FAO)
- Marine habitats (WCMC-UNEP, FAO)
- Genetic diversity (FAO). Vegetation and wildlife genetic resources
- Extension of agricultural ecosystems under sustainable management (FAO)
- Nutritional status of biodiversity (FA = international biodiversity): analyzes food consumption through sustainable use of species and ecosystems

The UNESCO Chair for Sustainable Development has produced a comprehensive guide of ecological indicators, based on the indicators previously established by the Commission on Sustainable Development (CSD) of the United Nations, 2001, shown in Table 1.1.

Carrying capacity and EF: The concept of carrying capacity, developed by Meadows in 1992, sets the maximum population of a species that a habitat can support indefinitely. It can be calculated according to a basic model developed by Ehlrich and Holdren in 1971, which is given by the following expression:

$$I = P \cdot A \cdot T \qquad (1.28)$$

where
I is the environmental impact
P is the population
A is the per capita income
T is the technology, waste quantity per unit of production

TABLE 1.1
Ecological Indicators of Sustainable Development

Chapter	Impulse Indicator	State Indicator	Reaction Indicator
18. Water resource protection	Water consumption	Water reserve and water quality	Water treatment and available networks
17. Sea and ocean protection	Protection of coastal areas and pollution level	Fish catches	—
10. Land use planning	Land use change	Changes in land condition	Natural resource management
12. Desertification and drought	Population evolution in arid zones	Evolution of rainfall and area affected by desertification	—
13. Management of mountain areas	Population evolution in mountain areas	Natural resource sustainable use	—
14. Sustainable agriculture	Pesticides, fertilizers, energy, and irrigation	Arable land and salty areas	Agricultural education
11. Deforestation	Forest harvesting level	Change in forest area	Forest area under a management plan
15. Biodiversity conservation	—	Number of endangered species	Protected areas
16. Sustainable technology management	—	—	Research expenditures
9. Protection of the atmosphere	Pollutant emissions	Concentration of pollutants in the air	Waste management expenditures
21. Waste management	Waste generation	Recycling level	Waste management expenditures
19. Management of toxic chemicals	—	Number of intoxications	Number of forbidden products
20. Dangerous waste management	Waste generation	Contaminated area	Treatment expenditures
22. Nuclear waste management	Waste generation	—	—

Source: United Nations, *Indicators of Sustainable Development: Guidelines and Methodologies*, New York, 2007.

The rate of technological change must equal the sum of the growth rates of population and per capita income to attain a sustainable economic growth.

In 1996, FAO calculated a carrying capacity based on food production by means of the following expression:

$$CC = \frac{Q}{M} \tag{1.29}$$

where

  CC is the maximum sustainable population for each country
  Q is the potential food production
  M is the minimum calories per person

The EF determines the corresponding area of productive land and aquatic ecosystems required to maintain a given level of output (Wackernagel and Rees, 1996):

$$ef = \Sigma\, a_i = \Sigma\, \frac{c_i}{p_i} \qquad\qquad (1.30)$$

and

$$EF = N \cdot ef \qquad\qquad (1.31)$$

where

  ef is the ecological footprint per person (ha/person)
  EF is the total ecological footprint (ha)
  c_i is the average annual consumption (kg/person)
  p_i is the average annual productivity (kg/ha)
  N is the population (number of people)

Other indicators have been developed from the concept of EF, such as "water footprint" that sets the annual volume of water required to sustain a population for a given living standard.

1.5.3.4 Energy

Energy indices: There are about 1800 million people in the world who do not have access to electricity, and their energy supply is coming from animal or vegetal sources. There are many other people who cannot afford fossil fuels despite their availability. These people are those with a lower level of socioeconomic development, and this circumstance highlights the uneven distribution of energy consumption in the world.

Moreover, the access of all people, present and future, to the use of energy makes the system unsustainable. For this reason, energy sustainability indicators are so important.

The United Nations CSD has defined 30 indicators classified into three dimensions:

- Social: Energy availability is an element of direct impact on poverty, employment, education, demography, and health. It is essential to ensure social equity.
- Economic: Based on industrialization, it is essential to know the energy intensity for each industry.
- Environmental: Gas emissions, waste, water changes, and landscape.

The following energy sustainability indicators can be highlighted:

- Index of renewability of the energy used $\alpha = RE/TE$
- Index of consumed energy cleanness $\beta = 1 -$ damage caused/maximum potential damage
- Energy self-sufficiency index $\gamma =$ own energy/total energy
- Energy efficiency index $\psi =$ output power/input power
- Energy Sustainability Index $EgSI = f(\alpha, \beta, \gamma, \psi)$

This index integrates the other four indices earlier and can be calculated as

$$EgSI = a_1 \cdot \alpha + a_2 \cdot \beta + a_3 \cdot \gamma + a_4 \cdot \Psi \tag{1.32}$$

where
$a_i = Ca_i/$Total cost
Ca_i being the energy cost associated to each a_i

Consumption measurement through the use of energy indices has the advantage of avoiding the problem of the scarcity of environmental statistics in monetary terms:

$$Rc = Rp\,(1 + \delta) \tag{1.33}$$

where
Rc is the energy resource consumption rate
Rp is the restocking fee
Δ is the sustainability indicator
$\delta < 0$ for the economy to be sustainable

1.5.3.5 Materials

Dematerialization indices: They measure the growth of the economy in terms of its reliance on the use of physical materials, both those manufactured in the country and those resulting of the balance of foreign trade. The following two indices can be highlighted:

Material flow index: which measures the amount of materials consumed per person per year. It shows the relationship between natural resource consumption of a product and the services it generates.

Human appropriation of net primary production (*HANPP*): It is measured by the ratio of the production appropriated by humans for consumption and total biomass production potential:

$$HANPP\,(\%) = \frac{HANPP}{NPP} \cdot 100 \dagger \tag{1.34}$$

where

    HANPP(%) measures the impact of the economic and social activity and assesses the economic and ecological sustainability of industrial societies

    HANPP is the human appropriation of net primary production, obtained as the sum of agricultural activities, energy consumption, and material consumption

    NPP is the net primary production

    NPP = Solar energy (photosynthesis) − respiration

It comes from the annual photosynthetic capacity of the planet and generates renewable natural resources, enabling the maintenance of all organisms including humans. Nowadays, *HANPP* is around 20%–40%.

1.6 PARTICIPATORY TECHNIQUES FOR THE SUSTAINABLE USE OF NATURAL RESOURCES

1.6.1 QUANTITATIVE TECHNIQUES TO SUPPORT THE SUSTAINABLE USE OF NATURAL RESOURCES

Natural resource management is a complex decision-making process involving a high number of stakeholders, each one of them having different objectives and imposing his or her own constraints, thus leading to a high number of conflicting alternatives.

According to Myllyviita et al. (2011), the techniques used to support the sustainable use of natural resources can be classified into four different groups: optimization, cost–benefit analysis (CBA) and monetary valuation methods (MVM), multicriteria decision analysis (MCDA), and other approaches.

Optimization deals with the election of the optimal alternative from a feasible set of options with respect to the considered objectives. Optimization methods include linear programming, goal programming, mixed integer programming, and heuristic optimization.

CBA and MVM are used to estimate the social net surplus (or the private profitability) of a decision. If the benefits of a decision are greater than its costs, the decision should be made, while it should not otherwise. The terms benefits and costs include not only direct financial benefits and costs (returns and expenses) but also social and environmental benefits and costs. CBA has to cope with the problem of selecting the appropriate discount rate, which reflects the time preference of the decision-maker. When environmental services and/or costs are included in the analysis, MVM such as choice experiment or contingent valuation are used to estimate values of nonmarket products and services. The necessary data are obtained through a survey in which the respondents are asked about their willingness to pay for a positive environmental change or the willingness to accept compensation for a negative one. There are many studies on the incentives to give a biased answer and the way to design the questionnaires to avoid that problem (Johnston and Swallow, 1999).

MCDA is based on the measurement of people's stated preferences. These preferences can then be transformed into single utility values to allow for comparison between alternatives. Multiattribute utility theory and multiattribute value theory are two methods within this group, the difference between them being that multiattribute value functions exclude decision-makers' risk preferences, while multiattribute utility functions include them.

Besides the three groups of methods mentioned earlier, some *other techniques* and tools have been used to support decision-making. Sometimes, more than one method is used giving rise to the so-called hybrid methods. They can be classified as follows:

- Soft operations research: It provides tools for problem structuring and includes several methods such as the soft systems methodology (SSM), strategic choice approach (SCA), and strategic options development and analysis (SODA).
- Cognitive mapping: It provides visualization of the problem in the form of loops, links, and relationships between the concepts, thus facilitating the understanding of the process by the participants.
- Strengths, weaknesses, opportunities, and threats (SWOT): The name of the method includes the aspects of the problem that are revealed and analyzed.
- Interviews and voting are two ways of obtaining feedback from the participants. Interviewing requires considerable time and effort, and the information obtained is useful for qualitative analysis and decision modeling, whereas voting provides quantitative data but can undervalue the opinion of minorities and is easier to manipulate.

1.6.2 PUBLIC PARTICIPATION IN THE METHODS TO SUPPORT THE SUSTAINABLE USE OF NATURAL RESOURCES

Most of the current methods for the sustainable use of natural resources allow different levels of public participation. The results of the study of Myllyviita et al. (2011), where the authors compared the level of public participation in 35 case studies applying the methods referred to in the previous section, are described in the following.

Regarding public participation, the information analyzed in the previously mentioned paper was the type of public involved (experts and/or stakeholders, or the general public) and the kind of participation (active or passive). Passive participation included stating preferences, filling out questionnaires, or answering questions, while active participation allowed participants to influence and actually modify the process.

The main results are highlighted in the following:

- Over half the study cases (54%) included some kind of public participation. In most of them (52.6%), the participants were experts and/or stakeholders, and in 31.6% of them, the feedback came from the general public, while the remaining 15.8% were studies with participants belonging to both groups.
- Regarding the decision support methods, only 11% of the optimization studies incorporated public participation, while 67% of the studies using CBA and MVM, 55% of the papers using the MCDA, and 87% of those using other hybrid approaches included some kind of public involvement.

- As for the type of participants, 59% of the papers using the MCDA and other methods included participation of experts and/or stakeholders for only 22% of the CBA/MVM case studies. On the contrary, 56% of these latter studies included participation of the general public for only 17.6% of the MCDA and other methods.
- The kind of participation of the general public was mainly passive, while experts and/or stakeholders seem to have a more active role in the participation process.

Other conclusions refer to the very few case studies using CBA and MVM that included indicators to assess the social aspects of sustainability. On the contrary, in the studies using MCDA and other hybrid methods, social sustainability indicators were considered.

The following consequences of the current situation can be deduced from the analysis of the levels of public participation:

- Since the social aspects of sustainability are given less attention than the economic and ecological ones, it seems that there is a need for regionally defined social indicators with stakeholders' participation. New insight to social indicators could be provided by increasing public participation through group decision support systems.
- The need to use both quantitative and qualitative data makes MCDA a powerful tool to assess social sustainability, probably in conjunction with other methods such as cognitive mapping and SWOT to help in the problem structuring phase.
- Whatever method is used, it is important that the feedback along the decision process is properly addressed and the effect of different decisions on the stakeholders and the general public is correctly communicated.

1.6.3 Suitability of the Tools for Participatory Decision-Making

SFM is an increasingly complex process, which must combine the conventional forestry background with the quantitative techniques described in Section 1.6.1 and the participation of new stakeholders. Sometimes there are social aspects that influence forest management and that are difficult to incorporate into a quantitative model. In such cases, a qualitative model may be more appropriate. In summary, both kinds of models are used, sometimes in combination.

An important part of any modeling effort is user needs assessment. In participatory decision-making, the users are not only foresters but a wide range of stakeholders, and usability is the key concept regarding the tools to be used, which must comply with a series of requirements.

According to Lawrence and Stewart (2011), usability is enhanced when the tool is easy to use, considers the needs of the intended users, and has been developed and tested in collaboration with them.

A good design is basic since the tools must be accessible to all users, allow for iterative integration of experience, and have a self-explanatory user interface. User needs must be understood and taken into account by forest planners and decision-makers. Both users and planners must work together in the development of forest management tools, for example, by participating in the selection of the indicators that are included in decision support systems. Stakeholders' involvement during the development process ensures that the resulting tools are adapted to local conditions and produce credible outcomes. As stated in Section 1.3.2, when stakeholders have the opportunity to express their opinions and criticize model assumptions from the beginning of the process, they perceive the model as theirs and tend to accept the results obtained.

Once the model has been developed, the next step is testing it with the users. Testing is usually carried out as a pilot study, which allows observing how users get familiar with the tools and the problems they encounter and assesses the effectiveness of the designed tools. The testing process provides some useful feedback to improve the tools. Lawrence and Stewart (2011) mention the testing of two models based on MCDA in Canada. One of them taught the developers that they could simplify the process because users did not need to be trained to use the tool properly.

Voting methods have been extensively tested in Finland. They have a series of advantages, that is, many people can participate, it is a usually transparent and familiar process, it is easy to understand, and its results are easy to interpret. However, depending on the voting method, different results may be obtained leading to different decisions. In addition, voting can be manipulated if some people's votes have a greater impact than others. This problem is not exclusive of this method, but it is typical of any analysis where group preferences are assembled.

The goal of this participatory process is to make better forest management decisions. A better decision may be a decision with a higher acceptance degree, a decision that avoids conflict, or a decision that takes into account more information from local sources leading to a more efficient outcome.

The way to assess the impact of public participation on the quality of a management decision is to carry out a testing process of the methodology used by looking at the effect that the model has on the outcome. There is a lack of practical case studies that examine whether these participatory tools have actually led to better management decisions.

1.6.4 HIGHLIGHTS OF THE PROPOSED METHODOLOGY

In view of the current techniques to incorporate public participation in forest management (Sections 1.6.1 through 1.6.3) and their main deficiencies (Section 1.1.4), the methodology we propose has two main features to recommend it. First, it is open to participation by all interested parties. Second, the choice is based on direct comparison of representative sustainability scenarios through a process of pair-wise comparison. Both these characteristics determine the type of quantitative technique that we propose to apply.

The first of these features is achieved by adopting a web-based collective decision-making system (Watkins and Rodriguez, 2008). On the one hand, web-based decision-making provides two potential benefits. One, new knowledge is generated statistically by merging independent and individual judgments and is freely accessible to numerous potential users. Two, the final solution obtains benefits from expertise.* On the other hand, by using the term "collective decision," we emphasize the need to get away from the concept of group decision support systems—as the decisions we seek are not necessarily collaborative in nature (DeSanctis and Gallupe, 1987)—and to move closer to social software.

To fulfill the second characteristic, rather than using the participatory process to choose between different forest management alternatives (which in any case must represent every possible management scenario), we ask each individual directly for their sustainability preferences from pairs of representative real scenarios. The answer to these questions produces a representation of individual preferences. Now, the application of algorithms to arrive at a numerical representation of preferences gives a sustainability assessment, which is consistent with each individual system of preferences. This procedure allows the classic process of multicriteria analysis (MCA) to be reversed and permits a statistical model to be fitted in order to calculate the overall assessment of sustainability from the measures of as many criteria as possible. It is therefore possible to determine how each evaluator combines sustainability indicators to achieve his or her own assessment of sustainability.

The reverse use of MCA reduces the importance of structuring objectives and therefore makes the application of quantitative techniques† a less critical part of the assessment process. However, these techniques still play a role in defining the way the public is informed about the complexity of the system, although other concepts derived from applying quantitative techniques are even more important in designing the information: the type of rationality or coherence in the opinions of each individual and the extent of the individual's knowledge of the system to be evaluated.

Hence, each individual's assessment is compared to the sustainability assessment derived from the preferences of other evaluators, and the relative location of each individual assessment can also be seen on a scale of null-maximum sustainability (obtained from a structured model of sustainability for the study area). Both comparisons make it easy for each individual either to modify his or her preferences or to reinforce his or her opinions.

---

* Condorcet's jury theorem (Condorcet, 1785) states that if each individual in a collective is more likely to be correct than not to be, then as the size of the group scales, the probability of the collective decision being correct moves toward certainty. If the participation of those who are more likely not to be correct is discouraged, then the probability of making a right decision increases.

† Like *cognitive* or *causal mapping* (CM), a visual approach to thinking where ideas are shown as nodes and where the links between nodes represent causality or influence, or SODA, which builds on cognitive mapping, is used to aid understanding and structuring subjective concerns and competing objectives through workshops, interviews, and analysis (Tikkanen et al, 2006). The main use of these techniques is to provide tools for problem structuring and defining criteria and decision alternatives.

Collective decision-making tools can also be applied further down the line. In general, public participation involves weighting stakeholders, whose relative level of influence can be aggregated into scores by applying quantitative MCA. Alternatively, voting models using social choice theory can be used (Kangas et al., 2006; Laukkanen et al., 2002) to transform these individual preferences into a collective choice. This requires the voting model, the voting procedure, and the voting method to be outlined.

Finally, specific techniques make it possible to design the forest management alternative that maximizes the individual or collective concept of sustainability. Here, Martins and Borges (2007) suggest that heuristic approaches may be more appropriate to deal with the complexity of multiobjective and multiowner scenarios, although linear and goal programming is used to characterize the best forest management alternative.

Transparency, rigor, and robustness are key requirements throughout the whole process. Transparency is crucial for the social acceptance of the decision-making tools, methods, and their ultimate outcomes (Martins and Borges, 2007). And in order to predict the assessment of sustainability for each system of preferences, modeling must accommodate the scope and complexity of natural resource management (Mendoza and Prabhu, 2006).

REFERENCES

Aledo, A., L.R. Galanes y J.A. Ríos, 2001. Éticas para una sociología ambiental, en *Sociología Ambiental*, capítulo VII. Grupo Editorial Universitario, Granada, Spain.

Allan, C. and A. Curtis, 2005. Adaptive management of natural resources. In: *Proceedings of the 5th Australian Stream Management Conference*. Australian Rivers: Making a Difference, Albury, 21-25 May. Institute for Land, Water and Society, Charles Sturt University, Thurgoona. New South Wales. Australia.

André, P., B. Enserink, D. Connor, and P. Croal, 2006. *Public Participation International Best Practice Principles*. Special Publication Series No. 4. Fargo, ND: International Association for Impact Assessment.

Bass, S., 2004. Certification. *Encyclopedia of Forest Sciences*. Academic Press, Oxford, U.K., pp. 1350–1357.

Bertalanffy, L., 1968. *General System Theory*. New York: George Braziller Publisher.

Bolt, K., M. Matete, and M. Clemens, 2002. *Manual for Calculating Adjusted Net Savings*. Environment Department, World Bank.

Bousquet, F. and C. Le Page, 2004. Multi-agent simulations and ecosystem management: A review. *Ecological Modelling*, 76 (3–4): 313–332.

Brenes, A., 2002. The earth charter principles: Source for an ethics of universal responsibility. In: Miller, P. and L. Westra (eds.), *Just Ecological Integrity: The Ethics of Maintaining Planetary Life*, pp. 26–36. Oxford, U.K.: Rowman & Littlefield.

Buchy, M. and S. Hoverman, 2000. Understanding public participation in forest planning: A review. *Forest Policy and Economics*, 1(1): 15–25.

Caparrós Gass, A., 2009. *Contabilidad nacional verde en el sector forestal: de la teoría a la práctica*. 5° Congreso forestal español. Centro Municipal de Congresos y Exposiciones de Ávila. Spain.

Catton, W.R. and R.E. Dunlap, 1978. Environmental sociology: A new paradigm. *The American Sociologist*, 13: 41–49.

Clark, M.R. and J.S. Kozar, 2011. Comparing sustainable forest management certifications standards: A meta-analysis. *Ecology and Society*, 16(1): 3.

Clark, B. and R. York, 2005. Dialectical materialism and nature: An alternative to economism and deep ecology. *Organization and Environment*, 18 (3): 318–337.

Condorcet, 1785. *Essai sur l'application de l'analyse à la probabilité des decisions rendus à la probabilité des voix*. Paris, France: De l'imprimerie royale.

Conte, R., R. Hegselmann, and P. Terna, 1997. Simulating social phenomena. *Lecture Notes in Economics and Mathematical Systems*, p. 456. Berlin, Germany: IEEE Computer Society Press.

Costanza, R., 1991. The ecological economics of sustainability: Investing in natural capital. In: Goodland, R., H. Daly, S. El Serafy, and B. von Droste (eds.), *Environmentally Sustainable Economic Development: Building on Brundtland*, pp. 83–90. New York: Unesco.

Daly, H., 1989. *Toward a Measure of Sustainable Net National Product. Environmental Accounting for Sustainable Development*. UNEP, World Bank Symposium, Washington, DC.

Daly, H.E. and J. Cobb, 1989. *For the Common Good: Redirecting the Economy towards Community, the Environment, and a Sustainable Future*. Boston, MA: Beacon Press.

DeSanctis, G. and R.B. Gallupe, 1987. A foundation for the study of group decision support systems. *Management Science*, 33 (5): 589–609.

Dietz, T., E. Rosa, and R. York, 2007. Driving the human ecological footprint. *Frontiers in Ecology and the Environment*, 5: 13–18.

Durán, G., 2000. *Medir la sostenibilidad: Indicadores económicos, ecológicos y sociales*. Albacete, España: VII Jornadas de Economía Crítica.

Edmonds, B., S. Moss, and P. Davidson, 2001. The Use of Models—making MABS actually work. In: *Multi-Agent-Based Simulation, Lecture Notes in Artificial Intelligence 1979*, pp. 15–32. Berlin, Germany: Springer-Verlag.

Ehlrich, P.R. and J.P. Holdren, 1971. Impact of population growth. *Science*, New Series, 171(3977): 1212–1217.

El Serafy, S., 1989. *The Proper Calculation of Income from Depletable Natural Resources*. Washington, DC: UNEP/World Bank on Environmental Accounting.

Falconí, F., 2001. *An Integrated Economic-Environmental Assessment of the Ecuadorian Economy*. Tesis doctoral. Universidad Autónoma de Barcelona, Barcelona, Spain.

FAO (Food and Agriculture Organization of the United Nations), 2001. *State of the World's Forests 2001*. FAO, Rome, Italy, 181pp.

Foster, J.B., 2005. The treadmill of accumulation. *Organization & Environment*, 18: 7–18.

Freese, L., 1997. *Environmental Connections*. Greenwich, CT: JAI Press.

Friedman, T., 2008. *Hot, Flat, and Crowded: Why we Need a Green Revolution and How it Can Renew America*. New York: Farrar, Straus, and Giroux.

Frittaion, C.M., P.N. Duinker, and J.L. Grant, 2010. Suspending disbelief: influencing engagement in scenarios of forest futures. *Technological Forecasting and Social Change*, 78(3): 421–430.

Forrester, J.W., 1961. *Industrial Dynamics*. Cambridge, MA: The MIT Press. Reprinted by Pegasus Communications, Waltham, MA.

Gafo Gómez-Zamalloa, M., A. Caparrós, and A. San-Miguel Ayanz, 2011. Fifteen years of Forest Certification in the European Union. Are we doing things right? *Forest Systems*, 20(1): 81–94.

Germain, R.H., D.W. Floyd, and S.V. Stehman, 2001. Public perceptions of the USDA Forest Service public participation process. *Forest Policy and Economics*, 3: 113–124.

Gilbert, N., 2002. Varieties of emergence. In: En Macal, C. and Sallach, D. (eds.), *Social Agents: Ecology, Exchange, and Evolution. Agent 2002 Conference*, pp. 41–50. Chicago, IL: University of Chicago and Argonne National Laboratory.

Gilbert, N., 2008. *Agent-Based Models*. London, U.K.: Sage Series in Quantitative Applications, vol. 153.

Gilbert, N. and P. Terna, 2000. How to build and use agent-based models in social science. *Mind and Society*, 1 (1): 57–72.

Gilbert, N. and K.G. Troitzsch, 1999. *Simulation for the Social Scientist*. Buckingham, U.K.: Open University Press.

Grober, U., 2007. Deep roots—A conceptual history of 'sustainable development' (Nachhaltigkeit) Best.-Nr. P 2007–002 Wissenschaftszentrum Berlin für Sozialforschung (WZB). Berlin.

Gunderson, L.H. and C.S. Holling, 2002. *Panarchy: Understanding Transformations in Human and Natural Systems*. Washington, DC: Island Press Center for Resource Economics.

Hamilton, K., 2000. *Genuine Saving as a Sustainability Indicator*. World Bank Environmental Economics Series No. 77, Washington, DC.

Harich, J., 2010. Change resistance as the crux of the environmental sustainability problem. *System Dynamics Review*, 26(1): 35–72.

Hartwick, J.M., 1977. Intergenerational equity and the investment of rents from exhaustible resources. *The American Economic Review*, 67: 972–974.

Hicks, R., 1939. *Value and Capital: An Inquiry into Some Fundamental Principles of Economic Theory*. Oxford, U.K.: Clarendon Press.

Holland, J.H., 1998. *Emergence. From Chaos to Order*. Reading, MA: Addison-Wesley.

Holling, C.S., 1973. Resilience and stability of ecological systems. *Annual Review of Ecology and Systematics*, 4: 1–23.

Hotelling, H., 1931. The economics of exhaustible resources. *Journal of Political Economics*, 35 (2): 137–175.

IAP2 (International Association for Public Participation), 2000. *Planning Effective Public Participation (Module 1)*. International Association for Public Participation, Denver, CO.

IAP2 (International Association for Public Participation), 2007. *IAP2 Core Values of Public Participation*. [online] N.p., July 12, 2007. Web August, 15 2012. http://www.iap2.org/associations/4748/files/CoreValues.pdf

International Association for Public Participation (IAP2), 2000. *Victorian Parliament Outer Suburban/Interface Services and Development Committee Report—Building New Communities*.

International Association for Public Participation (IAP2), 2007. *IAP2 Core Values of Public Participation*, http://www.iap2.org/associations/4748/files/CoreValues.pdf

Johnson, S., 2001. *Emergence: The Connected Lives of Ants, Brains, Cities, and Software*. New York: Scribner.

Johnston, R.J. and S.K. Swallow, 1999. Asymmetries in ordered strength of preference models: Implications of focus shift for discrete choice preference estimation. *Land Economics*, 75(2): 295–310.

Jorgenson, A.K. and T.J. Burns, 2007. The political-economic causes of change in the ecological footprints of nations, 1991–2001: A quantitative investigation. *Social Science Research*, 36: 834–853.

Jorgenson, A.K. and B. Clark, 2009. The economy, military and ecologically unequal exchange relationships in comparative perspective: A panel study of the ecological footprints of nations, 1975–2000. *Social Problems*, 56(4): 621–646.

Kangas, A., S. Laukanen, and J. Kangas, 2006. Social choice theory and its applications in sustainable forest management: A review. *Forest Policy and Economics*, 9: 77–92.

Kates, R.W., W.C. Clark, R. Corell, J.M. Hall, C.C. Jaeger, I. Lowe, J.J. McCarthy et al., 2001. Sustainability science. *Science*, 292(5517): 641–642.

Kidd, C.V., 1992. The evolution of sustainability. *Journal of Agricultural and Environmental Ethics*, 5(1): 1–26.

Laukkanen, S., A. Kangas, and J. Kangas, 2002. Applying voting theory in natural resource management: A case of multiple-criteria group decision support. *Journal of Environmental Management*, 64: 127–137.

Lavandeira, X., C. León, and y M. Vázquez, 2007. *Economía ambiental*. Madrid, Spain: Pearson Prentice Hall.

Lawrence, A. and A. Stewart, 2011. Sustainable forestry decisions: On the interface between technology and participation. *Mathematical and Computational Forestry & Natural-Resource Sciences*, 3(1): 42–52.

Linkov, I., F.K. Satterstrom, G. Kiker, C. Batchelor, T. Bridges, and E. Ferguson, 2006. From comparative risk assessment to multi-criteria decision analysis and adaptive management: Recent developments and applications. *Environment International*, 32: 1072–1093.

López Paredes, A. and C. Hernández Iglesias, 2008. *Agent Based Modelling in Natural Resource Management*. Insisoc. España. ISBN 978-84-205-4560-8.

Mackey, B.G., 2004. The earth charter and ecological integrity—Some policy implications. *Worldviews: Environment, Culture, Religion*, 8: 76–92.

Maestu, J., D. Tabara, and D. Sauri, 2003. *Public Participation in River Basin Management in Spain. Reflecting Changes in External and Self-Created Contexts*. Workpackage 4 HarmoniCOP Project. University of Alcalá de Henares. Madrid. Spain.

Martins, H. and J.G. Borges, 2007. Addressing collaborative planning methods and tools in forest management. *Forest Ecology and Management*, 248: 107–118.

MCPFE (Ministerial Conference on the Protection of Forests in Europe), 2003. *Improved Pan-European Indicators for Sustainable Forest Management*. MCPFE Liaison Unit, Vienna, Austria, 6pp.

Mendoza, G.A. and R. Prabhu, 2006. Participatory modeling and analysis for sustainable forest management: Overview of soft system dynamics models and applications. *Forest Policy and Economics*, 9(2): 179–196.

Miller, P. and L. Westra (eds.), 2002. *Just Ecological Integrity: The Ethics of Maintaining Planetary Life*. Lanham, MD: Rowman & Littlefield Publishers, Inc.

Mol, A.P.J. and G. Spaargaren, 2000. Ecological modernization theory in debate: A review. *Environmental Politics*, 9(1): 17–49.

Mol, A.P.J., G. Spaargaren and D.A. Sonnenfeld, (eds.), 2009. *The Ecological Modernisation Reader: Environmental Reform in Theory and Practice*. New York: Routledge Press.

Moore, F.C., 2011. Toppling the tripod: Sustainable development, constructive ambiguity, and the environmental challenge consilience. *The Journal of Sustainable Development*, 5(1): 141–150.

Mortensen, L.F., 1997. The driving force-state-response framework used by CSD. In: Moldan, B. and S. Billharz (eds.), *Sustainability Indicators*. A Report on the Project on Indicators of Sustainable Development. SCOPE Report 58. John Wiley & Sons: Chichester. UK.

Myllyviita, T., T. Hujala, A. Kangas, and P. Leskinen. 2011. Decision support in assessing the sustainable use of forests and other natural resources—A comparative review. *The Open Forest Science Journal*, 4: 24–41.

Neumayer, E., 2003. *Weak versus Strong Sustainability: Exploring the Limits of Two Opposing Paradigms*, 2nd edn. Cheltenham, U.K.: Edward Elgar Publishers.

North, J.M. and C.M. Macal, 2007. *Managing Business Complexity: Discovering Strategic Solutions with Agent-Based Modeling and Simulation*. New York: Oxford University Press.

Ostrom, E., R. Gardner, and J. Walker, 1994. *Rules, Games, and Common-Pool Resources*. Ann Arbor, MI: University of Michigan Press

Parkins, J.R., 2006. De-centering environmental governance: A short history and analysis of democratic processes in the forest sector of Alberta, Canada. *Policy Sciences*, 39: 183–203.

Parkins, J.R., 2010. The problem with trust: Insights from advisory committees in the forest sector of Alberta. *Society and Natural Resources*, 23: 822–836.

Paton, R., R. Gregory, C. Vlachos, J. Saunders, and H. Wu, 2004. Evolvable social agents for bacterial systems modeling. *IEEE Transactions on Nanobioscience*, 3(3): 208–216.

Pearce, D. and G. Atkinson, 1992. *Are National Economies Sustainable? Measuring Sustainable Development*. CSERGE GEC Working Paper 92–11. University College London, London, U.K., p. 18.

Pearce, D.W. and R.K. Turner, 1990. *Economics of Natural Resources and the Environment*. Baltimore, MD: Johns Hopkins University Press.

Pearce, D. and G. Atkinson, 1995. Measuring sustainable development. Chapter 8 In D.W. Bromley ed., *The Handbook of Environmental Economics*. Oxford, U.K.: Blackwell.

Pearce, D.W. and J.J. Warford, 1993. *World Without End: Economics, Environment and Sustainable Development*. New York: Oxford University Press for the World Bank, 440pp.

Prabhu, R., C.J.P. Colfer, and R.G. Dudley, 1999. Guidelines for developing, testing and selecting criteria and indicators for sustainable forest management. Center for International Forestry Research, Jakarta, Indonesia, p. 186.

Reed, M.S., 2008. Stakeholder participation for environmental management: A literature review. *Biological Conservation*, 141: 2417–2431.

Reed, M.G. and J. Varghese, 2007. Gender representation on Canadian Forest Sector advisory committees. *Forestry Chronicle*, 83: 515–525.

Reynolds, C.W., 1987. Flocks, herds, and schools: A distributed behavioral model. *Computer Graphics*, 21(4): 25–34.

Richardson, G.P., 1991/1999. *Feedback Thought in Social Science and Systems Theory*. Philadelphia, PA: University of Pennsylvania Press. Reprinted by Pegasus Communications, Waltham, MA.

Richardson, G.P. and D.F. Andersen, 2010. Systems thinking, mapping, and modeling for group decision and negotiation. In: Eden, C. and D.N. Kilgour (eds.), *Handbook for Group Decision and Negotiation*. Dordrecht, the Netherlands: Springer, pp. 313–324.

Sadler, B., 1996. *Environmental Assessment in a Changing World: Evaluating Practice to Improve Performance*. Final Report of the International Study of the Effectiveness of Environmental Assessment. Canadian Environmental Assessment Agency and International Association for Impact Assessment, p.263, Ottawa, Canada.

Schnaiberg, A. and K.A. Gould, 2000. *Environment and Society: The Enduring Conflict*. West Caldwell, NJ: Blackburn Press.

Selin, S. and D. Chavez, 1995. Developing a collaborative model for environmental planning and management. *Environmental Management*, 19(2): 189–195.

Sen, A.K., 1987. *On Ethics and Economics*. London, U.K.: Blackwell Publishing Ltd., 131pp.

Smith, C.L., V.L. Lopes, and F.M. Carrejo, 2011. Recasting paradigm shift: "True" sustainability and complex systems. *Human Ecology Review*, 18(1): 67–75.

Solow, R.M., 1974. The economics of resources or the resources of economics. *The American Economic Review*, 64: 1–14.

Solow, R., 1986. On the intergenerational allocation of natural resources. *Scandinavian Journal of Economics*, 88(1): 141–145.

Squazzoni, F., 2008. The micro-macro link in social simulation. *Sociologica*, 1/2008, doi: 10.2383/26578.

Sterman, J.D., 2000. *Business Dynamics: Systems Thinking and Modeling for a Complex World*. Boston, MA: McGraw Hill.

Sugden, R., 1993. Welfare, resources, and capabilities: A review of inequality reexamined by Amartya Sen. *Journal of Economic Literature*, 31(4): 1947–1962.

Tikkanen, J., T. Isokaanta, J. Pykalainen, and P. Leskinen, 2006. Applying cognitive mapping approach to explore the objective-structure of forest owners in a Northern Finnish case area. *Forest Policy and Economics*, 9: 139–152.

Tuler, S. and T. Webler, 2010. How preferences for public participation are linked to perceptions of the context, preferences for outcomes, and individual characteristics. *Environmental Management*, 46: 254–267.

UNDP, 1992. *Agenda 21, Programme of Action for Sustainable Development*. Rio de Janeiro, Brazil.

UNECE, 1998. *Convention on Access to Information, Public Participation in Decision-Making and Access to Justice in Environmental Matters*, Aarhus, Denmark.

United Nations, 1993. *International Trade Statistics Yearbook*, Vol. 1. New York: Department for Economic and Social Information and Policy Analysis.

United Nations, 2007. *Indicators of Sustainable Development: Guidelines and Methodologies*. New York: United Nations.

Vennix, J.A.M., 1996. *Group Model Building: Facilitating Team Learning Using System Dynamics*. Chichester, England: Wiley.

Wackernagel, M. and W.E. Rees, 1996. *Our Ecological Footprint: Reducing Human Impact on the Earth*. Philadelphia, PA: New Society Publishers, 160pp.

Walker, D.C., G. Hill, R.H. Smallwood, and J. Southgate, 2004a. Agent-based computational modelling of wounded epithelial cell monolayers. *IEEE Transactions on Nanobioscience*, 3: 153–163.

Walker, D.C., J. Southgate, G. Hill, M. Holcombe, D.R. Hose, S.M. Wood, S. Mac Neil, and R.H. Smallwood, 2004b. The epitheliome: Agent-based modelling of the social behaviour of cells. *Biosystems*, 76(1–3): 89–100.

Watkins, J.H. and M.A. Rodriguez, 2008. A survey of web-based collective decision making systems. In: Nayak, R., N. Ichalkaranje, and L.C. Jain (eds.), *Studies in Computational Intelligence: Evolution of the Web in Artificial Intelligence Environments*, pp. 245–279. Berlin, Germany: Springer-Verlag.

Weitzman, M.L., 1976. On the welfare significance of national product in a dynamic economy. *Quarterly Journal of Economics*, 90: 156–162.

Weitzman, M.L., 2001. A contribution to the theory of welfare accounting. *Scandinavian Journal of Economics*, 103(1): 1–23.

World Bank, 2001. *Making Sustainable Commitments: An Environment Strategy for the World Bank*. Washington, DC: World Bank, 274pp.

World Commission on Environment and Development (WCED), 1987. *Our Common Future*. Oxford, U.K.: Oxford University Press.

World Economic Forum (WEF), 2001. *Indicators of Sustainable Development: Framework and Methodologies*. Department of Economics and Social Affairs, New York.

York, R., E. Rosa, and T. Dietz, 2009. A tale of contrasting trends: Three measures of the ecological footprint in China, India, Japan, and the United States, 1961–2003. *Journal of World-Systems Research*, 25: 134–146.

2 Inventory Techniques in Participatory Forest Management

Cristina Pascual, Francisco Mauro,
Ana Hernando, and Susana Martín-Fernández

CONTENTS

2.1 INTRODUCTION

The participatory approach allows the representation of a broad range of interests and enables participants to be fully involved in the whole planning process from its inception to the implementation of the decision and the monitoring process. All participants acquire and share information, and this becomes a key feature of the process (Moote et al., 1997). Within the scope of sustainable forest management, traditional methods such as surveying and sampling (plots, trees) are supplemented by spatial information (mapping). Sampling techniques, geographic information systems (GISs), and remote sensing are thus essential tools for collecting, generating, and integrating personal perceptions and the environmental, economic, and social data required in the planning and decision-making process (Pfeiffer et al., 2008).

This approach, integrating geospatial tools and methods designed to represent people's spatial knowledge using physical or virtual media to help in the learning, discussion, and exchange of information, is known as "public participation GIS" (PPGIS) in the analysis and decision-making process (Bernard et al., 2011).

2.1.1 Chapter Content

This chapter is divided into four main sections. Section 2.2 describes the classic techniques for sampling static populations (i.e., sampling by means of surveys, plots, or trees). Sections 2.3, 2.4 and 2.5 discuss information technologies (IT) such as GIS and remote sensing. Specifically, Section 2.3 examines the main algorithms for spatial interpolation of data. Section 2.4 reviews the concept of PPGIS, and Section 2.5 explores the foundations of remote sensing and its potentialities for capturing and generating spatial information. This section pays special attention to object-based image classifications and light detection and ranging (LiDAR) data and their applications in forest management. Finally, Section 2.6 presents and discusses three applications of these geospatial tools, namely, (i) object-oriented classification approach for mapping habitats, (ii) forest structure characterization maps using *LiDAR* data and expert opinion, and (iii) current issues in the *LiDAR* area-based approach (ABA): co-registration error, accuracy of predictions, and modeling diametric distributions.

2.2　CLASSIC SAMPLING TECHNIQUES IN STATIC POPULATIONS

Forest inventory may be defined as the technique of collection, analysis, presentation, and interpretation of forest data. Since the size of the populations is very high, data are obtained through different sampling techniques. Sampling can be defined as the procedure used to choose a representative subset of individuals from a population. This information enables the population and the decision-making process to be characterized. There are three main types of sampling:

Probabilistic sampling is when the probability of every element to be chosen is previously known.

Intentional non-probabilistic sampling (or purposive sampling) is a type of non-probability sampling in which the researcher consciously selects specific elements or subjects for inclusion in a study. The objective of this method is to ensure that the elements will have certain characteristics that are relevant to the study.

Finally, *accidental sampling* is a type of non-probability sampling in which the population selected is easily accessible to the researcher; available subjects are simply entered into the study without any attempt at randomization.

In forestry, the sampling elements are usually trees or plots of different shapes and sizes. It is frequently necessary to carry out a prior pilot sampling to determine the variability of the data in order to select the best plot size.

The most commonly estimated population parameters are population mean, population total, proportion of population elements that possess a certain qualitative characteristic, and sample variance.

There are various types of sampling methods according to the characteristics of the population and the information available on the variables that have a wide application in forestry. The sampling error and the size of the sample will depend on the method chosen. These methods differ if the probability of choosing the elements of a population is the same or not and if the sampling involves the replacement of the elements. These most commonly applied methods are the following:

- Simple random sampling: Every element of the population has the same probability of being chosen. Therefore, all the samples of size *n* that can be chosen in a population have the same probability of being selected.
- Stratified random sampling: This method is applied when the population can be split into various subpopulations or strata that are more homogeneous than the population as a whole. Each stratum is sampled.
- Ratio and regression sampling.
- Double sampling, also known as double sampling with regression or ratio estimator, is a type of forest sampling in which the auxiliary variable is the same as the primary variable measured in a previous period.
- Probability sampling methods. In forest inventories, it is common to apply a method in which the probability of choosing each element of the population changes. This probability assigned to each element is often proportional to the size of some characteristic of the element. This is the reason they are known as probability proportional to size (PPS).

2.2.1 SIMPLE RANDOM SAMPLING

In this method, the sampling elements are selected as an independent random sample from the population. Each element of the population has the same probability of being selected. Likewise, each combination of *n* sampling elements has the same probability of being eventually selected.

The advantages of this type of sampling are its simple design and its clearly known estimators.

The main disadvantages are the difficulties in allocating the sampling elements in the field; if the populations are heterogeneous, the sampling errors are high, and the whole population may not be represented in the sample.

Some drawbacks can be avoided by applying systematic sampling methods; these are applied when the elements of the population can be ordered in a list or on the terrain, and the variable is uniformly distributed throughout the whole population.

In this case, the first element of the sample is drawn from among the first *k* elements of the population when they are listed. The next elements of the sample are chosen from every *k* elements.

The advantage of this type of sampling is that the selection process is very easy; the drawback is that all the possible samples need to be known in order to calculate the errors of the estimators, which is often impossible.

TABLE 2.1

Population Parameters and Estimators in Simple Random Sampling

Parameter	Population Value	Sample-Based Estimator
Mean	$\mu = \dfrac{\sum_{i=1}^{N} x_i}{N}$	$\bar{x} = \dfrac{\sum_{i=1}^{n} x_i}{n}$
SD	$\sigma = \sqrt{\dfrac{\sum_{i=1}^{N} (x_i - \mu)^2}{N}}$	$S_{n-1} = \sqrt{\dfrac{\sum_{i=1}^{n} (x_i - \bar{x})^2}{n-1}}$
Standard error (without replacement or from a finite population)	$\sigma_{\bar{x}} = \sqrt{\dfrac{N-n}{N-1}} \dfrac{\sigma}{\sqrt{n}}$	$S_{\bar{x}} = \sqrt{\dfrac{N-n}{N-1}} \dfrac{S_{n-1}}{\sqrt{n}}$
Standard error (with replacement or from an infinite population)	$\sigma_{\bar{x}} = \dfrac{\sigma}{\sqrt{n}}$	$S_{\bar{x}} = \dfrac{S_{n-1}}{\sqrt{n}}$

where

N is the population size

n is the sample size

x_i is the observed value of the ith sampling element

They are used in forest inventories when the sampling elements are plots distributed like a grid or in equidistant strips in the field.

The main estimators of simple random sampling method are shown in Table 2.1.

2.2.1.1 Estimation of the Population Total

The total amount of a population total (τ) can be useful in forestry, for example, when values such as the total volume of the timber in the forest have to be estimated.

This estimator is obtained from the sampling mean:

$$\hat{\tau} = N \bar{x} \tag{2.1}$$

The expressions of the standard error and deviation are the same as for the mean but multiplied by N.

2.2.1.2 Estimator of the Population Proportion

When estimating the proportion of individuals or elements in a population that have a specific characteristic, the best estimator of the population parameter is

$$\hat{\mu} = \bar{x} = \hat{p} = \frac{a}{n} \tag{2.2}$$

where a is the number of elements in the sample with the target characteristic.

The standard error is

$$\sigma_p = \sqrt{\frac{pq}{n}\left(1 - \frac{n}{N}\right)} \tag{2.3}$$

2.2.1.3   Sampling Error

The sampling error, e, shows the accuracy of the estimator. It is computed from the probability:

$$P\big[|T - \theta| \le e\big] = 1 - \alpha \tag{2.4}$$

where

   T is any sampling estimator
   θ the population parameter to be estimated

According to Chebyshev's theorem, this probability depends on the variance of the estimator and on its probability distribution.

   If the estimator follows the normal distribution or is unbiased and the size of the sample is greater than 30, the confidence interval of the population parameter will be

$$\hat{\theta} \pm z_{\alpha/2}\sqrt{V(\hat{\theta})} \tag{2.5}$$

The estimation error (e) cannot usually be calculated with the previous expression as the population variance is unknown.

   Applying the estimator of the variance, the approximate value of the error is

$$e \cong z_{\alpha/2}\sqrt{\hat{V}(\hat{\theta})} \tag{2.6}$$

When the estimator does not have a normal distribution, it is computed from Markov's inequality that the limit error of the sampling with a confidence level of $1-\alpha\%$ is

$$e \cong \sqrt{(1/\alpha)\hat{V}(\hat{\theta})} \tag{2.7}$$

So its general expression is

$$e \cong d\sqrt{\hat{V}(\hat{\theta})} \tag{2.8}$$

where d has a different value depending on the probability distribution of the estimator or if it is unknown.

   The expression of the relative error can be also obtained from this expression as

$$e_r = \frac{e}{x}100 \tag{2.9}$$

2.2.1.4   Size of the Sample

The size of the sample will depend on the value of the error e and on the cost of the sampling.

- Size of the sample for the estimation of the population mean
  The estimation of the size of the sample can be obtained from the expression of the standard error of the mean:

$$e^2 = d^2 \frac{N-n}{N} \frac{s^2}{n} \tag{2.10}$$

So, without a replacement sampling, we obtain

$$n = \frac{(ds/e)^2}{1+(ds/e)^2 (1/N)} = \frac{n_0}{1+n_0/N} \tag{2.11}$$

where

$$n_0 = \left(\frac{ds}{e}\right)^2 \tag{2.12}$$

The value of d is initially chosen as $1.96 \approx 2$ (value of the normal distribution for a confidence of 95%). If in case the sample value is less than 30, it must be recalculated, obtaining d from the t-Student distribution with $n'-1$ degrees of freedom, where n' is the sample size obtained from expression (2.11).

Another way to calculate this size is using the relative error:

$$n_0 = \left(\frac{\widehat{cv}(x)d}{e_r}\right)^2 \tag{2.13}$$

where

$$\widehat{cv}(x) = \frac{s}{\bar{x}} 100 \tag{2.14}$$

- Size of the sample for the estimation of the population total
  The process is the same as explained earlier:

$$e^2 = d^2 \frac{N(N-n)}{n} s^2 \tag{2.15}$$

From the previous expression, the value of n is

$$n = \frac{N^2 n_0}{1+n_0 N} \tag{2.16}$$

- Size of the sample for the estimation of the population proportion
  In this case, the expression of the size of the sample is

$$n = \frac{n_0}{1 + (n_0 - 1)/N} \tag{2.17}$$

where n_0 is

$$n_0 = \left(\frac{d}{e}\right)^2 pq \tag{2.18}$$

2.2.2 STRATIFIED SAMPLING

Stratified sampling is a procedure applied when the population is subdivided into separate and more homogeneous subpopulations. The total sample is formed by the stratum subsamples. These subsamples are obtained by independent sampling studies in each stratum. Stratified sampling is efficient especially in those cases where the variability inside the strata is low and the difference of means between the strata is large (Akca, 2001).

These strata must fulfill the condition of nonoverlapping strata (de Vries, 1986), and each one of the strata must have sufficient observations. The main advantages of this method are that subpopulations can be studied separately and the estimations of the population parameters are more accurate.

The criteria used to choose the strata, their number, and the type of sampling inside each one all depend on the specific objectives of the study. In general, simple random sampling is the method most usually applied in the strata, although the systematic method is also widely used in environmental and forest studies.

The stratification criteria can be related to geographic criteria such as ecozones, forest structure, and topographical conditions or to other criteria associated to forest management such as tree sociological classes, age classes, and species. The main population parameters and their estimators are shown in Table 2.2.

The variance of this estimator for a simple random sampling inside every stratum is

$$\hat{V}\left[\bar{x}_{est}\right] = \sum_{h=1}^{L} W_h^2 \hat{V}\left(\bar{x}_h\right) = \sum_{h=1}^{L} W_h^2 \left(\frac{N_h - n_h}{N_h}\right) \frac{s_h^2}{n_h} \tag{2.20}$$

The estimator of the population total is

$$\hat{\tau}_{est} = N\bar{x}_{est} \tag{2.21}$$

TABLE 2.2
Population Parameters and Estimators in Stratified Sampling

Parameter	Population Value	Sample-Based Estimator
Stratum mean	$\mu_h = \dfrac{\sum_{i=1}^{N_h} x_{ih}}{N_h}$	$\bar{x}_h = \dfrac{\sum_{i=1}^{n_h} x_{hi}}{n}$
Stratum variance	$\sigma_h^2 = \sum_{i=1}^{N_h} \dfrac{\left(x_{ih} - \mu_h\right)^2}{N_h}$	$S_{h,n-1} = \sqrt{\dfrac{\sum_{i=1}^{n_h}\left(x_{hi} - \bar{x}_h\right)^2}{n_h - 1}}$
Mean	$\mu = \dfrac{\sum_{h=1}^{L} N_h \mu_h}{N}$	$\bar{x}_{est} = \dfrac{\sum_{h=1}^{L} N_h \bar{x}_h}{N} = \sum_{h=1}^{L} W_h \bar{x}_h$
Variance	$\sigma^2 = \dfrac{\sum_{h=1}^{L} N_h \sigma_h^2}{N} + \dfrac{\sum_{h=1}^{L} N_h \left(\mu_h - \mu\right)^2}{N}$	$S^2 = \dfrac{\sum_{h=1}^{L} N_h S_h^2}{N} + \dfrac{\sum_{h=1}^{L} N_h \left(\bar{x}_h - \bar{x}\right)^2}{N}$

where

N is the size of the population

$$N = \sum_{h=1}^{L} N_h \tag{2.19}$$

L is the total number of strata in the population

N_h the population size of the hth stratum

x_{ih} is the value of the variable for the ith element in strata h

n is the size of the sample and n_h the size of the sample in stratum h

W_h is N_h/N

whose variance, for independent samples, is

$$V\left[\hat{\tau}_{est}\right] = N_1^2 V\left[\bar{x}_1\right] + N_2^2 V\left[\bar{x}_2\right] + \cdots + N_L^2 V\left[\bar{x}_L\right] \tag{2.22}$$

The estimation of a population proportion is

$$\hat{p}_{est} = \frac{1}{N} \sum_{h=1}^{L} N_h \hat{p}_h \tag{2.23}$$

where $\hat{p}_k$ is the proportion in stratum k.

For independent samples, the variance of the estimator is

$$V\left[\hat{p}_{est}\right] = \sum_{h=1}^{L} W_h^2 V\left(\hat{p}_h\right) \tag{2.24}$$

This expression, when the sampling method in every stratum is the simple random method, is

$$V[\hat{p}_{est}] = \sum_{h=1}^{L} W_h^2 V(\hat{p}_h) = \sum_{h=1}^{L} W_h^2 \left(\frac{N_h - n_h}{N_h - 1}\right) \frac{p_h q_h}{n_h} \tag{2.25}$$

If the proportions are unknown, their estimators should be used in the previous expression.

2.2.2.1 Error of Estimates

The standard errors are the typical deviations of the estimators.

With regard to the sampling errors for this type of sampling, the central limit theorem can be applied only if in every stratum $n_h > 30$, then $d = z_{\alpha/2}$ or $d^2 = 1/\alpha$; Markov's theorem must be applied in all other cases.

The sampling error for the mean is

$$e \cong d\sqrt{\hat{V}\left[\bar{x}_{est}\right]} \tag{2.26}$$

The sampling error for the population proportion is

$$e \cong d\sqrt{\hat{V}\left[\hat{p}_{est}\right]} \tag{2.27}$$

The sampling error for the population total is

$$e \cong d\sqrt{\hat{V}\left[\hat{\tau}_{est}\right]} \tag{2.28}$$

2.2.2.2 Size of the Sample

In this type of sampling, the variances of the estimators depend on the size of the sample in the stratum, so it is necessary to establish a relationship between stratum size and total size of the sample, n.

If $n_h = w_h n$, for $h = 1, 2, \ldots, L$, n can be related to error, e, as

$$e \cong d\sqrt{\hat{V}\left[\bar{x}_{est}\right]} = d\sqrt{\sum_{h=1}^{L} W_h^2 \left(\frac{N_h - nw_h}{N_h}\right) \frac{s_h^2}{nw_h}} \tag{2.29}$$

but w_h must be previously allocated, so

$$n = \frac{\sum_{h=1}^{L} \left(N_h^2/w_h\right) s_h^2}{\left(Ne/d\right)^2 + \sum_{h=1}^{L} N_h s_h^2} \tag{2.30}$$

For the population total, the sample size is

$$n = \frac{\sum_{h=1}^{L} \left(N_h^2 / w_h \right) s_h^2}{\left(e/d \right)^2 + \sum_{h=1}^{L} N_h s_h^2}$$

(2.31)

Finally, for the population proportion,

$$n = \frac{\sum_{h=1}^{L} \left(N_h^2 / w_h \right) \hat{p}_h \hat{q}_h}{\left(Ne/d \right)^2 + \sum_{h=1}^{L} N_h \hat{p}_h \hat{q}_h}$$

(2.32)

The total size of the sample must be assigned to the different strata. This process is the allocation of the sample. We can distinguish the following:

Uniform allocation: In this case, the expression of the parameter w_h is

$$w_h = \frac{1}{L} \rightarrow n_h = \frac{n}{L}$$

(2.33)

Proportional allocation:

$$w_h = \frac{N_h}{N} \rightarrow n_h = n \frac{N_h}{N}$$

(2.34)

Optimum allocation: In this case, either the variance or the sampling costs are minimized. When the cost is previously fixed, the total size of the sample is

$$n = \frac{(C - C_0) \sum_{h=1}^{L} N_h S_h / \sqrt{C_h}}{\sum_{h=1}^{L} N_h S_h \sqrt{C_h}}$$

(2.35)

Considering that the cost function is

$$C = c_0 + \sum_{h=1}^{L} C_h n_h$$

(2.36)

where
  C_0 is the fixed costs

C_h the sampling costs in stratum h, then the size of the sample in each stratum is

$$n_h = n \frac{N_h S_h / \sqrt{C_h}}{\sum_{h=1}^{L} N_h S_h / \sqrt{C_h}}$$ (2.37)

When the sampling error is previously fixed, the expression of the size of the sample in each stratum is the same as in the case when the costs are fixed. The expression of the size of the sample is

$$n = \frac{\sum_{h=1}^{L} \frac{N_h}{N} S_h \sqrt{C_h} \sum_{h=1}^{L} \frac{N_h}{N} S_h / \sqrt{C_h}}{V + \sum_{h=1}^{L} \frac{N_h}{N} S_h^2 / N}$$ (2.38)

2.2.3 RATIO ESTIMATORS

Indirect methods of estimation may allow better results than the methods explained earlier when the value of the target variable is known (or suspected) to be highly correlated to another variable (called co-variable). The sampling error and the cost may decrease applying these methods.

The co-variable may be

- Information from the same population at different dates. In this case, the co-variable is the target variable measured before.
- A highly correlated variable that is easier to measure than the target variable.

The ratio estimator does not introduce a new sampling technique, but it incorporates new elements: two variables have to be measured now, and the ratio estimator integrates the co-variable into the estimator.

2.2.3.1 Estimators

The ratio R is the relation between the target variable and the co-variable, for example, number of trees per hectare. From this, the total of the target variable can be estimated.

The population ratio is

$$R = \frac{\mu_Y}{\mu_X}$$ (2.39)

where

μ_Y is the population mean of the target variable

μ_X is the population mean of the co-variable

The estimator of the ratio is

$$r = \frac{\bar{y}}{\bar{x}} \tag{2.40}$$

And the estimated variance of the estimated ratio r is

$$\mathrm{var}(r) = \frac{N-n}{N} \frac{1}{n} \frac{1}{\mu_x^2} \frac{\sum_{i=1}^{n} (y_i - rx_i)^2}{n-1} \tag{2.41}$$

2.2.4 REGRESSION ESTIMATOR

The regression estimator explicitly establishes a simple linear regression between target variable and co-variable. The mean of the target variable from the sample can be estimated as

$$\bar{y}_{rl} = \bar{y} + b_0(\mu_x - \bar{x}) \tag{2.42}$$

where
b_0 is the estimated regression coefficient
μ_x is the population mean of the co-variable that must be known

The estimated variance of the estimated mean is (Cochran, 1977)

$$v(\bar{y}_{rl}) = \frac{N-n}{N} \frac{1}{n} \left(S_y^2 - 2b_0 S_{yx} + b_0^2 S_x^2 \right) \tag{2.43}$$

where $b_0 = S_{yx} / S_x^2$ is the population regression coefficient.
When the population parameters are unknown, they must be estimated using the sampling data.

2.2.5 DOUBLE SAMPLING

The aim of this sampling method is to determine as precisely as possible the population parameters of the co-variable X necessary for the ratio and the regression estimators.

The data collect is in two phases:

In *the first phase*, a sample of size n' is taken to estimate the mean or total of the co-variable X. The sample taken is usually large because measurement of X is cheap, fast, and easy. This sample provides accurate estimators of μ_x with simple random estimators:

$$\overline{x'} = \frac{\sum_{i=1}^{n'} x_i}{n'}$$

(2.44)

$$s_{x'^2} = \frac{\sum_{i=1}^{n'} (x_i - \overline{x'})^2}{n' - 1}$$

(2.45)

$$v(\overline{x'}) = \frac{s_{x'}^2}{n'} \frac{N - n'}{N}$$

(2.46)

In the *second phase*, a sample is selected on which both the target variable and co-variable are observed. Now the size of the sample n is much smaller than in the first phase. The estimators are

$$\overline{y} = \frac{\sum_{i=1}^{n} y_i}{n}$$

(2.47)

$$s_y^2 = \frac{\sum_{i=1}^{n} y_i^2 - \left(\sum_{i=1}^{n} y_i\right)^2 / n}{n - 1}$$

(2.48)

$$\overline{x} = \frac{\sum_{i=1}^{n} x_i}{n}$$

(2.49)

$$s_x^2 = \frac{\sum_{i=1}^{n} x_i^2 - \left(\sum_{i=1}^{n} x_i\right)^2 / n}{n - 1}$$

(2.50)

$$s_{xy} = \frac{\sum_{i=1}^{n} x_i y_i - \left(\sum_{i=1}^{n} x_i\right)\left(\sum_{i=1}^{n} y_i\right) / n}{n - 1}$$

(2.51)

For the *ratio estimator*, the mean of the target variable is estimated as

$$\hat{R}_{2-P} = \frac{\overline{y}}{\overline{x}}$$

(2.52)

$$\overline{y}_{2-P} = \hat{R}_{2-P} \overline{x'}$$

(2.53)

with an estimated variance of the estimated mean of

$$s_{\bar{y}2-p}^2 = \frac{s_y^2 + \hat{R}_{2-p}^2 s_x^2 - 2\hat{R}_{2-p} s_{xy}}{n} + \frac{-\hat{R}_{2-p}^2 s_x^2 + 2\hat{R}_{2-p} s_{xy}}{n'} - \frac{s_y^2}{N} \tag{2.54}$$

Finally, for the regression estimators,

$$b = \frac{s_{xy}}{s_x^2} \tag{2.55}$$

$$\bar{y}_{lr2-p} = \bar{y} - b\left(\bar{x} - \bar{x}'\right) \tag{2.56}$$

$$s_{\bar{y}lr2-p}^2 = \left(\frac{n-1}{n-2}\right)\left(\frac{s_y^2 - s_{xy}^2/s_x^2}{n}\right)\left(1 + \frac{1}{n} - \frac{1}{n'}\right) + \left(b^2 \frac{s_x^2}{n'}\right) \tag{2.57}$$

2.2.6   SAMPLING WITH UNEQUAL SELECTION PROBABILITIES

We would like to highlight from among these methods the sampling method with a selection *PPS* and the *3P* sampling method.

2.2.6.1   PPS Sampling

The requirements of this method are that the selection probabilities must be defined for each and every element of the population before sampling and none of the population elements must have a selection probability of 0.

Various sampling strategies of importance for forest inventory are based on the principle of unequal selection probabilities, such as Bitterlich sampling developed by Walter Bitterlich (1948). The main idea is to assign a higher probability of selection to the larger trees, of which there are usually fewer in a stand.

In the Bitterlich sampling method, from a selected sample point, the neighboring trees are selected strictly proportional to their basal area. It is necessary to have a device that produces a defined opening angle, such as a relascope. While standing at the sample point and aiming the relascope at the DBHs of the surrounding trees, and sweeping around 360°, all the trees that appear larger than the angle are counted. See Figure 2.1.

Clearly the larger trees have a greater probability of being chosen as a sample tree. From this count alone, an estimate of basal area per hectare is obtained.

The only additional information required is the "calibration factor" of the measurement device, as obviously the number of trees counted depends on the opening angle produced by the instrument.

If *S* is the area, the probability of selection of a tree *i*, with a basal area of g_i, measured with a device with a specific constant *C*, is

$$P_i = g_i \frac{1}{2500C^2 S} \tag{2.58}$$

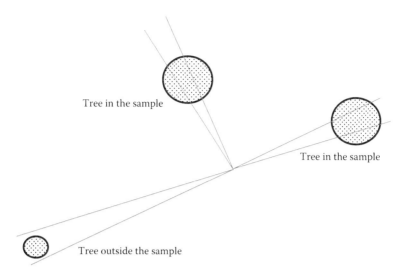

Tree in the sample

Tree in the sample

Tree outside the sample

FIGURE 2.1   Bitterlich sampling method.

2.2.6.2   3P Sampling

The operation of this method is as follows (see Rondeux (2010) for more information): First, the size of the sample is selected:

$$\frac{1}{n} = \frac{e_r^2}{t^2 CV^2} + \frac{1}{N} \qquad (2.59)$$

where
   e_r is the relative error
   t is the value of the t-Student distribution
   CV is the coefficient of variation
   N is the total number of trees in the set

Second, an *a priori* estimation is made of the total volume V_e of the trees in the set. Third, the parameter Z is calculated such that the minimum number of trees n is obtained from

$$n + 2\left(n - \frac{n^2}{N}\right) = \frac{V_e}{Z} \qquad (2.60)$$

Fourth, for every tree in the set, its volume x_i is estimated, and a number between 0 and Z is randomly obtained. If the number obtained is less than or equal to x_i, the volume of the tree is more accurately measured, y_i.

Therefore, the number of trees selected, in terms of probability, is

$$\sum_{i=1}^{N} P_i = \frac{V_c}{Z} \tag{2.61}$$

As a result, the estimated value of the total volume of the N trees in a lot, when the size of the sample is n, is

$$V_i = \sum_{i=1}^{N} x_i \frac{\sum_{i=1}^{N} Y_i / x_i}{n} \tag{2.62}$$

2.3   ESTIMATION OF SPATIAL DATA IN GIS

According to Burrough and McDonnell (1998), GISs comprise a powerful set of tools for collecting, storing, retrieving at will, transforming, and displaying spatial data from the real world.

The most common approaches used to represent the location of geographic objects are *raster* (grid cell) and *vector* (polygon) formats. In raster format, the location is defined by the row and column position of the cells the object occupies. The value stored for each cell indicates the type of object or condition that is found at that location over the entire cell. In the vector format, the feature boundaries of the objects are converted to straight-sided polygons that approximate the original regions. These polygons are encoded by determining the coordinates of their vertices, called *nodes*, which can be connected to form *arcs* (Lillesand et al., 2008).

A common question when using GIS arises when the user needs spatially continuous information on one variable, but there is only a limited set of measurements available for it, for example, to generate a digital elevation model (DEM) from GPS measurements, traverse surveys, or *LiDAR* data. This is also the problem we encounter when we need to create weather forecast maps or maps of climatic variables. Another common situation occurs when a map indicating the concentration of a pollutant in a specific area is needed. It is impossible to measure the concentration of this substance at every point in the area of interest, and only a limited number of measurements are taken. It is then necessary to create a map from these measurements. The aforementioned are three typical examples, but this same problem appears whenever a set of measurements are collected on the variables of interest over a discrete set of points, and the gaps between these points must be "filled" with the estimated values of the variable of interest, and no auxiliary information is available for the whole study area. If only the measurements and their positions are known, then there are a wide variety of methods for estimating or interpolating a variable at a point in the space where no measurement exists. We shall illustrate the interpolation problem with the following example (this is an artificial example generated simply to illustrate the interpolation).

Mean annual temperatures were measured at 200 randomly selected locations of a forest area. First, it is necessary to obtain a map of mean annual temperatures in the area of interest. This map should provide the mean temperature for any point in the study area; however, as these measurements are discrete, it is necessary to estimate the mean annual temperature in the gaps between the measured points. To estimate the temperature in the gaps, we need to use an algorithm. To illustrate this problem, we will use the simplest algorithm for interpolation. The estimated temperature at a gap point will be the temperature of the closest measured point. This interpolation is known as nearest neighbor (NN) interpolation, Voronoi or Thiessen polygon map. This method was very popular in the past, when computing power was limited and the cartography was managed in paper format. In Figure 2.2a, we can observe the temperatures measured. In Figure 2.2b, every point in the area of interest was assigned the temperature of the closest measured point.

Each point can be characterized by three variables; two describe its position (x- and y-coordinates), and one is the variable of interest (v). At those points for which we have measurements, we know the values of these three variables. For other points in the space, we can only know their x- and y-coordinates and the value of v must be estimated. The method for estimating v will need the coordinates x, y and the value v of the variable at the measured points as input data. These methods are usually classified based on two alternative criteria.

If it is based on the estimated value at the measured points

- *Exact methods*: A method is exact if the estimated value at a point where a measurement exists equals the measured value.
- *Non-exact methods:* A method is non-exact if the estimated value at a point where a measurement exists may be different from the measured value.

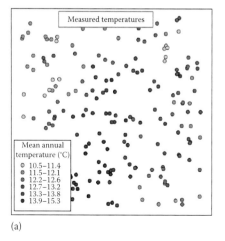
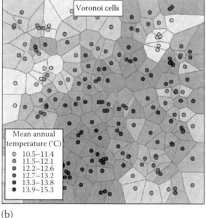

(a)   (b)

FIGURE 2.2   (**See color insert.**) Voronoi or Thiessen interpolation algorithm. (a) Measured temperatures; (b) Voronoi cells.

TABLE 2.3
Classification of Interpolation Algorithms

		Estimated Value at the Measurement Points	
		Equals the Measured Value (Exact Methods)	Different to the Measured Value (Non-Exact Methods)
For each point of the space, we have…	Only one value of the variable of interest (deterministic)	TIN, splines, IDW, NN, or Voronoi cells	Polynomial interpolation
	A random variable (stochastic)	Kriging	—

This terminology can be misleading. The terms exact and non-exact do not refer to the accuracy of the resulting estimates; they only indicate whether or not the area created by a particular method has to contain the measurement points.

Another important classification is commonly used. This alternative classification groups the interpolation methods based on how the variable of interest is interpreted.

- *Deterministic methods:* Deterministic methods are those in which one and only one value of the variable of interest is linked to every point in the space.
- *Stochastic methods:* Stochastic methods are those in which a random variable is linked to each point. The values observed are realizations of these random variables. These methods are based on a study of the spatial correlation. The spatial correlation is modeled, and then the model is used to generate predictions. The prediction at one point is the most likely value for the corresponding location, once we know the realized values at the measured points.

These two classifications are complementary. We will use this second classification as the main criteria to organize this section. Table 2.3 shows how different methods are classified according to these criteria.

2.3.1 Deterministic Methods to Estimate Spatial Data

Deterministic methods are those in which the variable of interest takes only one value at every point. The variable is not conceived as random. In this category, we find the following methods.

2.3.1.1 Thiessen Polygons or Voronoi Cells

As explained, this is probably the oldest interpolation method. It consists of determining the area of influence of each measured point and assigning to each area the value of the corresponding measurement. Each cell or area of influence groups the points in the space that are closer to it than to any other measured point. The method is exact and the area generated is discontinuous. Discontinuities are at the edges of the cells.

2.3.1.2 Triangulations

If we define triangles by linking the centers of three adjacent Voronoi cells, we obtain a Delaunay triangulated irregular network (TIN) (Figure 2.3). In a Delaunay TIN, each triangle defines the area in which any point is closer to one of the three triangle vertices than to any other measured point. This is not the only way of defining a network of triangles from the measured points, but it is optimal in the sense that three points for which we know x, y, and v define a plane in a 3D space X, Y, V. The vertices in a TIN are the measured points, but there are multiple combinations to link the points and create a network of triangles. When using TINs to estimate the value of v at a query point, we first locate the triangle that contains this point, then the estimated value is the height of the point on the 3D triangle. This method is exact and the surface produced is continuous, but not smooth. The slope changes at the edges of the triangles. This method is commonly used to interpolate digital terrain models (DTM).

2.3.1.3 Polynomial Interpolation

The estimated value at a new point is a polynomial function of grade n of the coordinates x and y. The polynomial function is the same for the whole area of interest. We obtain a smooth surface and the estimate at a point is the height of this surface. The coefficients of the polynomial are obtained by least squares. The order

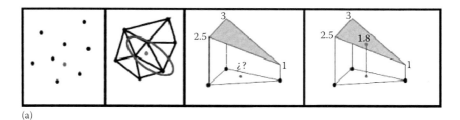

(a)

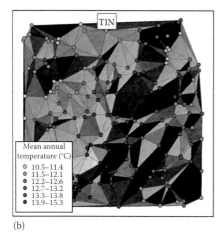
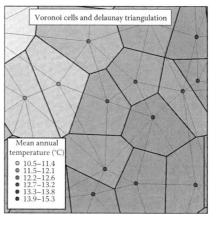

(b)

FIGURE 2.3 (See color insert.) Delaunay TIN. (a) Algorithm and (b) triangles obtained after interpolation.

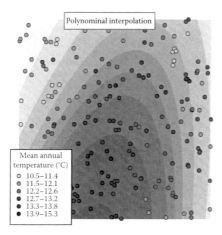

FIGURE 2.4 Polynomial interpolation.

of the polynomial controls the curvature of the surface. This interpolation technique is used to obtain general geographic patterns of change in a variable, so it is not advisable to use a large polynomial order—higher than 2 or 3—otherwise general patterns can be hidden. This is a non-exact method, and the surface generated is continuous and smooth. In Figure 2.4, we can see a hot area in the south. In general, the temperature decreases when moving away from this area. The temperature change is stronger when we move east or west than when we move north.

2.3.1.4 Inverse Distance Weighted

The estimated value at a query point is the weighted average of the measured points that are relatively close to the query location. The number of neighboring points that are considered to compute the weighted average is determined by establishing either a maximum number of measures to be averaged, a maximum distance in which to search for measures, or a combination of both conditions. The estimated value for a query point is the weighted average of the selected neighbors. The weighting factor $w_{j,q}$ attached to a neighbor j decreases when the distance from this point to the query point q $d_{j,q}$ increases. The function f that controls the weight is a negative power of $d_{j,q}$:

$$f\left(d_{j,q}\right) = K * \left(d_{j,q}\right)^{a} \tag{2.63}$$

The exponent a is called power parameter and K is a normalizing constant.

If power parameter a is smaller than 2, the interpolated surface is dominated by the influence of distant points. The larger the power parameter, the stronger the influence of the closest point in the interpolated value. If the power parameter takes a very high value, the interpolated surface becomes similar to the Voronoi cells. The method is exact and the interpolated surface is smooth. Figure 2.5 shows how the importance of the neighboring points increases for larger power parameters.

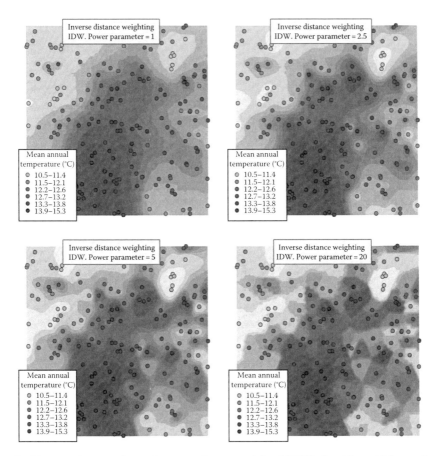

FIGURE 2.5   **(See color insert.)** Inverse distance weighted (IDW) algorithm with increasing values for power parameter a.

2.3.2 Stochastic Methods to Estimate Spatial Data

Stochastic methods differ from the previous methods with regard to how the variable of interest is understood. In these methods, a random variable is linked to every point in the space. The phenomenon studied is seen as a stochastic process or random field. A stochastic process is a set of indexed random variables. In geographic applications, each variable is indexed by its coordinates. These variables are called regionalized variables. Each random variable has a probability distribution. The variables linked to a set of n points have a joint n-dimensional probability distribution.

A key concept when dealing with these methods is that the random variables of different points in the space can be correlated. This correlation can be expected to be strong for points that are close and weak for points separated by a great distance. Stochastic methods involve studying this spatial correlation and how it changes.

It is also important to keep in mind that the variable of interest is treated as a random variable. When generating predictions for one point, we predict an outcome for this variable, and some uncertainty will remain. If we look at two points, we will have two random variables. These two random variables will have a joint probability distribution. Let us imagine that we are interested in two points. We observe the outcome of the variable at one of both points and then guess what the outcome of the variable would be at the other point. If the points are "close," both variables will be strongly correlated, and if we know the outcome of the variable at one point, the uncertainty as to the outcome of the other variable will be reduced. If both points are not "close," the correlation between the variables of interest will be weak or even null. Then the knowledge of the outcome of the variable at one point provides no information about the possible outcome of the other variable. This example illustrates the role of spatial correlation when we use stochastic methods. Our predictions for locations for which the variable of interest has not been measured will be based on the outcome of the variable of interest that has been measured and on the correlation between the variables at the point of interest and at the query point.

Example 2.1

Let us imagine that we are in the center of a room. The temperature in this room remains approximately constant at 18°C, except for a small variation at each point. We guess what the temperature would be at three points (Figure 2.6), a (left point), b (middle point), and c (right point), located at a distance of 4, 2, and 6 m, respectively, from the center. For now we only know that for every point in the room, the marginal distribution of the temperature follows a normal distribution with a mean of 18°C and a standard deviation (SD) of 1°C.

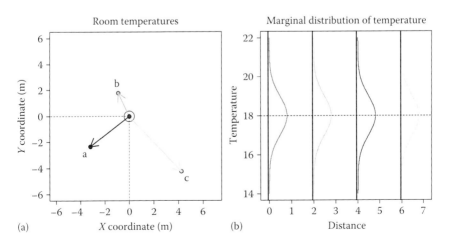

FIGURE 2.6 (a) Points at different temperature inside a room (center, a, b, and c). (b) Marginal distribution of temperature of the points in a room (center, a, b, and c).

We also know how the spatial correlation changes. The joint probability distribution of the temperature at two points ($p1$ and $p2$) follows a bivariate normal with means (18.18), and the variance–covariance matrix is

$$\Sigma = \begin{bmatrix} 1 & \sigma^2_{p1,p2}(d) \\ \sigma^2_{p2,p1}(d) & 1 \end{bmatrix} \qquad (2.64)$$

where $\sigma^2_{p1,p2}(d) = \sigma^2_{p2,p1}(d) = \begin{cases} 1 - \dfrac{d}{6} & \text{if } d < 6 \\ 0 & \text{otherwise} \end{cases}$ being d the distance between

$p1$ and $p2$ (Figure 2.7)

Based on this, we can find the joint distribution of the temperature in the center and at point a, the center and point b, and the center and point c. However, if we do not know the outcome of the temperature in the room center, we can only base our guess as to the temperature at a, b, or c on their marginal distributions (Figure 2.8).

However, once we have observed that the actual temperature in the center of the room is 20°C, we can estimate the temperatures at a, b, and c using their conditional distributions $p(t_a|t_{center}=20)$, $p(t_b|t_{center}=20)$, and $p(t_c|t_{center}=20)$.

As we can see (Figure 2.9), the conditional distributions $p(t_a|t_{center}=20)$, $p(t_b|t_{center}=20)$ are different from the marginal distributions that we would have used to make our guesses if we had not observed the temperature in the center. The expected values are different, and the variances of these distributions are also smaller than the variances of the marginals. This occurs because the temperatures of points separated by 2 or 4 m are correlated. The correlation becomes smaller as the

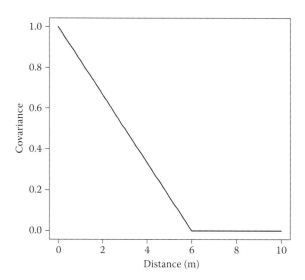

FIGURE 2.7   Covariance function for the temperature in the room.

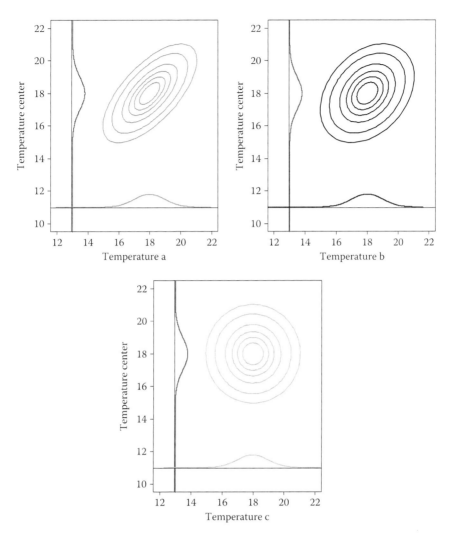

FIGURE 2.8   Joint probability distribution of the temperature at the room center and point a (left), point b (middle), and point c (right). Marginal distribution of the temperature in the room center (left border) and at points a, b, and c (lower border).

distance increases, and this is why the variance of $p(t_b|t_{center}=20)$ is larger than the variance of $p(t_a|t_{center}=20)$. When the distance is 6 m, the correlation becomes 0 and that is why $p(t_c|t_{center}=20)$ equals the marginal $p(t_c)$. This means that if we measure the temperature at one point, we obtain no information about the temperature at points located at distances of equal or greater than 6 m (Figure 2.9).

Although this example is not a real case, it illustrates the basics of stochastic interpolation methods. When using these methods, we based our inferences of the value at a given point on both the observed values at the measurement points and the spatial correlation. Another property of these methods is that they provide a

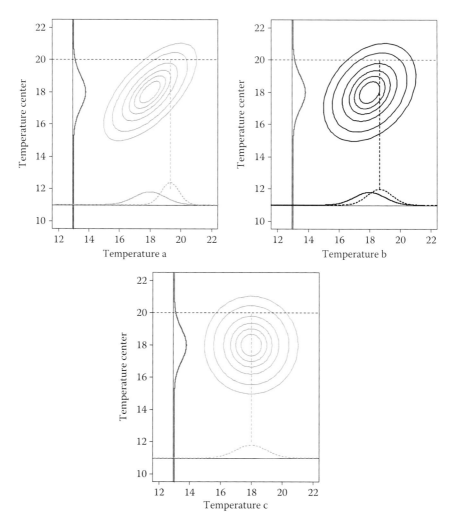

FIGURE 2.9   Solid lines: joint and marginal probability distributions of the temperature in the room center and a (left), room center and b (middle), and room center and c (right). Dashed lines: conditional probability distributions of the temperatures at a, b, and c if the temperature at the plot center is 20°C.

measure of reliability of the predictions. In real applications, we know neither the shape of the distributions nor how the correlation changes, and we need to model them from our measurements.

2.3.2.1   Characterization of the Stochastic Process

Example 2.1 described the basic stochastic interpolation with an artificial example in which the mean and spatial correlations were known beforehand. This is not the case in real applications, and for this reason, the stochastic process must be analyzed and modeled prior to estimation using the observations. Further predictions at query points will be based on the model fitted. As predictions are based on a model, an incorrect specification of this model will result in invalid predictions. Certain properties of a stochastic process are interesting from the modeling perspective: these properties are stationarity and isotropy. When a process is stationary and isotropous, it is easier to model.

2.3.2.2   Stationarity

A stochastic process is called stationary if for any set of points $(x_1, x_2, \ldots x_n)$ being $n \geq 1$, the joint probability distribution function, or the probability distribution function if $n = 1$, is invariant under translations.

This means that at every point in the area of interest, the distribution of the studied variable is always the same; that is, the joint distribution of the variables linked to a set of points $f(Z(x_1), Z(x_2), \ldots Z(x_n))$ is equal to the joint distribution of the variables at $f(Z(x_1 + h), Z(x_2 + h), \ldots Z(x_n + h))$, where h is any displacement vector. This is a very restrictive property and can be relaxed. Best linear unbiased predictors (BLUPs) are only based on means and covariances. Stationarity can be required only for these properties. Stochastic processes in which mean and covariance are stationary are known as second-order stationary processes.

A stochastic process is called second-order stationary if

1. $E(Z(x)) = m$ exists and does not depend on the location x.
2. $Cov(Z(x_a), Z(x_b)) = Cov(Z(x_a + h), Z(x_b + h))$

Covariance is invariant under translations, which implies that it is the only function of the vector $v = x_i - x_j$. The function $C(v)$ provides the covariance for points separated by a vector v. $C(0)$ is the variance.

An important relation in stationary processes is that

$$C\left(x_a - x_b\right) = Cov\left(Z\left(x_a\right), Z\left(x_b\right)\right) =$$

$$= E\left(\left(Z\left(x_a\right) - m\right)\left(Z\left(x_b\right) - m\right)\right) = E\left(Z\left(x_a\right)Z\left(x_b\right)\right) - m^2 \qquad (2.65)$$

$E(Z(x_a)Z(x_b))$ is invariant under translations because both m^2 and $Cov(Z(x_a), Z(x_b))$ are stationary. This relation will be used in the derivation of the kriging equations.

Second-order stationarity only requires the mean and the covariance to remain invariant. Only these moments need to be constant, not the whole probability distribution function. We will limit this section to second-order stationary processes (see Bivand et al. (2008) for an extended review of geostatistical methods and practical examples with R).

Another function closely related to the covariance function in second-order stationary processes and used to describe a stochastic process is the semivariogram.

This function provides the expectation of the square of the difference in the variable for two points divided by 2:

$$\gamma(x_a - x_b) = \frac{1}{2} E\left(\left(Z(x_a) - Z(x_b)\right)^2\right) \tag{2.66}$$

The semivariogram does not depend on the mean m of the variable. This is an important issue because in practical applications, the mean m of the stochastic process is unknown.

For second-order stationary processes, the variogram can be expressed in terms of the covariance function:

$$\gamma(x_a - x_b) = \frac{1}{2} E\left(\left(Z(x_a) - Z(x_b)\right)^2\right) = \frac{1}{2} E\left(\left(\left(Z(x_a) - m\right) - \left(Z(x_b) - m\right)\right)^2\right)$$

$$\times \gamma(x_a - x_b) = C(0) - C(x_a - x_b) \tag{2.67}$$

2.3.2.3 Isotropy

Another important property of a stochastic process is isotropy. A process is isotropous if it is uniform in all directions. A process that changes depending on the direction is called anisotropous.

For second-order stationary processes, we are interested in changes in the covariance function (v). If the process is isotropous, changes in $C(v)$ will not depend on the direction of the vector v. Only the modulus of v, which represents the distance between points, will be important for $C(v) = C(|v|)$. This property also holds for the semivariogram $\gamma(v) = \gamma(|v|)$.

2.3.2.4 Kriging

Kriging is an estimation procedure in which the value of a variable of interest at a point x is estimated using a weighted average of the observed values. The weight attached to each observation depends on the spatial correlation. Estimates are linear combinations of the observations, so this method provides linear estimators. Kriging provides BLUPs. Nonlinear estimators may be better than those provided by kriging, but if the variables are normally distributed, kriging provides the best possible predictors.

This section deals with only second-order stationary processes. The equations derived are only valid for these processes. Several alternatives exist to obtain estimators in other types of stochastic processes (see Bivand et al. (2008) for further information).

2.3.2.5 Best Linear Unbiased Predictor Derivation

The objective is to predict the value of the variable of interest in x_q using a linear combination of the values observed.

The prediction $Z^*\left(x_q\right)=\sum_{i=1}^{n}\lambda_i Z\left(x_i\right)$ should be unbiased, and the expectation of the squared estimation error should be minimized. If the estimator is unbiased, the expectation of the squared estimation error is the variance of the error. Assuming a second-order stationary process in which the spatial correlation is known, we can obtain the weights λ_i. The problem can be seen as a constrained minimization of the error variance. We will seek the weights that minimize the error variance subject to the condition of providing unbiased predictions.

Then

$$E\left(Z\left(x_q\right)\right)=m=E\left(Z^*\left(x_q\right)\right) \tag{2.68}$$

If the process is stationary for the mean $E(Z(x_i))=m$ for every i,

$$E\left(Z^*\left(x_q\right)\right)\doteq E\left(\sum_{i=1}^{n}\lambda_i Z\left(x_i\right)\right)=\sum_{i=1}^{n}\lambda_i E\left(Z\left(x_i\right)\right)=\sum_{i=1}^{n}\lambda_i m=m\sum_{i=1}^{n}\lambda_i \tag{2.69}$$

This reduces the unbiasedness condition to

$$E\left(Z\left(x_q\right)\right)=m=E\left(Z^*\left(x_q\right)\right)=m\sum_{i=1}^{n}\lambda_i \tag{2.70}$$

Canceling m from the previous equations yields

$$\sum_{i=1}^{n}\lambda_i=1$$

The objective now turns into finding the linear combination $Z^*\left(x_q\right)=\sum_{i=1}^{n}\lambda_i Z\left(x_i\right)$ such that $E(Z((x_q)-Z^*(x_q))^2)$ is minimum subject to the zero-bias constraint $\sum_{i=1}^{n}\lambda_i=1$.

If the covariance function is known, $E((Z(x_q) - Z^*(x_q))^2)$ can be expressed as follows:

$$E\left(\left(Z\left(x_q\right)-Z^*\left(x_q\right)\right)^2\right)=E\left(Z\left(x_q\right)^2\right)+E\left(Z^*\left(x_q\right)^2\right)-2E\left(Z^*\left(x_q\right)Z\left(x_q\right)\right)$$

$$E\left(\left(Z\left(x_q\right)-Z^*\left(x_q\right)\right)^2\right)=E\left(Z\left(x_q\right)^2\right)+\sum_{i=1}^{n}\sum_{j=1}^{n}\lambda_i\lambda_j E\left(Z\left(x_i\right)Z\left(x_j\right)\right)$$

$$-2\sum_{i=1}^{n}\lambda_i E\left(Z\left(x_q\right)Z\left(x_i\right)\right)$$

$$E\left(\left(Z(x_q)-Z^*(x_q)\right)^2\right)=C(0)+\sum_{i=1}^{n}\sum_{j=1}^{n}\lambda_i\lambda_j C(x_i-x_j)-2\sum_{i=1}^{n}\lambda_i C(x_i-x_q)$$

$$+m^2\left(1+\sum_{i=1}^{n}\sum_{j=1}^{n}\lambda_i\lambda_j-2\sum_{i=1}^{n}\lambda_i\right) \quad (2.71)$$

$$E\left(\left(Z(x_q)-Z^*(x_q)\right)^2\right)=C(0)-2\sum_{i=1}^{n}\sum_{j=1}^{n}\lambda_i\lambda_j C(x_i-x_j)+\dot{+}\sum_{i=1}^{n}\lambda_i C(x_i-x_q)$$

Minimizing Equation 2.1 subject to $\sum_{i=1}^{n}\lambda_i=1$ yields the following system of linear equations where the last equation and the parameter μ are introduced in the minimization of $E((Z(x_q)-Z^*(x_q))^2)$ using Lagrange multipliers:

$$\begin{bmatrix} C(0) & C(x_1-x_2) & \cdots & C(x_1-x_n) & 1 \\ C(x_2-x_1) & C(0) & \cdots & C(x_2-x_n) & 1 \\ \vdots & \vdots & \ddots & \vdots & \vdots \\ C(x_n-x_1) & C(x_n-x_2) & \cdots & C(0) & 1 \\ 1 & 1 & \cdots & 1 & 0 \end{bmatrix}\begin{bmatrix} \lambda_1 \\ \lambda_2 \\ \vdots \\ \lambda_n \\ \mu \end{bmatrix}=\begin{bmatrix} C(x_1-x_q) \\ C(x_2-x_q) \\ \vdots \\ C(x_n-x_q) \\ 1 \end{bmatrix} \quad (2.72)$$

In second-order stationary processes, equations can be expressed in terms of a semivariogram instead of using the covariance function:

$$E\left(\left(Z(x_q)-Z^*(x_q)\right)^2\right)=2C(0)-2\sum_{i=1}^{n}\sum_{j=1}^{n}\lambda_i\lambda_j C(x_i-x_j)+\dot{+}\sum_{i=1}^{n}\lambda_i C(x_i-x_q)-C(0)$$

$$E\left(\left(Z(x_q)-Z^*(x_q)\right)^2\right)=-2\sum_{i=1}^{n}\lambda_i\sum_{j=1}^{n}\lambda_j\gamma(x_i-x_j)+2\sum_{i=1}^{n}\lambda_i\gamma(x_i-x_q) \quad (2.73)$$

Solving for $(\lambda_1,\lambda_2,\ldots\lambda_n)$ subject to $\sum_{i=1}^{n}\lambda_i$ using Lagrange multipliers yields the following system of equations:

$$\begin{bmatrix} 0 & \gamma(x_1-x_2) & \cdots & \gamma(x_1-x_n) & 1 \\ \gamma(x_2-x_1) & 0 & \cdots & \gamma(x_2-x_n) & 1 \\ \vdots & \vdots & \ddots & \vdots & \vdots \\ \gamma(x_n-x_1) & \gamma(x_n-x_2) & \cdots & 0 & 1 \\ 1 & 1 & \cdots & 1 & 0 \end{bmatrix}\begin{bmatrix} \lambda_1 \\ \lambda_2 \\ \vdots \\ \lambda_n \\ \mu \end{bmatrix}=\begin{bmatrix} \gamma(x_1-x_q) \\ \gamma(x_2-x_q) \\ \vdots \\ \gamma(x_n-x_q) \\ 1 \end{bmatrix} \quad (2.74)$$

For second-order stationary processes, the optimal weights can be obtained using either the covariance function or the semivariogram. Once the weights are obtained, the variance of the error $E\left(\left(Z\left(x_q\right)-Z^*\left(x_q\right)\right)^2\right)=$ $C\left(0\right)-\sum_{i=1}^{n}\lambda_iC\left(x_i-x_q\right)+\mu=\sum_{i=1}^{n}\lambda_i\gamma\left(x_i-x_q\right)+\mu$. This measure of reliability of the estimators is an interesting property of stochastic interpolator methods.

Kriging equations are based on the covariance function or the semivariogram, although both functions are unknown and must be modeled. The semivariogram does not depend on the mean—which may be unknown—and this is why it is used in more applications. Due to this equivalence, we will explain the following steps in kriging interpolation in terms of semivariograms. Hypotheses of stationarity and isotropy should also be tested.

2.3.2.6 Empirical Semivariogram

An approximation to $\gamma(v)$ based on the observations is the empirical semivariogram

$$\gamma^*\left(v\right)=\sum_{i=1}^{N(v)}\left(Z\left(x_i\right)-Z\left(x_i+v\right)\right)^2 \tag{2.75}$$

where $N(v)$ is the number of pairs of observations $Z(x_i)$, $Z(x_j)$ such that $x_i=x_j+v$. It is important to note that the empirical semivariogram depends on the orientation of v and that we will only have a finite number of observations. A very small—or even null—number of pairs of observations will be separated by a specific vector v. For these reasons, the range of modulus and orientations of v is discretized, and tolerances for both are defined. Then, $\gamma^*(v)$ is a discontinuous representation of $\gamma(v)$ and it is constructed using pairs of observations $Z(x_i)$, $Z(x_j)$ such that $x_i=x_j+v\pm\text{tolerance}$.

2.3.2.7 Isotropy Testing

If we restrict the orientation of v, we obtain directional semivariograms. These semivariograms provide γ for every distance when moving in the considered direction. For isotropous processes, γ only depends on the modulus of v and not on its orientation $\gamma=\gamma(|v|)$. Therefore, directional semivariograms do not change. If the empirical semivariogram is computed for different directions, its shape should be constant for isotropous processes. If we hypothesize that a process is isotropous and the shape of $\gamma^*(v)$ changes depending on the direction, we should reject our hypothesis. If no changes are observed, we can assume isotropy. Under this assumption, only γ is only a function of the distance, and then calculate $\gamma^*(|v|)$ using observations separated a distance $|v|\pm\text{tolerance}$.

2.3.2.8 Semivariogram Modeling

A previous step to perform when using kriging to interpolate the values of a variable is to obtain a model for the semivariogram from the observations. We will use $\gamma^*(v)$ to obtain an approximation to $\gamma(v)$ and then search for a function that provides a good representation of the variations observed in $\gamma^*(v)$. The following example illustrates this process. This is an example in which we know beforehand that the process is isotropous. For this reason, only $\gamma^*(|v|)$ will be calculated and the model will be fitted for $\gamma(|v|)$.

Example 2.2

The mean annual temperature has been measured in 200 random locations of a forest area. We need to obtain a map of this variable and we decide to use kriging to create the map. Figure 2.10 represents the observations in the area of interest. Symbols are proportional to the mean annual temperature. Small and large circles appear everywhere. Small circles are usually surrounded by small circles and large circles are surrounded by large circles. This is due to the spatial correlation.

Assuming isotropy and using the field measurements, we computed the empirical semivariogram. The empirical semivariogram increases until the distance is close to 0.5 km, when it appears to remain stable (Figure 2.11). If we use kriging to predict the value of the mean annual temperature in a new location, we need to know the elements of Equation 2.74. Distances between observations and query points can take any value, and for this reason, we need a function (a model) that provides $\gamma(|v|)$ for any distance.

For modeling the pattern observed in $\gamma^*(|v|)$, we chose the following function:

$$\gamma\left(|v|\right) = \begin{cases} \sigma\left(1 - 1.5 * \left(|v|/\varphi\right) + 0.5\left(|v|/\varphi\right)^3\right) & \text{if} \quad |v| \ge \varphi \\ 0 & \text{otherwise} \end{cases} \tag{2.76}$$

Then, we search for the parameters (σ, φ) that best fit $\gamma^*(|v|)$. Once σ and φ are determined, we obtain a model for $(|v|)$. This model (Figure 2.12) is then used to fill the matrix and the right-hand side vector of Equation 2.74.

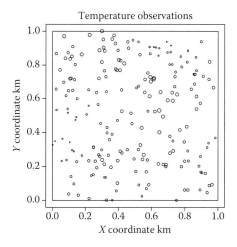

FIGURE 2.10 Mean annual temperature in 200 random locations. Symbols are proportional to the temperature.

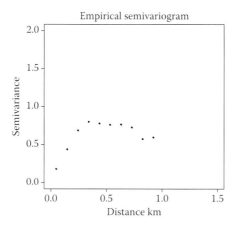

FIGURE 2.11   Empirical semivariogram for the mean annual temperatures.

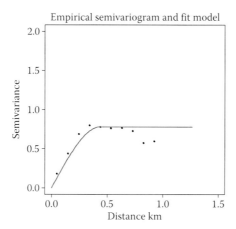

FIGURE 2.12   Semivariogram model. Model fit to the empirical semivariogram for the mean annual temperatures.

In Example 2.2, we obtained a model that provided an analytical expression for $\gamma(|v|)$. The model used in this example is called a spherical semivariogram model. This is one of the most commonly used models, but not the only one; there are a wide variety of semivariogram models. Table 2.4 shows the expression for the most widely used models.

Interesting properties of a stochastic process can be derived from the semivariogram. For second-order stationary processes, $\gamma(|v|) = C(0) - C(|v|)$. An expected property for the stochastic process is that $C(|v|)$ will decrease when $|v|$ increases. If this happens, the semivariogram should present an asymptotic increase. The value of this asymptote is called the *Sill* (Figure 2.13). The range is the distance at which the semivariogram reaches the *Sill*. If the *Sill* equals $C(0)$, then the range is the distance $|v|$ from which $C(|v|) = 0$. In these cases, the observation of the variable of interest at

TABLE 2.4

Theoretical Semivariogram Models

Pure nugget effect

$$\gamma\left(|v|\right) = \begin{cases} 0 & \text{if } |v| = 0 \\ \sigma & \text{otherwise} \end{cases}$$

Spherical semivariogram model

$$\gamma\left(|v|\right) = \begin{cases} \sigma\left(1.5 * \left(\dfrac{|v|}{\varphi}\right) - 0.5\left(\dfrac{|v|}{\varphi}\right)^3\right) & \text{if } |v| \geq \varphi \\ \\ \sigma & \text{otherwise} \end{cases}$$

Exponential semivariogram model $\gamma(|v|) = \sigma(1 - e^{-3(h/\varphi)})$

Gaussian semivariogram model

$$\gamma\left(|v|\right) = \sigma\left(1 - e^{-3(h/\varphi)^2}\right)$$

Combination of models

Nugget effect and spherical semivariogram model

$$\gamma\left(|v|\right) = \begin{cases} 0 & \text{if } |v| = 0 \\ \sigma_n + \left(\sigma - \sigma_n\right)\left(1.5 * \left(\dfrac{|v|}{\varphi}\right) - 0.5\left(\dfrac{|v|}{\varphi}\right)^3\right) & \text{if } 0 < |v| \geq \varphi \\ \sigma & \text{otherwise} \end{cases}$$

Nugget effect and exponential semivariogram model

$$\gamma\left(|v|\right) = \begin{cases} 0 & \text{if } |v| = 0 \\ \sigma_n + (\sigma - \sigma_n)\left(1 - e^{-3(h/\varphi)}\right) & \text{otherwise} \end{cases}$$

Nugget effect and Gaussian semivariogram model

$$\gamma\left(|v|\right) = \begin{cases} 0 & \text{if } |v| = 0 \\ \sigma_n + (\sigma - \sigma_n)\left(1 - e^{-3(h/\varphi)^2}\right) & \text{otherwise} \end{cases}$$

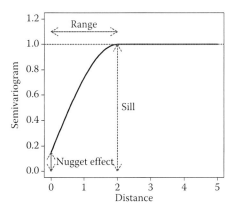

FIGURE 2.13   Semivariogram properties.

one point would provide no information about the possible outcome at another point separated by a distance greater than |v|.

The semivariogram must be 0 for |v|=0. When the empirical semivariogram is significantly different from 0 for small distances, we see an effect called the *nugget effect*. If two points were very close, the covariance would tend to equal $C(0)$. This would mean that knowing the outcome of one would determine the outcome of the other without any uncertainty. There may be a limitation on the uncertainty reduction for very close measurements, and this limitation is considered by the *nugget effect*. On the other hand, in a stochastic process with the following pure nugget-effect semivariogram

$$\gamma\left(|v|\right) = \begin{cases} 0 & \text{if} \quad |v| = 0 \\ \sigma = C(0) & \text{otherwise} \end{cases} \tag{2.77}$$

the observation of the variable of interest at one point does not provide any information about the possible outcome at any other point. This is a special case of a process without spatial correlation.

2.4 PUBLIC PARTICIPATION GIS

PPGIS was conceived in 1996 at the meeting of the National Center for Geographic Information and Analysis (NCGIA). However, the formal definition of PPGIS remains nebulous (Tulloch, 2007), and in literature, the terms PPGIS and "participation GIS" (PGIS) are often used interchangeably, although significant differences can be found between them.

According to Brown and Weber (2011), the term PPGIS emerged in developed countries as an intersection of public participatory *planning* and information technologies, while PGIS is often used to describe the result of a spontaneous merger of participatory *development* (i.e., participatory learning and action (PLA) methods) with geographic information technologies.

PPGIS combines the practice of GIS and mapping at local levels to produce knowledge of place; it is therefore conceived as a tool that allows the merger of geospatial and socioeconomic data. In short, it is a way of communicating findings between different stakeholders, initiates learning processes, and identifies key areas of interventions.

In contrast, PGIS is often used to describe participatory planning/development approaches in rural areas of developing countries (Brown and Weber, 2011). PGIS originated in mental maps that give insights into locally constructed positive or negative connotations of space, important landmarks, or the perceived size of the geographic areas covered. The use of participatory mental maps began in the late 1980s in order to elicit indigenous knowledge at a time when interaction between communities and policy makers was scarce. This changed in the 1990s with the introduction of GIS.

Based on the specific case studies in Brown and Weber (2011) and Bernard et al. (2011), we propose and describe the main phases for the implementation of PPGIS

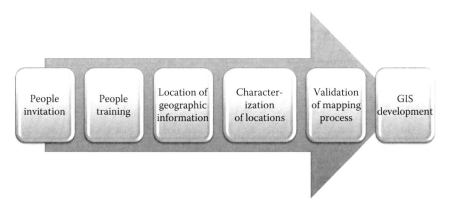

FIGURE 2.14   Phases in a PGIS project.

and PGIS projects. In the case of a PGIS system process (Figure 2.14), the main phases can be summarized as follows:

Phase 1: In this stage, local people are invited to attend the mapping activities in the villages.

Phase 2: The population is trained in the basic concepts of the project, cartography and remote sensing.

Phase 3: Location of reference points such as rivers, isolated houses, plantations, communities, and hunting areas. In this phase, the local people show the project managers the reference points in the field for their location in the cartography information.

Phase 4: Characterization of the reference points. Qualitative information is also gathered on the reference points, such as economic, descriptive, or social information.

Phase 5: Validation of the mapping process. The reference points are visited again to test whether or not they are correctly located on the maps.

Phase 6: GIS development. The information gathered in the previous phases is dumped and organized in a PGIS.

The implementation of a PPGIS project involves the performance of five different phases (Figure 2.15):

Phase 1: Formulation and description of the problem to be solved. The main objectives are defined, followed by a hierarchical structuring of the objectives. This structure determines the main variables or attributes to measure. For example, in a participatory study about the landscape value of a specific territory, variables such as aesthetics, recreation, biodiversity, and economics are considered in the process.

Phase 2: Data collection. Different data collection methods can be found: mail surveys with a mapping exercise (e.g., Brown, 2005); structured interviews, panels of experts, and workshops (e.g., Donovan et al., 2009; Raymond et al., 2009); or the currently most widely used GIS or Internet-based applications (e.g., Beverly et al., 2008; Brown and Reed, 2009; Simão et al., 2009; Brown and Weber, 2011; Clement and Cheng, 2011; Pocewicz et al., 2012).

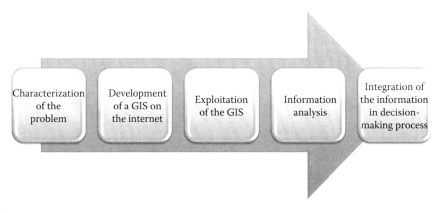

FIGURE 2.15 Phases in a PPGIS project.

The development of GIS on the Internet includes the modules:

1. Information about the application itself, including the time needed to complete it, and technical requirements such as operating system, memory space, and registration process.
2. Explanation of the study. The objectives of the study must be explained, together with any technical information on the problem that may be required by any individual in the participatory process. All these explanations are given in everyday language.
3. Demonstration module. This module can show the user how the program works.
4. Presentation of iterative maps. The user can interact with these maps to introduce the required information in the survey. This capacity of the program enables the user to draw polygons, lines, and points, as well as to introduce categorical and multimedia information.
5. Survey information. This information includes all the users' interfaces with the questionnaire, interactive maps, and project-specific explanations.

Phase 3: Exploitation of the website. Depending on the aim of the survey, various mechanisms are developed and triggered to invite people to participate. These mechanisms may be asking people to participate after an activity related to the area under study, e-mailing people of interest, or via the social networks (i.e., if the issue involves the use of a protected area, visitors may be asked to participate after their visit).

Phase 4: Analysis of the information. The information gathered from the users is statistically analyzed, indicators are developed and georeferenced, and groups of opinions are formed according to their personal characteristics such as gender, age, level of formal education, livelihood, income, association with the area or issue under study, and considering the problem-specific information provided through the application.

Phase 5: Integration in the decision-making process. The results of Phase 4 are integrated in the decision-making process.

An overview of the literature focused on environmental and natural resource management and planning indicates that research using PPGIS has been conducted to identify the location of highway corridor values (Brown, 2003), to identify preferences for tourism and residential development (Brown, 2006), to manage recreation resources on public lands (McIntyre et al., 2008), to formulate natural area plans (Brown and Weber, 2011; Gil et al., 2011b), for municipal transport planning (sustainable mobility plans) (Gil et al., 2011a), etc.

2.5   REMOTE SENSING

Remote sensing is the science and technique of acquiring information about the Earth's surface without actually being in contact with it. It therefore comprises both (i) the use of satellite-borne sensors to observe, measure, and record the electromagnetic radiation reflected or emitted by the Earth and (ii) the subsequent processing, analysis, and extraction of information (Chuvieco, 2008; Lillesand et al., 2008).

Any remote sensing process involves having an energy source (generally the sun) that provides electromagnetic energy to the surface of interest (i.e., the forest, the sea) (Figure 2.16(A)). When this electromagnetic energy reaches the surface of interest (target), it interacts with it, and depending on the characteristics of the target, the radiation will be partially absorbed (Figure 2.16) (Ab), transmitted (Tr), and/or reflected (Rf). A sensor on board a satellite collects and records the reflected radiation (B), which is finally transmitted in electronic form to a receiving and processing station (C). The receiver station converts the data into digital images (D). The incoming (A) and reflected radiation (Rf) travels through the atmosphere (E). Particles and gases in the atmosphere partially absorb and scatter the radiation. These effects have to be partially corrected in the preprocessing stage of the digital images.

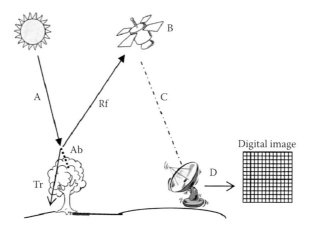

FIGURE 2.16   Elements of the remote sensing process for extraction of information.

According to wave theory, electromagnetic radiation consists of two perpendicular fields (electric and magnetic) that travel at the speed of light. Electromagnetic waves can be described in terms of their

- Wavelength (λ): the length of one wave cycle (nanometers (nm) or micrometers (μm))
- Frequency (ν): the number of cycles of a wave passing a fixed point per unit of time (hertz)

Wavelength and frequency are related by the speed of light (c):

$$c = \lambda \cdot \nu \qquad (2.78)$$

The electromagnetic spectrum ranges from the shorter wavelengths (gamma rays) to the longer wavelengths (radio waves). However, the regions or bands of the electromagnetic spectrum of practical interest in remote sensing are (Chuvieco, 2008) visible (blue, green, and red) (0.4–0.7 μm), near infrared (NIR) (0.7–1.3 μm), short-wave infrared (SWIR) (1.3–2.5 μm), thermal infrared (or the radiation emitted from the Earth's surface (8–14 μm)), and microwaves (over 1 mm).

Electromagnetic reflected radiation is collected by the sensor and transmitted in electronic form to a receiving and processing station. There, the electronic signal is converted to a matrix of numerical values, that is, a digital image. Digital images are integrated by small cells or pixels. The value stored for each cell, that is, the digital number (DN), is the reflected energy encoded into 8-bit, 11-bit, etc. A digital image consists of multiple layers corresponding to the different bands of the electromagnetic spectrum registered by the sensor (Figure 2.17)

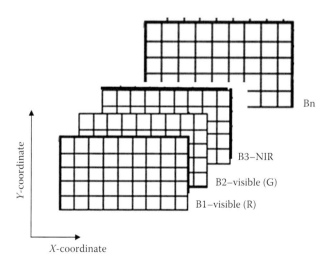

FIGURE 2.17 Digital image is a multilayer stack of grid cells (multiband dataset). Each band contains information from a specific region of the electromagnetic spectrum. Reflected energy is encoded into a DN.

2.5.1 Sensors and Platforms

2.5.1.1 Resolution of a Sensor

Sensors on board satellite or airborne platforms have different features that configure the characteristics of the image they provide. The resolution of the sensor can be set as the ability to discriminate information in detail (Estes and Simonett, 1975). Sensor resolution is one of the main criteria used for selection. This resolution involves (Chuvieco, 2008)

1. *Spatial resolution*: This is the smallest object that can be distinguished on an image; pixel size is the most common reference.
2. *Spectral resolution*: This indicates the number and width of the regions or bands of the electromagnetic spectrum that can be discriminated on the sensor.
3. *Radiometric resolution*: This refers to the sensitivity of the sensor to detect the variations of spectral radiance received. Typically this resolution is expressed in the number of bits required for each picture element to be stored.
4. *Temporal resolution*: This is the frequency coverage provided by the sensor.

These four aspects of the resolution are closely related. A higher spatial resolution can decrease the time needed to reduce the spectral resolution. Therefore, each detection system offers particular features and may be chosen or not depending on the goal to be attained.

2.5.1.2 Overview of Different Sensors and Platforms

Sensors can be classified as passive or active sensors. The former records the electromagnetic radiation reflected or emitted by the Earth's surfaces. In contrast, active sensors are able to emit a specific wavelength of electromagnetic radiation that is collected by the same sensor, after it has been reflected over the surface of interest (Chuvieco, 2008). Currently, the main active sensors are radar and *LiDAR*. *LiDAR* technology is discussed later in this chapter.

An overview of the currently available passive sensors based on Labrador García et al. (2012) is summarized in Table 2.5 and described as follows.

Disaster Monitoring Constellation (*DMC*) is a remote sensing satellite constellation owned by multiple nationalities, initially designed to monitor natural disasters and covering more than one daily visit to any point on the globe.

The Earth Observation (*EO-1*), an experimental satellite of NASA of the so-called "New Millennium Program" (NMP) was released on November 21, 2000. It has been tested and validated to apply new technologies in future LANDSAT programs, to reduce the high costs of current ones. In order to compare the images obtained spatially and temporally, the orbit of EO-1 is designed to pass 1 or 2 min after the LANDSAT-7.

Earth Resources Observation Satellite (*EROS*) is a series of Israeli commercial satellites designed by Israel Aircraft Industries. The satellites are operated by the company ImageSat International. The sensors of these satellites are panchromatic cameras with a lateral vision capability of up to 45° to the vertical, resulting in a corridor potential for the imaging of about 960 km.

TABLE 2.5
Overview of Main Remote Sensing Sensor and Satellites

Satellite	Sensor	Year	Country	Spatial Resolution (m)	Band Number	Temporal Resolution
DCM (several)	SLIM-6	2002–2009	Great	32 m	3	<1 day
	SLIM-6-22		Britain	22 m	3	<1 day
EO-1	ALI	2000	United	9 m	9	16 days
	Hyperion		States	220 m	220	16 days
EROS-A-B	EROS-A	2000	Israel	1,8 m	1	4 days
	EROS-B	2006		0,7 m	1	4 days
FORMOSAT-2	FORMOSAT-2	2004	Taiwan	8 m	4	1 day
GeoEye-1	GeoEye-1	2008	United States	2 m	4	3 days
IKONOS	IKONOS	1999	United States	4 m	4	3–5 days
KOMPSAT-2	KOMPSAT-2	2006	Korea	4 m	4	3 days
Landsat-7	ETM +	1999	United States	30 m	8	16 days
NOAA	AVHRR		United States	1100 m	5	12 h
QuickBird	QuickBird	2001	United States	2.44 m	4	2–4 days
RapidEye	RapidEye	2008	Germany	6.5 m	5	1 day
Resourcesat-2	LISS-IV	2011	India	5.8 m	3	5 days
	LISS-III			23.5 m	4	24 days
SPOT-5	HRG	2002	France	10	4	2.4–3.7 days
	HRS			PNA: 10	1	26 days
Terra	ASTER	1999	United States/ Japan/ Canada	15–90 m	14	16 days
	MODIS	1999	United States	250–1000 m	36	1–2 days
THEOS	THEOS	2008	Thailand	15 m	4	1–5 days
WorldView-2	WorldView-2	2009	United States	2 m	8	1–3 days

FORMOSAT-2, initially called ROCSAT-2, is one of the few satellites that combine good spatial resolution with a daily revisit period, although this feature—based on geosynchronous orbit—is assumed not to be able to cover the entire surface.

GeoEye-1 is one of the commercial satellites that provides the highest spatial resolution today. The main investor and customer of this satellite is the National Geospatial Intelligence Agency (NGA), and its second most famous investor and client is Google, which has direct access to the images that it uses to update its Google Earth mapping.

IKONOS was the first commercial satellite to provide satellite images of very high spatial resolution (VHSR) (1 m in the panchromatic channel and 4 m in the multispectral) and was an important milestone in the history of Earth observation from space. The current owner of this satellite is the company GeoEye.

Landsat-7, launched into space on April 15, 1999, is the latest satellite so far in the series, which began with the launch of LANDSAT-1 in 1972. The satellites that followed this first release yielded the largest series to date of existing commercial satellite images of Earth observation and are able to track the major changes occurring on the surface of our planet.

NOAA-6 was launched in 1979 to acquire meteorological and small-scale Earth observation data with high temporal resolution (12 h).

The U.S. commercial satellite of very high-resolution, QuickBird, operated by the company DigitalGlobe, and satellites WorldView-1 and WorldView-2, which also belong to DigitalGlobe, and form a constellation with very high resolution and a high revisit frequency.

RapidEye is a constellation of five satellites with the trademark of RapidEye AG, a German company providing geospatial information. The constellation is characterized by the small size of the satellites (about 1 m^3).

The *Resourcesat-2* satellite, launched on April 20, 2011, is the 18th Indian Remote Sensing (IRS) series national satellite. Resourcesat-2 improves and continues the work of the IRS-P6 (Resourcesat-1), launched into orbit in 2003 and still operating.

The French program Systeme d'Observation Probatoire de la Terre (*SPOT*), approved in 1978 and developed by the Centre National d'Etudes Spatiales (CNES) in collaboration with Belgium and Sweden, has spawned a total of five satellites for civilian use until the present day.

Terra is a scientific satellite sent into orbit by NASA on December 18, 1999, with the participation of the space agencies of the United States, Japan, and Canada. The main objective of this satellite is the study of carbon and energy cycles, to contribute to analyzing the "health" of the Earth as a whole.

Thailand Earth Observation Satellite (*THEOS*) is the first Thai national satellite for the observation of the Earth's surface, launched into space on October 1, 2008.

The *WorldView-2* commercial satellite is a very high-resolution U.S. satellite operated by the company DigitalGlobe. It was launched on October 8, 2009, and was the first commercial satellite capable of capturing eight spectral bands with a resolution of 2 m/pixel.

2.5.2 Digital Image Processing

The ultimate aim of remote sensing technology is to extract meaningful information from the imagery. This objective can be addressed by combining the visual interpretation (i.e., human interpretation of the image) and the digital analysis of the image. The most common image processing functions for the digital analysis and extraction of information are

- Preprocessing
- Image enhancement
- Image transformation
- Image classification

2.5.2.1 Preprocessing

These functions involve all the operations that are normally required prior to the main data analysis and extraction of information. According to Chuvieco (2008), any image acquired by a remote sensor presents a series of radiometric and geometric alterations due to several factors. Therefore, the final image captured does not exactly match the tone, position, shape, and size of the objects it includes. The most common image alterations are distortions caused by the platform or the sensor and as a consequence of the absorption and scattering of energy when passing through the atmosphere.

Some of these problems are routinely solved by the receiver stations. However, others persist, leading to a need for preprocessing techniques and particularly when seeking to

- Compare images from different sensors and/or dates.
- Perform environmental, ecological, or geophysical analyses.
- Improve classification results.

Preprocessing functions are generally grouped as radiometric or geometric corrections:

1. *Radiometric corrections* seek to modify the original DN values of the image to bring them as close as possible to the DN values the image would have in the case of ideal data reception (without distortions or atmosphere effects) (Chuvieco, 2008). These corrections also include the calibration of the reflected energy into DN.
2. *Geometric corrections* (*co-registration*) include any change in the position of the pixels in the image. Contrary to radiometric corrections, the co-registration process does not aim to modify the DN values of the pixels but their position. Often the geometric correction consists of transforming the geometric coordinates (longitude–latitude) to Cartesian plane coordinates. The final aim is to allow multiple source data integration in a GIS platform or the overlapping of two or more images (Chuvieco, 2008).

The co-registration of a digital image is based on regression functions:

$$f(c') = f_1(c, r); f(x, y) \tag{2.79}$$

$$f(r') = f_2(c, r); f(x, y) \tag{2.80}$$

Thus, column (c') and row (r') of the corrected image are a function of the coordinates' column and row (c, r) of the original image or a function of the projected map coordinates (x, y) to which the image will be overlapped (Chuvieco, 2008).

2.5.2.2   Image Enhancement

Enhancement functions improve the appearance of the imagery to assist in visual interpretation and analysis. The most useful techniques include contrast stretching to increase the tonal distinction between various features in a scene and spatial filtering to enhance (or suppress) specific spatial patterns in an image (CCRS, 2000).

2.5.2.3   Image Transformation

This group typically involves multiple transformations designed to extract information and features from satellite imagery. Arithmetic operations are performed to combine and transform the original bands into "*new*" images that better display or highlight certain features in the scene (CCRS, 2000). We will focus on three main groups of transformations: (i) vegetation indices (VIs), (ii) principal components analysis (PCA), and (iii) multitemporal analysis.

2.5.2.3.1   Vegetation Indices

VIs are ratios of two or more bands of the image, designed to enhance the contribution of vegetation properties (Huete et al., 2002). VIs have been highly successful in assessing vegetation condition, foliage, cover, phenology, and processes such as evapotranspiration and primary productivity (Glenn et al., 2008).

Normalized difference vegetation index (NDVI), a normalized ratio of the NIR and red (visible) bands of the image (Huete et al., 2002), is the VI most widely used:

$$NDVI = \frac{\rho_{NIR} - \rho_{Red}}{\rho_{NIR} + \rho_{Red}} \tag{2.81}$$

where

   ρ_{NIR} is the reflected energy in the NIR band of the electromagnetic spectrum
   ρ_{RED} is the reflected energy in the red (visible) band of the electromagnetic spectrum

Vegetated areas have a relatively high reflection in the NIR and a low reflection in the visible range of the spectrum. Clouds, water, and soil reflect more or similar energy in the visible region than in the NIR. Therefore, densely vegetated areas have high NDVI values, while cloud, soil, or water present low or negative NDVI values. Thus, NDVI improves the discrimination of vegetation from the Earth's other surface coverings.

Specifically, NDVIs have been widely applied to show high correlations with many vegetation parameters such as chlorophyll content, water foliage content, net primary productivity, leaf area index (LAI), and evapotranspiration (Chuvieco, 2008).

2.5.2.3.2   Principal Components Analysis

PCA is a statistical analysis that transforms a number of possibly correlated variables into a smaller number of uncorrelated variables.

Different bands of digital images are often highly correlated and thus contain similar information. Therefore, PCA transformation is often applied to multiband images to reduce the dimensionality (i.e., the number of bands) in the data and compress much of the information in the n-original bands into fewer bands. The "new" bands that result from this statistical procedure are called *principal components* and are obtained as a linear combination of the n-original bands. This process aims to maximize the amount of information (or variance) from the original data into the least number of new components (CCRS, 2000).

2.5.2.3.3   Multitemporal Analysis

Sensors are able to capture repetitive information from the same area over time, making digital images a valuable source of information for monitoring the dynamic processes of the Earth's cover.

The subtraction of DN or NDVI from two images from two different dates to form a new image is a particularly simple transformation for assessing changes in the territory. Applying a PCA transformation over a multiple dataset (multiband image) from the two dates is also a useful technique for detecting change. Multitemporal analysis usually requires the images to be accurately co-registered and radiometrically corrected in order not to consider misregistration or different atmospheric effects as changes when in fact they are errors.

2.5.2.4   Image Classification

Image classification has so far been considered one of the main remote sensing applications. The final output of the classification is a thematic map of categorical pixels or polygons. In the classification process, the multiple dataset becomes another single-band image of the same size and characteristics as the original. However, in the classified image, the pixel values are not related to reflected energy in different electromagnetic spectrum bands, but are a categorical value that represents a category (i.e., type of vegetation, land use) (Chuvieco, 2008) (Figure 2.18).

2.5.2.4.1   Pixel-Based versus Object-Based Image Classification

In the past, most digital image classifications were based on processing the entire scene, in a process known as pixel by pixel. This is commonly referred to as a *pixel-based classification*. The recent availability and accessibility of VHSR images at a reasonable price has increasingly opened up the applications for these images. Higher spatial resolution has allowed the improvement of the field classifications made up to the present day. However, new problems have emerged in the pixel-based classifications of VHSR images. The high spatial resolution also increases the spectral variability, in contrast to the integration effect of earlier sensors; pixel-based

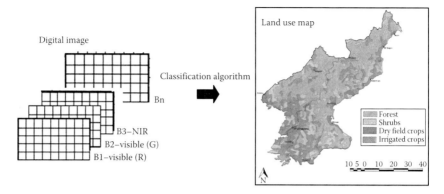

FIGURE 2.18   (**See color insert.**) Classification of digital image process.

classification methods have shown their inability to process this additional spectral variability, generating too many or poorly defined classes (Arroyo, 2006), thus decreasing the accuracy of the classification.

In addition, pixels do not make use of spatial concepts or take into account contextual information (Benz et al., 2004). Nevertheless, much information is contained in the relationship between adjacent pixels—including information on texture and shape—that allows identification of individual objects as opposed to single pixels (Laliberte et al., 2004). Finally, the domain of remote sensing always assumes a certain scale based on pixel resolution, although the objects of interest often have their own inherent scale.

These issues highlighted the fact that the new classification approaches need to overcome the difficulties of traditional pixel-based classification methods and take advantage of the potentialities of VHSR images. The challenge was to produce proven man–machine methods that externalized and improved on human interpretation skills. Some of the most promising results came from the adoption of image segmentation algorithms and the development of the so-called *object-based classification methodologies* (Blaschke et al., 2006).

2.5.2.4.2 Object-Based Image Classification

Object-based image analysis (OBIA) technology examines pixels, not as individual cells but in a group context. It works by imitating the human mind, using a combination of pixel color, shape, texture, and size. Therefore, in OBIA classifications, pixels are aggregated before and not after classification (Arroyo et al., 2006).

When human beings make use of their eyes, they are performing a complex mental procedure. This procedure is called image understanding. When we survey a region with our eyes, we register that certain areas have a particular size, form, and color. Thus, in our vision, it becomes an object. For example, we see a triangle area and we classify it as triangle, object. The same thing happens with the round object and the rectangle object (Figure 2.19).

If we immediately combine all these figures and relate them to each other in a fourth big rectangular object, we can recognize a tree in a landscape (Figure 2.20). These objects are meaningful. This cognition process compares our view of the object and its relationships with the knowledge existing in our memory.

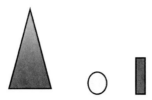

FIGURE 2.19 Different objects.

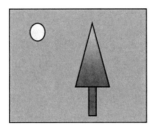

FIGURE 2.20 Landscape image.

Similar to human vision, the concept of image understanding is based on a correct segmentation of the visual image content of interest, against other visual image content. Segmentation is performed in zoned partial areas of different characteristics, and the segments are called image objects. From these image objects, we can produce a classification according to particular criteria (Definiens, 2010).

OBIA allows the analyst to decompose the images into many relatively homogeneous objects (referred to as patches or segments) using a multi-resolution image segmentation process. The various statistical characteristics of these homogeneous image objects are then subjected to traditional statistical or fuzzy logic classification. Therefore, OBIA classification methods consist of two basic steps: (1) a segmentation of pixels into objects and (2) a fuzzy classification of these objects using their spectral and contextual information.

OBIA has provided positive results when applied to VHSR imagery for mapping biotopes, fuels, trees, riparian zones, rangelands or wildland–urban interface areas, forest vegetation, and forest stands, among other examples (Laliberte et al., 2007; Mallinis, 2008; Tiede et al., 2008; Arroyo et al., 2010; De Chant et al., 2010; Petr et al., 2010; Tiede et al., 2010).

Both *pixel- and object-based classifications* can be classified using the following listed methods and the algorithms described in the next sections.

2.5.2.4.3 Classification Methods

There are two main methods to classify pixels or objects:

1. *Unsupervised classification*: The computer or algorithm automatically groups pixels with similar spectral characteristics (means, SDs, covariance matrices, correlation matrices, etc.) into unique clusters according to some statistically determined criteria. The analyst then relabels and combines the spectral clusters into information classes.

2. *Supervised classification:* Identification of units like *training sites* known a priori through a combination of fieldwork, map analysis, and personal experience; the spectral characteristics of these sites are used to train the classification algorithm for eventual land-cover mapping of the remainder of the image. Each pixel, both inside and outside the training sites, is then evaluated and assigned to the class of which it has the highest likelihood of being a member.

2.5.2.4.4   Algorithms

The most common *types of algorithms* used for pixels or object classifications are (1) NN, (2) membership functions, and (3) user-controlled threshold:

1. The NN uses the values of a series of samples of different classes and assigns membership values (MVs). For this process, it is first necessary to train the system with the samples that define each class. Thus, the NN classifier returns an MV between zero and one, based on similarity to the samples given for that class (Figure 2.21). The result of the classification process is presented in two ways: as a fuzzy classification and as a classical classification (rigid). In the first, an allocation is provided to each of the categories for each object. In the second, each object is assigned to a single category, in which one class has the highest probability of assignment (Schowengerdt, 1983).

2. The *membership functions* transform the values of the variable considered in assigning grades. The type assignment is blurred (fuzzy); that is, each object takes a degree of assignment, from 0 (zero allocation) to 1 (maximum allocation). Figure 2.22 shows the value of the variable 100 is set to 0, and as this value is approached, the variable will take values of

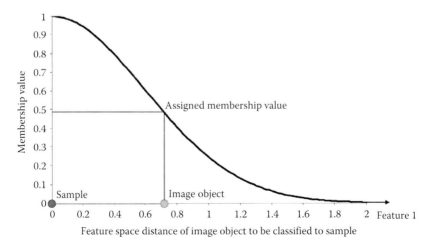

FIGURE 2.21   MV based on NN. (From Definiens, *Definiens eCognition Developer 8.0.1. Reference book*, Definiens AG, Munich, Bavaria, Germany.)

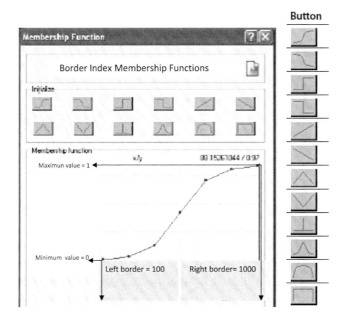

FIGURE 2.22   **(See color insert.)** Types of membership functions.

greater than 1000–1. We can choose different types of functions as shown in the same figure.

3. *Thresholds* are set using one or more conditions. If the object fulfills them, it is assigned the value 1 and is classified; if it does not comply, it is assigned the value 0 and is not classified. Thresholds are typically used when classes can be clearly separated by a variable.

2.5.3   LiDAR Data

Laser altimetry, or *LiDAR*, is an active remote sensing technology, analogous to radar, but using laser light (electromagnetic energy in the optical range). These systems measure the distance between a sensor and a target surface (range) based on the time between the emission of a pulse and the detection of a reflected return. Therefore, laser systems provide 3D coordinates of target surfaces (Figure 2.23). In addition, some physical properties of the target object can be derived from the interaction of the radiation with the target. *LiDAR* has many scientific applications, for example, detection of pollutants and chemical agents in atmosphere or water, 3D mapping of topography, bathymetry, and forest structure (Baltsavias, 1999; Chuvieco, 2008). *LiDAR* sensors have been supported on terrestrial, airborne, and satellite platforms.

LiDAR systems are classified as either discrete-return or full-waveform recording. Full-waveform recording *LiDAR* systems digitize the entire reflected energy from a return, resulting in complete submeter vertical vegetation profiles. In contrast, discrete-return systems record single or multiple returns from a given laser pulse.

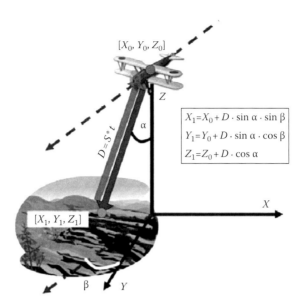

$$X_1 = X_0 + D \cdot \sin \alpha \cdot \sin \beta$$
$$Y_1 = Y_0 + D \cdot \sin \alpha \cdot \cos \beta$$
$$Z_1 = Z_0 + D \cdot \cos \alpha$$

FIGURE 2.23   (**See color insert.**) LiDAR systems provide the distance between a sensor and a target surface based on the time between the emission of a pulse and the detection of a reflected return. D = distance of target surface; S = speed of light; t = time recorded by the lidar sensor. GNSS receivers and INSs allow the source of the return signal to be located in three dimensions.

Thus, within a forest environment, full-waveform systems record the entire waveform for analysis, while discrete-return systems record clouds of points representing intercepted features (Figure 2.24) (Dubayah and Drake, 2000; Lefsky et al., 2002; Wulder et al., 2012). *LiDAR* footprint is the diameter of a laser pulse's circle of illumination on the ground (Figure 2.24). According to this, *LiDAR* sensors may be small footprint (typically 0.1–2 m) or large footprint (typically 10–100 m) (Wulder et al., 2008a). Most often, discrete-return systems are usually small footprint, while waveform sensors provide large footprint data. Currently, airborne small-footprint discrete-return sensors are used for virtually all operational applications (Næsset, 2004b; Wulder et al., 2008a, 2012) and will be discussed at greater length in this chapter.

Airborne *LiDAR* systems (either discrete-return or waveform sampling sensors) are typically used in combination with two complementary technologies for locating the source of the return signal in three dimensions. These are global navigation satellite system (GNSS) receivers to record the position of the platform and inertial navigation systems (INSs) to measure the attitude (roll, pitch, and yaw) of the *LiDAR* sensor. Combining this information with accurate time referencing of each source of data yields the absolute position of the reflecting surface for each laser pulse (Lefsky et al., 2002; Wulder et al., 2012).

2.5.3.1 LiDAR Preprocessing: LiDAR Models

Small-footprint laser scanning provides 3D coordinates (x–y–z) of any intercepted surface (i.e., terrain topography, vegetation, or buildings). For useful applications in

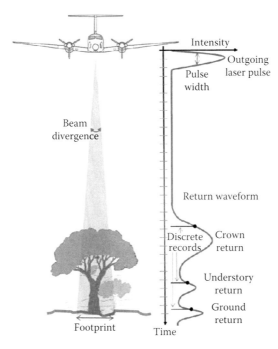

FIGURE 2.24 **(See color insert.)** Differences between waveform recording and discrete-return lidar devices. At the left is the intersection of the laser illumination area (footprint). In the center, the hypothetical return signal of a waveform recording sensor. To the right, the signal recorded by discrete-return lidar sensors. (From Fernández-Díaz, J.C., *Imag. Notes*, 26(2), 31, 2011.)

forest environments, this 3D point cloud is processed into different elaborated models such as DEM to represent the Earth's surface and digital surface models (DSMs) to represent any protruding object or surface (vegetation, buildings, etc.).

The first step to process a *LiDAR*-derived DEM is to filter or extract bare-earth (ground) returns from the point clouds. Different algorithms have been developed for this purpose (see Sithole and Vosselman (2004) for a comparison). Then, interpolation algorithms (such as those previously reviewed in Section 2.3.1) are applied over *LiDAR* points previously classified as ground returns, to provide a continuous earth surface. The quality of DEMs is influenced by data characteristics (e.g., point density, flight height, or scan angle). The selection of an interpolation algorithm and the spatial resolution of the models may also influence the accuracy of DEM generation (Bater and Coops, 2009). Finally, external factors, such as canopy cover, land use, and slope, also involve significant differences in the vertical accuracy of *LiDAR*-derived DEMs (Hyyppa et al., 2008; González-Ferreiro, 2012).

Extracting forest attributes from small-footprint discrete-return *LiDAR* data usually involves generating the DSM. The DSM is obtained by means of the interpolation of first *LiDAR* returns. The subtraction of the *LiDAR*-derived DEM from the corresponding DSM provides the continuous surface of the height of vegetation, often referred to as the digital canopy model (DCM). When working with point clouds, the processing of the point cloud data into canopy heights or normalized heights is simply calculated as the difference between the elevation values of laser hits and the estimated terrain elevation values (DEM) at the corresponding location (Hyppa et al., 2008; González-Ferreiro et al., 2012). Therefore, when processing *LiDAR* data for vegetation height estimation, DEM and DSM errors may be propagated to vegetation height estimation.

2.5.3.2   Forest Inventory Using LiDAR Data

There are two main types of approaches for estimating forest attributes from *LiDAR* data (Hyyppa et al., 2008): (i) methods based on statistical canopy height distribution (i.e., empirical regression methods also known as the ABA) and (ii) methods focusing on individual tree detection (i.e., physical methods).

This latter approach consists of segmenting individual tree crowns from the DCM or normalized tree heights to derive biophysical parameters and measures such as crown size, individual tree height, and location of individual trees. Allometric equations allow the estimation of individual tree parameters such as volume and normal section. Finally, forest stand attributes are obtained by simply aggregating individual parameters of the segmented trees within a stand (Hyyppa et al., 2008). Identification of individual trees from *LiDAR* data requires a high point density during the data acquisition process (at least 4–5 point/m^2) (Wulder et al., 2008a; Reutebuch et al., 2005).

ABA approaches consist of deriving several metrics from *LiDAR* cloud points within established field plots (i.e., percentiles, density, or variability metrics). Additionally, field forest attributes (i.e., mean tree height, dominant height, basal area, volume) are measured in the same field plots as in traditional forest inventory. Finally, stand- or plot-level forest attributes are estimated by regression analysis, where field biophysical parameters and *LiDAR*-derived metrics are dependent and independent variables, respectively. Multiple forest biophysical parameters have been estimated by applying this empirical approach (Table 2.6).

The accuracy of tree detection approaches from *LiDAR* data is more influenced by forest structure than by tree detection algorithms. Thus, many tree-segmentation algorithms fail to identify understory and suppressed trees or to delineate trees under certain canopy conditions such as dense forests and grouped trees (Goodwin et al., 2006; Zhao et al., 2009; Vaukonen et al., 2012).

ABA or empirical regression approaches (i.e., regressing *LiDAR*-derived variables with field data) are effective methods for estimating forest structure attributes, although there is a large set of assumptions and site-specific considerations that must be made for each study (Gleason and Im, 2012). Most of the study areas are located in boreal forest, and thus, additional studies should be conducted in order to assess the potential of this approach in other areas (González-Ferrero, 2012). According to Zhao et al. (2009), regression models are also scale dependent; that is, models are built to estimate forest attributes for a specific plot size, and changing this plot size may affect the accuracy of the results. To reduce all these effects, new approaches based on machine learning techniques offer promising estimation approaches for the near future (Breidenbach et al., 2010; Vauhkonen et al., 2010; Gleason and Im, 2012).

TABLE 2.6
Main Attributes Estimated Using *LiDAR* Data and the ABA.

Forest Attribute	References
Mean height	Næsset (1997a, 2002, 2004a,b); Magnussen et al. (1999); Næsset and Bjerknes (2001); Holmgren et al. (2003); Hall et al. (2005); Stephens et al. (2007); Heurich and Thoma (2008); Treitz et al. (2010); González-Ferreiro et al. (2012); Mauro et al. (2012)
Mean diameter	Næsset (2002, 2004b)
Quadratic mean diameter	Treitz et al. (2010); Mauro et al. (2012)
Basal area	Means et al. (2000); Næsset (2002, 2004a,b); Hall et al. (2005); Stephens et al. (2007); Heurich and Thoma (2008); Treitz et al. (2010); González-Ferreiro et al. (2012); Mauro et al. (2012)
Volume	Næsset (1997b, 2002, 2004a,b); Means et al. (2000); Holmgren et al. (2003); Hollaus et al. (2007); Heurich and Thoma (2008); Rombouts et al. (2008); Treitz et al. (2010); González-Ferreiro et al. (2012); Mauro et al. (2012)
Dominant height	Næsset (2002, 2004b); Lovell et al. (2005); Stephens et al. (2007); Heurich and Thoma (2008); Rombouts et al. (2008); Treitz et al. (2010); González-Ferreiro et al. (2012); Mauro et al. (2012)
Stem number	Næsset (2002, 2004b); Hall et al. (2005); Heurich and Thoma (2008); Treitz et al. (2010); Mauro et al. (2012)

2.6 APPLICATIONS OF IT TO FOREST MANAGEMENT

2.6.1 Mapping Natura 2000 Habitats Using OBIA

Habitats mapping is necessary to ensure the good conservation status required in the Natura 2000 network. VHSR satellite images and OBIA are operational tools for this purpose. We proposed a "methodology for habitats mapping" using QuickBird images and eCognition network language (Hernando et al., 2012).

2.6.1.1 Introduction

The lack of consistent information on type, location, size, and quality of habitats has been identified as a major constraint for the implementation of the Natura 2000 network (Weiers et al., 2004). This network has the primary goal of guaranteeing the favorable conservation status of natural habitats and species in order to ensure European biodiversity (Hernando et al., 2010). OBIA has emerged as an alternative to pixel-based classification that largely neglects shape and context aspects of the image information—among the main clues for the human interpreter (Baatz et al., 2008; Lang, 2008a; Blaschke, 2010)—and also to optimize the current "VHSR" satellite images. Additional expert knowledge for the object-building process or the inclusion of auxiliary datasets has been proved to enrich not only the classification but also the entire information extraction workflow (Bock et al., 2005; Tiede et al., 2010). Multi-scale segmentation was introduced (Benz et al., 2004) and implemented in the software package eCognition (Blaschke, 2010).

With the aim of developing a methodology for creating detailed habitats mapping, we propose using high spatial resolution—*QuickBird* image—and OBIA using *Definiens Developer* software.

2.6.1.2   Study Area and Data

The study was carried out in a Natura 2000 site in Avila, region of Castile–León, in central Spain. This forest territory is included in the Mediterranean biogeographic region designated Special Protection Area ES0000186: "Pinares del bajo Alberche" for the existence of nine habitats listed in the first annex of the Habitats Directive (92/43/EEC).

These habitats are *H. 4090* (*Genista*), endemic oro-Mediterranean heaths with gorse; *H. 5120* (*Cytisus purgans*), Mountain Cytisus purgans formations; *H. 5210* (*J. oxycedrus*), arborescent shrubland with Juniperus spp.; *H. 6220* (*Thero-Brachyp.*), pseudo-steppe with grasses and annuals of *Thero-Brachypodietea; H. 8220* (*Rocks*), siliceous rocky slopes with chasmophytic vegetation; *H. 9340* (*Q. ilex), Quercus ilex* and *Quercus rotundifolia* forests; *H. 6230* (*Nardus*), species-rich *Nardus* grasslands on siliceous substrates in mountains; *H. 9230* (*Q. pyrenaica*), Galicio-Portuguese oak woods with *Quercus pyrenaica*; and *H. 9540* (*P. pinaster*), Mediterranean pine forests with endemic Mesogean pines.

For the present investigation, a *QuickBird* scene was acquired on August 5, 2005. Data include three visible and one infrared spectral channel with a resolution of 2.44 m. The scene covers 17,283 ha and has an 11-bit radiometric resolution. The DTM-(1 m resolution) and the thematic data (Avila forest map (*AVFM*), urban and river areas) were supplied by the Forestry Department of the government of Castile–León. AVFM provides forest species mapping within a 1:5000 scale and is very accurate in the species location, although coarse in the delineation. NDVI was computed with the original red and infrared bands. Orthorectification was carried out using a 1 m DTM with bilinear interpolation implemented in ENVI©.

2.6.1.3   Methods

The methodology for habitats mapping aims to delineate the habitats listed for this study area. Five consistent levels were created in eCognition to provide thematic preknowledge, classify objects, and finally extract the target classes (Hernando et al., 2012). OBIA offers a methodological framework for machine-based interpretation of complex classes, defined by spectral, textural, spatial, and hierarchical object properties (Benz et al., 2004; Lang, 2008b). This new approach, OBIA, interlinks two main phases: (1) segmentation, which creates objects considering a scale parameter and one or more criteria of homogeneity, and (2) classification, encoding, and relating the relevant intrinsic spectral and spatial properties in the image (Tiede et al., 2010). These were created according to the following strategies for the purpose of the final classification of habitats:

- *Level 5 "thematic layers"* was generated to delineate similar objects to the thematic layers provided for further classification. The 5700 objects were classified into eight categories: *A. Juniperus oxycedrus, A. Pinus*

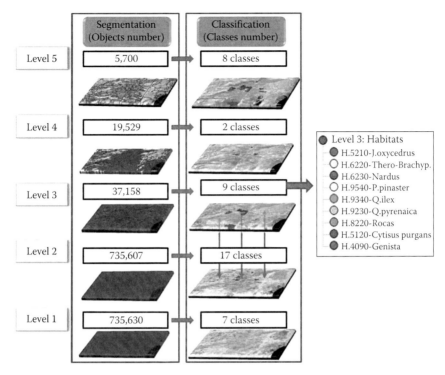

FIGURE 2.25 **(See color insert.)** Segmentation levels (left) and their corresponding classifications (right).

nigra, A. Pinus pinaster, A. Pinus sylvestris, A. Q. ilex, A. Q. pyrenaica, A. river, and *A. urban* (Figure 2.25). It was necessary to introduce this information as the four QuickBird bands are not able to distinguish species and the river.

- *Level 4 "arable"* was created to extract *arable* areas that can easily be confused with *H.6220 (Thero-Brachypodietea)*, due to their spectral features. We extracted *arable* from the 19,529 segmented objects, taking into account certain features such as brightness (>490), merging potential areas, and refining them with area size (<13,000 pixels), DTM (<1000 m), and rectangular fit (>0.65) for the study area.

- *Level 1 "vegetation"* was segmented to distinguish the land cover in greater detail; 735,630 objects were created. Seven categories (*c. tree, c. transition, c. pasture, c. shrubs, c. pasture-shrub, c. rocks, c. road*) were classified using NDVI thresholds that were set up visually using the visualization and information tools provided by eCognition.

- *Level 2 "species"* was segmented with exactly the same parameters as level 1 and was classified using the upper and lower levels described previously. The objective of this scale was to achieve the species. For this purpose, we established class-related features between the super- and subobjects previously created. For example, for tree species, an object in level 2 would be classified as *s. Q. pyrenaica* if this object had a lower object *c. tree* and an upper object *A. Q. pyrenaica*. After classification, we obtained the following 17 categories with a highly detailed resolution: *s. Q. pyrenaica, s. Q. ilex, s. P. nigra, s. P. pinaster, s. P. sylvestris, s. Juniperus, s. shrub Genista, s. shrub C. purgans, s. shrub-rocks, s. pasture, s. mountain pasture, s. arable, s. oak shrub, s. river, s. urban, s. transition, and s. rock.*

- Finally, *level 3 "habitats"* was created for the target classification. The level was segmented into 37,158 objects favoring medium-size polygons, which were classified into the nine targeted habitats considering the relative area of the subobject of level 2 with fuzzy membership functions.

2.6.1.4 Results and Discussion

Level 3 was created for the final classification, which was habitats. After segmentation, 57.99% of the 37,158 created objects belonged to the unclassified category, and the rest were classified in the target habitats. Considering the whole area—17,291.28 ha—the habitats with the largest percentage of relative area were H. 6220 (Thero-Brachyp.) (10.46%) and H. 9340 (*Q. ilex*) (9.04%), covering 1,808.7 ha and 1,562.85 ha, respectively. They were followed by H. 9230 (*Q. pyrenaica*) (4.51%), H. 8220 (*Rocks*) (4.10%), and H. 9540 (*P. pinaster*) (3.21%). Some habitats cover less than 2.5% of the whole area, such as H. 6230 (*Nardus*) (2.25%) and H. 5210 (*J. oxycedrus*) (1.85%). In the last place, H. 4090 (*Genista*) covers 0.21% of the total area. Integration via fuzzy logic is useful for habitats mapping as there is not always a clear threshold, and the class with the highest probability will be assigned to the object. For the five segmentation scales used in this study, the fine scales (level 1 and level 2) define smaller objects, while the medium scales (level 3) define the target habitats. In any case, all scales may be useful for forest management and monitoring.

To support image interpretation and mapping, extensive field references were collected after classification. Field validation of randomly classified potential habitats—30 samples for each of the nine categories and the unclassified category (Hawth's tools ArcGIS® 9.3)—was conducted in July and August 2009. The study area is quite large and difficult to access, and for this reason, 300 samples were confirmed in the field (no. of plots inside the three public lands) and visually (no. of plots outside the three public lands) with the PNOA orthophotos.

The classification accuracy of the habitats mapping was assessed by means of a *confusion matrix*. The overall accuracy was 86.3%, and the overall kappa statistic was 0.84.

Pasture producer's accuracy (PA), *H. 6220 (Thero-Brachyp.)* and *H. 6230 (Nardus)*, had values of over 90%, as did some forest habitats—*H. 9540 (P. pinaster)* and *H. 9230 (Q. pyrenaica)* (Table 2.7).

However, some forest habitats, *H. 5210 (J. oxycedrus)* (82.4%) and specially *H. 9340 (Q. ilex)* (60%), had lower PA values. *H. 5210 (J. oxycedrus)* was misclassified with *H. 6220 (Thero-Brachyp.)*, as *Juniperus* has a small crown and is sometimes not segmented separately but mixed with pasture, leading to its misclassification. Otherwise, *H. 4090 (Genista)* is the understory of *H. 9340 (Q. ilex)*, and if covered

TABLE 2.7
Accuracy Assessments

Classes	PA (%)	UA (%)	MV	St
H. 5210—J. oxycedrus	82.4	93.3	0.94	0.82
H. 6220—Thero-Brachyp.	92.3	80	0.68	0.67
H. 6230—Nardus	96.6	93.3	0.95	0.95
H. 9540—P. pinaster	100	100	0.99	0.85
H. 9340—Q. ilex	60	90	0.92	0.79
H. 9230—Q. pyrenaica	96.7	96.7	0.94	0.94
H. 8220—Rocks	92.9	86.7	0.95	0.94
H. 5120—Cytisus purgans	80.6	83.3	0.30	0.30
H. 4090—Genista	89.5	56.7	0.48	0.40

PA, producer's accuracy; UA, user's accuracy; MV, membership value; St, classification stability.

by tree crowns, satellite images are unable to detect whether it actually exists. Regarding shrub habitats, *H. 5120* (*Cytisus purgans*) and *H. 4090* (*Genista*) had values of between 80% and 90% PA and also the lowest user's accuracy (UA) 83.3% and 56.7%. We had no previous thematic layers to support the shrub species; therefore, the results for *H. 5120* (*Cytisus purgans*) can be considered good, taking into account that it only relies on spectral features from NDVI and DTM data. The results for *H. 4090* (*Genista*) were not very successful (very fragmented habitats), but at least we were able to map them due to their class-related features. The problem is that both habitats—*H. 4090* (*Genista*) and *H. 9340* (*Q. ilex*)—are mixed, and one of them is the understory that is finally neglected. Other inaccuracies come from the coarse delineation of the AVFM layer, but we considered its inclusion necessary in order to distinguish the species; we could simply separate coniferous and broad leaves, but we need more specific species information for habitats mapping. The confusion matrix and its derived statistics provide information about the quality of the thematic maps, but the precision of the delineated boundaries still remains, which could be a subjective task.

Furthermore, due to the use of fuzzy membership functions for habitats mapping, another approach to accuracy assessment is also reported by the software eCognition. The image object is assigned to the class with the highest MV, from 0 to 1 for the best classification results. The best classification value for most of the objects is high (>0.9) and significantly lower (0.68) in the case of *H. 6220* (*Thero-Brachyp.*) (Table 2.7). There are a couple of classes—*H. 5120* (*Cytisus purgans*) *and H. 4090* (*Genista*)—with a low assignment value (0.3 and 0.48). A high MV to a certain class does not necessarily indicate definite membership in this class. If there is only a low difference between the best and the second best MV, the classification result is relatively unclear. This fact occurs in the habitats mentioned with 0.48 and 0.4 classification stabilities, as well as lower UA. This correspondence shows a relation between both accuracy assessment approaches.

2.6.1.5   Conclusions

The monitoring and forest management of Natura 2000 sites for assuring a good conservation status requires cost-efficient and time-consistent practices. As VHSR satellite images become more easily available, they can be used for this objective. OBIA with multi-scale levels, ancillary data, and class-related features performs well for this purpose. Therefore, forest managers are provided not only with habitats maps but also with very accurate species maps for taking decisions to assure biodiversity as required in the Habitats Directive. The proposed methodology for habitats mapping combines all the new operational tools for forestry improvements.

2.6.2   Application of LiDAR Data to Forest Structure Characterization Mapping in Forest Inventory

2.6.2.1   Introduction

In Spain, forest inventories follow a traditional procedure, consisting of three basic stages: First (1) stands (i.e., contiguous group of trees sufficiently uniform in species composition, arrangement of age classes, site quality, and condition to be a distinguishable unit (Smith et al. 1997)) are delineated over the whole forest area; (2) a systematic sampling design of field plots is extended over all the study area (for each plot, basic tree attributes—diameter at breast height (DBH), height, crown size, etc.—are measured); (3) stand-level mean values of biophysical variables such as dominant height, basal area, stem number, volume, and growth are calculated as average field plot measurements. For management purposes, the stands are grouped in broader forest structure types according to their stand-level attributes, to be treated as homogeneous areas.

Remote sensing via image segmentation, and to a lesser extent *LiDAR* data, has been used to assist in the aforementioned stage (1) In addition, *LiDAR* data have been widely applied for (2) and (3) forest inventory stages. The statistical approaches based on *LiDAR*-derived metrics provide information about *LiDAR* height distributions in homogeneous areas. However, this classification may present certain limitations within a forest management approach. Thus, the automated algorithm can provide a solution for homogeneous units based on statistical results, although the opinion of experts in forest management may modify this automated delineation according to their personal experience, knowledge of ecological interactions, and/or based on specific forest management criteria (i.e., species conservation, recreational use, or wood production).

Pascual et al. (2008; 2013) implemented various approaches for the forest structure characterization of heterogeneous *P. sylvestris*, L stands, ranging from null to high incorporation of expert opinion. The aim of the current work is twofold: to map homogeneous forest areas for forest management purposes in the mountain area of the Madrid region using *LiDAR* data and to evaluate the role of forest expert opinion in this mapping process.

2.6.2.2   Study Site and LiDAR Data

The study site is located in the municipality of Cercedilla, in the Guadarrama mountains, about 50 km north of the city of Madrid, Spain (40° 45′ N, 4° 5′ W), and has an area of 127 ha. The predominant forest is Scots pine (*P. sylvestris*, L.) with abundant shrubs of *Cytisus scoparius* (L.) Link., *Cytisus oromediterraneus* (Rivas Mart. et al.), and *Genista florida* (L.). There are also small pastures and an extensive rocky area in the north of the study site. Elevations range from 1310 to 1790 m above sea level, with slopes of between 20% and 45%. The average aspect of the study site is east.

An airborne laser scanning (ALS) survey was conducted over the study area in August 2002 using a Falcon sensor (http://www.toposys.com). First and last returns were recorded with a nominal height above ground of 1000 m, leading to an average point density of 4.5 points m^{-2}.

The raw data delivered by the sensor (*x*-, *y*-, and *z*-coordinates) were processed into two 1 m pixel digital models by the data provider. The DSM was processed using the first pulse reflections, and the DTM was constructed using the last returns and filtering algorithms. To obtain a digital canopy height model (DCHM), the DTM was subtracted from the DSM. Both the DTM and DCHM were validated before use by land surveying and ground-based tree-height measurements.

2.6.2.2.1 Forest Structure Types in the Study Area

The main forest attributes of the five forest structure types of the spatially heterogeneous *P. sylvestris* L. forest were described in the study area (Table 2.8). DBH and height of all the trees were collected from ten plots in the study site to describe the five forest structure types.

Forest type 1: Uneven-aged forest (multilayered canopy) with very high crown cover. These forest stands are located in the lowest part of the study area. This forest type corresponds to a multilayered, uneven-aged Scots pine formation. Crown cover ranges between 75% and 85%. This forest type includes the tallest trees in the study area.

Forest type 2: Multi-diameter forest with high crown cover. These stands are distributed between 1310 and 1600 m in the southern part of the study area with some discontinuous polygons in the north sector. This forest type can be described as having a multi-diameter distribution and a two-story vertical distribution. Canopy cover is over 65%–70%. Trees included in this forest type have a slightly lower height and diameter than in the previous one (Table 2.8).

TABLE 2.8

Forest Stand Attributes for the Five Forest Structure Types in the Study Area

Forest Type	Mean Height (m)	SD of Height (m)	Lorey's Height (m)	Basal Area (m²/ha)	Density (Tree/ha)
1	9.9	6.2	17.4	39.9	850
2	14.7	4.6	17.3	40.7	640
3	11.4	5.4	15.4	35.3	378
4	11.4	4.1	13.1	26.2	175
5	8.7	3.5	9.7	6.6	76

Forest type 3: Multi-diameter forest with medium crown cover. This type occurs discontinuously across the elevation gradient of the study area (1310–1790 m). Crown cover is over 55%. This type of forest has a multi-diameter distribution but is less dense than type 2 earlier. In some polygons, the pines form clumps of trees.

Forest type 4: Even-aged forest (single story) with low crown cover. These stands are distributed throughout the higher elevations of the study area (1500–1790 m), which has a predominantly eastern orientation. The distribution of diameter classes is close to an even-aged formation, and the height distribution represents a single-story condition. This forest type includes mature trees of greater diameter but with a slightly lower height and larger crown diameters than other types. Crown cover is relatively low and is generally less than 40%.

Forest type 5: Zones with scarce tree coverage (Table 2.8). This type consists of dense coverage of shrubs under isolated pine trees. Crown cover is between 10% and 15%. These stands are located at the highest elevations (1550–1750 m), with a mean slope of up to 40% and a predominantly northern or eastern aspect.

2.6.2.3 Mapping Forest Structure

We implemented four approaches for forest structure characterization, ranging from null to high incorporation of expert opinion and from fully automated to fully manual approaches that we designated Aut-I, SAut-II, SAut-I, and M-I, respectively. These approaches consisted of three basic stages: (1) forest stand identification, (2) forest stand classification into forest structure classes, and (3) validation. The methodological steps in the three stages for all the approaches (Table 2.9) are briefly described as follows:

2.6.2.3.1 Definition of Height Classes by Forestry Experts and Statistical Validation

Local forest managers were asked to define up to 10 height classes according to their expert knowledge and based on the ecological factors of the study site. The expert opinion height classes were validated using an analysis of variance (ANOVA) test of the trees' DBH with height class as a factor. An ANOVA test was carried out for 1600 trees from 10 plots located throughout the study area and surroundings. The expert opinion height classes were used for the SAut-I, SAut-II, and M-I approaches (Table 2.9).

2.6.2.3.2 Binning LiDAR-Derived DCHM into Expert Opinion Height Classes

The DCHM cells were binned into one of the height classes defined by the forestry experts, thereby transforming the continuous *LiDAR*-derived DCHM into a categorical canopy height model. This process was carried out for the SAut-I, SAut-II, and M-I approaches only (Table 2.9). This binned DCHM image provided information on forest canopy height, gaps, and forest cover, which are commonly used parameters in forest structure photo interpretation (Franklin, 2001).

2.6.2.3.3 Manual Delineation of Binned DCHM

For the SAut-I and M-I approaches (Table 2.9), polygons corresponding to forest stands were manually digitized on-screen from the binned DCHM. The forest managers established the perimeter of the polygons based on their experience, according

TABLE 2.9

Synopsis of the Four Approaches to Forest Structure Characterization Compared in this Study

	Increasing Level of Expert Opinion and Manual Tasks			
Stage	Automated Segmentation of LiDAR-Derived DCHM and Statistical Classification (Aut-I)	Automated Segmentation of Binned LiDAR-Derived DCHM and Statistical Classification (SAut-II)	Manually Delineated Binned LiDAR-Derived DCHM and Statistical Classification (SAut-I)	Manually Delineated Binned LiDAR-Derived DCHM and Manual Classification (M-I)
1. Forest stand identification		Definition of height classes by forestry experts and statistical validation		
		Binning the LiDAR-derived DCHM into expert opinion height classes		
	Object segmentation of LiDAR DCHM	Object segmentation of binned DCHM	Manual delineation of binned DCHM	
2. Forest stand classification into forest structure classes	Cluster of forest stands using LiDAR-derived metrics			Manual assignment of forest stands into forest classes
3. Validation	Validation (based on hypsographs and percentiles)			

to the spatial distribution pattern of the height classes (i.e., texture and color of the binned DCHM), and aided by a thorough knowledge of the area.

2.6.2.3.4　Object Segmentation of LiDAR-Derived DCHM

Polygons corresponding to forest stands were segmented from the LiDAR-derived DCHM, implementing an OBIA approach with eCognition 4.0 software. The scale and homogeneity parameters for segmentation were obtained from the *LiDAR* DCHM (i.e., canopy height as a continuous variable). We applied three consecutive segmentations. The polygons obtained for each segmentation were later aggregated at higher levels. This segmentation was used only in the Aut-I approach (Table 2.9).

2.6.2.3.5　Object Segmentation of Binned DCHM

For the SAut-II approach, we developed a segmentation procedure that worked with the binned DCHM using eCognition 4.0 software. This time the scale and homogeneity parameters were obtained from the binned DCHM (i.e., the categorical height classes established by the forestry experts). Three consecutive segmentations were also applied. The polygons obtained for each segmentation were later aggregated at higher levels.

2.6.2.3.6 Cluster of Forest stands Using LiDAR-Derived Metrics

Manually delineated (SAut-I approach) and automatically segmented (SAut-II and Aut-I approaches) polygons were grouped into five structure types by K-means cluster analyses (Table 2.9). Individual polygons were assigned to the different clusters using the sequential threshold method, where distances in cluster seeds were sorted, and observations of the distances between them were taken at constant intervals. Median and SD of *LiDAR*-derived height within each polygon were the entry variables for cluster analysis. The ANOVA test was used to test the statistical significance of the forest structure types derived from the cluster analysis. Euclidean distances between the cluster's centroids were also calculated in order to determine the proximity of the statistical clusters (Hair et al., 1995).

2.6.2.3.7 Manual Assignment of Forest Stands into Forest Classes

For the manual approach (M-I), forestry experts visually inspected the manually delineated polygons and assigned each one into a forest structure class. Their decision was based on a management approach, considering the spatial distribution of texture patterns and colors of the binned DCHM (i.e., spatial distribution of heights and forest covers). Experts also incorporated their personal experience of forest management in this area. An ANOVA test was applied to assess the separability of the manually assigned forest structure types.

2.6.2.3.8 Validation Based on Percentiles and Hypsographs

Forest structure type mapping is considered to be an arbitrary and subjective process, and it is therefore impossible to compare the results against any one correct stand delineation (Wulder et al., 2008b; Koch et al., 2009). According to Koch et al. (2009), Mustonen et al. (2008), and Falkowski et al. (2009), any reasonable stand classification should provide a separation of stands that differ from each other with respect to quantitative parameters. Assuming this criterion, we validated and compared the performance of the approaches developed by assessing the statistical separability of quantitative parameters such as *LiDAR*-derived hypsographs and hypsograph percentiles.

Hypsographs are the cumulative distribution of canopy heights as a function of proportional area within each polygon. These graphs were also summarized as percentiles, that is, heights at which 10%, 25%, 50%, 75%, and 90% of the polygon surface area occurs within each polygon (H10%, H25%, H50%, H75%, and H90%, respectively) (Figure 2.26). *LiDAR*-derived hypsographs are widely used both to analyze stand structure and to synthesize the 3D distribution of forest canopies (Lefsky et al., 1999; Harding et al., 2001; Parker and Russ 2004; Maltamo et al., 2005).

Hypsographs of each polygon were generated to validate and compare the classifications of the four approaches. ANOVA tests with Tukey's honestly significant difference (HSD) method for post hoc analyses were used to assess whether indices describing canopy height distributions (H10%, H25%, H50%, H75%, and H90%) varied significantly among forest structure types.

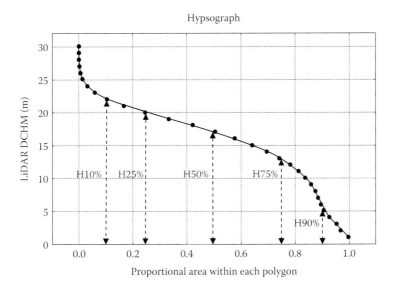

FIGURE 2.26   Hypsograph from a lidar polygon and its percentiles.

2.6.2.4   Results

Local forest managers defined eight height classes to aid manual forest stand delineation. Class 1 (0–1 m) represents areas with little to no vegetation; class 2 (1–3 m)
represents areas with shrubs (<3 m height); classes 3, 4, 5, 6, 7, and 8 considered
the following DCHM height intervals: 3–6, 6–10, 10–15, 15–20, 20–25, and 25–32
m, respectively. Robustness of height classes was validated by an ANOVA and post
hoc Tukey HSD test of tree diameters with height class as a factor. The results
indicated that the height classes were all statistically independent ($p < 0.001$ in all
cases). The *LiDAR*-derived DCHM was then binned into expert opinion height
classes (Figure 2.27).

For the Aut-I and SAut-II approaches, 112 and 103 polygons were automatically
segmented from *LiDAR* (continuous variable) and binned DCHM, respectively
(Figure 2.27a and b). For the SAut-I and M-I approaches, 64 polygons were manually delineated from the binned DCHM (Figure 2.27c and d).

For the Aut-I, SAut-I, and SAut-II approaches, polygons were clustered into five
forest structure types using K-means algorithm (Figure 2.27a and c). Clustering of
polygons was based on median and SD of DCHM within the heights in each polygon. The ANOVA F-ratios between cluster centers revealed that cluster analysis was
able to separate all five forest structure types in the three automated approaches
(Aut-I, SAut-I, and SAut-II). Thus, the ANOVA results for the Aut-I approach were
median $F = 526.91$, $p < 0.001$, and SD $F = 3.67$, $p < 0.001$; similarly, the ANOVA
results for SAut-I and SAut-II approaches were median $F = 110.0161$, $p < 0.001$, and
SD $F = 8.7116$, $p < 0.001$ and median $F = 346.8224$, $p < 0.001$, and SD $F = 3.7262$,
$p < 0.001$, respectively.

For the M-I approach, manually delineated polygons were manually assigned
to five forest structure types (Figure 2.27d). We also applied an ANOVA test to

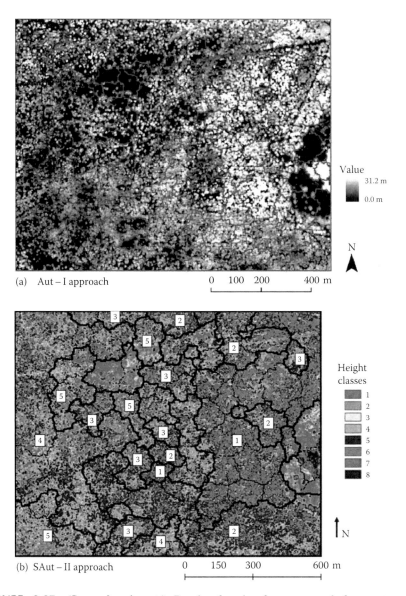

(a)   Aut – I approach

(b) SAut – II approach

FIGURE 2.27 (See color insert.) Results for the four proposed forest structure characterization approaches: (a) Aut-I approach, (b) SAut-II approach, and (c) SAut-I approach. The numbers inside the polygons indicate the forest structure type (1, 2, 3, 4, or 5) to which each polygon was assigned. Polygons with no number indicate that they were not forest stands and were not included in the forest structure classification. Dashed line indicates the automated segmented (a) and (b) and manually delineated polygons (c) and (d).

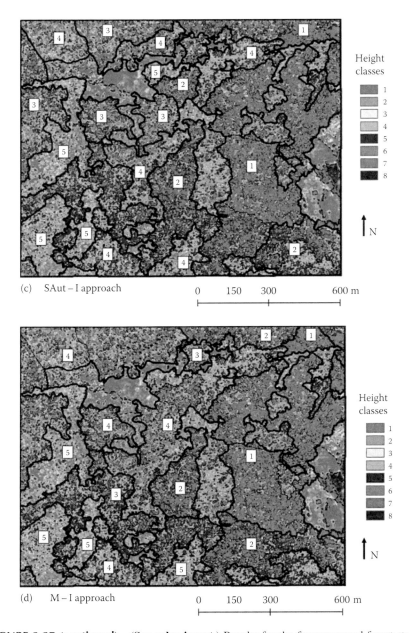

(c) SAut – I approach

0   150   300   600 m

(d) M – I approach

0   150   300   600 m

FIGURE 2.27 (continued) **(See color insert.)** Results for the four proposed forest structure characterization approaches: (a) Aut-I approach, (b) SAut-II approach, and (c) SAut-I approach. The numbers inside the polygons indicate the forest structure type (1, 2, 3, 4, or 5) to which each polygon was assigned. Polygons with no number indicate that they were not forest stands and were not included in the forest structure classification. Dashed line indicates the automated segmented (a) and (b) and manually delineated polygons (c) and (d).

TABLE 2.10

Euclidean Distance (m) among Cluster (i.e., Forest Structure Types) Centers for Aut-I, SAut-I, and SAut-II Approaches

	Aut-I					SAut-I					SAut-II			
	1*	**2***	**3***	**4***		**1***	**2***	**3***	**4***		**1***	**2***	**3***	**4***
2*	2.08	—	—	—	**2***	2.51	—	—	—	**2***	2.19	—	—	—
3*	3.97	1.93	—	—	**3***	3.82	1.77	—	—	**3***	3.52	1.47	—	—
4*	6.02	3.99	2.06	—	**4***	5.05	2.65	1.43	—	**4***	4.95	2.83	1.43	—
5*	9.42	7.37	5.45	3.41	**5***	8.17	5.72	4.49	3.13	**5***	6.84	4.72	3.31	1.89

1*, 2*, 3*, 4*, and 5* stand for forest types 1, 2, 3, 4, and 5, respectively.

check whether the manual groups achieved maximum variability among groups and minimum variability within a group and also for the purpose of comparison. The ANOVA results for the M-I approach (median $F = 36.120$, $p < 0.001$, and SD $F = 9.537$, $p < 0.001$) indicated that forest experts were able to provide homogeneous forest structure groups.

Euclidean distances among cluster centroids were also considered to evaluate cluster results. Thus, Euclidean distances provided better separabilities for the fully automatic approach (Aut-I) than for both semiautomatic approaches (SAut-I, SAut-II) (Table 2.10). Euclidean distances for the cluster analysis of segmented polygons from the binned DCHM (SAut-II approach) presented the worst separability among forest structure types (Table 2.10).

The validation analyses using hypsograph percentiles revealed marked differences among forest structure types for the four approaches. Tukey HSD tests of hypsograph percentiles (Table 2.11) indicate that the best separability among forest structure types was achieved for the Aut-I approach. Significant differences ($p < 0.05$) for at least four percentiles for each pair of forest structure types were obtained. The SAut-I approach produces slightly poorer results. This approach provided acceptable separability among forest structure types, with significant differences ($p < 0.05$) for at least two hypsograph percentiles for each pair of forest structure types (Table 2.11). This was similar to the M-I approach, which showed limited discrimination between forest structure types 1 and 2 ($p < 0.05$ for H10% only). Finally, the SAut-II approach was unable to discriminate forest structure types 2 and 3 and forest structure types 4 and 5 ($p > 0.05$ in all hypsograph percentiles), showing the worst separability among forest structure types. In summary, both the hypsographs and the Tukey HSD tests for hypsograph percentiles highlighted the Aut-I and SAut-I approaches as being the best able to discriminate among forest structure types.

2.6.2.5 Discussion

In this work, we developed and compared four approaches for forest structure characterization that incorporate expert opinion in a progressive manner. Our results (ANOVA test and Tables 2.10 and 2.11) indicate that all four procedures are valid

TABLE 2.11
Tukey HSD Test

	Type 1	Type 2	Type 3	Type 4
Aut-I Approach				
Type 2	$H_{25\%}$ $H_{50\%}$ $H_{75\%}$ $H_{90\%}$			
Type 3	$H_{10\%}$ $H_{25\%}$ $H_{50\%}$ $H_{75\%}$ $H_{90\%}$	$H_{10\%}$ $H_{25\%}$ $H_{50\%}$ $H_{75\%}$ $H_{90\%}$		
Type 4	$H_{10\%}$ $H_{25\%}$ $H_{50\%}$ $H_{75\%}$ $H_{90\%}$	$H_{10\%}$ $H_{25\%}$ $H_{50\%}$ $H_{75\%}$ $H_{90\%}$	$H_{10\%}$ $H_{25\%}$ $H_{50\%}$ $H_{75\%}$ $H_{90\%}$	
Type 5	$H_{10\%}$ $H_{25\%}$ $H_{50\%}$ $H_{75\%}$ $H_{90\%}$	$H_{10\%}$ $H_{25\%}$ $H_{50\%}$ $H_{75\%}$ $H_{90\%}$	$H_{10\%}$ $H_{25\%}$ $H_{50\%}$ $H_{75\%}$ $H_{90\%}$	$H_{25\%}$ $H_{50\%}$ $H_{75\%}$ $H_{90\%}$
SAut-II Approach				
Type 2	$H_{10\%}$ $H_{25\%}$ $H_{50\%}$ $H_{75\%}$ $H_{90\%}$			
Type 3	$H_{10\%}$ $H_{25\%}$ $H_{50\%}$ $H_{75\%}$ $H_{90\%}$	—		
Type 4	$H_{10\%}$ $H_{25\%}$ $H_{50\%}$ $H_{75\%}$ $H_{90\%}$	$H_{25\%}$ $H_{50\%}$ $H_{75\%}$ $H_{90\%}$	$H_{50\%}$	
Type 5	$H_{10\%}$ $H_{25\%}$ $H_{50\%}$ $H_{75\%}$ H_{90}	$H_{25\%}$ $H_{50\%}$ $H_{75\%}$ $H_{90\%}$	$H_{50\%}$ $H_{75\%}$	—
SAut-I Approach				
Type 2	$H_{10\%}$ $H_{25\%}$ $H_{50\%}$ $H_{75\%}$			
Type 3	$H_{10\%}$ $H_{25\%}$ $H_{50\%}$	$H_{50\%}$ $H_{90\%}$		
Type 4	$H_{10\%}$ $H_{25\%}$ $H_{50\%}$ $H_{75\%}$	$H_{10\%}$ $H_{25\%}$ $H_{50\%}$	$H_{10\%}$ $H_{25\%}$ $H_{50\%}$ $H_{75\%}$ $H_{90\%}$	
Type 5	$H_{10\%}$ $H_{25\%}$ $H_{50\%}$ $H_{75\%}$	$H_{10\%}$ $H_{25\%}$ $H_{50\%}$ $H_{75\%}$ $H_{90\%}$	$H_{10\%}$ $H_{25\%}$ $H_{50\%}$	$H_{10\%}$ $H_{25\%}$ $H_{50\%}$ $H_{75\%}$ $H_{90\%}$
M-I Approach				
Type 2	$H_{10\%}$			
Type 3	$H_{10\%}$ $H_{25\%}$ $H_{50\%}$	$H_{10\%}$ $H_{25\%}$ $H_{50\%}$		
Type 4	$H_{10\%}$ $H_{25\%}$ $H_{50\%}$ $H_{75\%}$ $H_{90\%}$	$H_{10\%}$ $H_{25\%}$ $H_{50\%}$ $H_{75\%}$ $H_{90\%}$	$H_{75\%}$ $H_{90\%}$	
Type 5	$H_{10\%}$ $H_{25\%}$ $H_{50\%}$ $H_{75\%}$ $H_{90\%}$	$H_{10\%}$ $H_{25\%}$ $H_{50\%}$ $H_{75\%}$ $H_{90\%}$	$H_{10\%}$ $H_{25\%}$ $H_{50\%}$ $H_{75\%}$ $H_{90\%}$	$H_{10\%}$ $H_{25\%}$ $H_{50\%}$

Significant differences between forest types ($p < 0.05$) hypsograph percentiles.

and acceptable, although quantitative attributes for validation (i.e., percentiles) (Table 2.11) indicated that the fully automated approach (Aut-I) provided a slightly higher degree of separability for the five forest structure classes than the mixed procedures with increasing expert participation. Therefore, our results demonstrate that the incorporation of expert opinion does not imply any improvement in precision nor does it represent a significant loss.

The fact that the M-I and SAut-I methods correctly discriminated forest structure types suggests that it may be advantageous to incorporate expert opinion and manual procedures in order to establish structure typologies where specific software or trained users are not available. In fact, the quantitative attributes (Euclidean distance and percentiles) associated to the different forest structures (Table 2.11) showed a greater degree of statistical separation in the procedures with greater expert participation (M-I and SAut-I) than in the approach that included the segmentation with eCognition (SAut-II). These results are consistent with previous findings that considered that manual (based on expert opinion) and automated approaches should be mutually complementary, especially in heterogeneous forest areas (Wulder et al., 2008b).

The reclassification of the *LiDAR*-derived DCHM into expert opinion height classes is a suitable approach for the manual delineation of forest stands. Although binning the *LiDAR*-derived DCHM into height classes implies the simplification of the data and the loss of information, this prior step aided the manual delineation of forest stands. The reclassification of DCHM into height classes allowed the transformation of a grayscale map of a continuous variable (i.e., *LiDAR* heights) into a map where the spatial pattern of colors and textures aided the identification of forest stands (Figure 2.27). The color distribution and texture of the binned *LiDAR* height categories accurately synthesized the spatial distribution of crown cover and gaps, attributes that describe forest structure (Maltamo et al., 2005; Poage and Tapeiner 2005). Falkowski et al. (2009) indicate that the *LiDAR* height and the degree of forest coverage are the *LiDAR* parameters that best discriminate the forest successional types in their study area. Thus, binning the *LiDAR*-derived DCHM allowed experts to distinguish and digitalize polygons in a similar way to traditional methods (i.e., based on photo interpretation) but more easily, as no stereoscopic restitution equipment was required. In addition, compared to the more individual work of the photo interpreter, this procedure facilitates team discussion among forest managers in order to delimit structure units according to the required management focus (productive, conservational, etc.), as well as based on their personal experience and knowledge of ecological interactions and other ancillary data. The procedures developed in this work offer the following advantages: (1) They allow greater expert participation, (2) they make it possible to give a specific management focus in each case, and (3) they provide accessibility by the users (forest managers) to the source of *LiDAR* information.

2.6.3 CURRENT ISSUES IN LiDAR AREA-BASED APPROACH: CO-REGISTRATION ERROR, ACCURACY OF PREDICTIONS, AND MODELING TREE SIZE DISTRIBUTIONS

2.6.3.1 Introduction

The use of LiDAR data and ABA is becoming increasingly popular for forest practitioners and natural resource managers. This methodology focuses on finding empirical relationships between predictors and forest properties (Erdody and Moskal 2010; Maltamo et al., 2006; Næsset, 2002). In the ABA approach, LiDAR-derived predictors are regressed against the variable of interest, thereby obtaining a model for relating the variable of interest and the LiDAR predictors. Predictions at points where only LiDAR data is available are then based on the model obtained. The advantages of this approach over the traditional field-based inventory method are

- The possibility of obtaining realistic maps that provide a better illustration of the spatial variability of the forest variables.
- The predictions of the aggregated values of the variables of interest are enhanced by using the auxiliary LiDAR information.

This method is sometimes criticized because the models function like black boxes. Even though the models are obtained without considering the physical interactions between LiDAR pulses and the forest, the flexibility and good results provided by this methodology have made it a standard for operational applications.

The ABA approach, in combination with regression analysis, is probably the most widespread technique. Most of the applications developed during the last decades using this methodology share three common features:

1. The methodology relies on a correct co-registration between field measurements and LiDAR data.

   Georeferencing field observations requires the use of GPS equipment, and this equipment does not perform well in forest environments. Positioning errors can be within the range of 5–10 m when operating in dense forest environments. This lack of co-registration introduces an additional uncertainty into the estimation of forest variables.

2. The accuracy assessment usually focuses on studying the marginal distribution of prediction errors at the pixel or plot level.

   This assessment is not sufficient for management planning, which usually relies on aggregated estimates made for areas that include a large number of pixels. Moreover, inventory estimates of spatially aggregated values often need to fulfill accuracy requirements that are not directly derived from the model's accuracy assessment at the pixel level. Spatially aggregated predictions and assessments of their accuracy are clearly required, although they have received less attention in the literature than studies regarding accuracy at the pixel level.

3. The variables of interest summarize variables that do not give information on the variability of tree sizes.

   Summarizing variables—for example, basal area, quadratic mean diameter, mean height, or dominant height—is not informative enough for applications such as timber value assessments that require knowledge about the distribution of tree sizes. Very few studies have explored the possibilities for estimating tree size distributions from LiDAR.

   Several questions arise from the issues mentioned. The following examples aim to illustrate how these questions can be answered.

Example 2.3: Co-Registration Errors

Georeferencing of field plots for remote sensing applications is usually based on the positioning of GNSSs. These techniques involve measuring distances between satellites and receivers. GNSS may refer to the U.S. Global Positioning System (GPS) or the Russian GNSS (GLONASS). This technology offers a number of advantages over traditional survey methods in terms of time, cost, and ease of use. Several factors affect the accuracy of GNSS, depending on either observation conditions or device characteristics and processing mode. Forested environments are far from optimal conditions for GNSS positioning, as the forest canopy blocks and reflects the satellite signal, causing multipath effects and signal losses that reduce accuracy. Nominal accuracies are difficult to achieve, and ad hoc experiments are required to evaluate the real accuracy of a specific device and processing mode under given canopy conditions. Some

TABLE 2.12

Nominal and Undercanopy Accuracies for Different Types of GNSS Devices and Processing Modes

Observable	Processing Mode	[a]Nominal Accuracy (m)	GNSS Type	Error Ranges (m) Reported under Forest Canopies	
				Min–Max	Reference
C/A code	Autonomous	5–15 m	GPS	2.95–6.72	Hasegawa and Yoshimura (2003)
			GPS/GLONASS	5.70–21.60	Wing et al. (2005)
	Differential	0.5–3 m	GPS	5.60–7.16	Tucek and Ligos (2002)
			GPS	0.77–5.02	Hasegawa and Yoshimura (2003)
			GPS	5.20–8.40	Rodriguez-Perez et al. (2007)
			GPS/GLONASS	1.05–14.01	Andersen et al. (2009)
Carrier phase	Differential	0.003 ±	GPS	na–3.61	Næsset (1999)
		0.5 ppm x D	GPS/GLONASS	0.01–2.21	Næsset et al. (2000)
			GPS/GLONASS	0.01–1.79	Næsset (2001)
			GPS/GLONASS	0.00–1.29	Andersen et al. (2009)
			GPS/GLONASS	na–2.5	Valbuena et al. (2010)

[a]   Nominal accuracies have been obtained from three different manufacturers (Topcon Positioning System Inc. 2006, Topcon Positioning Systems Inc. 2009; Leica Geosystems AG 2010a, Leica Geosystems AG 2010b; Trimble Navigation Ltd 2010a, Trimble Navigation Ltd 2010b). A detailed description of these processing modes can be found in Mauro (2011).

previous cases are summarized in Table 2.12. GNSS devices can be classified into three groups depending on the observable used for ranging and on the processing mode.

Greater accuracy of the GNSS equipment can be expected when the observation time is increased, although this gain in accuracy is limited and becomes practically null when the observation time reaches a certain limit. The accuracy of different GNSS equipment when extending the observation time from 5 to 30 min was analyzed by Valbuena et al. (2010), comparing survey-grade and handheld phase differential devices. Significant differences were found between both types of devices for observation times shorter than 20 min. After 20 min, the performance of both types of receivers was similar, leading to an important practical conclusion. The use of phase differential handheld devices, which are less expensive than survey-grade receivers, should be avoided unless considerable time is available for positioning the plot center. The reference of 20 min was obtained in a specific forest environment and should therefore be considered with caution, as it is not a universal reference. In any case, we recommend using the best receiver to hand and extending the observation time as much as possible. This is not always a problem, as in many situations the collection of the ground truth data requires considerable time and can be done while the GNSS receiver is collecting observations.

Following the recommendations earlier, the user can minimize the positioning errors. However, the positioning error itself is not the main problem for the ABA methodology. The final products of this methodology are estimates of forest variables. On the one hand, even when the positioning error is minimized, its effects may be strong, and on the other hand, major positioning errors may have little influence on the results of this methodology. We can therefore conclude that positioning errors are important if their effects are significant.

The effect of co-registration errors on tree-height estimates from LiDAR data was studied by Gobakken and Næsset (2009). These authors analyzed both the influence of GNSS positioning errors on various LiDAR metrics and the estimation of Lorey's height, basal area, and volume, based on the aforementioned metrics when using plots of different sizes. Frazer et al. (2011) performed a similar study using simulated stands and LiDAR datasets. In these studies, the uncertainty introduced by the positioning errors is transmitted to the predictors. The influence of positioning errors on the tree-height distribution observed in the field for plots of different sizes was analyzed by Mauro et al. (2011). This study proposes a methodology for integrating changes in a variable of interest and positioning errors. The difference in this case is that the uncertainty introduced by the positioning errors is transmitted to the dependent variable. Analyzing the problem from this perspective requires less computation effort, as changes are not observed in the large LiDAR dataset, and only changes in the field data are analyzed. However, approaching the problem from this perspective requires intense data collection in the field. All the trees within different relatively wide areas must be georeferenced using a total station. Then the variable of interest must be compared in simulated plots within these wide areas. The studies by Gobakken and Næsset (2009), Frazer et al. (2011), and Mauro et al. (2011) verified that increasing the size of the plots helps to reduce the effects of positioning errors. These studies provide general references for estimating the effect of positioning errors for consideration when dimensioning the field plots for training models in the ABA approach. However, a similar ad hoc analysis would be needed for specific applications where the effect of the positioning errors must be monitored.

Example 2.4: Accuracy for Spatially Aggregated Predictions

Introduction
Traditional inventory methods use only field-data information, that is, single-stage sampling (SSS). Double-stage sampling (DSS) techniques use information available from field samples and from auxiliary variables for a larger sample of the population of interest (see Section 2.2). In most cases, this auxiliary information is available for the whole population or the study area. The strong correlation between LiDAR-derived variables and variables of interest for forest planning has been demonstrated in many studies (Næsset 1997a; Næsset 1997b; Næsset and Økland 2002; Magnussen et al., 1999; Maltamo et al., 2006). This fact suggests that

LiDAR may provide a good source of auxiliary information for making estimates with a high level of precision, using a reduced number of plots (Andersen and Breidenbach 2007).

The use of LiDAR data and ABA is becoming increasingly popular for forest practitioners and natural resource managers. The advantages of this approach over traditional field-based inventory methods include the possibility of obtaining realistic maps illustrating the spatial variability of the forest variables and enhancing predictions of aggregated values of the variables of interest by using auxiliary information in a DSS. Most of the applications that use this approach provide general estimators of the accuracy of the predictive models at the pixel level. This assessment is not sufficient for management planning, which usually relies on aggregated estimates made for areas that include a large number of pixels. Moreover, inventory estimates of aggregated values often need to fulfill accuracy requirements that are not directly derived from the assessment of the model's accuracy at the pixel level. Spatially aggregated predictions and assessments of their accuracy are clearly required, although they have received less attention in the literature than studies regarding accuracy at the pixel level.

Questions about trade-offs using either DSS with LiDAR data as auxiliary information or SSS to estimate aggregated values can arise when planning a sampling for developing forest management plans in relatively small areas, ranging from hundreds to thousands of hectares. Potential DSS reduction of fieldwork activities, which can be very expensive, may offset the additional cost derived from ALS data acquisition and processing.

Objectives

This study aims to investigate the possibilities for fieldwork reduction when using DSS with ALS auxiliary information and compares the sampling intensities that are needed to reach a relative error of 5%, 10%, and 15% when using SSS or DSS.

Methods

The study area is a 300 ha *P. sylvestris* L. forest located on the northern slopes of the Valsaín valley in the Guadarrama mountains (central Spain). Elevations range between 1310 and 1450 m above sea level, with slopes of between 10% and 45%. The general aspect of the study site is northwest.

Tree height (H) and DBH were measured in a total of 37 georeferenced circular plots with a radius of 20 m (1256.64 m^2). Plots were systematically distributed in three lines starting in randomly selected locations close to points with easy access. Basal area (G), stand density (N), dominant height (Ho), mean tree height (Hm), and quadratic mean diameter (Dg) were computed for each plot directly from the field measurements. Local models for *P. sylvestris* L. were used to estimate stem volume (Rojo and Montero, 1996) and total tree biomass (Montero et al., 2006). These models used DBH and H, respectively, and DBH as predictors. Tree attributes were aggregated to obtain plot-level values of volume and total biomass.

For the DSS, ALS-derived variables were obtained for each plot using Fusion software. In a previous step, irregularities in pulse densities were removed, obtaining a final number of two pulses m^{-2}. Generalized regression (GREG) estimators are model-assisted estimators. A linear model is fitted using both field data at plot level and predictors associated to the plots. The model is then applied to every pixel or unit in the study area and aggregated. An additional correction term that considers the residuals of the modeling stage is included.

For each variable, the arithmetic mean from SSS and the GREG estimators for the mean from DSS were computed using a different number of field plots. The number of field plots was gradually lowered from 37 to 10. Two hundred and fifty bootstrap replications were computed for each number of field plots. The mean of the sample means and the GREG estimator for the whole area mean computed using 37 plots was used as a reference value for each method. Then, for each number of plots, the width of 95% CI and the percentage of replicates that was different to these values in at least 5%, 10%, and 15% were computed. Differences in precision between the estimators based only on field data and the GREG estimators can be observed in Figure 2.28.

Results and conclusion

A considerable reduction in fieldwork is achieved when auxiliary information is considered. The accuracy of the sample mean and GREG estimators appears to be similar for the stand density. Several studies (Næsset 2002) have found that this variable is poorly correlated with LiDAR predictors, and this may account for this pattern.

Example 2.5: Prediction of Diameter Distribution

Diameter distribution is probably the most relevant variable for forest managers. Many forest attributes can be related to it, such as basal area, volume, biomass, number of stems, and their distributions by diameter classes. Various studies (Gobakken and Næsset 2004, 2005; Maltamo et al., April 2006; Maltamo et al., 2007; Breidenbach et al., 2008) have explored the possibility of estimating diameter distributions using LiDAR data. These studies have been developed in boreal forest areas and have shown that diameter distribution can be accurately predicted. Basically three different methods exist for estimating diameter distributions.

The first method, known as the parameter prediction method, consists of modeling the diameter distribution of field plots using parametric models (usually Weibull models). Model parameters in training plots are then regressed against LiDAR predictors. Models for each parameter are applied to each pixel to obtain estimates of the diameter distribution parameters. Examples of this method can be found in Maltamo et al. (2006), Gobakken and Næsset (2004), and Breidenbach et al. (2008).

The second method is called the percentile prediction method (Gobakken and Næsset, 2005; Maltamo et al., 2007). In this method, diameter distributions in training plots are modeled from a set of percentiles. These percentiles are usually the 10, 20…, and 100 percentile. This is a nonparametric approach that allows a very flexible definition of the diameter distribution. Its main disadvantage is that the number of models to fit adds up to the number of percentiles considered plus an additional model for a scaling parameter. This means fitting eleven models to predict the diameter distribution.

The third method is called the parameter recovery method. In this method, models are obtained to predict several variables closely related to the diameter distribution. A parametric model for the diameter distribution is assumed. Then a series of equations are established between the predicted variables

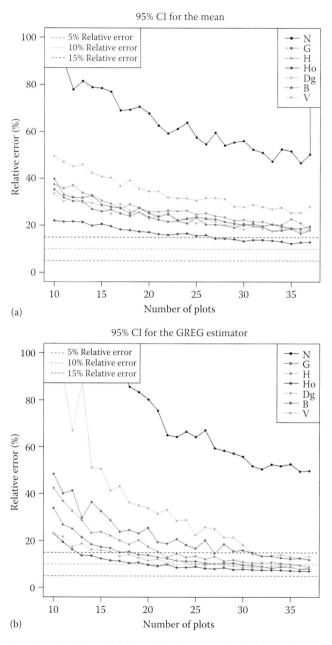

FIGURE 2.28 **(See color insert.)** (a) 95% relative error for SSS sampling (sample mean). (b) 95% relative error for DSS sampling (GREG estimator).

and the parameters of the diameter distribution. The diameter distribution parameters are obtained by solving the system of equations. An example of this methodology applied to obtain tree-height distributions can be found in Mauro et al. (2012).

REFERENCES

Akca, A. 2001. *Waldinventur*. J.D. Sauerländer's Verlag. Frankfurt am Main, 193 S.

Andersen, H.E. and J. Breidenbach 2007. Statistical properties of mean stand biomass estimators in LiDAR based double sampling survey design. *ISPRS International Society for Photogrammetry and Remote Sensing*. Istanbul, Turkey, 8p. http://www.fs.fed.us/pnw/pubs/journals/pnw_2007_andersen002.pdf

Andersen, H.E., T. Clarkin, K. Winterberger, and J. Strunk 2009. An accuracy assessment of positions obtained using survey- and recreational-grade global positioning system receivers across a range of forest conditions within the Tanana valley of interior Alaska. *Western Journal of Applied Forestry* 24(9): 128–136.

Arroyo, L.A. 2006. Cartografía de combustibles forestales a partir de imágenes de alta resolución espacial y clasificadores de contexto. Doctoral thesis. E.T.S.I. Montes, U.P.M, Madrid, Spain.

Arroyo, L.A., S.P. Healey, W.B. Cohen, D. Cocero, and J.A. Manzanera 2006. Using object-oriented classification and high-resolution imagery to map fuel types in a Mediterranean region. *Journal of Geophysical Research-Biogeosciences* 111: G04S04.

Arroyo, L.A., K. Johansen, J. Armston, and S. Phinn 2010. Integration of LiDAR and QuickBird imagery for mapping riparian biophysical parameters and land cover types in Australian tropical savannas. *Forest Ecology and Management* 259: 598–606.

Baatz, M., M. Hoffmann, and G. Willhauck 2008. Progressing from object-based to object-oriented image analysis. In: *Object Based Image Analysis*, eds. T. Blaschke, S. Lang, and G.J. Hay. Heidelberg, Germany: Springer.

Baltsavias, E.P. 1999. Airborne laser scanning: Basic relations and formulas. *ISPRS Journal of Photogrammetry and Remote Sensing* 54: 199–214.

Bater, C.W. and N.C. Coops 2009. Evaluating error associated with lidar-derived DEM interpolation. *Computers and Geosciences* 35: 289–300.

Benz, U.C., P. Hofmann, G. Willhauck, I. Lingenfelder, and M. Heynen 2004. Multi-resolution, object-oriented fuzzy analysis of remote sensing data for GIS-ready information. *ISPRS Journal of Photogrammetry and Remote Sensing* 58: 239–258.

Bernard, E., L. Barbosa, and R. Carvalho 2011. Participatory GIS in a sustainable use reserve in Brazilian Amazonia: Implications for management and conservation. *Applied Geography* 31: 564–572.

Beverly, J.L., K. Uto, J. Wilkes, and P. Bothwell 2008. Assessing spatial attributes of forest landscape values: An internet-based participatory mapping approach. *Canadian Journal of Forest Research* 38: 289–303.

Bitterlich, W. 1948. Die Winkelzahlprobe. *Allgemeine Forst und Holtz Zeit-schrift* 59: 4–5.

Bivand, R. S., E. J. Pebesma, and V. Gomez-Rubio 2008. *Applied Spatial Data Analysis with R*. New York: Springer.

Blaschke, T. 2010. Object based image analysis for remote sensing. *ISPRS Journal of Photogrammetry and Remote Sensing* 65: 2–16.

Blaschke, T., C. Burnett, and A. Pekkarinen 2006. Image segmentation methods for object-based analysis and classification. In: *Remote Sensing Image Analysis: Including the Spatial Domain*, eds. S.M. de Jong and F.D. Van der Meer. Dordrecht, the Netherlands: Springer.

Bock, M., P. Xofis, J. Mitchley, G. Rossnerc, and M. Wissenc 2005. Object-oriented methods for habitat mapping at multiple scales—Case studies from Northern Germany and Wye Downs, UK. *Journal for Nature Conservation* 13: 75–89.

Breidenbach, J., C. Gläser, and M. Schmidt 2008. Estimation of diameter distributions by means of airborne laser scanner data. *Canadian Journal of Forest Research* 38(6): 1611–1620.

Breidenbach, J., E. Næsset, V. Lien, T. Gobakken, and S. Solberg 2010. Prediction of species specific forest inventory attributes using a nonparametric semi-individual tree crown approach based on fused airborne laser scanning and multispectral data. *Remote Sensing of Environment* 114: 911–924.

Brown, G. 2003. A method for assessing highway qualities to integrate values in highway planning. *Journal of Transport Geography* 11: 271–283.

Brown, G. 2005. Mapping spatial attributes in survey research for natural resource management: Methods and applications. *Society and Natural Resources* 18: 17–39.

Brown, G. 2006. Mapping landscape values and development preferences: A method for tourism and residential development planning. *International Journal of Tourism Research* 8: 101–113.

Brown, G. and P. Reed 2009. Public participation GIS: A new method for use in national forest planning. *Forest Science* 55: 166–182.

Brown, G. and D. Weber 2011. Public participation GIS: A new method for national park planning. *Landscape and Urban Planning* 102: 1–15.

Burrough, P.A. and R.A. McDonnel 1998. *Principles of Geographic Information Systems*. Oxford, U.K.: Oxford University Press.

CCRS (Canadian Centre for Remote Sensing) 2000. Fundamentals of remote sensing, (http://www.nrcan.gc.ca/earth-sciences/geography-boundary/remote-sensing/fundamentals/1430) (last accessed on 20 October 2012)

Chuvieco, E. 2008. *Teledetección Ambiental*, La observación de la Tierra desde el Espacio, Barcelona, Spain: Ariel.

Clement, J.M. and A.S. Cheng 2011. Using analyses of public value orientations, attitudes and preferences to inform national forest planning in Colorado and Wyoming. *Applied Geography* 31: 393–400.

Cochran 1977. *Sampling Techniques*. New York: John Wiley & Sons.

De Chant, T., A. Hernando-Gallego, J. Velázquez-Saornil, and M. Kelly 2010. Urban influence on changes in linear forest edge structure. *Landscape and Urban Planning* 96: 12–18.

Definiens 2010. *Definiens eCognition Developer 8.0.1. Reference book*. Munich, Germany: Definiens AG.

Donovan, S.M., C. Looney, T. Hanson, Y.S. de León, J.D. Wulfhorst, and S.D. Eigenbrode 2009. Reconciling social and biological needs in an endangered ecosystem: The palouse as a model for bioregional planning. *Ecology and Society* 14(1): 9.

Dubayah, R.O. and J.B. Drake 2000. Lidar remote sensing for forestry. *Journal of Forestry* 98: 44–52.

Erdody, T.L. and L.M. Moskal 2010. Fusion of LiDAR and imagery for estimating forest canopy fuels. *Remote Sensing of Environment* 114(4)(4/15): 725–737.

Estes, J.E. and D.S. Simonett 1975. Fundamentals of the image interpretation. In *Manual of Remote Sensing*, ed. R.G. Reeves. Falls Church, VA: American Society of Photogrammetry, pp. 869–1076.

Falkowski, M.J., J.S. Evans, S. Martinuzzi, P.E. Gessler, and A.T. Hudak 2009. Characterizing forest succession with lidar data: An evaluation for the Inland Northwest, USA. *Remote Sensing of Environment* 113: 946–956.

Fernández-Díaz, J.C. 2011. Lifting the canopy veil. *Imaging Notes* 26(2): 31–34.

Franklin, S.E. 2001. *Remote Sensing for Sustainable Forest Management*. Boca Raton, FL: Lewis Publisher.

Frazer, G.W., S. Magnussen, M.A. Wulder, and K.O. Niemann 2011. Simulated impact of sample plot size and co-registration error on the accuracy and uncertainty of LiDAR-derived estimates of forest stand biomass. *Remote Sensing of Environment* 115(2)(2/15): 636–649.

Gil, A., H. Calado, and J. Bentz 2011a. Public participation in municipal transport planning processes—The case of the sustainable mobility plan of Ponta Delgada, Azores, Portugal. *Journal of Transport Geography* 19: 1309–1319.

Gil, A., H. Calado, L.T. Costa, J. Bentz, C. Fonseca, and A. Lobo 2011b. A methodological proposal for the development of Natura 2000 sites management plans. *Journal of Coastal Research* 2011: 1326–1330.

Gleason, C.J. and J. Im 2012. Forest biomass estimation from airborne LiDAR data using machine learning approaches. *Remote Sensing of Environment* 125: 80–91.

Glenn, E.P., A.R. Huete, P.L. Nagler, and S.G. Nelson 2008. Relationship between remotely-sensed vegetation indices, canopy attributes and plant physiological processes: What vegetation indices can and cannot tell us about the landscape. *Sensors* 8: 2136–2160.

Gobakken, T. and E. Næsset 2004. Estimation of diameter and basal area distributions in coniferous forest by means of airborne laser scanner data. *Scandinavian Journal of Forest Research* 19(6): 529–542.

Gobakken, T. and E. Næsset 2005. Weibull and percentile models for lidar-based estimation of basal area distribution. *Scandinavian Journal of Forest Research* 20(6): 490–502.

Gobakken, T. and E. Næsset 2009. Assessing effects of positioning errors and sample plot size on biophysical stand properties derived from airborne laser scanner data. *Canadian Journal of Forest Research* 39(5): 1036–1052.

González-Ferreiro, E. 2012. Aplicación de lidar aerotransportado en la generación de modelos digitales del terreno y en la estimación de variables dasométricas. PhD thesis. University of Santiago de Compostela, La Coruña, Spain.

González-Ferreiro, E., U. Diéguez-Aranda, and D. Miranda. 2012. Estimation of stand variables in *Pinus radiata* D. Don plantations using different LiDAR pulse densities. *Forestry* 85: 281–292.

Goodwin, N.R., N.C., Coops, and D.S. Culvenor 2006. Assessment of forest structure with airborne LiDAR and the effects of platform altitude. *Remote Sensing of Environment* 103: 140–152.

Hair, J.F., R.E. Anderson, R.L. Tatham, and W.C. Black 1995. *Análisis Multivariante*. London, U.K.: Prentice Hall International.

Hall, S.A., I.C. Burke, D.O. Box, M.R. Kaufmann, and J.M. Stoker 2005. Estimating stand structure using discrete-return lidar: An example from low density, fire prone ponderosa pine forests. *Forest Ecology and Management* 208: 189–209.

Harding, D.J., M.A. Lefsky, G.G. Parker, and J.B. 2001. Laser altimeter canopy height profiles: Methods and validation for closed-canopy, broadleaf forests. *Remote Sensing of Environment* 76: 283–297.

Hasegawa, H. and T. Yoshimura 2003. Application of dual-frequency GPS receivers for static surveying under tree canopies. *Journal of Forest Research* 8(2): 103–110.

Hernando, A., L. Arroyo, J. Velazquez, and R. Tejera 2012. Object based image analysis for mapping Natura 2000 habitats to improve forest management. *Photogrammetric Engineering and Remote Sensing* 78(9): 991–999.

Hernando, A., R. Tejera, J. Velázquez, and M.V. Nuñez 2010. Quantitatively defining the conservation status of Natura 2000 forest habitats and improving management options for enhancing biodiversity. *Biodiversity and Conservation* 19: 2221–2233.

Heurich, M. and F. Thoma 2008. Estimation of forestry stand parameters using laser scanning data in temperate, structurally rich natural European beech (*Fagus sylvatica*) and Norway spruce (*Picea abies*) forests. *Forestry* 81: 645–661.

Hollaus, M., W. Wagner, B. Maier, and K. Schadauer 2007. Airborne laser scanning of forest stem volume in a mountainous environment. *Sensors* 7: 1559–1577.

Holmgren, J., M. Nilsson, and H. Olsson 2003. Simulating the effects of lidar scanning angle for estimation of mean tree height and canopy closure. *Canadian Journal of Remote Sensing* 29: 623–632.

Huete, A., K. Didan, T. Miura, E.P. Rodriguez, X. Gao, and L.G. Ferreira 2002. Overview of the radiometric and biophysical performance of the MODIS vegetation indices. *Remote Sensing of Environment* 83: 195–213.

Hyyppä, J., H. Hyyppä, D. Leckie, F. Gougeon, X. Yu, and M. Maltamo 2008. Review of methods of small-footprint airborne laser scanning for extracting forest inventory data in boreal forests. *International Journal of Remote Sensing* 29: 1339–1366.

Koch, B., C. Straub, M. Dees, Y. Wang, and H. Weinacker 2009. Airborne laser data for stand delineation and information extraction. *International Journal of Remote Sensing* 30: 935–963.

Labrador-García, M., J.A. Évora-Brondo, and M. Arbelo-Pérez 2012. *Satélites de Teledetección para la Gestión del Territorio*. La Laguna, Spain: Consejería de Agricultura, Ganadería, Pesca y Aguas del Gobierno de Canarias.

Laliberte, A.S., E.L. Fredrickson, and A. Rango 2007. Combining decision trees with hierarchical object-oriented image analysis for mapping arid rangelands. *Photogrammetric Engineering and Remote Sensing* 73: 197–207.

Laliberte, A.S., A. Rango, K.M. Havstad, J.F. Paris, R.F. Beck, R. McNeely, and A.L. Gonzalez 2004. Object-oriented image analysis for mapping shrub encroachment from 1937 to 2003 in southern New Mexico. *Remote Sensing of Environment* 93: 198–210.

Lang, S. 2008a. Object-based image analysis for remote sensing applications: Modeling reality-dealing with complexity. In *Object-Based Image Analysis*, eds. T. Blaschke, S. Lang, and G.J. Hay. Berlin, Germany: Springer.

Lang, S. 2008b. Object-based image analysis for remote sensing applications: Modeling reality-dealing with complexity. In *Object-Based Image Analysis*, eds. T. Blaschke, S. Lang, and G.J. Hay. Berlin, Germany: Springer.

Lefsky, M.A., W.B. Cohen, S.A. Acker, G.G. Parker, T.A. Spies, and D. Harding 1999. Lidar remote sensing of the canopy structure and biophysical properties of Douglas-Fir Western Hemlock forests. *Remote Sensing of Environment* 70: 339–361.

Lefsky, M.A., W.B. Cohen, G.G. Parker, and D.J. Harding 2002. Lidar remote sensing for ecosystem studies. *Bioscience* 52: 19–30.

Leica 2010a. *Leica viva GNSS GS10 receiver datasheets*.

Leica 2010b. *Leica zeno 10 & zeno 15 datasheet*.

Lillesand, T.M., R.W. Kiefer, and J. Chipman. 2008. *Remote Sensing and Image Interpretation*. Hoboken, NJ: John Wiley & Sons.

Lohr, S. 1999. *Sampling Design and Analysis*. Pacific Grove, CA: Brooks/Cole Publishing Co.

Lovell, J.L., D.L.B. Jupp, G.J. Newnham, N.C. Coops, and D.S. Culvenor 2005. Simulation study for finding optimal lidar acquisition parameters for forest height retrieval. *Forest Ecology and Management* 214: 398–412.

Magnussen, S., P. Eggermont, and V.N. LaRiccia 1999. Recovering tree heights from airborne laser scanner data. *Forest Science* 45(16): 407–422.

Mallinis, G., N. Koutsias, M. Tsakiri-Strati, and M. Karteris 2008. Object-based classification using Quickbird imagery for delineating forest vegetation polygons in a Mediterranean test site. *ISPRS Journal of Photogrammetry and Remote Sensing* 63: 237–250.

Maltamo, M., K. Eerikäinen, P. Packalen, and J. Hyyppä 2006. Estimation of stem volume using laser scanning-based canopy height metrics. *Forestry* 79(2): 217–229.

Maltamo, M., P. Packalen, X. Yu, K. Eerikainen, J. Hyyppa, and J. Pitkanen 2005. Identifying and quantifying structural characteristics of heterogeneous boreal forests using laser scanner data. *Forest Ecology and Management* 216: 41–50.

Maltamo, M., A. Suvanto, and P. Packalén 2007. Comparison of basal area and stem frequency diameter distribution modelling using airborne laser scanner data and calibration estimation. *Forest Ecology and Management* 247(1–3): 26–34.

Mauro, F., R. Valbuena, A. García-Abril, E. Ayuga-Téllez, and C. Pascual 2012. Tree height distribution estimation from ALS data in central Spain. Paper presented at *Silvilaser 2012*, Vancouver, British Columbia, Canada. First return. http://silvilaser2012.com/wp-content/uploads/2011/11/Silvilaser2012_Full_Proceedings.pdf

Mauro, F., R. Valbuena, J.A. Manzanera, and A. García-Abril 2011. Influence of global navigation satellite system errors in positioning inventory plots for tree-height distribution studies. *Canadian Journal of Forest Research* 41(1): 11–23.

Mauro, F., R. Valbuena, S. Martín Fernandez, J.A. Manzanera, and E. Peral 2012. Uncertainty of mean values of Forest variables, calculated combining field plots and ALS data for a 300 ha area. Estimation of possibilities for fieldwork reduction. Paper presented at *Forests at 2012*, Corvallis, OR. U.S.A.http://forestsat2012.files.wordpress.com/2012/04/forestsat2012_titles_and_abstracts_final1.xlsx

McIntyre, N., J. Moore, and M. Yuan 2008. A place-based, values-centered approach to managing recreation on Canadian crown lands. *Society and Natural Resources* 21: 657–670.

Means, J.E., S.A. Acker, J.F. Brandon, M. Renslow, L. Emerson, and C.J. Hendrix 2000. Predicting forest stand characteristics with airborne scanning lidar. *Photogrammetric Engineering and Remote Sensing* 66: 1367–1371.

Montero, G., R. Ruiz-Peinado, and M. Muñoz 2006. *Producción de Biomasa y Fijación de CO2 por los bosques españoles*. Madrid, Spain: INIA.

Moote, M.A., M.P. McClaran, and D.K. Chickering 1997. Theory in practice: Applying participatory democracy theory to public land planning. *Environmental Management* 21: 877–889

Mustonen, J., P. Packalen, and A. Kangas 2008. Automatic segmentation of forest stands using canopy height model and aerial photography. *Scandinavian Journal of Forest Research* 23: 534–545.

Næsset, E. 1997a. Estimating timber volume of forest stands using airborne laser scanner data. *Remote Sensing of Environment* 61(2): 246–253.

Næsset, E. 1997b. Determination of mean tree height of forest stands using airborne laser scanner data. *ISPRS Journal of Photogrammetry and Remote Sensing* 52: 49–56.

Næsset, E. 1999. Point accuracy of combined pseudorange and carrier phase differential GPS under forest canopy. *Canadian Journal of Forest Research* 29(5): 547–553.

Næsset, E. 2002. Predicting forest stand characteristics with airborne scanning laser using a practical two-stage procedure and field data. *Remote Sensing of Environment* 80(1): 88–99.

Næsset, E. 2004a. Effects of different flying altitudes on biophysical stand properties estimated from canopy height and density measured with a small-footprint airborne scanning laser. *Remote Sensing of Environment* 91(2): 243–255.

Næsset, E. 2004b. Practical large-scale forest stand inventory using a small-footprint airborne scanning laser. *Scandinavian Journal of Forest Research* 19(2): 164–179.

Næsset, K. and O. Bjerknes 2001. Estimating tree heights and number of stems in young forest stands using airborne laser scanner data. *Remote Sensing of Environment* 78(3): 328–340.

Næsset, E. and T. Økland 2002. Estimating tree height and tree crown properties using airborne scanning laser in a boreal nature reserve. *Remote Sensing of Environment* 79(1): 105–115.

Næsset, E., B. Trygve, Ø. Ola, and H.R. Lorentz 2000. Contributions of differential GPS and GLONASS observations to point accuracy under forest canopies. *Photogrammetric Engineering and Remote Sensing* 66(4) 403–407.

Parker, G.G. and M.E. Russ 2004. The canopy surface and stand development: Assessing forest canopy structure and complexity with near-surface altimetry. *Forest Ecology and Management* 189(1–3): 307–315.

Pascual, C., A. García-Abril, L.G. Garcia-Montero, S. Martin Fernandez, and W.B. Cohen 2008. Object-based semi-automatic approach for forest structure characterization using LIDAR data in heterogeneous *Pinus sylvestris* stands. *Forest Ecology and Management* 255(11): 3677–3685.

Pascual, C., L.G. Garcia-Montero, L.A. Arroyo, and A. García-Abril 2012. *Annals of Forest Science*. DOI 10.1007/s13595-012-0232-1 (published online).

Pascual, C., L.G. Garcia-Montero, L.A. Arroyo, and A. García-Abril 2013. Increasing the use of expert opinion in forest characterization approaches based on LiDAR data. *Annals of Forest Science* 70(1): 87–99.

Petr, M., M. Smith, and J. Suarez 2010. Object-based approach for mapping complex forest structure phases using LIDAR data. In *The International Archives of Photogrammetry, Remote Sensing and Spatial Information Science*, *GEOBIA 2010*, Ghent, Belgium.

Pfeiffer, C., S. Glaser, J. Vencatesan, E. Schliermann-Kraus, A. Drescher, and R. Glaser 2008. Facilitating participatory multilevel decision-making by using interactive mental maps. *Geospatial Health* 3: 103–112.

Poage, N.J. and J.C. Tappeiner 2005. Tree species and size structure of old-growth Douglas-fir forests in central western Oregon, USA. *Forest Ecology and Management* 204(2–3): 329–343.

Pocewicz, A., M. Nielsen-Pincus, G. Brown, and R. Schnitzer 2012. An evaluation of internet versus paper-based methods for Public Participation Geographic Information Systems (PPGIS). *Transactions in GIS* 16: 39–53.

Raymond, C.M., B.A. Bryan, D.H. MacDonald, A. Cast, S. Strathearn, and A. Grandgirard, A. 2009. Mapping community values for natural capital and ecosystem services. *Ecological Economics* 68: 1301–1315.

Reutebuch, S.E., H. Andersen, and R.J. McGaughey 2005. Light detection and ranging (LIDAR): An emerging tool for multiple resource inventory. *Journal of Forestry* 103(6): 286–292.

Rodriguez-Perez, J.R., M.F. Alvarez, and E. Sanz-Ablanedo 2007. Assessment of low-cost GPS receiver accuracy and precision in forest environments. *Journal of Surveying Engineering* 133(4): 159–167.

Rojo, A. and G. Montero 1996. *El Pino Silvestre en la Sierra de Guadarrama: historia y selvicultura de los pinares de Cercedilla, Navacerrada y Valsain*. Madrid, Spain: Ministerio de Agricultura, Pesca y Alimentación.

Rombouts, J., I.S. Ferguson, and J.W. Leech 2008. Variability of LiDAR volume prediction models for productivity assessment of radiata pine plantations in South Australia. *Proceedings of SilviLaser 2008: 8th International Conference on LiDAR Applications in Forest Assessment and Inventory*, Edinburgh, U.K., September 17–19, 2008).

Rondeux, J. 2010. Medición de árboles y masas forestales. Translated by A. Díaz Barrionuevo. Mundi Prensa. 521 pp. Madrid.

Schowengerdt, R.A. 1983. *Techniques for Images Processing and Classification in Remote Sensing*. New York: Academic Press.

Simão, A., P.J. Densham, and M. (Muki) Haklay 2009. Web-based GIS for collaborative planning and public participation: An application to the strategic planning of wind farm sites. *Journal of Environmental Management* 90(6): 2027–2040.

Sithole, G. and G. Vosselman 2004. Experimental comparison of filter algorithms for bare-Earth extraction from airborne laser scanning point clouds. *ISPRS Journal of Photogrammetry and Remote Sensing* 59: 85–101.

Smith, D.M., Larson, B.C., Kelty, M.J., and Ashton, P.M.S. 1997. *The Practice of Silviculture. Applied Forest Ecology*. New York: John Wiley & Sons.

Stephens, S.L., R.E. Martin, and N.E. Clinton, 2007. Prehistoric fire area and emissions from California's forests, woodlands, shrublands and grasslands. *Forest Ecology and Management* 251(3): 205–216.

Tiede, D., S. Lang, F. Albrecht, and D. Holbling 2010. Object-based class modeling for cadastre-constrained delineation of geo-objects. *Photogrammetric Engineering and Remote Sensing* 76: 193–202.

Tiede, D., S. Lang, and C. Hoffmann 2008. Domain-specific class modelling for one-level representation of single trees. In *Object-Based Image Analysis*, eds. T. Blaschke, S. Lang, and G.J. Hay. Berlin, Germany: Springer.

Topcon 2006. *Topcon hiper pro operator's manual*.

Treitz, P., L. Kevin, W. Murray, P. Doug, D. Nesbitt, and D. Etheridge 2010. LiDAR data acquisition and processing protocols for forest resource inventories in Ontario, Canada. *In Silvilaser 2010: The 10th International Conference on LiDAR Applications for Assessing Forest Ecosystems*, eds. B. Koch and G. Kendlar. Freiburg, Baden-Württemberg, Germany.

Trimble 2010a. Trimble GeoXT datasheet.

Trimble 2010b. Trimble R8 GNSS datasheet.

Tucek, J. and J. Ligos 2002. Forest canopy influence on the precision of location with GPS. *Journal of Forest Science* 48(39): 399–407.

Tulloch, D. 2007. Public participation GIS (PPGIS). *Encyclopedia of Geographic Information Science* SAGE Publications (http://www.sage-ereference.com/geoinfoscience/Article_n165.html) (accessed October 20, 2012).

Valbuena, R., F. Mauro, R. Rodriguez-Solano, and J.A. Manzanera 2010. Accuracy and precision of GPS receivers under forest canopies in a mountainous environment. *Spanish Journal of Agricultural Research* 8(4): 1047–1057.

Vauhkonen, J., I. Korpela, M. Maltamo, and T. Tokola 2010. Imputation of single-tree attributes using airborne laser scanning-based height, intensity, and alpha shape metrics. *Remote Sensing of Environment* 114: 1263–1276.

de Vries, P.G. 1986. *Sampling Theory for Forest Inventory. A Teach-Yourself Course*. Berlin, Germany: Springer.

Weiers, S., M. Bock, M. Wissen, and G. Rossner 2004. Mapping and indicator approaches for the assessment of habitats at different scales using remote sensing and GIS methods. *Landscape and Urban Planning* 67(1–4): 43–65.

Wing, M.G., A. Eklund, and D. Kellogg 2005. Consumer-grade global positioning system (GPS) accuracy and reliability. *Journal of Forestry* 103(5): 169–173.

Wulder, M.A., C.W. Bater, N.C. Coops, T. Hilker and J. C. White 2008a. The role of LiDAR in sustainable forest management. *Forestry Chronicle* 84: 807–826.

Wulder, M.A., J.C. White, G.J. Hay, and G. Castilla 2008b. Towards automated segmentation of forest inventory polygons on high spatial resolution satellite imagery. *Forestry Chronicle* 84: 221–230.

Wulder, M.A., J.C. White, R.F. Nelson, E. Næsset, H.O. Ørka, and N.C. Coops 2012. Lidar sampling for large-area forest characterization: A review. *Remote Sensing of Environment* 121: 196–209.

Zhao, K., S. Popescu, and R. Nelson. 2009. Lidar remote sensing of forest biomass: A scale-invariant estimation approach using airborne lasers. *Remote Sensing of Environment* 113(1): 182–196.

3 Criteria and Indicators for Sustainable Forest Management

M. Victoria Núñez, Rosario Tejera,
Antonio García-Abril, Esperanza Ayuga-Téllez,
and Eugenio Martínez-Falero

CONTENTS

3.1   INTRODUCTION

3.1.1   CHAPTER CONTENT

This chapter explores the different approaches to assess criteria and indicators (C&I) for sustainable forest management (SFM) as a result of the United Nations Conference on Environment and Development (UNCED) 1992 and presents a case study of computing indicators at the local scale, based on light detection and ranging (LiDAR) survey and yield tables in *Pinus sylvestris* forests in Central Spain. These indicators are measures for tree height distribution, timber yield, and biomass.

Section 3.2 describes C&I for SFM at the regional level. It focuses on the international processes and provides an overview of national initiatives for the C&I and forest certification. This section ends with the evolution of sustainability during the last decade, paying attention to the countries with most accomplished processes and some conclusions about the implementation of SFM.

Section 3.3 relates the importance of C&I for SFM at the forest management unit (FMU) level. This section includes the proposal of a methodology that uses information from LiDAR airborne system to assess three SFM indicators designed for this purpose and its application and meanings at the FMU scale.

3.1.2   STATE OF THE ART OF C&I FOR SFM AND PARTICIPATORY FOREST MANAGEMENT

SFM is a concept specifically designed to embrace and reconcile the different interests on forests, including the maintenance of biodiversity. However, the interests of different stakeholders are rarely fully mutually reinforcing. Interests normally require trade-offs and some are simply mutually exclusive. Certification of good or SFM has to deal with these diverging values of different stakeholders, including the importance placed on biodiversity maintenance relative to other aspects (Rametsteiner and Simula 2003).

C&I are tools used to define, assess, and monitor periodic progress toward SFM in a given country or in a specified forest area, over a period of time (Prabhu et al. 1999). The ultimate aim of C&I is to promote improved forest management practices over time and to further the development of a healthier and more productive forest conditions, taking into account the social, economic, environmental, cultural, and spiritual needs of the full range of stakeholder groups in countries concerned.

Through sustainable management, forests can contribute to the resilience of ecosystems, societies, and economies while also safeguarding biological diversity and providing a broad range of goods and services for present and future generations.

Criteria define the essential elements against which sustainability is assessed, with due consideration paid to the productive, protective, and social roles of forests and forest ecosystems. Each criterion relates to a key element of sustainability and may be described by one or more indicators (FAO 2001).

Indicators are parameters that can be measured and correspond to a particular criterion. They measure and help monitor the status and changes of forests in quantitative, qualitative, and descriptive terms that reflect forest values as seen by those who defined each criterion (FAO 2001).

The C&I are considered as monitoring instruments by which progress toward implementation of SFM may be evaluated and reported (Kotwal et al. 2008; Khadka and Vacik 2012).

The multiple C&I involved, the variety of underlying goals and objectives of different interest groups and the possibility of nontransparency of the decision-making process, can hinder the adoption of C&I or may even result in the failure to gain public acceptance of the results of the C&I assessments (Mendoza and Prabhu 2000a). Therefore, the new context of sustainable forestry places demands on forest planning processes, in terms of integrating science with participatory decision support (Mendoza and Prabhu 2000b; Sheppard 2005).

C&I can be applied at a range of spatial scales. Early emphasis was on the development of national level, under the international processes for the purpose of raising awareness, of gaining commitment, and of assisting in measuring broad progress toward achieving SFM (Raison et al. 2001). Therefore, C&I are developed at three different levels: international, national, and FMU levels.

At the international level, there are nine ongoing international C&I processes:

1. African Timber Organization (ATO) Process
2. Dry Forest in Asia Process
3. Dry-Zone Africa Process
4. International Tropical Timber Organization (ITTO) Process
5. Lepaterique Process of Central America
6. Montreal Process (Temperate and Boreal Forests)
7. Near East Process
8. Pan-European Forest Process
9. Tarapoto Proposal for the Sustainability of the Amazon Forest

While each of these processes differs in specific content or structure, all of them center around seven globally agreed thematic areas corresponding to criteria:

a. Extent of forest resources
b. Biological diversity
c. Forest health and vitality
d. Protective functions of forests
e. Productive functions of forests
f. Socioeconomic functions
g. Legal policy and institutional framework

C&I for SFM processes at the international level are closely linked to a number of international forest-related and cross-sectorial processes such as the Forest Resources Assessment (FRA) program of FAO, the United Nations Forum on Forests (UNFF), and the Convention on Biological Diversity (CBD).

At the national level, more than 150 countries are taking part in one or more of the nine international processes (FAO 2001) (Table 3.1). Other 63 countries not members of these international processes are developing their own national C&I. All these countries are somehow supported by partner institutions such as FAO, Centro Agronómico Tropical de Investigación y Enseñanza (CATIE—Tropical Agronomic Research and Training Center), Center for International Forestry Research (CIFOR), International Union of Forest Research Organizations (IUFRO), ITTO, and United Nations Environment Programme (UNEP).

National Forest Programmes (NFPs) evolved from this international forum on forests. Since 2002, over 70 countries attended the NFP *Facility* (FAO 2006). This is a response to intergovernmental dialogue, which has recognized the essential role of NFP in addressing forest sector issues. Its main objective is to assist countries in developing and implementing NFPs that effectively address local needs and national priorities and reflect internationally agreed principles.

Other important initiatives aimed at *forest certification* evolved from the requested C&I implementation. Certification is the process whereby an independent third party (called a certifier or certification body) assesses the quality of forest management in relation to a set of predetermined requirements (the standard). The certifier gives written assurance that a product or process conforms to the requirements specified in the standard (Rametsteiner and Simula 2003).

C&I in international processes are used among national governments to monitor and exchange information on their implementation of SFM, while forest certification schemes are used by forest management organizations to establish proof of SFM in the forest product markets (Holvoet and Muys 2004).

In 1993, concerned business representatives, social groups, and environmental organizations got together and established the *Forest Stewardship Council* (FSC) with the purpose of supporting environmentally appropriate, socially beneficial, and economically viable management of the world's forests. FSC brings together people, organizations, and businesses to develop consensus-based solutions that promote responsible stewardship of the world's forests. The 10 FSC principles and criteria form the basis for all FSC forest management standards and policies.

TABLE 3.1
International Processes on SFM, Number of C&I, and Countries

ITTO

Number of criteria	7
Number of indicators	66
Countries	*Producers:* Bolivia, Brazil, Cambodia, Cameroon, Central African Republic, Colombia, Congo, Cote-d'Ivoire, Democratic Republic of the Congo, Ecuador, Fiji, Gabon, Ghana, Guatemala, Guyana, Honduras, India, Indonesia, Liberia, Malaysia, México, Myanmar, Nigeria, Panama, Papua New Guinea, Peru, Philippines, Suriname, Thailand, Togo, Trinidad and Tobago, Vanuatu, and Venezuela
	Consumers: Australia, Austria, Belgium, Canada, China, Denmark, Egypt, Finland, France, Germany, Greece, Ireland, Italy, Japan, Luxembourg, Nepal, New Zealand, Norway, Poland, Portugal, Republic of Korea, Spain, Sweden, Switzerland, the Netherlands, United Kingdom, United States of America, and the European Union

Dry-zone Africa process

Number of criteria	7
Number of indicators	47
Countries	Burkina Faso, Cape Verde, Chad, Gambia, Guinea-Bissau, Mali, Mauritania, Niger and Senegal. Djibouti, Eritrea, Ethiopia, Kenya, Somalia, Sudan, and Uganda. Angola, Botswana, D. R. of Congo, Lesotho, Malawi, Mauritius, Mozambique, Namibia, Seychelles, South Africa, Swaziland, United Republic of Tanzania, Zambia, and Zimbabwe

Montreal Process

Number of criteria	7
Number of indicators	67
Countries	Argentina, Australia, Canada, Chile, China, Japan, Mexico, New Zealand, Republic of Korea, Russian Federation, Uruguay, and United States of America

Pan-European Forest Process

Number of criteria	6
Number of indicators	35
Countries	Albania, Austria, Belarus, Belgium, Bosnia-Herzegovina, Bulgaria, Croatia, Czech Republic, Denmark, Estonia, European Community, Finland, France, Georgia, Germany, Greece, Hungary, Iceland, Ireland, Italy, Latvia, Liechtenstein, Lithuania, Luxembourg, Monaco, Netherlands, Norway, Poland, Portugal, Republic of Andorra, Romania, Russian Federation, San Marino, Slovakia, Slovenia, Spain, Sweden, Switzerland, Turkey, Ukraine, and United Kingdom

(continued)

TABLE 3.1 (continued)
International Processes on SFM, Number of C&I, and Countries

ATO

Number of criteria	28
Number of indicators	60
Countries	Angola, Cameroon, Central African Republic, Congo, Cote-d'Ivoire, Democratic Republic of Congo, Equatorial, Guinea, Gabon, Ghana, Liberia, Nigeria, Sao Tome et Principe, and United Republic of Tanzania

Tarapoto

Number of criteria	7
Number of indicators	47
Countries	Bolivia, Brazil, Colombia, Ecuador, Guyana, Peru, Suriname, and Venezuela

Lepaterique Process of Central America

Number of criteria	8
Number of indicators	53
Countries	Belize, Costa Rica, El Salvador, Guatemala, Honduras, Nicaragua, and Panama

Dry Forests in Asia

Number of criteria	8
Number of indicators	49
Countries	Bangladesh, Bhutan, China, India, Mongolia, Myanmar, Nepal, Sri Lanka, and Thailand

Near East Process

Number of criteria	7
Number of indicators	65
Countries	Afghanistan, Algeria, Azerbaijan, Bahrain, Cyprus, Djibouti, Egypt, Iraq, Islamic Republic of Iran, Jordan, Kuwait, Kingdom of Saudi Arabia, Kyrgyzstan, Lebanon, Libya, Malta, Mauritania, Morocco, Oman, Pakistan, Qatar, Somalia, Sudan, Syria, Tajikistan, Tunisia, Turkey, Turkmenistan, United Arab Emirates, and Yemen

Source: http://www.fao.org/forestry/16435-091114c04e64187ce8caa8299fcd3fa8c.pdf

Since 1999, the *Programme for the Endorsement of Forest Certification* (PEFC), an international nonprofit, nongovernmental organization, is dedicated to promoting SFM through independent third-party certification. It works by endorsing national forest certification systems developed through multistakeholder processes and tailored to local priorities and conditions. PEFC supplements the principles, C&I derived from the international processes with additional requirements, developed through multi-stakeholder processes to make them operational as performance measures in the forest.

PEFC is an umbrella organization that endorses national schemes, some of which were developed within the PEFC framework, while others existed as independent schemes for several years before PEFC was formed (e.g., American Tree Farm System (ATFS), Canadian Standard Association (CSA), or Sustainable Forest Initiative (SFI)).

FSC- and PEFC-endorsed schemes together account for almost 100% of the world's certified forest. The total worldwide area of forests certified by these schemes is estimated about 375 million ha in May 2011 (UNECE-FAO 2011).

3.2 PROGRESS TOWARD SFM

The application of SFM during the last decades has improved the state of forests in several ways. The full-detail data and information about the seven thematic elements of SFM and data of trends are best available at www.fao.org/forestry/fra2010. Here we present a summary of them at the global level, by regions and for some important countries, as well as some conclusions related with the implementation of SFM.

3.2.1 Progress toward SFM at the Global Level, 2000–2010

A brief summary of the key findings in SFM by main criteria (themes) are exposed in the following, based on the *Global FRA 2010, Main Report* (FAO 2010):

3.2.1.1 Extent of Forest Resources

The change in forest area is negative (−0.13% annual rate) in the period 2000–2010. The net change in forest area is estimated at −5.2 million hectares per year, down from −8.3 million hectares per year in the period 1990–2000.

However, deforestation—mainly the conversion of tropical forests to agricultural land—shows signs of decreasing in several countries but continues at a high rate in others. Around 13 million hectares of forest was converted to other uses or lost through natural causes each year in the last decade compared to 16 million hectares per year in the 1990s. Both Brazil and Indonesia, which had the highest net loss of forest in the 1990s, have significantly reduced their rate of loss. Nevertheless, Africa, South America, and Oceania continue to have the largest net loss of forest. This may denote a low level of implementation of SFM in these areas.

3.2.1.2 Forest Biological Diversity

The area of forest where conservation of biological diversity is designated as the primary function has increased by more than 95 million hectares since 1990. However, not all of them are located inside protected areas, which might mean that SFM is not applied.

3.2.1.3 Forest Health and Vitality

Forest fires are severely underreported at the global level, with information missing from many countries, especially in Africa.

Outbreaks of forest insect pests damage some 35 million hectares of forest annually, primarily in the temperate and boreal zone. Severe storms, blizzards, and earthquakes have also damaged large areas of forest since 2000. Around 0.2 million hectares of forest by year was lost by these causes in the last decade.

3.2.1.4   Productive Functions of Forest Resources

At the global level, reported wood removals amounted to 3.4 billion cubic meters annually, similar to the volume recorded for 1990 and equivalent to 0.7% of the total growing stock. Though informally and illegally removed, wood, especially wood fuel, is not recorded, so the actual amount of wood removals is undoubtedly higher.

3.2.1.5   Protective Functions of Forest Resources

Around 330 million hectares of forest is designated for protective functions (8% of the world's forests). The area of these forests increased by 59 million hectares between 1990 and 2010, primarily because of large-scale planting in China.

3.2.1.6   Socioeconomic Functions of Forests

Information availability for the social and cultural functions of forest is scarce. The only subregions and regions with fairly good data are East Asia and Europe.

The value of wood removals has fallen since 2005, while the value of non-wood forest products remains underestimated due to information still missing from many countries in which non-wood forest products are highly important, and the true value of subsistence use is rarely captured.

3.2.1.7   Legal, Policy, and Institutional Framework

The area of forest covered by a management plan is steadily increasing, yet information is only available for 80% of the total forest area.

Close to 75% of the world's forests are covered by an NFP, that is, a participatory process for the development and implementation of forest-related policies and international commitments at the national level.

3.2.2   Progress toward SFM by Regions

3.2.2.1   Progress in Africa, 2000–2010

On the whole, progress toward SFM in Africa has improved when comparing the last decade to the 1990s. The net loss of forest area has slowed down and the areas of forest designated for the conservation of biological diversity included in protected areas have increased slightly. There is also a positive increase in the area of forest with a management plan over the last 10 years. However, there is a continued, rapid loss of forest area and of primary forest (3.4 million hectares and 0.572 million hectares per year, respectively, since 2000–2010) (FAO 2010).

3.2.2.2   Progress in Asia, 2000–2010

Overall, the forest area in Asia is about 16 million hectares larger in 2010 than it was in 1990 as a result of large-scale afforestation efforts during the last 10–15 years, particularly in China. The decrease in area of primary forest reached 0.342 million hectares, while there was an increase in the forest area designated for conservation of biological diversity (annual rate of 1.4 million hectares in the period 2000–2010), the area of forest in protected areas (annual rate of 1.5 million hectares in the period 2000–2010), and forests designated for protective functions (annual rate of

2.6 million hectares in the period 2000–2010). The area affected by fire decreased, while that affected by insects increased slightly. Variables representing the legal, policy, and institutional framework are largely positive or stable, and information availability in the region is generally good (FAO 2010).

3.2.2.3 Progress in Europe, 2000–2010

The status of forest resources in Europe has essentially been stable over the last 10 years. While the area of forest is expanding (annual change of 0.676 million hectares in the period 2000–2010), the focus of forest management in Europe has converged toward conservation of biological diversity, protection, and multiple uses (FAO 2010).

Moreover, the proportion of old plus uneven-aged forest increased slightly in most regions. Both these categories are valuable for biodiversity and recreation. While these increases are small, they are still noteworthy because change in forest structure is generally a very slow process. Also, in Europe, the most common form of involvement in SFM was through NFP workshops, followed by consultation (FOREST EUROPE-UNECE and FAO 2011).

3.2.2.4 Progress in North and Central America, 2000–2010

Progress toward SFM was generally positive in North and Central America as a whole during the period 2000–2010, with the notable exception of the significant negative trends noted for the area of forest affected by fire and by insect pests (annual rate of 4.1 million hectares in the period 2000–2010) and the slight decrease in the level of employment (FAO 2010). There was, however, considerable variation among subregions. More detailed information can be seen in Section 3.2.3.1 for the United States and Section 3.2.3.2 for Canada.

3.2.2.5 Progress toward SFM in Oceania, 2000–2010

Data availability is largely determined by Australia, since it accounts for 78% of the forest area in this region. It is impossible to assess long-term trends in this region for most of the themes due to the low reporting level. An increase in the net loss of forest area (annual rate of loss 0.7 million hectares) was reported, despite the fact that part of the latter may be a temporary loss of forest cover due to an extensive drought in Australia (FAO 2010). Extensive information about SFM in this country can be found in Section 3.3.2.3.

3.2.2.6 Progress in South America, 2000–2010

Overall, progress toward SFM was mixed in South America. The rate of net forest loss continues to increase (annual change, −3.997 million hectares in the period 2000–2010) although significant progress has been made, particularly in the last 5 years. The rate of loss of primary forest also remains alarmingly high (nearly an average of 3 million hectares per year in the period 2000–2010). Nonetheless, there were also positive signs, for example, in the increased areas of forest designated for conservation of biological diversity and in protected areas (annual change, 3.1 million hectares and 2.4 million hectares in the period 2000–2010). The decrease in removals of wood fuel may reflect a reduced demand for this product in the region, but this was partly offset by an increase in removals of industrial wood since 2000.

The area of planted forests increased (3.76 million hectares in the period) and may meet a larger proportion of the demand for wood in the future. The increase in the area of forest with a management plan (19.37 million hectares in the period) is also a positive sign (FAO 2010).

3.2.3 SFM Trends in the United States, Canada, EU plus Russian Federation, and Australia

3.2.3.1 Trends in the United States

Related to the Montreal Process, in 2010, the Forest Service of the U.S. Department of Agriculture published a National Report on Sustainable Forest. One of the report's key findings is the fact that the United States is richly endowed with forests (751 million acres). That area has remained remarkably stable over the last 50 years, and the amount of wood in these forests is increasing. At the same time, however, forests in the United States face a number of threats, ranging from fragmentation and loss of forest integrity due to development to an alarming increase in the area and severity of forest disturbances. Sustained capacity and willingness to manage forests sustainably are evidenced by a growing number of public–private collaborations on projects devoted to landscape-scale conservation (Forest Service 2010). Detailed information regarding the current evaluation of the seven Montreal Process criteria is exposed in the following*:

3.2.3.1.1 *Criterion 1: Conservation of Biological Diversity*

The total area of forests in the United States currently stands at 751 million acres. This number has been stable to slightly increasing in recent decades. The area of forests impacted by fragmentation has been increasing at a steady rate.

The indicators covering species richness and genetic diversity do not yield a clear signal regarding changes in richness and diversity since 2003.

Changes in richness and diversity are highly variable across geographic regions and general species categories (vascular plants, mammals, birds, and so on), with declines in species counts in some areas or categories being offset by gains in others.

The area of forests that is formally protected by government designation totals some 106 million acres; this number has changed little since 2003. At the same time, alternative ways of protecting forests through land trusts and conservation easements have grown rapidly.

3.2.3.1.2 *Criterion 2: Maintenance of Productive Capacity of Forest Ecosystems*

The current use of the forests is sustainable from the perspective of timber production capacity; the area of timberland is stable, and timber stocking on these lands has been increasing.

In the case of non-wood forest products, the data are not sufficient to reach a definitive conclusion about the sustainability of productive capacity.

---

* Source: National Report on Sustainable Forest, 2010. Available at http://www.fs.fed.us/research/ sustain/docs/national-reports/2010/2010-sustainability-report.pdf

3.2.3.1.3   Criterion 3: Maintenance of Ecosystem Health and Vitality

The findings for the indicators in this criterion point to a substantial increase in the levels of biotic disturbance and an increase in fire extent and intensity relative to the 1997–2002 reference period.

3.2.3.1.4   Criterion 4: Conservation and Maintenance of Soil and Water Resources

For Indicator 4.19, which measures soil degradation, trends over time cannot yet be determined.

Measures of water conditions (Indicator 4.21) are limited by the data on hand and improvements in reporting are expected.

Indicators 4.17, 4.18, and 4.20 measure forest areas subject to certain land-use designations or management practices. They rely largely on state-level reports of management activity and land-use designations. The lack of consistency in these reports presents considerable challenges in addressing the indicators. None of these three indicators were included in the 2003 report, and relevant comparisons could not be drawn with past activities to determine significant trends.

3.2.3.1.5   Criterion 5: Maintenance of Forest Contribution to Global Carbon Cycles

Forested ecosystems in the United States currently contain an amount of carbon equivalent to more than 165 billion metric tons of CO_2, a figure close to 27 times the 5.9 billion tons of CO_2 emitted nationally every year through the burning of fossil fuels and similar sources. Live trees and forest soils account for the bulk of forest-based carbon stocks. In terms of flows, forests sequester approximately 650 million metric tons of additional CO_2 every year, offsetting close to 11% of total U.S. annual carbon emissions. This rate of sequestration has been relatively stable for several decades.

A carbon equivalent to around 8 billion metric tons of CO_2 is currently stored in long-lived forest products and in discarded forest products in landfills. Annual rates of sequestration are approximately 100 million tons, substantially less than 650 million tons annually sequestered by forests but still a significant number.

Annual production of energy from the combustion of wood in the United States is around 2100 trillion British thermal units (BTUs) (about 2% of the 101 quadrillion BTUs consumed in 2007). When converted to avoid carbon emissions, this number translates to between 100 and 200 million metric tons of carbon depending on the energy source used for comparison. This number has been slightly falling since the mid-1990s.

3.2.3.1.6   Criterion 6: Maintenance and Enhancement of Long-Term Multiple Socioeconomic Benefits to Meet the Needs of Society

The criterion includes 20 indicators divided into 5 subcriteria. These are

   a. Production and consumption

   The indicators covering timber and wood products (Indicators 6.25, 6.28, 6.30, 6.32, and 6.33) show that both timber harvest and wood product production are down slightly relative to 2003. Production and trade figures for non-timber forest products are down 30% relative to 1998, while exports are up 38% since 2003.

b. Investment in the forest sector

Investments in the wood products and pulp and paper sectors totaled $10.9 billion in 2006, up from $7.5 billion in 2003 but still substantial lower than the $13.6 billion reported for 1997.

Investments in research, extension services, and education totaled $608 million in 2006, an increase of 18% in inflation-adjusted terms since 2000.

c. Employment and community needs

Forest product industry employment, which currently stands at 1.3 million employees, decreased by about 15% since 1997, with much of the drop concentrated in the pulp and paper sector. Vitality and adaptability of forest-dependent communities are new measures for indicator 6.38, which will rely on survey and community assessment techniques to characterize the resiliency of individual communities.

d. Recreation and tourism

Although the area of public forest lands has increased to a very slight degree since 2003, the falling percentage of private lands that are accessible for recreation use points to an overall decline in forest land available to recreation.

The number of recreational activity days has increased by 25% since 2000 and currently stands at 83 billion days. The number of people participating in these activities has increased at a slower pace (4.4%).

e. Cultural, social, and spiritual needs and values

Due to the more intangible values and attachments people have to forests, a pilot approach was explored. It relied on survey techniques to assess the various dimensions of people's relationship to forests and the importance they attach to them.

Results highlight the diversity of feelings people have for forests and the fact that these are largely determined by cultural background.

3.2.3.1.7   Criterion 7: Legal, Institutional, and Economic Framework for Forest Conservation and Sustainable Management

A wide variety of legal, institutional, and economic approaches exist that encourage SFM in the United States, at all levels of government.

Many new market-based mechanisms, including forest certification, wetland banks, payments for environmental services, conservation easements, and environmental incentives, are also being developed to implement SFM in the United States.

3.2.3.2   Trends in Canada

Harvest rates across Canada are set at levels to ensure long-term ecosystem sustainability. The rate of deforestation in Canada has declined, with the annual rate dropping from just 64,000 ha in 1990 to some 45,000 ha in 2009 (Canadian Forest Service 2011). As a result, the country's forests are able to support species diversity and resilience over vast landscapes with dynamic, ever-evolving ecosystems (Canadian Council of Forest Ministers 2010).

Moreover, one of the most notable achievements of the Montreal Process has been the establishment of mutual trust and confidence, which has encouraged the 12 member countries to develop a "network of knowledge" through discussion, research, cooperation, and communication (The Working Group for the conservation and sustainable management of temperate and boreal forests 2009).

A summary of the main current themes related with SFM status is exposed in the following*:

3.2.3.2.1   Status of Forest-Associated Species at Risk

From 1990 to 2011, nearly 70 of 215 species have moved to a higher-risk category, while 135 did not change or moved to a lower-risk category.

3.2.3.2.2   Addition and Deletions of Forest Area

The rate of deforestation has declined from just 64,000 ha in 1990 to 45,000 ha in 2009.

3.2.3.2.3   Area of Forest Disturbed by Fire, Insects, Disease, and Harvesting

Fires: Although the number of fires was the same, the area burned was much higher in 2010: 3 million hectares—nearly double (86%) the 10-year average.

Insects: In 2009, 15.2 million hectares of forest was defoliated by insects or contained beetle-killed trees, an increase from 13.7 million hectares in 2008.

Deceases: Native forest pathogens have evolved to exist in equilibrium with natural communities.

Harvesting: Each province and territory sets an allowable annual cut based on the sustainable growth rate of the particular forest area. In 2009, approximately 612,000 hectares of forest was harvested (9.5% less than in 2008).

3.2.3.2.4   Proportion of Timber Harvest Area Regenerated by Artificial and Natural Means

Between 2008 and 2009, naturally regenerated area decreased by 3.5% and artificially regenerated area decreased by 13.3%. This reflects the steep decline (42%) in annual harvest area over the previous 5 years, from a 10-year high in 2005 to a 20-year low in 2009.

3.2.3.2.5   Carbon Emissions/Removals

Forest acted as net carbon sinks in 12 of the 20 years from 1990 to 2009.

3.2.3.2.6   Forest Sector Carbon Emissions

A changing energy mix and greater energy efficiency are clearly reducing greenhouse gas (GHG) emissions in the sector.

---

* Source: the State of Canada's Forests, Annual Report 2011. Available at http://cfs.nrcan.gc.ca/ pubwarehouse/pdfs/32683.pdf

*3.2.3.2.7   Annual Harvest of Timber Relative to the Level
            of Harvest Deemed to be Sustainable*

Canada's aggregated allowable annual cut in 2009 is estimated to be 207 million cubic meters. Total annual wood supply has been relatively stable since 1999, at about 240 million cubic meters, although in recent years it has increased modestly, reaching 246 million cubic meters in 2009.

3.2.3.2.8   Certification

As of December 2012, Canada had 149.8 million hectares of forest certified, up from 142.8 million hectares in 2009. Also, Canada has the largest area of certified forest in the world, with 42% of the total worldwide.

3.2.3.2.9   Forest Industry Employment

In 2010, direct employment in the Canadian forest industry fell 6.6% from 2009 levels and over 4.4% from the previous 10 years.

3.2.3.2.10   Forest Product Exports

In 2010, the value of Canada's forest product exports increased to $26 billion from $23.6 billion in 2009 but decreased by 5.9% from the previous 10-year average.

3.2.3.2.11   Forest-Independent Communities in Canada

The number of forest-independent communities is down from approximately 300 recorded in the 2001 census to fewer than 200 in 2006.

3.2.3.3   Trends in Europe and Russian Federation

A summary of the main current themes related with SFM status is exposed in the following*:

3.2.3.3.1   Ecosystem Health and Vitality

Sulfur deposition has decreased over the last decade.

The development of pH and base saturation of soils did not show a uniform pattern. However, increased pH and base saturation were found in acid forest soils.

The rate of defoliation of most tree species varied moderately during the last decade, and the level showed a mean defoliation of 25% or more.

3.2.3.3.2   Productive Functions of Forests

In the Russian Federation, the felling rate has decreased from 41% in 1990 and stabilized around 20% since 2000. In Europe, without the Russian Federation, the felling rate increased from 58% in 1990 to 62% in 2010.

More than 578 million cubic meters of roundwood were produced and reached 21.1 billion € in 2010.

The total reported value of marketed non-wood goods amounts to 2.7 billion € and has almost tripled since the 2007 assessment—although some of the increase may be due to improved reporting.

---

* Available at: http://www.unece.org/forests/fr/outputs/soef2011.html

Finally, most of the forest area in Europe is covered by a forest management plan or its equivalent.

3.2.3.3.3   Biological Diversity in Forest Ecosystems

Europe has increased by around half a million hectares annually over the last 10 years due to policies to improve biodiversity.

Forest management practice has changed toward greater integration of biodiversity aspects.

For instance, deadwood components and important vulnerable small biotopes are kept in forests managed for wood production. There is an increasing use of natural regeneration, and mixed tree species stands. In several countries, long-term monitoring of threatened forest species has indicated that adoption of new forest management measures has reduced the decline of threatened species.

3.2.3.3.4   Protective Functions in Forest Management

There is growing awareness of the importance of forest management for protection of water, soil, and infrastructure. More than 20% of Europe's forest fulfills these functions.

3.2.3.3.5   Socioeconomic Functions and Conditions

The general trend is a decrease in occupation, but there are substantial differences between regions, which reflect the mechanization level and the potential for increased productivity. The importance and recognition of other forest services, as source of energy, recreation, and cultural and spiritual values, are increasing.

NFPs are the most widely applied approach by countries to develop sound forest policy frameworks. They are usually based on and elaborated through participatory processes. In many countries, NFPs contribute to consistent and broadly supported policies and strategies for putting SFM into practice. However, particular effort is needed to keep such processes relevant for key stakeholders and flexible, to effectively respond to emerging issues, and keep related costs low. While NFP principles are more widely followed than before, there is a need to strengthen substantive participation and the link to overall national development goals and forest-related sectors.

3.2.3.3.6   Policies, Institutions, and Instruments by Policy Area

Countries have highlighted the need for improved forest information and monitoring to implement NPF. This is a response to the growing multiple requirements placed on forests by society and global markets and is reflected in the concept of SFM.

3.2.3.4   Trends in Australia

A summary of the main current themes related with SFM status is exposed in the following*:

3.2.3.4.1   Criterion 1: Conservation of Biological Diversity

Genetic resource conservation plans exist for more than 40 native timber and oil-producing species, a 70% increase on the number in 2003. Since the 2003, the area

---

* Source: Australia's State of the Forest Report 2008 Executive Summary Available at: http://adl.brs. gov.au/forestsaustralia/publications/execsummary.html

of Australia's native forest in formal nature conservation reserves has increased by about 1.5 million hectares to 23 million hectares, from 13% to 16%.

3.2.3.4.2 Criterion 2: Maintenance of Productive Capacity of Forest Ecosystems

The area of plantations increased from 1.63 million hectares to 1.82 million hectares from 2001 to 2006.

3.2.3.4.3 Criterion 3: Maintenance of Ecosystem Health and Vitality

Large areas of Australia were affected by severe drought over 2003–2008 periods, with significant regional impacts on tree health. Several exotic organisms that pose a threat to Australian forests moved closer to Australia's shores during the reporting period, increasing the importance of effective quarantine. Fire, including some very intense fires in southern Australia, burnt an estimated 24.7 million hectares of forest in the period from 2001–2002 to 2005–2006. Of that total, an estimated 20 million hectares was burnt by unplanned fire (wildfire) and 4.7 million hectares by planned fire (e.g., prescribed burning).

3.2.3.4.4 Criterion 4: Conservation and Maintenance of Soil and Water Resources

Over 30 million hectares of public forests (20% of the total forest area) is managed primarily for protection, including of soil and water values; most is in nature conservation reserves. In most jurisdictions, codes of practice or other instruments are applied.

3.2.3.4.5 Criterion 5: Maintenance of Forest Contribution to Global Carbon Cycles

Plantations offset about 3.5% and managed native forests about 5.5% of total national GHG emissions in 2005. Additional storage in wood products offset a further 1% of emissions. The net amount of carbon sequestered by managed native forests in 2005 was 43.5 million tons (carbon dioxide equivalent). GHG emissions from deforestation declined from about 70 million tons carbon dioxide equivalent in 2002 to an estimated 53.3 million tons in 2005, which was about 9% of total national GHG emissions. The removal of carbon from native forests by timber harvesting stayed relatively constant and was compensated about three times over by sequestration.

3.2.3.4.6 Criterion 6: Maintenance and Enhancement of Long-Term Multiple Socioeconomic Benefits to Meet the Needs of Societies

Total direct employment in wood and wood product industries increased marginally between 2001–2002 and 2006–2007. Total national employment in businesses dependent on growing and using timber in 2006 was estimated to be about 120,000 people.

Indigenous-managed land includes more than 21 million hectares of forest, which is 13% of Australia's total forest area.

3.2.3.4.7 Criterion 7: Legal, Institutional, and Economic Framework for Forest Conservation and Sustainable Management

The use of forest certification has grown from 2.3 million hectares to over 9 million hectares of native forests and plantations by 2007.

3.2.4 CERTIFIED FOREST EVOLUTION

By May 2011, the global area of certified forest, endorsed by one or the other of the international frameworks—the FSC and the PEFC—amounted to 375 million hectares, up to 7% (23.5 million hectares) since May 2010. There is a rough overlap of 3.75 million hectares due to double certification. The rate of increase of certified forest area has slowed during the past decade. Since 2009, two certification schemes (PEFC and FSC) have been dominant, since all smaller schemes have been endorsed by PEFC. The area of forest certified by FSC increased by 11% and that certified by PEFC by 5%, between 2010 and 2011. However, the trends for both systems have been similar over the past decade (UNECE-FAO 2011).

Globally, the certified area is not evenly distributed. More than half (54%) the certified forest area is in North America, just under one-quarter (23%) in the European Union (EU)/European Free Trade Association (EFTA) region and 12% in other Europe and the Commonwealth of Independent States (CIS). The remaining 11% is split across the southern hemisphere (UNECE-FAO 2011).

After 10 years of implementation, the original intention to save tropical biodiversity through certification has largely failed to date. Most of certified areas are in the temperate and boreal zone. Only around 2% of the total forest areas in Oceania, Africa, Latin America, and Asia together are certified (44.2 million ha of 243.9 million ha).

While the quality of actual audits of the standards is of varying quality, there are indications that independent audits are an incentive for improving forest management. Regardless of many difficulties, forest certification has been very successful in raising awareness and disseminating knowledge on a holistic SFM concept, embracing economic, environmental, and social issues, worldwide (Rametsteiner and Simula 2003).

3.2.5 UTILITY AND EFFICIENCY OF C&I TO ASSESS SFM

From these data, we can conclude that, at least, there is a deficiency in collecting information for several indicators related with these thematic areas: forest health and vitality, socioeconomic functions of forests, and legal, policy, and institutional framework.

This may be due to difficulties in data collection, for example, indicators of forest health and vitality need extensive and expensive inventories or a network of permanent plots. Another reason is the unclear methodology of evaluation, for example, for non-wood products, part of them may be collected from areas outside forests (other wooded land and trees outside forests) and some may come from forests designated for multiple use—including community forests—rather than from forests designated primarily for productive purposes. Moreover, other forest products and services like cultural values remain difficult to measure.

Besides, the area of forest with a management plan is not necessarily an adequate indicator of the area of forest under SFM. For example, plans may not be effective, or forests may be conserved and sustainably used without a plan.

On the other hand, nowadays the implementation of SFM needs improvement since the results of C&I during the last decade show some negative trends for key subjects like area of forest and area of primary forest at the global level.

Especially in Africa, Oceania, and South America, there is a net loss of forest, which indicates SFM is not the predominant type of forest management. Another subject that may point out this problem is the scarce area of protected forest or forest under SFM plans. Even if these areas are improving their efforts toward SFM, broader implementation of SFM is needed.

It is widely assumed that C&I frameworks can facilitate international reporting and agreements while still reflecting national differences (Hall 2001). Moreover, there is a recognized need to define a collaborative approach to C&I research and monitoring frameworks that will improve communication, reduce duplication, increase efficiency, and make more effective use of investment funds (Wolfslehner and Vacik 2012).

However, indicators should be designed for considering their potential interactions and feedbacks within a given set. This would help to gain more insight into systemic cause–effect relationships and—by identifying key processes and indicators—help to make data collection and analysis more efficient (Requardt 2007). This means a change from "monitoring and reporting" to "assessment" of sustainability. For example, in Europe, a new, experimental method to assess sustainability was designed using official data, objective and transparent parameters, and thresholds, in addition to detailed comments to put the situation in context (MCPFE 2011). The assessment aims to give policy- and decision-makers as well as the general public a clear overview of complex issues and facilitate balanced strategic and operational decision-making, as well as communication and dialogue with the general public and other relevant sectors.

3.3 C&I AT THE FOREST MANAGEMENT UNIT LEVEL: COMPUTING INDICATORS IN A CASE STUDY

C&I provide a framework for the formulation of policy options, help to advance international cooperation, and also provide an assessment of the positive and negative changes in forest conservation and management at different levels (Kondrashov 2004). Thus, there is a need to develop and examine C&I for SFM at the FMU level.

C&I at the FMU level provide a science-supported framework upon which national policy decisions can be based (Hall 2001). The objective of SFM evaluation at the FMU level is to support the framework for the sustainable conservation at higher scales and to apply the measurements for the management and development of forests. The progress accomplished needs to be followed up and measured. Moreover, FRA activities are to be taken into account as they reflect the state and change of forest resources. They allow to (i) reply to certain indicators of SFM with a numerical value and (ii) note if an intention to follow the situation of the forestry domain exists in order to better control the development of a forestry policy.

The FMU level indicators depend on local, often site-specific, environmental factors such as forest type and topography, local economic and social considerations, and priorities. These indicators may thus differ between individual forest areas in any one country, at any one time, in accordance with prevailing conditions, priorities, and objectives of management.

The criteria at FMU level are likely to be identical or very similar to those defined at national level, although they are more flexible. Thus, they must be mutually compatible to help ensure complementarity over the country.

Methodological developments have largely operated on the assumption that the relationship between management and indicators of sustainability is well understood and less attention has been paid to the actual derivation of the indicators from the state of the stand (Annikki et al. 2012).

Here we present a methodology and a case study for evaluating three SFM indicators related with forest structure, timber yield, and biomass, assessed with information from LiDAR airborne system that may promote a reliable cost-effective methodology.

The idea of using LiDAR in SFM is not new (Wulder et al. 2008). LiDAR-based forest variables, in particular height-related variables, have been shown to be precise and more cost-effective than field measurements (Nelson et al. 2003; Wulder and Seemann 2003; Lovell et al. 2005). A number of studies have found significant relationships between LiDAR variables and field-measured canopy variables, such as crown dimensions (Lovell et al. 2003, 2005; Coops et al. 2007; Dean et al. 2009; Véga and St-Onge 2009), canopy volume (Lefsky et al. 1999, 2006; Coops et al. 2007), diameter at breast height (DBH), basal area (Lefsky et al. 2002; Chen et al. 2007), and growth rates (Yu et al. 2008). Consequently, LiDAR imagery is useful in forest inventory taking and forest sustainability and ecosystem quality assessments (Lefsky et al. 2002; Wulder et al. 2008; Zhao et al. 2009).

Recent publications confirm the utility of LiDAR to characterize forest structure (Kane et al. 2010; Miura and Jones 2010) and estimate stand volume at plot level (Ioki et al. 2010) or forest biomass (Dubayah et al. 2010). Akay et al. (2009) point out that LiDAR can be used in wide-scale forestry activities such as stand characterizations, forest inventory and management, fire behavior modeling, and forest operations. Castillo et al. (2012) demonstrate that changes in the forest vertical structure (such as height) associated with principal successional stages (early, intermediate, and late) of tropical dry forest secondary growth can be effectively identified from LiDAR data. Overall, there is much potential for automated approaches and ancillary data sets to aid analysis and classification of LiDAR images (Morgan et al. 2010).

Here we present a new application of LiDAR forest data: the assessment of SFM indicators at stand level. These indicators can be used for the assessment of variables for Criteria 1, 3, and 4 of Pan-European Process (MCPFE 2002); Criteria 1, 2, and 5 of Montreal Process; or Criteria 2 and 4 of ITTO Process (ITTO 2005), among others. This approach to evaluate these indicators can also facilitate public participation as the objectives can be easily designed and modified by current computational means.

3.3.1 STUDY AREA

The study area is located in the Fuenfría Valley (Madrid, Spain).* It covers an area of 127.10 ha (1293 m×983 m) (40°45′N, 4°5′W), with elevations ranging from 1310 to 1790 m. The average annual temperature of the area is 9.4°C, the average annual rainfall is 1180 mm, and the predominant tree species is Scots pine (*P. sylvestris*). The study site falls within phytoclimatic subregion IV (VI), that is, subhumid Mediterranean with a Central European trend (Allué 1990). The potential vegetation is supra-Mediterranean Carpetan–Iberian–Leonese and subhumid siliceous Alcarrian series of *Quercus pyrenaica* (Rivas-Martínez 1987). Some forest characteristics and a brief description of the five forest structure types classified by Pascual (2008) in this study area are shown in Table 3.2.

TABLE 3.2
Forest Structure Type, Number of Trees ha^{-1}, Mean Height, Basal Area, and Description Based on Pascual et al. (2008)

Forest Structure Type	Density (Tree. ha^{-1})	H Mean (m)	Basal Area (m^2 ha^{-1})	Description
Type 1 (T1)	850	8.7	39.9	Multilayered, uneven-aged Scots pine formation that includes the tallest trees in the study area. Very high crown cover and density
Type 2 (T2)	640	14.7	40.7	Multidiameter forest with high crown cover, two-story vertical distribution. Trees with lower height and diameter than type 1
Type 3 (T3)	380	11.4	35.3	Multidiameter forest with medium crown cover, less dense than type 2, and trees also smaller in diameter and height
Type 4 (T4)	175	11.4	26.2	Even-aged forest (single story) with low crown cover. Includes mature trees of greater diameter but with a slightly lower height and larger crown diameters than other forest types
Type 5 (T5)	76	8.7	6.6	Dense coverage of shrubs under isolated pine trees

Source: Pascual, C. et al. 2008. *Forest Ecol. Manag.*, 255, 11, 3677.

---

* Section 2.3.4 of this book (Chapter 2) shows an application of *IT* techniques to the inventorying of forest structures and other forest characteristics, in this same study area. There also appears a deeper description of this zone.

3.3.2 LiDAR Data

In August 2002, TopoSys GmbH surveyed the study area with a LiDAR TopoSys II sensor, and a digital canopy height model (DCHM) was obtained after image processing,* as described in detail by Pascual et al. (2008) and in Section 2.5 (Chapter 2). The final DCHM was a raster on map with a pixel of 1 meter wide (Figure 3.1).

The information provided by the LiDAR image is the main source of information for the calculation of indicators of sustainability. Other sources were the potential evolution of the forest (reference scenario), the information in the current management plan, and inventories carried out in the field on permanent plots.

UTM-coordinates: 408108,5–4512228,5 ▨▨▨▨ Boundary between two types of forest structure

0 m   500 m   ↑ N

FIGURE 3.1 LiDAR image of the study area showing the limits of the forest structure zones defined.

---

* The TopoSys II LiDAR system recorded first and last returns with a footprint diameter of 0.95m; average point density was 5 points/m²; the raw data (x, y, z coordinates) were processed into two digital elevation models by TopoSys using as interpolation algorithm a special local adaptive median filter developed by the data provider. The digital surface model (DSM) was processed using the first pulse reflections, and the digital terrain model (DTM) was constructed using the last returns. The DSM and DTM horizontal positional accuracy was 0.5m and vertical accuracy was 0.15m. To obtain a DCHM, the DTM was subtracted from the DSM. The vertical accuracy for the DCHM under forest canopy was 1.3m.

TABLE 3.3
Plot Location (UTM) and Area (m²)

Forest Structure Type	Plot	UTM X	UTM Y	Area (m²)
T1	T1P1	408,830.5	4,512,473.5	2400
	T1P2	408,869.5	4,512,754.5	2400
T2	T2P1	408,890.5	4,512,195.5	1256.6
	T2P2	408,854.5	4,512,186.5	1256.6
T3	T3P1	408,283.5	4,512,580.5	1256.6
	T3P2	408,283.5	4,512,620.5	1256.6
T4	T4P1	408,052.5	4,512,613.5	1256.6
	T4P2	408,054.5	4,512,573.5	1256.6
T5	T5P1	408,196.5	4,512,532.5	1256.6
	T5P2	408,195.5	4,512,492.5	1256.6

3.3.3 FIELD MEASUREMENTS

Field data were compiled to validate the indicators. A point was randomly selected in each of the five forest structure types. These points were used to select two sets of five plots by systematic sampling. For this systematic sampling, a value of slope (west–east direction) and a distance were chosen. The two equidistant points in the straight line with this slope, passing on the random selected point, identify the center of the two sample plots for each forest structure (Table 3.3).

The full height and the height of the first living branch were recorded for each tree in order to calculate the crown height. The DBH and maximum radius of each crown were also recorded.

3.3.4 YIELD TABLES

Variable density yield tables for *P. sylvestris* in the central mountain (García Abejón and Gómez Loranca 1984) were consulted. These tables contain information about mean tree height (H_g), stand top height (H_0) as an average of the height of the 100 highest trees per hectare, quadratic mean diameter (D_g), basal area per hectare (G), stem number per hectare, mean increment, and current annual increment. The yield table for the study area contains these values for even stands of *P. sylvestris* from 20 to 120 years of age at 10-year intervals.

3.3.5 METHODS

To calculate the values of the indicators, the tree height distribution in the study area was compared with the height distribution in the reference scenario designed for the present study.

3.3.5.1 Reference Scenario
We designed a reference scenario to form one extreme for the indicators. Our reference scenario involves a situation of high structural diversity. It fits with a

complex horizontal and vertical forest structure at stand level where (i) all height classes are the same area and follow a Gaussian distribution and (ii) the height distribution curve is maintained constant for species and site quality. The set of stands contains all the possible random values for tree height.

Forest stand structures with a wide range of canopy layers and age classes are more favorable for biodiversity than simple or coetaneous stand structures (Pelissier and Goreaud 2001; De Warnaffe and Devillez 2002; Hernando et al. 2010). Increasing heterogeneity of horizontal and vertical stand structure is linked to a higher number of species and stands with greater ecological stability (Pommerening 2002). Moreover, biodiversity is a key element for evaluating the stability of the system (Kimmins 1997). Therefore, our reference scenario is assumed to represent a sustainable forest. We suppose that it can either maintain itself or be subject to silvicultural actions.

3.3.5.1.1 *Distribution Function of Tree Height in the Reference Scenario*

While yield tables are traditionally based on diameter–stem number relations, in this study, height–stem number relations were used.

The data for mean tree height in each age class are provided by the yield table ($\mu = Hg$). However, the standard deviation must be determined in order to obtain the height distribution curve. This estimator was made from the stand top height (H_0), which has been taken as the average of the height variable truncated by the percentile $1–100/n$ (Equation 3.1):

$$H_0 = \frac{1}{\sigma\sqrt{2\pi}}\left(\frac{\int_b^\infty t \cdot e^{-\frac{(t-H_g)^2}{2\sigma^2}}\,dt}{100\,/\,n}\right) \tag{3.1}$$

where
  b is the $\Phi^{-1}(1-100/n)$

$$\Phi(x) = \int_{-\infty}^x \frac{1}{\sigma\sqrt{2\pi}}\,e^{-\frac{u^2}{2\sigma^2}}\,du$$

  n is the total stem number—in the plot or age class

Therefore,

$$\sigma = \left(H_0 - H_g\right) \times \frac{100}{n} \times \sqrt{2\pi} \times e^{0.5 \times b^2} \tag{3.2}$$

where
  $\pi = 3.1416$
  $e = 2.7173$

TABLE 3.4

Standard Deviation and Mean Height of Trees by Age Class

Age (Years)	Before Thinning		After Thinning	
	h (m)	σ	h (m)	σ
0	0	0.4	0	0.4
10	1	0.47	1	0.47
20	3	0.5	3	0.5
30	6.1	0.553	6.5	0.578
40	8.6	0.566	9	0.591
50	11	0.584	11.4	0.607
60	13.1	0.601	13.5	0.638
70	15.1	0.616	15.5	0.658
80	16.8	0.635	17.2	0.657
90	18.4	0.651	18.8	0.674
100	19.9	0.67	20.3	0.694
110	21.2	0.689	21.6	0.713
120	22.4	0.71		

Note: Calculations were based on the quality I yield table and a moderate thinning schedule (before and after thinning values are provided). Age (years), h = mean height (m), and σ = standard deviation of the height.

Table 3.4 shows the main distribution parameters for the variable tree height obtained by Equations 3.1 and 3.2. The values for the age classes 0, 10, and 20 years were obtained by extrapolation from the available data (Ayuga Téllez et al. 2006).

The tree height distribution curve was achieved by convolution of the tree height distribution in even stands. However, as the estimated density functions for the heights include a probability distribution for trees of height <0 m, it was necessary to first set the truncated distribution.

The tree height distribution was calculated by spatially clustering the height distributions corresponding to 13 age classes. These ranged from the <5-year class up to the >115-year class in 10-year intervals. Each age class was represented by the mean and standard deviation of the tree heights. Eleven of these age classes are represented in the yield tables used (20–120 years). The remaining classes were obtained by extrapolation.

Since it was assumed that all age classes occupied the same area, it was also assumed that dividing the number of stems per area by the number of age classes would provide the stand density. If it is assumed that the density function for tree height in each age class has a normal distribution, the total number of trees, and therefore the number of trees per hectare in each age class, can be determined with Equation 3.3:

$$N_i = \frac{N_i^0}{N_{cl}} \left(1 - \int_{-\infty}^{0} \frac{1}{\sigma_i\sqrt{2\pi}} e^{-\frac{(x-\mu_i)^2}{2\sigma_i^2}} \, dx \right) \tag{3.3}$$

where

N_i is the actual stem number per ha in class i
N_i^0 is the stem number per ha in class i (according to yield tables)
N_{cl} is the number of age classes
μ_i is the mean tree height in class i
σ_i is the standard deviation of tree height in class i
$\pi = 3.1416$
$e = 2.7173$

The estimator of stem number in each height class, for this reference scenario, was derived from the yield tables and Equation 3.3 before and after thinning. Based on the density function for each age class, the distribution function of tree heights in the reference scenario was obtained with Equation 3.4:

$$F_{id}(h) = \sum_{i=1}^{N_{cl}} \int_{-\infty}^{h} \frac{1}{\sigma_i\sqrt{2\pi}} e^{-\frac{(x-\mu)^2}{\sigma_i\sqrt{2\pi}}} \, dx \tag{3.4}$$

where

$F_{id}(h)$ is the probability that a tree has a height $\leq h$
N_{cl} is the number of age classes

However, the computed distribution function showed irregularity between 1.5 and 9.5 m, compromising its functionality within this range (Figure 3.2). To solve this problem, tree height distribution function was recalculated to include age classes at 5-year intervals. The values of the mean, standard deviation, and stem number for the intermediate age classes were obtained by direct interpolation of the available data. The result was a height distribution based on 25 age classes (Figure 3.3).

3.3.5.1.2 Ten-Year Period Evolution of Number of Stems per Height Class in the Reference Scenario

Another condition for the reference scenario is that the tree height distribution function remains indefinitely constant for a single species and site quality (García-Abril et al. 1999). The characteristics of the thinning at the 10-year intervals between operations were calculated in order to constantly recover the initial tree height distribution function. This took into account the number of stems per height class, which, over the 10-year period, (A) naturally died or were felled, (B) grew to the next height class, or (C) remained in the same class (Table 3.5).

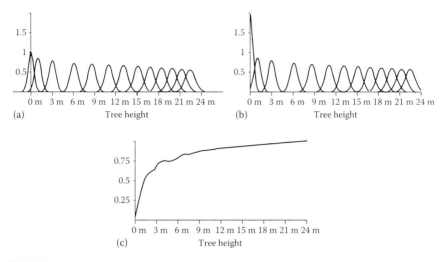

FIGURE 3.2   Height distribution function obtained by aggregating height density functions for age classes (10-year intervals). (a) Probability density functions (pdfs). (b) Truncated pdfs for 13 age classes. (c) Distribution function for 13 age classes.

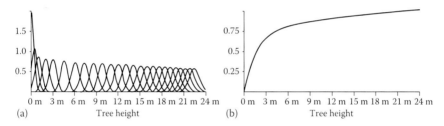

FIGURE 3.3   Height distribution function obtained by aggregating height density functions for age classes of 5-years interval. (a) Truncated pdfs for 25 age classes. (b) Distribution functions pdfs for 25 age classes.

3.3.5.2   LiDAR Forest

In this study, the term "LiDAR forest" refers to the position and height of the trees derived from the LiDAR data set. Tree locations were identified with a relative local maximum in the DCHM.* The local maximum filter represents a group of filtering methods that have been successfully used in different studies (Wulder et al. 2000; Nelson et al. 2005) to identify individual tree locations. The precision of this method is dependent, however, on the forest structure, and can be adjusted through the window size and the smoothing function. In this study, an 8 m window was used since this was the largest crown radius. Kernel estimation for small samples (Martínez-Falero 1992) was used for the nonparametric smoothing function.

---

* Chapter 2 (Section 2.3.4) shows a more detailed procedure to identify the spatial position of trees of a forest. However, the procedure we have applied in this chapter is faster and with enough precision for the purpose of this application (see Section 3.3.6).

TABLE 3.5

Stem Number ha⁻¹ Increase Rate from One Height Class to Another in the Reference Scenario

Height Class (m)	Stem Number/ha	A	B	C
0–3	2,300.7	5.0	11.4	83.6
3–6	476.1	13.4	41.5	45.1
6–9	308.4	35.0	29.0	36.0
9–12	176.5	22.3	28.4	49.3
12–15	123.6	13.0	27.5	59.5
15–18	101.3	8.5	25.1	66.4
18–21	85.6	5.8	23.8	70.3
21–24	45	45.1	0.2	54.7
24–27	0.1	100	0	0

Note: Percentage of stem number (A) extracted due to felling or natural death, (B) which moved to the following height class, or (C) remaining in the same class for the reference scenario (interval = 10 years).

The algorithm used for identifying the relative local maxima in the DCHM is summarized in Figure 3.4. It has been applied to each point in a regular 1×1 m grid superimposed on the study area (in agreement with the DCHM's spatial resolution).

If it is accepted that there is a tree at the coordinates analyzed, it is assigned the height of the DCHM at those coordinates. Figure 3.5 shows the trees' identification from the LiDAR image in a part of the study area.

Tree height distribution is computed, based on tree height and location, and stored, on each one of the points in a square grid—1 m wide—superimposed on the study area. Figure 3.6 shows the information used for computing the value of the indicators on each territorial point.

3.3.5.3 SFM Indicators

Three SFM indicators—tree height distribution, timber yield, and biomass—were designed and computed from the information provided by the LiDAR image and the yield tables.

Indicator values were calculated within a 30 m radius around each node of the grid described earlier. Therefore, information concerning all three SFM indicators was available for the whole study area.

3.3.5.3.1 Tree Height Distribution

In this study, tree height distribution indicator (I_1) expresses a statistical distance between the tree height distribution at each analyzed point and the tree height distribution in the reference scenario (Equation 3.5):

For x_0 = 408108.5 to 409338.5 *step* 1 (UTM *coordinates and steps of 1 meter*)
For y_0 = 4513211.5 to 4512251.5 *step* 1 (UTM *coordinates and steps of 1 meter*)
    For α = 0 to π step $\pi/8$
        Calculate: $R_\alpha \to S_\alpha \to P_\alpha \to PS_\alpha$
            R_α, *straight line that crosses through* (x_0, y_0) *and forms* α *angle with direction E-W.*
            S_α, *segment on* R_α *centred in* (x_0, y_0) *and 16 meters length both sides the centre.*
            P_α, *profile on* S_α *of tree height obtained from the* **DCHM**.
            PS_α, *nonparametric smoothing of* P_α *using Kernel estimation for small samples*
            (Martinez-Falero et al. 1992):
        Calculate: $J_{1\alpha}, J_{2\alpha}, J_{3\alpha}$
            $J_{1\alpha}$: *Existence of a relative maximum in* (x_0, y_0) *for the* PS_α *profile* ($J_{1\alpha}$=1 *if it exists*
            the maximum and 0 otherwise).
            $J_{2\alpha}$: *Existence of a higher height in* (x_0, y_0) *when comparing this point with the*
            height of the points at a distance of 1 meter from it on the R_α *straight line:*
            $[x_0+x_0\cos(\alpha),\ y_0+y_0\sin(\alpha)]$ *and* $[x_0-x_0\cos(\alpha),\ y_0-y_0\sin(\alpha)]$ *(the range of variations for*
            $J_{2\alpha}$ *is from 0 to 2).*
            $J_{3\alpha}$: *Existence of a higher height in* (x_0, y_0) *when comparing this point with the*
            height of the points at a distance equal to R_0 *(the average radius of the crown for*
            the trees with a maximum height equal to the **DCHM** *in the analysed point) on*
            the R_α *straight line:* $[x_0+x_0\cos(\alpha),\ y_0+y_0\sin(\alpha)]$ *and* $[x_0-x_0\cos(\alpha),\ y_0-y_0\sin(\alpha)]$ *(the*
            range of variations for $J_{3\alpha}$ *is also from 0 to 2).*
    Next α
    Calculate: J = 100 × (Σ_α [$J_{1\alpha} + J_{2\alpha}/2 + J_{3\alpha}/2$] / 24) (*as a percentage of the full certainty for the*
    existence of a maximum in the analyzed point).
    If J>90%, *then*
        The existence of a tree is accepted in (x_0, y_0) *with a height defined by the value of*
        DCHM *in* (x_0, y_0).
    Else
        The existence of a tree is not accepted in (x_0, y_0).
    End if
Next y_0
Next x_0

FIGURE 3.4 Outline of the algorithm for spatial identification of trees at each point of the study area.

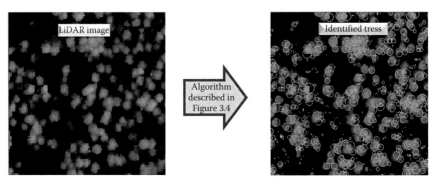

FIGURE 3.5 Example of spatial identification of trees in one zone in the study area.

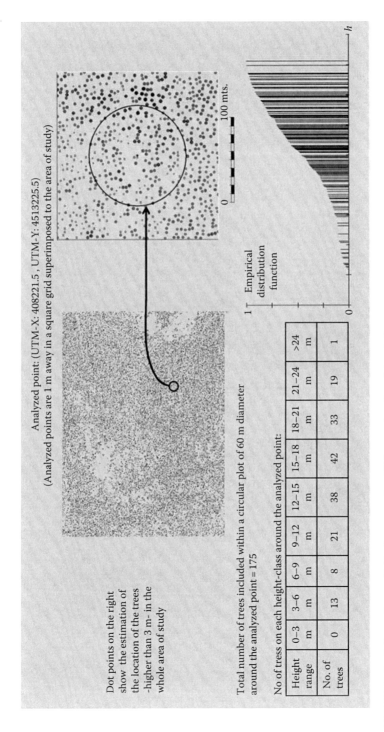

FIGURE 3.6   Information available for computing C&I on each point of the study area.

$$I_1 = 100 \times \left(1 - \frac{\sum_{i=1}^{h_{max}} D(i)}{\text{MaxDiv}} \right) \qquad (3.5)$$

where

    i is the height classes $(1, 2, \ldots, h_{max})$

    h_{max} is the maximum height class

$$D(i) = \frac{\left[p_{id}(i) - p(i) \right]^2}{p_{id}(i)}$$

with

$$p_{id} = F_{id}(i) - F_{id}(i-1), \left\{ F_{id}(i) \text{ computed as in Equation (3.4)} \right\}$$

and

$$p_i(i) = F(i) - F(i-1), \left\{ F(i) = \sum_{j=1}^{i} NR_j / N \right\}$$

    NR_j is the stem number/ha in the jth height class in the analyzed point

    N is the stem number/ha in the jth height class at each analyzed point

MaxDiv is the maximum value of $\sum_{i=1}^{h_{max}} D(i)$ for all the points in the study area

I_1 ranges from 0 to 100, with higher values as it approaches the reference scenario.

As shown in the examples in Figure 3.7, the tree height distribution indicator adequately reflects the structural diversity of a forest. Stage 1 represents one uneven-aged stand of a forest with high diversity and similarity with the reference scenario (ideal forest). Stage 2 shows a stand with great abundance of trees between 12 and 18 m in height (possibly even-aged) that exhibits media diversity. Finally, Stage 3 is an aged stand, with large proportion of tall trees and reduced diversity, where natural regeneration (and thus sustainability) is difficult.

The application of this indicator to the whole study area is shown in Figure 7.8 (Chapter 7) as the "structural diversity" indicator.

3.3.5.3.2   Timber Yield

The term "timber yield" indicates the harvested volume (m³ ha⁻¹ year⁻¹) obtained from the forest when it is managed to converge with the reference scenario. This not only provides a quantitative assessment of the volume of timber that can be regularly extracted but also makes it possible to plan thinning and felling operations in 10-year period until the height classes are balanced.

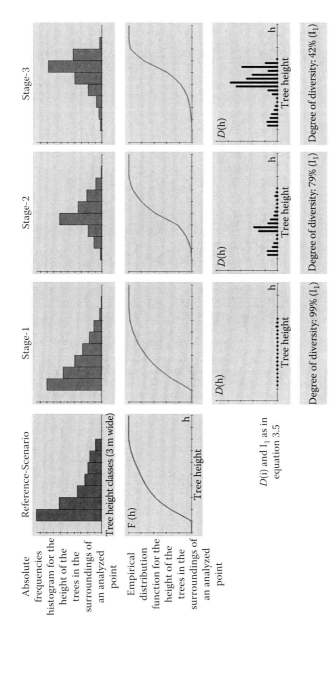

FIGURE 3.7 Sensitivity analyses for I_1 indicator.

To calculate the convergence with the height distribution in the reference scenario, the following features need to be considered:

i. The increase rate from one height class to another, calculated for the reference scenario, must be maintained.

ii. In each height class, a quantity of trees must be extracted every 10 years, which permits convergence with the height distribution in the reference scenario, expressed by Equation 3.6:

$$C_j = \begin{cases} N_j \times g_j & \text{if} \quad \dfrac{N_{j+1}}{NR_{j+1}} \leq 1 \\[2em] N_j \times g_j \times \dfrac{N_{j+1}}{NR_{j+1}} & \text{if} \quad \dfrac{N_{j+1}}{NR_{j+1}} > 1 \end{cases} \tag{3.6}$$

where

C_j is the stem number to be felled in height class j every 10 years

N_j is the stem number in height class j

g_j is the felling rate every 10 years in height class j

NR_j is the stem number in height class j for the distribution in the reference scenario

iii. The number of trees incorporated into the first height class is proportional to the gaps existing after thinning.

The timber yield indicator (I_2) is calculated using Equation 3.7:

$$I_2 = 100 \times \frac{(V_{150} - V_0)}{(VR_{150} - V_0)} \tag{3.7}$$

where

V_{150} is the average value of yield for 150 years in each analyzed point, managed for the purpose of attaining reference scenario (if $V_{150} > VR_{150}$, then $V_{150} = VR_{150}$)

V_0 is the average volume of yield starting from barren land after 150 years, managed during this period for the purpose of attaining the conditions of the reference scenario

VR_{150} is the average volume of yield of reference scenario over 150 years

Figure 3.8 shows the evolution of marketable wood existences from the current state and also for the reference scenario (constant) and for another stage representing a new reforestation.

The application of this indicator to the whole study area is shown in Figure 7.8 (Chapter 7).

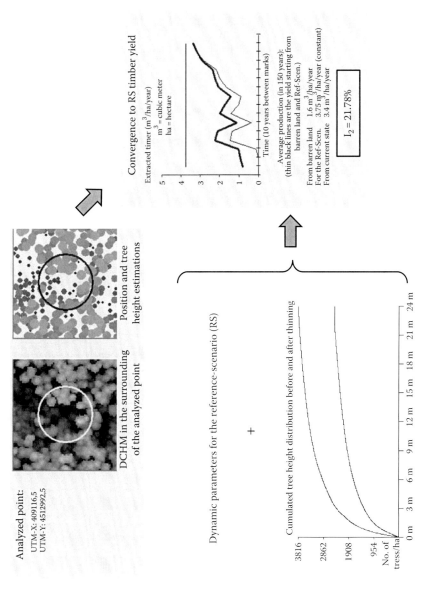

FIGURE 3.8   Example of convergence of current yield to the yield in the reference scenario and I_2 value.

3.3.5.3.3   Biomass Indicator

The biomass indicator (I_3) proposed assesses the proportion of total aerial biomass present in the LiDAR forest compared with the reference scenario. This is calculated with Equation 3.8:

$$I_3 = \frac{\sum_{i=1}^{h_{max}} B_i}{\sum_{i=1}^{h_{max}} BR_i} \tag{3.8}$$

where
  B_i is the total aerial biomass (kg of dry mass (d.m.)) of trees in height class i at each analyzed point
  BR_i is the total aerial biomass (kg d.m.) of trees in height class i in the reference scenario If $B_i > BR_i$, then $B_i = BR_i$.

The biomass of each tree was computed using the allometric expression for *P. sylvestris* in the central mountains (García Abejón and Gómez Loranca 1984) (Equation 3.9):

$$y = \exp\left(\frac{0.246887^2}{2}\right) \times \exp(-2.50275) \times dbh^{2.41194} \tag{3.9}$$

(adjusted determination coefficient, $r_{aj}^2 = 0.951$)
where
  y is the total aerial biomass of the tree (kg d.m.)
  dbh is the diameter at breast height of the tree, estimated for the whole study area based on the work of García-Abril (2007), relating dbh and height

The application of this indicator to the whole study area is shown in Figure 7.8 (Chapter 7).

3.3.5.4   Statistical Tests

The χ^2 goodness-of-fit test was used to confirm indicators are distributed according to a normal distribution; Student's t-test was used to verify the equality of the indicators calculated with field and LiDAR data.

Since the plot structure may make it more difficult to adequately measure the values of the indicators, a paired sample comparison was performed. This test is designed to identify differences between two data samples collected as pairs and determines whether the mean difference equals zero.

The ANOVA and Kruskal–Wallis tests were performed to verify whether the indictors showed differences between the central measurements of the five structure types. With the multiple range test (using Fisher's significant difference method; 95% confidence level), the structure groups differing from one another can be discerned.

3.3.6 RESULTS AND DISCUSSION

3.3.6.1 Suitability of LiDAR Data for Estimating Indicator Values

It was hypothesized that there were no significant differences between field- and LiDAR-derived indicators. Table 3.6 shows the indicators calculated for the 10 plots using both data sources. Table 3.7 shows the p values for goodness-of-fit test for a normal distribution and the Student's t-test of paired sample comparison. For all three indicators, no significant differences were seen between the results ($\alpha=0.05$). Hence, indicators calculated using the LiDAR forest data are reliable predictors of field-based indicators.

I_2 and I_3 have very similar values. For this reason, a t-test was performed to compare adjoining plots. The possibility that these indicators are of the same value cannot be rejected (p value $=0.2439$).

3.3.6.2 Assessment of the Results Obtained from the Entire Study Area

According to the earlier results, the LiDAR analysis is sufficient to extend the analysis to a large area.

TABLE 3.6
Field-Based and LiDAR-Based Indicator Results

Plot	I_1: Tree Height Distribution		I_2: Timber Yield		I_3: Biomass	
	Field	LiDAR	Field	LiDAR	Field	LiDAR
T1P1	63.29	49.26	53.49	49.82	53.15	47.01
T1P2	55.32	41.47	69.26	55.57	64.45	49.61
T2P1	49.08	41.06	88.4	89.18	83.8	82.38
T2P2	44.11	44.15	74.44	72.54	66.19	66.4
T3P1	58.41	54.47	33.68	33.68	33.88	33.9
T3P2	49.01	54.37	30.14	28.85	28.06	29.23
T4P1	57.93	56.16	10.68	15.23	10.99	19.01
T4P2	56.78	59.55	15.85	10.09	15.99	11.28
T5P1	61.47	56.77	2.49	8.74	2.69	9.03
T5P2	62.11	62.07	3.13	4.64	3.83	5.88

TABLE 3.7
p Values for Field-Based and LiDAR-Based Indicators

Indicators	Normality Test		Student's t-Test
	Field	LiDAR	p Values
I_1	0.1886	0.5397	0.0980
I_2	0.3326	0.5397	0.4761
I_3	0.3326	0.3326	0.6630

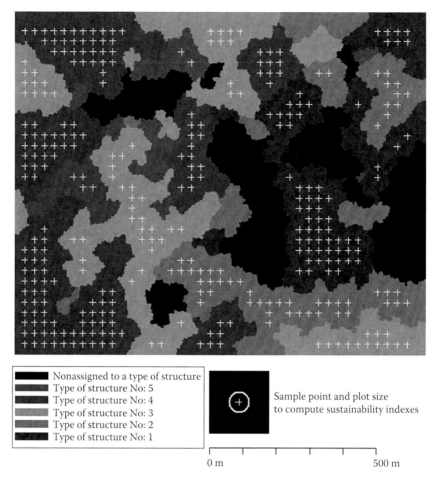

	Nonassigned to a type of structure
	Type of structure No: 5
	Type of structure No: 4
	Type of structure No: 3
	Type of structure No: 2
	Type of structure No: 1

Sample point and plot size
to compute sustainability indexes

0 m   500 m

FIGURE 3.9   **(See color insert.)** Sample points on the forest structures of the study area.

A stratified sampling was used, with proportional affixation to the area of each forest structure type (maximum standard error for the mean of each indicator <15%). A total of 337 points were selected: 64 in T1 (type of structure No. 1), 52 in T2, 62 in T3, 153 in T4, and 6 in T5 (see Figure 3.9).

Figure 3.10 shows the average results of applying the previously described indicators to the current forest structures. Indicators showed differences between the central measurements of the five structure types (ANOVA and Kruskal–Wallis test, $p < 0.0001$). They also varied significantly among the five structure types (multiple range tests $p < 0.0001$). Therefore, their use in forest structure classification is feasible (Figure 7.8).

3.3.6.3   Analysis of the Information Provided by the Indicators

I_1 expresses that, for forest structures T1 and T2, the proportion of high trees exceeds the height distribution for the reference scenario, while the proportion of small- and

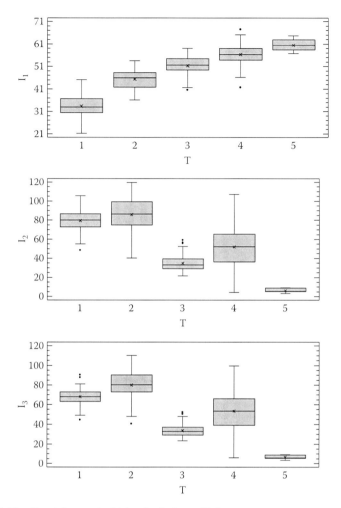

FIGURE 3.10 Graph box and whisker for I_1, I_2, and I_3 by structure type.

medium-height trees is lower than in the reference scenario. For the other forest structures, their height distribution is less distant from the height distribution in the reference scenario (Figure 3.11a).

I_2 graphs (Figure 3.11b) show that structure T1 is in an intermediate state between barren land and the reference scenario, while T2 is closer to the reference scenario and even exceeds the reference scenario yield for several periods due to the proportion of big trees. Structure T3 is nearer to barren land than to the reference scenario, whereas T4 and T5 are very similar to one another, and the yield is almost the same for barren land, in accordance with low crown cover structures.

In the case of the biomass indicator (I_3) (Figure 3.11c), structures T1 and T2 are nearer the reference scenario than the others due to their higher proportion of stem number. Structure T3 is intermediate, and structures T4 and T5 have a low tree density, quite far from the reference scenario in line to forest structures with few stem number.

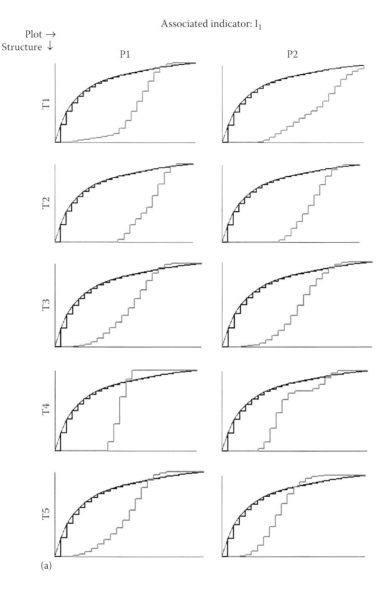

(a)

FIGURE 3.11   Graphics of the LiDAR-based indicators for each sample plot: (a) For tree height distribution (I_1), graphics indicate relative accumulated frequency histograms of stem number per height class in the LiDAR forest (grey line) and in the reference scenario (dotted line).

3.3.6.4   Management Implications

Canopy cover density may be described by means of the biomass indicator, whereas tree distribution per height class can be regarded as a dynamic equilibrium for the height distribution. The timber yield indicator can be used in order to plan felling and clearing activities at 10-year intervals until reaching the reference scenario.

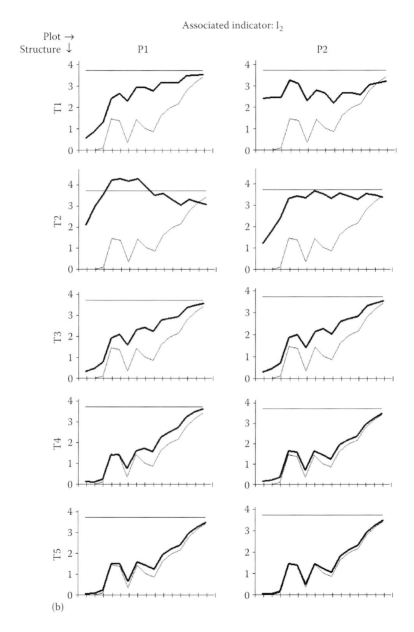

FIGURE 3.11 (continued) Graphics of the LiDAR-based indicators for each sample plot: (b) For I_2 graphics, ordinates are timber yield (m^3 ha^{-1} year^{-1}); X-axis = time (marks every 10 years). The bold line is the plot yield; the thin lines are the yield starting from barren land and reference scenario (3.75 m^3 ha^{-1} year^{-1}).

(*continued*)

Associated indicator: I_3

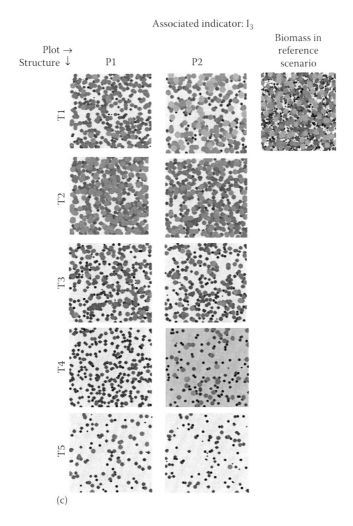

(c)

FIGURE 3.11 (continued) Graphics of the LiDAR-based indicators for each sample plot: (c) For I_3, horizontal representations of tree crowns are given. The darker the color of the circumference, the smaller the size of the tree crown.

Moreover, vertical forest structure was defined as the distribution of tree heights within a forest stand and horizontal structure as the distribution of percent canopy closure (Zimble et al. 2003). Hence, I_1 and I_3 may represent two important aspects for forest structure assessment.

The proximity to the reference scenario can be evaluated by means of these indicators, and areas can be classified according to this proximity. This affords a practical advantage since priority areas can be identified. Further, new values for the indicators can be recalculated after each treatment, so that their convergence on the reference scenario may be assessed. Therefore, they can be used for agreement on site-specific decisions.

Our methodology involves a landscape-sensitive approach that not only accounts for the predicted characteristics of forest stands but also allows different goals to be considered explicitly through different reference scenarios. It thus provides useful ways to evaluate management approaches with cost-effective inventory due to LiDAR.

REFERENCES

Akay, A. E., H. Oguz, I. R. Karas, and K. Aruga. 2009. Using LiDAR technology in forestry activities. *Environmental Monitoring and Assessment* 151(1–4):117–125.

Allué, J. L. 1990. *Mapa de Subregiones Fitoclimáticas de España Peninsular y Balear*. Madrid: Dirección General de Medio Natural y Política Forestal. Ministerio de Medio Ambiente, y Medio Rural y Marino.

Annikki, A., M. del Río, H. Jari, M. J. Hawkins, C. Reyer, P. Soares, M. van Oijen, and M. Tomé. 2012. Using stand-scale forest models for estimating indicators of sustainable forest management. *Forest Ecology and Management* 285:164–178.

Ayuga Téllez, E., A. J. M. Fernandez, C. G. García, and E. M. Falero. 2006. Estimation of non-parametric regression for dasometric measures. *Journal of Applied Statistics* 33(8):819–836.

Canadian Council of Forest Ministers. 2010. *Marking Canada's Progress in Sustainable Forest Management*. Ottawa, Ontario, Canada: Canadian Council of Forest Ministers.

Canadian Forest Service. 2011. *State of Canada's Forests*. Ottawa, Ontario, Canada: Natural Resources Canada.

Castillo, M., B. Rivard, A. Sanchez-Azofeifa, J. Calvo-Alvarado, and R. Dubayah. 2012. LIDAR remote sensing for secondary Tropical Dry Forest identification. *Remote Sensing of Environment* 121:132–143.

Chen, Q., P. Gong, D. Baldocchi, and Y. Q. Tian. 2007. Estimating basal area and stem volume for individual trees from lidar data. *Photogrammetric Engineering and Remote Sensing* 73(12):1355–1365.

Coops, N. C., T. Hilker, M. A. Wulder, B. St-Onge, G. Newnham, A. Siggins, and J. A. Trofymow. 2007. Estimating canopy structure of Douglas-fir forest stands from discrete-return LiDAR. *Trees-Structure and Function* 21(3):295–310.

De Warnaffe, G. D. and F. Devillez. 2002. Quantifier la valeur écologique des milieux pour intégrer la conservation de la nature dans l'aménagement des forêts: une démarche multicritères. *Annals of Forest Science* 59(4):369–387.

Dean, T. J., Q. V. Cao, S. D. Roberts, and D. L. Evans. 2009. Measuring heights to crown base and crown median with LiDAR in a mature, even-aged loblolly pine stand. *Forest Ecology and Management* 257(1):126–133.

Dubayah, R. O., S. L. Sheldon, D. B. Clark, M. A. Hofton, J. B. Blair, G. C. Hurtt, and R. L. Chazdon. 2010. Estimation of tropical forest height and biomass dynamics using lidar remote sensing at La Selva, Costa Rica. *Journal of Geophysical Research-Biogeosciences* 115:17.

FAO. 2001. Criteria and indicators for sustainable forest management: A compendium. FAO Corporate Document Repository. http://www.fao.org/DOCREP/004/AC135E/AC135E00.HTM

FAO. Criteria and indicators for sustainable forest management: A compendium 2001 cited. Available from http://www.fao.org/DOCREP/004/AC135E/AC135E00.HTM.

FAO 2006. *Understanding National Forest Programmes*. Rome: Food and Agriculture Organization of the United Nations.

FAO 2010. Global forest resources assessment 2010. Main report. Rome, Italy.

FOREST EUROPE, UNECE and FAO. 2011. *State of Europe's Forests 2011. Status and Trends in Sustainable Forest Management in Europe*. Aas, Norway: Ministerial Conference on the Protection of Forests in Europe-FOREST EUROPE Liaison Unit Oslo.

Forest Service. 2010. *National Report on Sustainable Forests*. Washington, DC: Unites States Department of Agriculture (USDA).

García Abejón, J. L. and Gómez Loranca, J. A. 1984. *Tablas de producción de densidad variable para Pinus sylvestris L. en el sistema central, Comunicaciones INIA. Serie recursos naturales*. Madrid: Instituto Nacional de Investigaciones Agrarias, Ministerio de Agricultura, Pesca y Alimentacion.

García-Abril, A., J. Irastorza, J. García Cañete, E. Marínez-Falero, J. Solana, and E. Ayuga. 1999. Concepts associated with deriving the balanced distribution of an uneven-aged structure from even-aged yield tables: Application to Pinus sylvestris in the central mountains of Spain. Ed. A. F. M. Olsthoorn, *Management of Mixed-Species Forest: Silviculture and Economics. Olsthoorn*.

García-Abril, A., J. Irastorza, J. García Cañete, E. Marínez-Falero, J. Solana, and E. Ayuga. 1999. Concepts associated with deriving the balanced distribution of an uneven-aged structure from even-aged yield tables: Application to Pinus sylvestris in the central mountains of Spain. In *Management of Mixed-Species Forest: Silviculture and Economics. Olsthoorn*, ed. A. F. M. Olsthoorn, DLO Institute for Forestry and Nature Research (IBN-DLO), Wageningen, the Netherlands.

García-Abril, A., S. Martin-Fernandez, M. A. Grande, and J. A. Manzanera 2007. Stand structure, competition and growth of Scots pine (*Pinus sylvestris* L.) in a Mediterranean mountainous environment. *Annals of Forest Science* 64:825–830.

Hall, J. P. 2001. Criteria and indicators of sustainable forest management. *Environmental Monitoring and Assessment* 67(1–2):109–119.

Hernando, A., R. Tejera, J. Velázquez, and M. V. Núñez. 2010. Quantitatively defining the conservation status of Natura 2000 forest habitats and improving management options for enhancing biodiversity. *Biodiversity and Conservation* 19(8):2221–2233.

Holvoet, B. and B. Muys. 2004. Sustainable forest management worldwide: A comparative assessment of standards. *International Forestry Review* 6(2):99–122.

Ioki, K., J. Imanishi, T. Sasaki, Y. Morimoto, and K. Kitada. 2010. Estimating stand volume in broad-leaved forest using discrete-return LiDAR: Plot-based approach. *Landscape and Ecological Engineering* 6(1):29–36.

ITTO. 2005. *Revised itto Criteria and Indicators for the Sustainable Management of Tropical Forests Including Reporting Format*. Yokohama, Japan: International Tropical Timber Organization.

Kane, V. R., R. J. McGaughey, J. D. Bakker, R. F. Gersonde, J. A. Lutz, and J. F. Franklin. 2010. Comparisons between field- and LiDAR-based measures of stand structural complexity. *Canadian Journal of Forest Research-Revue Canadienne De Recherche Forestiere* 40(4):761–773.

Khadka, C. and H. Vacik. 2012. Use of multi-criteria analysis (MCA) for supporting community forest management. *iForest—Biogeosciences and Forestry* 5(2):60–71.

Kimmins, J. P. 1997. *Forest Ecology: A Foundation for Sustainable Management*, 2nd edn. Upper Saddle River, NJ: Prentice Hall.

Kondrashov, L. 2004. Legal framework for forestry and hunting. *Report on the Study of Criteria and Indicators for Sustainable Forest Management in the Kyrgyz Republic*:1–32.

Kotwal, P. C., M. D. Omprakash, Sanjay Gairola, and D. Dugaya. 2008. Ecological indicators: Imperative to sustainable forest management. *Ecological Indicators* 8(1):104–107.

Lefsky, M. A., W. B. Cohen, S. A. Acker, G. G. Parker, T. A. Spies, and D. Harding. 1999. Lidar remote sensing of the canopy structure and biophysical properties of Douglas-fir western hemlock forests. *Remote Sensing of Environment* 70(3):339–361.

Lefsky, M. A., W. B. Cohen, G. G. Parker, and D. J. Harding. 2002. Lidar remote sensing for ecosystem studies. *Bioscience* 52(1):19–30.

Lefsky, M. A., D. J. Harding, M. Keller, W. B. Cohen, C. C. Carabajal, F. D. Espirito-Santo, M. O. Hunter, R. de Oliveira, and P. B. de Camargo. 2006. Estimates of forest canopy height and aboveground biomass using ICESat (vol. 33, art no L05501, 2006). *Geophysical Research Letters* 33(5):1.

Lovell, J. L., D. L. B. Jupp, D. S. Culvenor, and N. C. Coops. 2003. Using airborne and ground-based ranging lidar to measure canopy structure in Australian forests. *Canadian Journal of Remote Sensing* 29(5):607–622.

Lovell, J. L., D. L. B. Jupp, G. J. Newnham, N. C. Coops, and D. S. Culvenor. 2005. Simulation study for finding optimal lidar acquisition parameters for forest height retrieval. *Forest Ecology and Management* 214(1–3):398–412.

Martínez-Falero, E., E. Ayuga-Téllez, and C. González-García. 1992. A comparative study of different kernel functions according to data type. *Qüestiió* 16(1–3):3–26.

MCPFE. 2002. *Improved Pan-European Indicators for Sustainable Forest Management as Adopted by the Ministerial Conferences on the Protection of Forests in Europe (MCPFE) Expert Level Meeting 7–8 October 2002.* Vienna, Austria: MCPFE Liaison Unit.

MCPFE. 2011. *Summary for Policy Makers—State of Europe's Forests 2011.* Aas, Norway: Forest Europe publications.

Mendoza, G. A. and R. Prabhu. 2000a. Multiple criteria decision making approaches to assessing forest sustainability using criteria and indicators: A case study. *Forest Ecology and Management* 131(1–3):107–126.

Mendoza, G. A. and R. Prabhu. 2000b. Development of a Methodology for Selecting Criteria and Indicators of Sustainable Forest Management: A Case Study on Participatory Assessment. *Environmental Management* 26(6):659–673.

Miura, N. and S.D. Jones. 2010. Characterizing forest ecological structure using pulse types and heights of airborne laser scanning. *Remote Sensing of Environment* 114(5):1069–1076.

Morgan, J. L., S. E. Gergel, and N. C. Coops. 2010. Aerial photography: A rapidly evolving tool for ecological management. *Bioscience* 60(1):47–59.

Nelson, T., B. Boots, and M. A. Wulder. 2005. Techniques for accuracy assessment of tree locations extracted from remotely sensed imagery. *Journal of Environmental Management* 74(3):265–271.

Nelson, R., M. A. Valenti, A. Short, and C. Keller. 2003. A multiple resource inventory of Delaware using airborne laser data. *Bioscience* 53(10):981–992.

Pascual, C., A. García-Abril, L. G. García-Montero, S. Martin-Fernandez, and W. B. Cohen. 2008. Object-based semi-automatic approach for forest structure characterization using lidar data in heterogeneous *Pinus sylvestris* stands. *Forest Ecology and Management* 255(11):3677–3685.

Pelissier, R. and F. Goreaud. 2001. A practical approach to the study of spatial structure in simple cases of heterogeneous vegetation. *Journal of Vegetation Science* 12(1):99–108.

Pommerening, A. 2002. Approaches to quantifying forest structures. *Forestry* 75(3):305–324.

Prabhu, R., Colfer, C. J. P., and Dudley, R. G. (1999). *Guidelines for Developing, Testing and Selecting Criteria and Indicators for Sustainable Forest Management.* Jakarta, Indonesia: Center for International Forestry Research (CIFOR)–The Criteria & Indicators Toolbox Series.

Raison, R. J., D. W. Flinn, and A. G. Brown. 2001. Application of criteria and indicators to support sustainable forest management: Some key issues. In *Criteria and Indicators for Sustainable Forest Management*, eds. R. J. Raison, A. G. Brown and D. W. Flinn. New York: CABI Publishing.

Rametsteiner, E. and M. Simula. 2003. Forest certification—An instrument to promote sustainable forest management? *Journal of Environmental Management* 67(1):87–98.

Requardt, A. 2007. Pan-European criteria and indicators of sustainable forest management: Networking structures and data potentials of international data sources. PhD Thesis, University of Hamburg, Hamburg, Germany.

Rivas-Martínez, S. 1987. *Mapa de series de vegetación de España 1:400.000 y memoria.* Madrid: ICONA Ministerio de Agricultura, Pescay Alimentación.

Sheppard, Stephen R. J. 2005. Participatory decision support for sustainable forest management: A framework for planning with local communities at the landscape level in Canada. *Canadian Journal of Forest Research* 35(7):1515–1526.

The working group for the conservation and sustainable management of temperate and boreal forests. 2009. *A Vital Process for Addressing Global Forest Challenges: The Montreal Process.* New Zealand: SCION.

UNECE-FAO. 2011. *Forest Products Annual Market Review 2010–2011*, ed. G. T. a. F. S. P. 27. Geneva, Switzerland.

Véga, Cédric, and Benoît St-Onge. 2009. Mapping site index and age by linking a time series of canopy height models with growth curves. *Forest Ecology and Management* 257(3):951–959.

Wolfslehner, B., and H. Vacik. 2012. Mapping indicator models: From intuitive problem structuring to quantified decision-making in sustainable forest management. *Ecological Indicators* 11(2):274–283.

Wulder, M. A., C. W. Bater, N. C. Coops, T. Hilker, and J. C. White. 2008. The role of LiDAR in sustainable forest management. *Forestry Chronicle* 84(6):807–826.

Wulder, M. A., K. O. Niemann, and D. G. Goodenough. 2000. Local maximum filtering for the extraction of tree locations and basal area from high spatial resolution imagery. *Remote Sensing of Environment* 73(1):103–114.

Wulder, M. A. and D. Seemann. 2003. Forest inventory height update through the integration of lidar data with segmented Landsat imagery. *Canadian Journal of Remote Sensing* 29(5):536–543.

Yu, X., J. Hyyppa, H. Kaartinen, M. Maltamo, and H. Hyyppa. 2008. Obtaining plotwise mean height and volume growth in boreal forests using multi-temporal laser surveys and various change detection techniques. *International Journal of Remote sensing* 29(5):1367–1386.

Zhao, K. G., S. Popescu, and R. Nelson. 2009. Lidar remote sensing of forest biomass: A scale-invariant estimation approach using airborne lasers. *Remote Sensing of Environment* 113(1):182–196.

Zimble, D. A., D. L. Evans, G. C. Carlson, R. C. Parker, S. C. Grado, and P. D. Gerard. 2003. Characterizing vertical forest structure using small-footprint airborne LiDAR. *Remote Sensing of Environment* 87(2–3):171–182.

4 Soil-Quality Indicators for Forest Management

Fernando Arredondo-Ruiz, Luis García-Montero,
Inmaculada Valverde-Asenjo, and Cristina Menta

CONTENTS

4.1   INTRODUCTION TO FOREST-SOIL QUALITY

4.1.1   IMPORTANCE OF THE SOIL AND THE RHIZOSPHERE

Soil is one of the most fundamental components for supporting life on Earth. Jeffery et al. (2010) and Menta (2012) point out that the processes occurring within soil (most of which are driven by its living organisms) perform ecosystem and global functions that help maintain life aboveground. Moreover, soil delivers numerous services, ranging from providing the food we eat to filtering and cleaning the water we drink. It is used as a platform for building, it provides vital products such as antibiotics, and it contains an archive of our cultural heritage in the form of archeological sites. However, life within the soil is hidden and often suffers from being "out of sight and out of mind" (Blum, 1993; Ebel and Davitashvili, 2007).

Soils may be characterized in terms of the properties they inherit from the underlying rock (the parent material) and the properties resulting from alteration of the original parent material by soil-forming or "pedogenic" processes, namely, climate, vegetation, time, and human activity. Pedogenic processes operate mainly in the surface and subsurface horizons normally found in the upper 2 m (Blum, 1993; Ebel and Davitashvili, 2007).

From the point of view of both agriculture and forestry, another essential aspect of the functioning of the soil, vegetation, and ecosystem is the importance of the rhizosphere (Akeem, 2012). The presence of roots is generally associated with greater microorganism density and soil fauna in the nearby soil when compared to soil devoid of roots; the term rhizosphere is used in a broader sense to refer to the portion of soil surrounding roots in which the soil organisms are influenced by their presence (Killham, 1994). The rhizosphere can be distinguished from the greater part of the soil on the basis of its chemical, physical, and biological characteristics. As the roots penetrate into the ground, they act on clay minerals and the particles of soil surrounding them, leading to the formation of an area in which the water pathway and the movement of nutrients and microflora are more heavily channeled than in the rest of the soil. For the same reasons, the organic matter (OM) released by the roots accumulates nearby. The chemical nature of the rhizosphere is significantly different from the rest of the soil; this is largely the result of the fact that the roots release carbon (C) and selectively capture ions in solution in the groundwater. The plants act as C pumps, fixing what is available in the atmosphere in the root exudates, which in turn are quickly captured by soil organisms; for this reason, the level of C available around the roots is never very high. Instead, the selective absorption of ions causes some to be depleted in the rhizosphere and the accumulation of others, which are not absorbed by the roots. The relationship between the roots and microflora can be very close and leads to (1) bacteria or fungi becoming an integral part of the roots as in mycorrhizal symbiosis and (2) the association of bacteria and legumes. The peculiar characteristics of the rhizosphere are also reflected in the selectivity of the animal element. The interaction between soil animals and the plant roots can take a variety of forms that either produce benefits or repress plant growth and often involve interactions with the microbial populations in the soil (Akeem, 2012).

As a result of all these factors and soil functions, soil health indicates the capacity of soil to function as a vital living system to sustain biological productivity, promote environmental quality, and maintain plant and animal health. However, soils are affected by human activity, which often results in their degradation and the loss of their functions. Growing pressure from an ever-increasing global population, as well as threats such as climate change and soil erosion, is placing increasing stresses on the ability of soil to sustain its important role in the planet's survival. Evidence suggests that while the increased use of monocultures, intensive agriculture, and forestry has led to a decline in soil quality and biodiversity in many areas, the precise consequences of this loss are not always clear. Jeffery et al. (2010) and Menta (2012) indicate that too rarely do we pause to reflect on the fact that soil is the foundation upon which society is sustained and evolves and that it is a vital component of ecological processes and cycles, as well as the basis on which our infrastructure rests. These authors insist that more importance should be given to the fact that soil quality and its protection contribute significantly to preserving the quality of life and that the nutrition and health of humans and animals cannot be separated from the quality of the soil.

Only by understanding soil in all its complexity while maintaining its functionality and quality through actions designed to protect its properties and acknowledging its importance in the quality of life worldwide can we embark on a truly sustainable use of soil perceived as a resource and build a proper human/soil relationship to be passed on to future generations.

4.1.2 Foundations for Using Soil Properties as Indicators of Forest-Soil Quality

Monitoring ecosystem components is essential for acquiring basic data to assess the impact of land-management systems and to plan resource conservation. For this reason, soil scientists have always sought to link soil type and soil variables to potentials or limitations of land use (Knoepp et al., 2000). Doran (1994) proposed three main functions of soil in view of this goal: (1) to act as a medium for plant growth, (2) to regulate and partition water flow, and (3) to serve as an environmental buffer. Knoepp et al. (2000) described these approaches and remarked that this is borne out by the estimates of forest and agricultural productivity, as well as by their uses for recreation, wildlife, construction, and other functions listed in county soil surveys.

During the last decade, numerous authors have highlighted a fourth soil function, indicating that soil represents one of the leading reservoirs of biodiversity, as reported by Menta (2012). All these authors relate these soil functions to the need to identify soil-quality indicators (SQI). The increasing pressure on available land and the debate as to its proper use have brought about a parallel escalation in the movement to identify and set quality standards for both agricultural and forest soils.

The Soil Science Society of America officially defines soil quality as "the capacity of a specific kind of soil to function, within natural or managed ecosystem boundaries, to sustain plant and animal productivity, maintain or enhance water and air quality, and support human health and habitation" (Karlen et al., 1997). Likewise, environmental indicators are "quantitative assessments informing of the state of conservation and health of the environment." Some environmental indicators have a broad scope: for instance, greenhouse gas emissions, protection areas, biodiversity indices, proportion of forest land area, net annual forest growing stock volume, average condition of health and vigor, etc. All are useful to characterize the current status and track or predict significant changes in the ecosystem (Blum, 1993; Ebel and Davitashvili, 2007).

Soil quality affects forestry, agricultural sustainability, and environmental quality and consequently plant, animal, and human health (Bloem, 2003; Menta, 2012). A common criterion for evaluating the long-term sustainability of ecosystems is to assess the fluctuations in soil quality (Schoenholtz et al., 2000). Soil reflects ecosystem metabolism; within soils, all the biogeochemical processes in the different ecosystem components are combined (Dylis, 1964). Maintaining soil quality is of the utmost importance for the preservation of biodiversity and for the sustainable management of renewable resources (Menta, 2012). Based on the concept of soil quality, Andrews et al. (2004) and Garrigues et al. (2012) provide a more detailed description of mean soil functions, including water flow and retention, solute transport and retention, physical stability and support, retention and cycling of nutrients, buffering and filtering of potentially toxic materials, and maintenance of biodiversity and habitat.

In summary, protecting the quality, biodiversity, and productive capacity of soil is a paramount goal for sustainable forest management. To support these goals, it is necessary to control or restrict agricultural and forestry activities that could reduce on-site agricultural and forest productivity and environmental quality in order to prevent soil degradation and restore its potential. Prerequisites for this objective include reliable soil data and the use of SQI, as well as an input for the design of soil usage systems and soil management practices for a truly sustainable forest management (FAO, 2009; Miller et al., 2010).

4.1.3 Fertility and Organic Matter Content in Forest Soils

The soil functions that determine vegetation composition/structure are of particular importance, as they serve as a medium for root development and provide moisture and nutrients for plant growth (Minnesota Forest Resources Council, 1999). Natural soil fertility derives from the combination of many of these soil properties, including OM content, clay content and mineralogy, the presence of weatherable minerals, pH, base saturation, and biological activity, mainly in the rhizosphere.

For reference purposes, soil groups in Europe can be classified into three levels of natural soil fertility as follows:(1) low (arenosols, planosols, acrisols, podzols), (2) moderate (regosols, andosols, calcisols, umbrisols, luvisols, albeluvisols, histosols), and (3) high (fluvisols, gleysols, vertisols, kastanozems, chernozems, phaeozems) (Soil Atlas of Europe).

In association with the concept of soil fertility, it is worth noting the importance of soil organic matter (SOM) and total organic carbon (TOC). Living SOM represents only a small percentage of soil TOC and includes soil micro-, meso-, and macroorganisms. In particular, soil microbial biomass is regarded as the most active and dynamic pool of SOM and plays an important role in driving soil mineralization processes. Three main aspects of soil microbial biomass are usually considered, due to their effects on soil functions: pool size, activity, and diversity. The determination of the total amount of C immobilized within microbial cells allows soil microbial biomass to be determined as a pool of SOM. Since this pool is responsible for the decomposition of plant and animal residues and the immobilization and mineralization of plant nutrients, it is ultimately responsible for maintaining soil fertility (Brookes, 2001). Thus, the concept of microbial biomass has developed to serve as an "early warning" of changing soil conditions and as an indicator of the direction of change.

C mineralization activity, a key process of the soil's C cycle, determines the speed of the SOM degradation process in soil. Numerous studies have recently been carried out in order to verify—directly or indirectly—the potential for increasing C storage in soil by manipulating C inputs with a view to minimizing the rate of C mineralization (Jans-Hammermeister, 1997). The C mineralization process has a low sensitivity to changes in soil management, as small microbial populations in degraded soil can mineralize OM to the same extent and at the same rate as large microbial populations in undegraded soils (Brookes, 1995).

A combination of the two measurements, relating to both the size and activity of the microbial biomass, is more sensitive to soil management changes and more helpful as a SQI. C mineralization activity/unit of biomass (biomass-specific respiration) and the mineralization coefficient (respired C/TOC) indicate efficiency in C utilization and energy demand. Finally, soil microbial diversity measurements have assumed increasing importance as indicators of community stability and the impact of stress on that community (Akeem, 2012).

4.1.4 STORING CARBON DIOXIDE IN FOREST SOILS

Among the soil functions cited and with regard to the soil's capacity to act as an environmental buffer, the last decade saw significant interest in forests—and especially in forest soil—as a means of storing carbon dioxide (CO_2) from the atmosphere and providing a mechanism to combat the increase in atmospheric CO_2 (Soil Atlas of Europe). Forest soils generally accumulate C. Estimates range from 17 to 39 million tons of C per year, with an average of 26 million tons per year in 1990 and an average of as much as 38 million tons per year in 2005 (EC, 2002 Policy Makers).

However, there is still limited knowledge concerning the long-term impact of an increase in global average temperatures and the real role that trees and SOM could play in absorbing C, in addition to the costs and benefits involved in using restoration as a mechanism to offset C emissions. For these reasons, C knowledge projects have been proposed as a way of testing these parameters in the context of landscape-based forest restoration activities (Mansourian et al., 2005).

The atmospheric concentration of CO_2 has increased by over one-third since the Industrial Revolution. This increase is primarily attributed to fossil fuel combustion and also significantly to changes in land cover and use (e.g., forest degradation or conversion of forests to agriculture). In order to curb atmospheric CO_2 concentrations, it is essential to reduce human dependence on fossil fuels and impose legally binding targets for reduced CO_2 emissions, and this should be the central focus of any policy program. However, in order to stabilize atmospheric CO_2 concentrations, the international community must also reduce the rate of destruction of natural ecosystems that serve as important stocks and sinks of C. In addition to slowing the rate of land conversion, other mitigation tools considered to stabilize the burgeoning concentration of CO_2 in the atmosphere include increasing the land coverage of C-absorbing vegetation and soil C sinks.

The concept of C sinks is based on the natural ability of trees, plants, and soil organisms to take up CO_2 from the atmosphere and store the organic and inorganic C forms in wood, roots, leaves, and soil. The theory behind land-based C trading is that governments or institutions that wish or are required to reduce their fossil fuel emissions can offset some of these emissions by investing in afforestation and reforestation activities to sequester C. Indeed, in some cases, private companies are voluntarily electing to offset some of their fossil fuel CO_2 emissions through the purchase of C credits from land-based C sequestration projects (Mansourian et al., 2005).

Recent studies under various agroforestry systems (AFSs) in a range of ecological conditions showed that the extent of C sequestered in AFSs largely depends on environmental conditions and system management. Trading of sequestered C is a viable opportunity for bringing economic benefit to agroforestry practitioners, who are mostly resource-poor farmers in developing countries (EC 2002 Policy Makers; Ramachandran et al., 2010). For example, in contrast with industrial plantation approaches, the Scolel Té project for rural livelihood and C management (Mexico) aims to demonstrate how C finance can allow low-income rural farmers to invest in forest conservation, sustainable land-use systems, and improvements in their livelihood that would otherwise be inaccessible to them (Mansourian et al., 2005). However, more rigorous research results are required for AFSs to be used in global agendas for C sequestration (EU, 2002 Policy Makers; Ramachandran et al., 2010).

4.1.5 BIODIVERSITY AND FOREST SOILS

Although global biodiversity was one of the focal points of the Rio Conference, in the 1990s, virtually no attention was paid to activities for the conservation of soil communities. However, with the new millennium, the conservation of soil biodiversity has become an important aim in international environmental policies, as highlighted in the European Union (EU) Soil Thematic Strategy (European Commission, 2006), the Biodiversity Action Plan for Agriculture (European Commission, 2001), the Kiev Resolution on Biodiversity (EU/ECE, 2003), and subsequently in the Message from Malahide (European Commission, 2004) that lays down the goals of the Countdown 2010.

Historically, the study of soil biodiversity started with a mapping of soil food webs, perhaps because the most fundamental integrating feature of soil communities is the feeding relationships between organisms (EC, 2002). Nature's biodiversity reflects all ecosystem metabolisms, since all the biogeochemical processes of the different ecosystem components are combined within it; therefore, soil-quality fluctuations are considered to be a suitable criterion for evaluating the long-term sustainability of ecosystems. Within the complex structure of soil, biotic and abiotic components interact closely in controlling the organic degradation of matter and nutrient recycling processes.

Moreover, maintaining a high soil biodiversity may be of vital importance in structurally diverse ecosystems, as soil biodiversity may promote ecological stability (Grime, 1997; Van Bruggen and Semenov, 2000). Because soil is an environment that is constantly subjected to fluctuations, the establishment of an unfavorable environmental condition can result in the inhibition of some populations that perform essential functions. In highly diverse communities, however, there is a higher probability of the occurrence of quiescent microorganisms that could perform the same functional role but that have different physical, chemical, and biological demands (van Bruggen and Semenov, 2000; Chaer, 2008).

Therefore, soil communities, aside from being an important reservoir of global biodiversity, also play an essential role in these ecosystem functions and for this reason are often used to provide SQI (Menta, 2012).

4.2 THREATS TO FOREST-SOIL QUALITY

4.2.1 Eight Main Threats to Soil on a Global Scale

Maintaining soil condition is essential for ensuring the sustainability of society and biodiversity. However, the soil is under an increasing number of threats. These threats are complex, and although unevenly spread, their dimension is global and they are frequently interlinked. When numerous threats occur simultaneously, their combined effects tend to compound the problem. Therefore, soil disturbances linked to natural forces and human activities alter the physical, chemical, and biological properties of soil, which in turn can impact its long-term productivity (Gupta and Malik, 1996).

Human activities are increasingly polluting soils and groundwater through applied agrochemicals, deposition of atmospheric pollutants, and spreading of sewage sludge and manures, and these can have adverse impacts on the ability of soil to perform its vital functions (Blum, 1993; Ebel and Davitashvili, 2007). Consequently, numerous soils around the world have lost their fertility or their capacity to perform their functions due to processes that are accelerated or triggered directly by human activities and that often act in synergy with each other, amplifying the negative effect.

Among these processes, the most widespread at the worldwide level are erosion, OM loss, compaction, salinization, flooding and landslide phenomena, contamination, and reduction in biodiversity (Menta, 2012). In this regard, the communication of the European Commission to the Council and the European Parliament, entitled "Towards a Thematic Strategy for Soil Protection," defined the following eight main soil threats (Soil Atlas of Europe):

1. Soil sealing occurs mainly through the development of technical, social, and economic infrastructures, especially in urban areas.
2. Erosion is mainly due to the inadequate use of soil by agriculture and forestry but also through building development and uncontrolled water runoff from roads and other sealed surfaces.
3. Loss of OM is mainly due to intensive use of the land by agriculture, especially when organic residues are not sufficiently produced or recycled back into the soil. Agronomists consider soil with less than 1.7% OM to be in the pre-desertification stage.
4. The decline in biodiversity is linked to the loss of OM, as biodiversity depends on OM, meaning that all soil biota live on the basis of OM.
5. Contamination can be either diffuse (widespread) or localized and may be caused by many human activities such as industrial production and traffic and predominantly through the use of fossil material such as ores, oils, coals, and salts or through agricultural activities.
6. Compaction of soil is a rather new phenomenon caused mainly by high pressures exerted on soil through heavily loaded vehicles used in agriculture and forestry. An estimated 4% of soil throughout Europe suffers from compaction.
7. Hydrogeological risks are complex phenomena that result in floods and landslides deriving partly from uncontrolled soil and land uses (e.g., sealing, compaction, and other adverse impacts), as well as from unregulated mining activities.
8. Salinization is mainly a regional problem, but in the areas where it occurs—such as the Mediterranean basin and in Hungary—it severely endangers agriculture, forestry, and the sustainable use of water resources.

Of these eight main threats to soil and insofar as forest management is concerned, the impact implied by the loss of OM, erosion, and loss of biodiversity in the soils of agricultural and forest systems should be highlighted on a global scale.

4.2.2 EROSION AND LOSS OF ORGANIC MATTER IN FOREST AND AGRICULTURAL SOILS

Soil degradation processes constitute a major worldwide problem with significant environmental, social, and economic consequences. As the world population increases, so does the need to protect soil as a vital resource for the production of food and natural resources. Growing awareness by the international community of the need for global responses has led to an increasing number of international initiatives.

In fact, erosion is regarded as one of the most widespread forms of soil degradation and, as such, poses potentially severe limitations on sustainable land use. Topsoil is the most productive soil layer, and the most dominant effect of erosion is the loss of topsoil, which may not be conspicuous but is nevertheless potentially very damaging. Erosion literature commonly identifies "tolerable" rates of soil erosion, but these usually exceed the rates that can be balanced by the natural weathering of parent materials to form new soil particles. Soil loss in some places may be considered acceptable from an economic standpoint, but some modern agricultural and forest production methods are causing overall erosion rates that are becoming increasingly unacceptable from a long-term point of view.

Soil erosion is a physical phenomenon involving the removal of soil and rock particles by water, wind, ice, and gravity. Therefore, climate, topography, and soil characteristics are important physical factors affecting the amount of erosion. The most effective strategy for reducing erosion is to increase the cover of vegetation or litter and preferably both. Plants—especially woody plants with strong, deep roots—greatly increase soil strength by providing a stabilizing effect on slopes, in addition to protecting the soil with their canopy and adding litter to the soil surface. In some cases, the plants also transpire significant quantities of water from the slope, thus reducing the weight that contributes to mass movements (Mansourian et al., 2005; Soil Atlas of Europe, 2005).

Loss of OM is another of the most widespread forms of soil degradation. Following the unprecedented expansion and intensification of agricultural and forest production during the twentieth century, there is clear evidence of a consequent decline in the OC contents of many soils. This decline in OC contents has important implications for agricultural and forest production systems, as well as for ecosystems, as OC is a major component of SOM (Ebel and Davitashvili, 2007). A wide range of agricultural and forest management practices influence the abundance of OM, biomass, and diversity of soil biota and litter. These management practices include variations in tillage, treatment of pasture and crop residues, crop rotation, and applications of pesticides, fertilizers, manure, sewage, and ameliorants such as clay and lime, drainage and irrigation, and vehicle traffic (Baker, 1998).

4.2.3   Biodiversity Loss in Agricultural and Forest Soils

Humans have caused a widespread reduction in biodiversity on a global scale. The Convention on Biological Diversity (CBD) was the first global agreement aimed at the conservation and sustainable use of biological diversity (Secretariat of the Convention on Biological Diversity, 2000). The CBD lies at the heart of biodiversity conservation initiatives. It offers opportunities to address global issues at a national level through locally grown solutions and measures. One important requirement is the development of national biodiversity strategies and action plans channeled into relevant sectors and programs, as a primary means of implementing the Convention at the national level (United Nations, 1992). The recent Conference of the Parties of the CBD (May 2008, Bonn) demonstrated that there is unanimous acknowledgment of the need for action to protect biodiversity. Biodiversity conservation is essential both for ethical reasons and especially for the ecosystem services provided for current and future generations by the complex of living organisms. These ecosystem services are essential for the functioning of our planet (Menta, 2012).

Changes in biodiversity alter ecosystem processes and affect their resilience to environmental change. Human activities are estimated to have increased the rates of extinction 100–1000 times (Lawton and May, 1995). In the absence of major changes in policy and human behavior, our effect on the environment will continue to alter biodiversity.

Land use is considered to be the main element of global change for the near future, and land-use change is projected to have the greatest impact on biodiversity by the year 2100. In a review on changing biodiversity, Chapin et al. (2000) consider that land use will be the main cause of change in biodiversity for tropical, Mediterranean, and grassland ecosystems.

Forests, tropical or temperate, generally represent the biomes with the largest soil biodiversity. Consequently, any land-use change resulting in the removal of perennial tree vegetation will produce a reduction in soil biodiversity. In some cases, pasture or perennial grasslands succeed forests, while in others, arable land replaces formerly wooded areas. The change in soil biodiversity will therefore be influenced by the subsequent use of the land after the forest. However, reduction of soil biodiversity as a result of urbanization can be even more severe. The urbanization process leads to the conversion of indigenous habitat to various forms of anthropogenic land use, the fragmentation and isolation of areas of indigenous habitat, and an increase in local human population density. The urbanization process has been identified, for example, as one of the leading causes of decline in soil arthropod diversity and abundance in some areas (Menta, 2012). These reductions in soil biodiversity result in artificial ecosystems that require constant human intervention and extra running costs, whereas natural ecosystems are regulated by communities of plant, animal, and soil organisms through flows of energy and nutrients, a form of control that is being progressively lost with agricultural intensification.

Furthermore, differences in agricultural and forest production systems, such as integrated, organic, or conventional systems, have been demonstrated to affect soil fauna in terms of numbers and composition (Hansen et al., 2001; Cortet et al., 2002). The impact of soil tillage operations on OM and soil organisms is highly variable, depending on the tillage system adopted and on the soil characteristics. Conventional tillage by plowing inverts and breaks up the soil, destroys soil structure, and buries crop residues, thereby causing the highest impact on soil fauna; the intensity of these impacts is generally correlated to soil tillage depth (Menta, 2012). Minimum tillage systems may be characterized by a reduced tillage area (i.e., strip tillage) and/or reduced depth (i.e., rotary tiller, harrow, hoe); crop residues are generally incorporated into the soil instead of being buried. The negative impact of these conservation practices on soil fauna is reduced in comparison with conventional tillage. Under no-tillage crop production, the soil remains relatively undisturbed, and plant litter decomposes at the soil surface, much like in natural soil ecosystems.

Taking as an example the communities of soil microarthropods, observations on the impacts of different forms of agricultural and forest management on these communities showed that the high input of intensively managed systems tends to promote low diversity, while lower input systems conserve diversity (Menta, 2012). The influence on soil organism populations is expected to be most evident when conservation practices such as no-till are implemented on previously convention-ally tilled areas, as the relocation of crop residues to the surface in no-till systems will affect the soil decomposer communities (Beare et al., 1992). No-till (Hendrix et al., 1986) and minimum tillage generally lead to an increase in microarthropod numbers (Menta, 2012). Akeem (2012) also observed that higher values of mite den-sity were associated with a decrease in tillage impact. Similarly, Cortet et al. (2002) reported that the mite community—and in particular oribatids—was more abun-dant in no-tillage as compared to conventional tillage. However, the differences were found only at certain periods of the year. Conventional tillage caused a reduction in microarthropod numbers as a result of exposure to desiccation, destruction of habitat, and disruption of access to food sources (House and Del Rosario, 1989). The influence of these impacts on the abundance of soil organisms will either be moder-ated or intensified depending on their spatial location, that is, in row where plants are growing, near the row where residues accumulate, or between rows where they are subjected to possible compaction from mechanized traffic (Fox et al., 1999).

Moreover, the observed impact of different forms of agricultural and forest man-agement on bacterial communities showed similar patterns to microarthropod com-munities (Siepel and Bund, 1988; Bardgett and Cook, 1998). In fact, it is also evident that high-input systems favor bacterial pathways of decomposition, dominated by labile substrates and opportunistic bacterial-feeding fauna. In contrast, low-input systems favor fungal pathways with a more heterogeneous habitat, leading to domi-nation by more persistent fungal-feeding fauna (Bardgett and Cook, 1998)

The effects of fertilizers on soil invertebrates and microbial communities are a consequence of their impact both on the vegetation and directly on the organisms themselves. Increases in the quantity and quality of the food supplied by vegetation are frequently reflected in greater fecundity, faster development, and increased pro-duction and turnover of invertebrate herbivores (Curry, 1994). The effects of organic and inorganic fertilizers in terms of nutrient enrichment may be comparable, but these two types of fertilizers differ in that organic forms provide additional food material for the decomposer community. Ryan (1999) concluded that the total soil microbial biomass and the biomass of many specific groups of soil organisms will reflect the level of SOM inputs.

Hence, organic or traditional farming practices that include regular inputs of OM in their rotation determine larger soil communities than conventional farming practices (Ryan, 1999). Generally, the responses of soil fauna to organic manure will depend on the characteristics of the manure and the rates and frequency of application. Herbivore dung, a rich source of energy and nutrients, is exploited initially by a few species of coprophagous dung flies and beetles and subsequently by an increasingly complex community comprising many general litter-dwelling species (Curry, 1994).

Pfotzer and Schuler (1997) also reported that the soil microbial and faunal feed-ing activity responded to the application of compost with higher activity rates than with mineral fertilization. Studies related to compost from sewage sludge appli-cation on agricultural and forest soils showed an increase in the abundance of

Collembola (Lüben, 1989), *Carabidae* (Larsen et al., 1986), *Oligochaeta* (Cuendet and Ducommun, 1990), soil nematodes (Bruce et al., 1999), and *Arachnida* (Bruce et al., 1999). In some cases, the application of sewage sludge to agricultural and forest systems can input toxic substances that accumulate in the soil and reach potentially toxic levels for soil fauna (Bruce et al., 1999). Field studies have suggested that metals contained in sewage sludge do not reduce the abundance of euedaphic (Lüben, 1989) and epigeic collembolans (Bruce et al., 1997) but may alter their population structure. Bruce et al. (1997) reported negative effects on collembolan communities in soil treated with sewage sludge, and these effects can be attributed to anaerobic conditions and high ammoniacal levels. Indeed, the knowledge gained in relation to the effects of sewage sludge showed that species that are more sensitive to the toxic substances contained in sewage sludge may disappear, while other more tolerant species may dramatically increase (Menta, 2012).

Organic wastes could also become an easily available and cheap source of OM after composting processes. The use of compost obtained from organic waste in agricultural and forest activity enables waste materials to be converted into a useful resource. This has led national authorities in recent years to promote both the use of compost to reduce soil fertility loss and the research aimed at assessing its effects on both agricultural and forest production and soil environment (Allievi et al., 1993; Pinamonti et al., 1997; Bazzoffi et al. 1998). However, the possible negative effects on soil fauna deriving from the use of organic waste include the accumulation of trace metals in the soil (Tranvik et al., 1993).

Pesticide application to the soil can affect the soil fauna by influencing the performance of individuals and modifying ecological interactions between species. When pesticides impact one or more ecosystem components, they also affect microarthropod communities in terms of number and composition. Pesticide toxicity on soil fauna is determined by various factors, such as the pesticide's chemical and physical characteristics, the species' sensitivity, and the soil type. In fact, among soil microarthropods, different taxa showed a variety of responses.

The physical and chemical characteristics of the soil, such as its texture, structure, pH, OM content and quality, and the nature of the clay minerals, are important factors in determining the toxic effects of pesticides and other xenobiotics. A study carried out by Joy and Chakravorty (1991) showed reduced toxic effects, as a function of soil type, in the following order: sand > sandy loam > clay > organic soil. Often, the toxicity of pesticides can be directly related to SOM content (Van Gestel and Van Straelen, 1994). However, pesticide application does not always cause negative impacts on the entire soil microarthropod community. For example, for certain types of soil, there is evidence that some taxa can obtain a competitive advantage from the application of certain specific pesticides (Menta, 2012).

Finally, in biodiversity recovery processes using forest restoration, it must be considered that after stimulating the natural regeneration processes that establish forest species, it is necessary to manage and direct succession processes toward the desired objectives. It is important to promote continued development of the vegetation to conserve soil, nutrients, and organic resources; to restore fully functional hydrological processes, nutrient-cycling processes, and energy flow processes; and to create self-repairing landscapes that provide the goods and services necessary for biophysical, biochemical, and socioeconomic sustainability.

Different stages of forest-soil degradation call for management actions that focus on different processes. Severely degraded sites require early repair of hydrological, nutrient-cycling, and energy capture and transfer processes. As the forest vegetation increases in biomass and stature, it reduces the abiotic limitations of the site by improving soil and microenvironmental conditions. Directing natural processes toward land-use goals requires an understanding of the processes driving succession. The rate and direction of succession are influenced by the availability of species and the availability of suitable sites and by differential species performance (Mansourian et al., 2005).

4.2.4 DIFFICULTIES IN DEVELOPING INTERNATIONAL STRATEGIES FOR FOREST-SOIL PROTECTION

The realization that all of the soil degradation processes described are an environmental problem of global significance and the recognition of the importance of protecting our soils have led to an increase in international initiatives (Akeem, 2012). The protection of the soil and the preservation of its biological health and overall quality have therefore become a key international goal.

In Europe, for example, the threat to soil health was reported, and its biological roots were recognized almost three decades ago (Filip, 1973; Kovda, 1975). Nevertheless, there was a strong belief in soil's capacity for self-remediation. In the 1980s, the first moves toward the development of well-targeted soil protection initiatives were made in Germany and the Netherlands and later also in the European Community (Barth and L'Heremite, 1987; Howard, 1993). Finally, on March 1, 1999, a Federal Soil Protection Act came into force in Germany. Its practical application, however, requires the availability of properly justified standards and reliable monitoring methods (Filip, 2002).

The legislative interest in soil protection must therefore be accompanied by a concerted effort to seek a common methodology for the estimation of soil quality, in which biological and biochemical soil properties will play a very important role due to their high sensitivity to distorting agents (Filip, 2002). However, Howard (1993) indicated that variations in the approaches adopted by different countries reflect discrepancies in the nature and perceived seriousness of soil problems.

Continuing with the example of the EU's soil policies, Howard (1993) and Gzyl (1999) reported that European countries considered the "extension of urbanization, pollution by heavy metals, organic contaminants (including pesticides), acidification, over-fertilizing and artificial radio nucleids, groundwater nitrates, loss of organic matter, deteriorating soil structure, soil compaction and water and wind erosion" to be major threats to their soils. However, soil biodiversity is only indirectly addressed in a few European countries through specific legislation on soil protection or regulations promoting environmentally friendly farming practices (Jeffery et al., 2010; Turbé et al., 2010).

Given the differences between belowground and aboveground biodiversity, policies aimed at aboveground biodiversity may not be very effective for the protection of soil biodiversity. In contrast, the management of soil communities could form the basis for the conservation of many endangered plants and animals, as soil biota steer plant diversity and many of the regulating ecosystem services. No legislation or regulation exists to date that is specifically targeted at soil biodiversity, whether at the international, national, or regional level. This reflects the lack of awareness of soil biodiversity and its value, as well as the complexity of the subject (Akeem, 2012; Menta, 2012).

Due to the complexity of these topics, there is so far no instrument that specifically addresses the protection of soil (European Commission, 2005) on any significant scale, such as at the level of the EU.

In connection to the issue of biodiversity loss, for example, there are several policy areas that directly affect and could address soil biodiversity, including policies for soil, water, climate, agriculture, and nature. This aspect could be taken into account or highlighted in future biodiversity policies and initiatives such as the new European strategy for biodiversity protection post-2010 (Jeffery et al., 2010; Turbé et al., 2010). Previously in this area, among the priorities set by the Sixth Environment Action Program of the EU (EC, 2002) for the conservation of biodiversity and natural resources, it assumed the commitment of addressing soil alongside water and air as an environmental medium and as a nonrenewable resource to be preserved, hence undertaking to develop a thematic strategy for the protection of the soil (UE, 2005).

Johnston and Crossley (2002) highlight the study and protection of soil ecology as an essential component of forest ecosystem recovery and protection. They indicate that for many years, there was a lack of concern for forest-soil conservation: Brauns (1955) cautioned that it would be impossible to continue ignoring soil biology when resolving forest management problems and urged more consideration for the improvement of soil biological status. Similarly, Wells (1984), reflecting upon the increasing demands for shorter rotations and faster-growing trees, concluded that it would not be sufficient to maintain the existing fertility of forest soils. Bloem et al. (2003) and Brussard et al. (1998) explain that cryptobiota—hidden soil life—plays a key role in life-support functions but is not part of any recognized list of endangered species. It is questionable whether a species-based approach is sufficient to attain a sustainable use of ecosystems inside—and especially outside—protected areas. Therefore, research networks have been established to monitor large areas, including agricultural and forest soils (Bloem, 2003).

Johnston and Crossley (2002) also remarked that the chemical and biological conditions of forest soils are not routinely monitored and studied, as they should be in view of the priority of improved regional soil fertility within the overarching goal of ecosystem management. They posed two questions: Can forest managers continue to manage forest ecosystems without striving to improve soil fertility over the long term? Does the field of soil ecology have little to offer to the practice of managing forest ecosystems, including forests that produce timber products?

In view of all these questions, it is necessary to conduct a review of the application of SQI in forest-soil conservation and protection and of the protocols that allow the integration of these factors in participatory forest management.

4.3 SOIL-QUALITY INDICATORS IN SUSTAINABLE FOREST MANAGEMENT

4.3.1 SUSTAINABLE FOREST MANAGEMENT AND SOIL QUALITY

Given the numerous terms and definitions for the concept of sustainable forest, Powers et al. (1998) and Page-Dumroesea et al. (2000) pointed out that soil productivity is a key factor for maintaining ecosystem function. Forestry management on a sustainable basis has consistently included the maintenance or enhancement of forest-soil quality as a criterion of sustainability (Burger and Kelting, 1999; Schoenholtz et al., 2000). Furthermore, the overall goal of soil protection involves the conservation of good soil quality and requires certain soil functions to be fulfilled.

Schoenholtz et al. (2000) also reported that international and national calls for management of forestry on a sustainable basis have consistently included maintenance or enhancement of forest-soil quality as a criterion of sustainability. However, the choice of a standard set of specific soil chemical, physical, and biological properties as indicators of soil quality can be complex and will vary among forest systems and management objectives. Moreover, despite the dramatic changes in soil properties and processes detected during cropping and forest regeneration, no attempts have been made to develop SQI to assist in the assessment of soil conditions during such changes (Akeem, 2012; Bautista et al., 2012).

However, although considerable activity is currently aimed at the development and evaluation of sustainable management systems for agricultural and forest soils, these efforts have been hampered by the lack of agreement as to what constitutes credible measures of sustainability. The means to evaluate soil sustainability in terms of both design and performance have yet to be fully determined (Larson and Pierce, 1991; Pierce and Larson, 1993; Doran, 1994), but in recent decades, several new institutional, technical, and scientific approaches have been and are being developed.

As an example of the evolution of institutional approaches for sustainable soil management, the European Commission in its Sustainable Development Strategy published in 2001 noted that soil loss and declining fertility were eroding the viability of agricultural land (COM 2001, 264). In 2002, the European Parliament and Council established the Sixth Environmental Action Program (Sixth EAP), which covers a period of 10 years (Decision 1600/2002). This program addresses the community's key environmental objectives and priorities to be met through a range of measures, including legislation and strategic approaches. In Article 6, "Objectives and priority areas for action on nature and biodiversity," the Sixth EAP foresees the development of a thematic strategy on soil protection, addressing the prevention of pollution, erosion, desertification, land degradation, and hydrogeological risks, taking account of regional diversity, and including the specificities of mountain and arid areas (Bloem, 2003).

As an example of the evolution of technical approaches for sustainable agricultural and forest-soil management, several authors indicate that the current aims of these approaches are to maintain good crop yields with minimal impact on the environment while at least avoiding deterioration in soil fertility and providing essential nutrients for plant growth. Furthermore, sustainable agricultural and forest systems and agriculture should support a diverse and active community of soil organisms, exhibit a good soil structure, and allow for undisturbed decomposition. Thus, current sustainable agricultural and forest practices are adjusting to integrate organic—or more extensive—management. The main principles involve restricting stocking densities, avoiding synthetic pesticides and mineral fertilizers, and using organic manure. This will ultimately result in an increased role of soil organisms, for example, decomposers, nitrogen fixers, and mycorrhizae, in plant nutrition and disease suppression.

Current sustainable agriculture and forestry should therefore seek to conserve natural resources based on a concept of productivity, which is closely linked to maintaining a system aimed at saving energy and resources in the mid to long term, through optimizing recycling and enhancing biodiversity and through biological synergy (Hansen et al., 2001; Máder et al., 2002; Bloem, 2003).

As examples of the evolution of technical approaches for sustainable forest-soil management, the two most common approaches for monitoring soil quality in forestry in the last decade may be (1) direct comparisons of biomass production in successive rotations and (2) soil-quality indices based on direct measurements of soil properties such as rooting volume, soil strength, and OM or through some integrated measure using multiple soil properties (Henderson et al., 1990; Powers et al., 1998; Fox et al. 2000).

Regarding the evolution of scientific approaches for sustainable agricultural and forest-soil management, Bloem (2003) highlights the need for standards for evaluating management systems that allow an assessment of their sustainability. In recent decades, two different approaches have frequently been employed to evaluate sustainable management systems: (1) comparative assessment and (2) dynamic assessment.

In the comparative approach (1), the performance of a system is determined in relation to alternatives. The characteristics and biotic and abiotic soil attributes of alternative systems are compared at a particular time, and the decision with regard to the relative sustainability of each system is based on the magnitude of the measured parameters. The main limitation of the comparative approach is that if only outputs are measured, it provides little information about the process that created the condition measured.

In contrast, in the dynamic approach, (2) a management system is assessed in terms of its performance determined over a period of time. The main disadvantage of this approach is that it needs measurements of indicators for at least two points in time and consequently does not provide an immediate assessment of soil quality. Moreover, it can be misleading in the case of soils that are functioning at their highest attainable level and cannot be improved or when they are functioning at their lowest attainable level and cannot deteriorate further. Both these cases would show a static trend—indicating sustaining systems—but would have a completely different quality. These two approaches to assessment are complementary, since they allow different scales of evaluation. While monitoring trends are more useful for evaluation at the farm level, comparative assessment appears to be more suited to a broader scale of evaluation (on a regional scale) (Seybold et al., 1997; Bloem, 2003).

Continuing with the topic of the evolution of scientific approaches for sustainable agricultural and forest-soil management, Bastida et al. (2008) point out that given the complex nature of soils, it is important to select adequate indicators depending on the task. In order to obtain a complete picture of soil quality, a range of different parameters (physical, chemical, and biological) must be included (Frankenberger and Dick, 1983; Nannipieri et al., 1990; Dick, 1994; Gelsomino et al., 2006). The reason for the need to include a wider range of indicators is that certain factors may affect some indicators but not others, and this can affect the accuracy of the overall picture of soil quality.

However, Bastida et al. (2008) explain that despite the wide diversity of indices, they have never been used on large scales, nor even in similar climatological or agronomic conditions. They attribute this lack of applicability of soil-quality indices to the deficient standardization of certain methodologies, to the fact that some methods are beyond reach in some parts of the world, to spatial scale problems (soil heterogeneity), to poor definition of soil natural conditions (climate and vegetation), and to poor definition of the soil function to be tested for soil quality. They report that the most straightforward index used in the literature is a biparametric index (metabolic quotient qCO_2 or respiration-to-microbial biomass ratio) that has been widely used to evaluate ecosystem development, disturbance, and system maturity. However, they conclude that indices that integrate only two parameters did not provide enough information on soil quality or degradation.

Lately, as opposed to biparametric indices, there has been widespread development of multiparametric indices, which clearly establish differences between (1) management systems, (2) soil contamination or density, and (3) type of vegetation. These indices integrate different parameters, of which the most important are biological and chemical parameters such as pH, OM, microbial biomass C, respiration, and enzyme activities. The majority of multiparametric indices have been established on the basis of either expert opinion (subjective) or using mathematical–statistical methods (objective).

4.3.2 SOIL-QUALITY INDICATORS

Soil-quality assessment is of the utmost importance for determining the sustainability of land-management systems in the near and distant future. SQI indices are needed to identify problem production areas, to make realistic estimates of food and natural resource production, to monitor changes in sustainability and environmental quality as related to agricultural and forest management, and to assist national and state or regional agencies in formulating and evaluating sustainable agricultural and forest land-use policies (Granatstein and Bezdicek, 1992; Doran, 1994).

The concepts of soil quality and SQI have evolved in recent decades. Soil scientists have always sought to link soil quality and soil variables to land use. Agronomists and farmers most commonly define soil quality as the suitability of a soil to function under different uses, illustrating a broader concept that highlights the fact that agriculture has traditionally been more focused on soil interaction than forestry. Foresters, by comparison, have traditionally linked soil quality to the measurement of soil productivity using tree growth or wood yield; they usually define soil productivity as the ability of a soil to produce biomass per area and per time units (Ford, 1983; Warkentin, 1995; Schoenholtz et al., 2000).

Among the definitions for the concept of general soil quality that have been suggested in the past, Doran (1994) highlights that soil quality is "The capacity of the soil to interact with the ecosystem in order to maintain biological productivity, environmental quality and to promote animal and plant health." This definition is similar to the three essential criteria for soil quality that were identified by the Rodale Institute in 1991: (1) Productivity is the soil's capacity to increase plant biological productivity; (2) environmental quality is the soil's capacity to attenuate environmental contamination, pathogens, and external damage; and (3) the health of living organisms is the interrelation between soil quality and animal, plant, and human health (Benedetti and Dilly, 2003).

Schoenholtz et al. (2000) reviewed the concept of soil quality and defined it as a site's quality for forest productivity. They explained that soil quality is a value-based concept related to the objectives of ecosystem management and hence it will be management and ecosystem dependent. Soil quality may be broadly defined to include water retention capacity, C sequestration, plant productivity, waste remediation, and other functions, or it may be defined more narrowly. For example, a forest plantation manager may define soil quality as the capacity of a soil to produce biomass.

Bastida et al. (2008) reported that the concept of soil quality arouses greater controversy than the concept of water or air quality. These authors point out that the maintenance of soil quality is critical for ensuring the sustainability of both the environment and the biosphere; for this reason, they reviewed SQI as well as the parameters comprising them and highlighted the lack of consensus concerning the use of these indicators.

Heightened concerns regarding the sustainability of agricultural and forestry practices and the influence of soil conditions on environmental sustainability have led to considerable research efforts into SQI and indices (Doran, 1994; Hailu and Chambers, 2012).

Basic SQI are useful for comparing quality among soil types and before and after certain management practices are imposed on a soil type. SQI may be simple state variables with a measurable unit or a complex construct of several soil variables known as "soil-quality indices" that may include a time or rate dimension, which makes them dynamic (Burger and Kelting, 1999; Schoenholtz et al., 2000). Whatever the case, SQI should give some measure of the capacity of a soil to function in terms of plant and biological productivity and environmental quality (Seybold et al., 1997).

Indicators of soil quality should also be as follows: (1) sensitive to long-term change in soil management and climate but sufficiently robust not to alter as a consequence of short-term variations in weather conditions, (2) well correlated with beneficial soil functions, (3) useful for understanding why a soil will or will not function as desired, (4) comprehensible and useful to land managers, (5) easy and inexpensive to measure, and (6), where possible, should also be components of existing soil databases.

Furthermore, the suitability of SQI depends on the kind of land, land use, and scale of assessment. Different land uses may require different soil properties, and in consequence, some SQI in a given situation can be more helpful than others for the purpose of the assessment (Karlen et al., 2001; Bloem, 2003).

In summary, an SQI is a measurable surrogate of a soil attribute that determines how well a soil functions (Burger and Kelting, 1999; Schoenholtz et al., 2000). Soil quality can be evaluated using a large number of indicators (chemical, physical, biological) depending on the scale and objective of the evaluation.

A review of SQI showed that there is heavy reliance on a few appraisals: (1) SOM among chemical indicators (Liebig and Doran, 1999; Bowman et al., 2000; Brejda et al., 2000; Kettler et al., 2000; Gilley et al., 2001; Li et al., 2001), (2) bulk density (Liebig and Doran, 1999; Kettler et al., 2000; Gilley et al., 2001; Li et al., 2001), and (3) aggregate stability (Bowman et al., 2000; Six et al., 2000) among physical indicators, which were the most frequently used. In contrast, there were very few examples of biological indicators of soil quality (Pankhurst, 1997; Liebig and Doran, 1999; Gilley et al., 2001).

However, biological monitoring is necessary to correctly assess soil degradation and its correlated risks (Turco et al., 1994). In particular, there is an urgent need to identify indicator systems with the capacity to express soil-quality criteria so they can be used as benchmarks in environmental remediation, as well as to assess and monitor soil quality in soils subjected to risk of degradation (van Straalen and Krivolutsky, 1996; ANPA, 2002).

Due to the complexity of the soil system, a specific soil management system cannot be assessed using a single indicator of soil quality but requires the selection of a minimum data set (MDS) of attributes regarding the physical, chemical, and biological properties of the soil (Karlen et al., 2001; Bloem, 2003). Hailu and Chambers (2012) summarized the scientifically sound procedures for building soil-quality indices that combine diverse soil-quality attributes or indicators into summary measures in order to enhance the evaluation of land-management strategies. Over the last decade and a half, several procedures for building soil-quality indices have been proposed and implemented in the soil-science literature. They describe an overview of the index construction procedures employed in soil-science research provided by Andrews et al. (2002, 2004), Karlen et al. (2003), and Bremer and Ellert (2004).

The standard approach is to generate the index using the following three steps. The first step (1) involves choosing an MDS of physical, chemical, and biological soil-quality variables (Larson and Pierce, 1991), based on either expert opinion or statistical data reduction methods such as principal component analysis (PCA). In the second step (2), the variables in the MDS are transformed into 0–1 scores using either linear or nonlinear transformation functions. In the final step (3), the indicator scores are combined into a soil-quality index using different integration techniques including simple addition (Andrews and Caroll, 2001a), weighted addition (Harris et al., 1996), or decision support systems employing min-max objective functions (Yakowitz et al., 1993). Both the second and third steps are ad hoc. The transformation of individual soil-quality variables or indicators into 0–1 scores does not recognize that the impact of a soil-quality variable on outcomes (e.g., crop yield) might depend on the level of other soil variables or production inputs (Hailu and Chambers, 2012).

In summary, variation-sensitive soil indicators in forest management are needed to compare the effects of a management practice on soil over time (Schoenholtz et al., 2000). The assessment of soil quality and the identification of key soil properties that serve as indicators of soil function are complicated by the many issues defining quality and the multiplicity of physical, chemical, and biological factors that control biogeochemical processes and their variation in time, space, and intensity. The practical assessment of soil quality therefore requires the consideration of these functions and their variations in time and space (Larson and Pierce, 1991; Doran, 1994).

4.3.3 Assessment of Soil-Quality Indicator Usage in Forestry

The assessment of sustainable forestry systems, which allow the combination of production targets and environmentally friendly management practices while protecting both soil and biodiversity, is essential in order to prevent the decline of forest landscapes (Menta, 2012). Despite these warnings, it does not currently appear that many forest soils are being managed and protected actively in the way this author suggests.

The process of forest sustainability assessment requires evaluators to be reasonably informed about the practical appraisal and interpretation of soil quality using soil indicators in order to increase their chances of achieving correct joint decision making (E. Martínez-Falero, pers. com.). Thus, to improve the evaluators' understanding of the current sustainable forest management status and the use of SQI in forestry, this chapter and the following describe the main scientific outcomes of SQI and their current applications in forest sustainability assessment.

Regarding forest soils, foresters have always relied on the knowledge of soil chemical and physical properties to assess the capacity of sites to support productive forests. However, Schoenholtz et al. (2000) pointed out that in the last few years the need to assess soil properties has expanded due to the growing public interest in understanding the consequences of management practices on the quality of soil relative to the sustainability of forest-soil functions (water flow, biodiversity conservation, and as an environmental buffer), in addition to plant productivity. Moreover, soil-quality assessment is fundamental for determining the sustainability of land-management systems in the near and distant future. The challenge for the future is to develop sustainable management systems that are at the vanguard of soil health; SQI are merely a means toward this end.

In this regard, with the aim of assessing the current usage of soil-quality indices in forestry, we analyzed the available scientific bibliography in all experimental areas related to the concepts of "soil quality and its indicators" and "SQI in forestry," by conducting a systematic study of all the available databases on the Web of Knowledge (Thomson Reuters, 2011). This analysis was done up until 2011, which was the last complete year.

This bibliographical study is structured into successive stages of analysis. The first stage involves consulting the ISI Web of Knowledge base using the keywords "soil quality + indicators," and in the second stage, we consulted the ISI Web of Knowledge base using the keywords "soil quality + forestry" to determine the temporal evolution and research contents of the scientific works associated to these keywords. The third stage revises the bibliographical information contents on the selected articles, such as soil indicator typologies and what they were used for in relation to the sustainable forestry approach.

The field of soil science has historically been concerned with soil quality; however, the results of our bibliographical analysis show that it was not until 1991 when

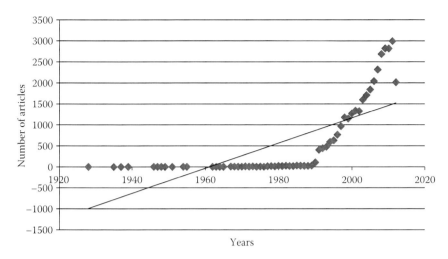

FIGURE 4.1   The number of articles studying soil quality has been rapidly increasing since 1990 to the present day (ISI WEB of Knowledge database. Keyword: "soil quality").

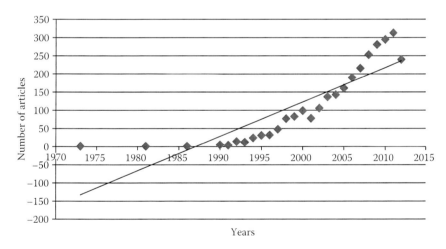

FIGURE 4.2   The number of articles using indicators to study soil quality has been rapidly increasing since 1990 to the present day (ISI WEB of Knowledge database. Keyword "soil-quality indicators").

significant research into "soil quality" and "SQI" began to appear. Figure 4.1 shows the trends for the articles incorporating "soil-quality" concepts, and Figure 4.2 shows the trends for the articles incorporating "SQI."

However, this interest in soil quality and its indicators has not been homogeneous across the various research and management areas. Figure 4.3 shows that out of the total articles incorporating the concept of "soil quality," only 4% refer to forestry, in contrast with 52% to agriculture and 32% to environmental sciences.

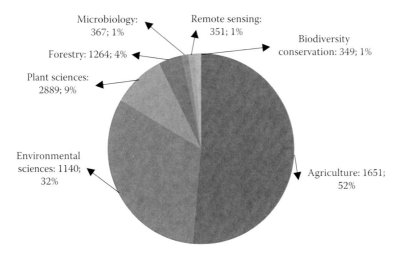

FIGURE 4.3   Article distribution by research and management areas studying soil quality (ISI WEB of Knowledge database: keyword "soil quality").

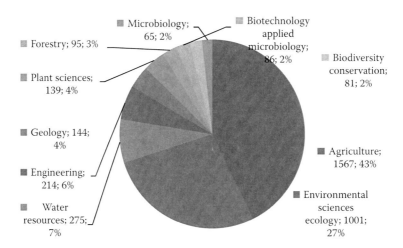

FIGURE 4.4   Article distribution by research and management areas using indicators of soil quality (ISI WEB of Knowledge database: keywords "soil-quality indicators").

Moreover, Figure 4.4 shows that out of the total articles incorporating "SQI," only 3% refer to forestry, as opposed to 43% to agriculture and 27% to environmental sciences.

Regarding forest production, soil fertility and soil quality determine vegetation composition, structure, and growth, as both serve as a medium for root development and provide moisture and nutrients for plant development. Both soil qualities would thus explain the reason that 55.1% of the forestry articles incorporating SQI correspond to research in applied forestry (51.7% articles on forestry production and 3.4% articles on forest management).

However, there has been a progressive decline in the number of articles incorporating SQI in the study of forest production and forest management in recent years. The study and protection of soil ecology are essential components of forest ecosystem conservation and recovery (Johnston and Crossley, 2002). However, our bibliographical analysis shows that only 25.2% of the forestry articles analyzed include SQI associated to edaphological and forest ecology studies (10.3% articles related to forest edaphology, 9.2% articles related to forest ecology, and 5.7% articles related to tree morphology and physiology). Moreover, only 6.4% of the forestry studies analyzed incorporate SQI relating forestry research to environmental issues. However, there has been a progressive increase in the number of articles incorporating SQI in the study of forest ecology in recent years.

The interest and use of SQI in forestry have not been homogeneous across all forest types. Our bibliographical analysis shows that 61.3% of the forestry articles using SQI are related to coniferous woods, which are usually associated to soils with limited fertility and limited biological productive capacity (29.3% articles related to *Pinus* spp. woods and 12.0% articles related to *Picea* spp. woods). In contrast, 28.0% of this kind of articles is associated to deciduous woods, and only 4.0% to tropical forests.

In the area of forest-soil elements and properties assessed using SQI, our bibliographical analysis showed that 72.1% of the articles are related to the study of soil fertility and soil biological production capacity (34.9% associated to soil nutrition, cationic exchangeable complex (CEC), N, C/N, P, and fertility; 16.3% associated to soil moisture and soil temperature; 14.5% associated to soil C, OM, and humus; and 6.4% associated to microbiology and soil biology). This kind of articles has increased over time.

In contrast, 25.0% of the forestry studies using SQI are related to soil profile characteristics and elements (8.1% are associated to geomorphology and soil depth, 4.7% are associated to pH, 4.7% are associated to texture and structure, 4.7% are associated to porosity and aeration, and 2.9% are associated to penetration resistance and bulk density). However, this kind of articles has increased over time. Finally, only 1.7% of forest SQI studies are related to soil degradation (salinity and contamination).

This last aspect is an important omission, as forest-soil degradation processes are an environmental problem of global significance with immediate consequences at both the economic and social level; recognition of the importance of protecting against this phenomenon has led to an increase in international initiatives (Akeem, 2012).

Foresters have always relied on the knowledge of the chemical and physical properties of soils to assess a site's capacity to support productive forests (Schoenholtz et al., 2000). Our bibliographical analysis shows that most of the soil-quality indices/indicator types used in forestry studies are either chemical (49.0%) or physical (43.5%). Moreover, both kinds of articles have increased over time.

However, in the field of forestry, SQI that integrate both physical and chemical variables have been poorly applied (2.0%), as have biological SQI (4.1%). Finally, only 12.6% of the forestry articles applying SQI have used statistical analysis and numerical models.

All these bibliographical analyses lead to the conclusion that there is a need for new procedures that integrate physical, chemical, biological, and other unforeseen elements into indices and indicators so they can perform seamlessly under any given forest-soil circumstance worldwide (Bastida et al., 2008). Current knowledge of forest-soil indicators has not benefited from the wealth of tools such as numerical models—as indicated by Hailu and Chambers (2012)—which incorporate the Luenberger indicator (currently available in economics), for example, into forest-soil-quality assessment.

4.4 BIOLOGICAL, CHEMICAL, AND PHYSICAL SOIL-QUALITY INDICATORS

4.4.1 FOUNDATIONS FOR UNDERSTANDING THE ROLE OF BIOLOGY IN SOIL-QUALITY ASSESSMENT

Some researchers have defined the soil biota as a "superorganism" that assumes a crucial significance due to the chemical–physical and biological processes that are rooted in the soil. A rapid survey of invertebrate and vertebrate groups reveals that at least a quarter of the living species described are strictly soil or litter dwelling. These large soil communities play an essential role in soil functions as they are involved in processes such as the decomposition of OM, the formation of humus, and the nutrient cycling of many elements (nitrogen, sulfur, C). Moreover, edaphic communities affect soil porosity and aeration as well as the infiltration and distribution of OM within its horizons (Maharning et al., 2008; Menta, 2012). It is therefore necessary to understand the role of biodiversity in soil functions as the foundation for identifying SQI based on soil biology.

With regard to biodiversity and soil function interactions, it must be noted that decomposition of OM by soil organisms is crucial for the functioning of an ecosystem, due to its substantial role in providing ecosystem services for plant growth and primary productivity (Menta, 2012). For example, soil fauna performs a mainly mechanical action, whereas fungi and bacteria, both free and intestinal symbionts of other organisms, essentially perform chemical degradation; furthermore, during digestion, organic substances are enriched by enzymes that are dispersed in the soil along with the feces, contributing to humification (Maharning et al., 2008; Menta, 2012).

However, at present, we are not sure of the extent of the importance of this biodiversity in the sustainability of a soil's functions, and it is extremely rare for the biological relationships between soil organisms to consist of a simple and clear interaction. Current ecological conditions are the result of many complex interactions that typically involve multiple participants in soil life such as plants, microbes, fungi, and animals (Killham, 1994).

Thus, the relationships between biodiversity and its functions are complex and somewhat poorly understood, even in aboveground situations. The exceptional complexity of belowground communities further confounds our understanding of soil systems. Four important mechanisms underlying relationships between biodiversity and function are responsible for (1) driving fundamental nutrient-cycling processes, (2) regulating plant communities, (3) degrading pollutants, and (4) helping to stabilize soil structure.

Some of the most significant functions—and the main biotic groups that carry them out—are (1) primary production [plants, cyanobacteria, algae]; (2) secondary production [herbivores]; (3) primary decomposition [bacteria, archaea, fungi, some fauna]; (4) secondary decomposition [some microbes, protozoa, nematodes, worms, insects, arachnids, mollusks]; (5) soil structural dynamics [bacteria, fungi, cyanobacteria, algae, worms, insects, mammals]; (6) suppression of pests and diseases [bacteria, actinomycetes, fungi, protozoa, nematodes, insects]; (7) symbioses [bacteria, actinomycetes, fungi—notably mycorrhizae]; (8) SOM formation, stabilization, and C sequestration [virtually all groups, directly or indirectly]; (9) atmospheric gas dynamics, including generation and sequestration of greenhouse gases [bacteria for nitrous oxides, methane, all biota for CO_2]; and (10) soil formation [bacteria, fungi] (Soil Atlas of Europe).

Rough estimates of soil biodiversity indicate several thousand invertebrate species per site, as well as relatively unknown levels of microbial and protozoan diversity. Soil ecosystems generally contain a large variety of animals including nematodes and microarthropods such as mites and collembolans, symphylans, chilopodans, pauropodans, enchytraeids, and earthworms. In addition, a large number of meso- and macrofauna species (mainly arthropods such as beetles, spiders, diplopods, chilopods, and pseudoscorpions, as well as snails) live in the uppermost soil layers, the soil surface, and the litter layer. Despite several decades of soil biological studies, it is still very difficult to provide average abundance and biomass values for soil invertebrates. This is partly due to their high variability in both time and space, as well as to differences in the sampling methods used (Jeffery et al., 2010).

In addition, most of the work has been conducted in the forest soils of temperate regions, while other ecoregions such as the tropics, or other land uses such as agriculture, have been seriously neglected (Jeffery et al., 2010). Soil fauna is very variable, and most is also highly adaptable regarding feeding strategies, ranging from herbivores to omnivores and including carnivores. Depending on the available food sources, many soil faunas are able to change their feeding strategies to a greater or lesser extent, and many carnivorous species are able to feed on dead OM at times of low food availability. The interactions between soil fauna are numerous, complex, and varied. As well as in the predator/prey relationships and in some instances parasitism, commensalism also occurs (Menta, 2012).

The degree of interaction between soil organisms and the soil itself can be highly variable among taxa and dependent on the part of the life cycle that is spent in the soil (Wallwork, 1970). Specifically—and in combination with the morphological adaptations and the ecological functions of organisms—soil fauna can be classified into four main groups: (1) temporarily inactive geophiles, (2) temporarily active geophiles, (3) periodical geophiles, and (4) geobionts. It should be noted that these groupings do not have any taxonomical significance but are useful when studying the life strategies of soil invertebrates. These different types of relationships between soil organisms and the soil environment determine a differentiated level of vulnerability among various groups in response to possible impact on the soil environment. For instance, if soil contamination occurs, any impact will be higher on geobionts (as they cannot leave the soil and must spend all their life there) and lower on temporarily inactive geophiles (Menta, 2012).

Due to the absence of light, which makes photosynthesis unfeasible, there are very few photoautotrophic organisms and very few real phytophages among the organisms populating the soil. Regarding soil heterotrophic organisms, the activity of animals (typically protozoa, nematodes, rotifers, certain springtails, and mites) that feed on microflora—namely, bacteria, actinomycetes, and fungi (both hyphae and spores)—is of crucial importance for both diffusing and regulating the density of these microorganisms. For example, through their feces, springtails—which feed on fungi—can spread fungal spores that are still viable to areas as far as a few meters away from their point of origin (Wallwork, 1970; Menta, 2012).

Therefore, the detritus food chain assumes an essential role within the soil, as it becomes the basis of the hypogean food web; in fact, numerous organisms such as isopods, certain myriapods, earthworms, springtails, many species of mites, and the larvae and adults of many insects feed on the vegetable and animal detritus that is deposited on the soil. For example, in the soil of a temperate forest—in which the contribution of litter each year can amount to 400 g/m^2—about 250 g/m^2 is ingested by earthworms and enchytraeids, 30–40 g/m^2 by mites, and 50–60 g/m^2 by springtails.

There are many extremely old groups of microarthropods in soils—including collembolans and mites—dating from the Devonian period (more than 350 million years ago). They are very useful as indicators of soil quality, as their biodiversity and density are influenced by numerous soil factors, particularly OM and water content, but also other factors such as pollution (Menta, 2012).

Earthworms are among the most important organisms in many of the soils of the world. The activity of earthworms produces a significant effect, not just on the structure, but also on the chemical composition of the soil, since a large part of the OM ingested by earthworms is returned to the soil in a form easily used by plants. While they are feeding, earthworms also ingest large quantities of mineral substances (minimally so in the case of the epigeic), which are then mixed with the OM they ingest and, after having been cemented with a little mucous protein, are expelled in small piles known as worm casts. In addition to being rich in nitrogen and other nutritive substances such as calcium, magnesium, and potassium, worm casts also contain a large quantity of nondigested bacteria that proliferate easily in this substratum and contribute to the humification and mineralization of OM (Zanella et al., 2001). In addition to their worm casts, earthworms contribute to increasing the amount of nitrogen present in the ground through the excretion of ammonia and urea, forms that are directly useable by plants; furthermore, a sizable quantity of nitrogen is returned to the soil on the death of animals, which have a 72% protein content (Dindal, 1990).

The continual burrowing activity of earthworms and other soil organisms contributes to the creation of spaces within the soil with a resulting rise in its porosity; the increase in the pores between the particles in turn enhances aerobic bacterial activity and the consequent demolition speed of organic substances. This bioturbation also has positive effects on water retention, percolation processes, and the development of the rhizosphere. The burrowing activity enables the soil to be mixed, thereby incorporating OM from the surface layers into the lower layers, while mineral substances are brought toward the surface (Menta, 2012).

The anthill is one of the most interesting and elaborate examples of the modification of the soil by cunicular burrowing organisms. It consists of a complex of chambers, generally constructed on several levels and linked together by tunnels and corridors. Some nests can reach a depth of more than 5 m and contain over 2000 chambers, some of which are set aside for the cultivation of fungi. The presence of channels and tunnels increases the porosity of the soil, aiding the penetration of air and water. In addition, a consequence of the movement of the fine material toward the surface by ants during the course of the construction and maintenance of the anthill is the creation of a surface layer with a fine particle size, which is more mineral than organic in nature (Bachelier et al., 1971).

Soil fauna—particularly mollusks and earthworms—affects the soil through the secretion of cutaneous mucous, which has a cementing effect on the particles in the ground, contributing to the stability and structure of the soil and making it less vulnerable to erosion processes. The animals' mucous secretions, feces (especially those of earthworms), and their own bodies (when they die) considerably influence the concentration of nutrients present in the soil—particularly potassium, phosphorous, and nitrogen—reducing the C/N ratio of the litter and facilitating decomposition (Menta, 2012).

The presence of roots is generally associated with a greater density of microorganisms in the nearby soil compared with soil devoid of roots; the term rhizosphere is used in a broader sense to refer to the portion of soil surrounding roots in which the microorganisms are influenced by their presence (Killham, 1994). The interaction between soil animals and plant roots may take a variety of forms that lead to benefits or repress the growth of the plant and often involve interactions with the microbial population of the soil. The dispersion of the inocula of mycorrhizal fungi by soil animals can have beneficial effects for the plants; this dispersion is particularly favored by burrowing organisms belonging to the mega- and mesofauna categories.

The hyphae of mycorrhizal fungi may comprise a significant proportion of the total microbial biomass in some soils and may become one of the most important sources of food for fungus-grazing animals such as springtails. Numerous soil animals feed directly on the roots of plants; these include a large number of springtail and myriapod species. It is still not entirely clear how much of the damage inflicted on plants can be attributed to the direct action caused by grazing on the roots or to the subsequent vulnerability of the roots to pathogens in the soil, especially fungi (Menta, 2012).

In summary, soil communities participate in most key soil functions, driving fundamental nutrient-cycling processes, regulating plant communities, degrading pollutants, and helping to stabilize soil structure. The ecosystem services provided by soil biota are thus one of the most powerful arguments for the conservation of edaphic biodiversity.

4.4.2   BIOLOGICAL INDICATORS TO ASSESS FOREST-SOIL QUALITY

The growing recognition of the problems deriving from soil degradation has led to the identification of soil fauna research as a priority for soil-quality assessment (Bongers, 1990, 1999; Pankhurst, 1997; Pankhurst et al., 1997; van Straalen, 1997).

Measurement of soil organisms meets many of the criteria for useful indicators of sustainable land management: (1) They respond sensitively to anthropogenic disturbances; (2) the area covered during their life cycle is representative of the site under examination; (3) their life histories permit insights into the soil ecological condition; (4) their abundance and diversity are well correlated with beneficial soil functions; (5) they are useful for elucidating ecosystem processes.

Recently, various authors have proposed new methods for soil-quality assessment based on soil fauna, using three levels of analysis: The simplest one is (1) the biomarker, which measures an individual's biochemical and physiological variability and its excretions; (2) bioindicators are organisms with specific ecological requirements, which serve as indicators of environmental change; and (3) the last type is community analysis, which has the greatest level of complexity and allows the totality of the information on zoocenosis and the relationships that characterize it to be assembled into one.

The obvious adaptations of soil fauna make edaphic organisms unable to leave the soil. This incapacity defines these organisms as being more sensitive to the variation in the physical and chemical parameters that can occur in the soil after tillage and other human activities. Moreover, the intricate relationships of soil invertebrates with their ecological niches in the soil, and the fact that many soil organisms live a rather sedentary life, serve as a good starting point for bioindicator changes in soil properties and the impact of human activities.

To be able to evaluate their role and function, it is important to use methodologies that highlight either the number of species present or the processes and roles they play in the soil environment. The key features for obtaining an assessment of soil disturbance are (1) species richness and diversity, (2) the relationship between species (rare vs. dominant species), (3) distribution of body sizes in the species analyzed, (4) classification of species' life-cycle attributes, (5) classification according to the ecophysiological preferences of the species analyzed, and (6) food chain structure.

The sample of the groups of organisms that can be used to assess soil ecosystems at the community level should meet the following requirements: (1) Their taxonomy must be well known and stable; (2) the main aspects of their natural history must be known; (3) their presence/absence should be easy to test and identify; (4) samples must contain individuals from a higher taxon level (genus, family, order) with a widespread diffusion both in geographic terms and in number of habitats; (5) they must contain individuals of lower taxa (species, subspecies) that are highly specialized and sensitive to environmental changes; (6) they must contain taxa with high ecological significance; (7) they must represent biodiversity models that reflect and represent other taxa, whether or not they are neighbors (Menta, 2012).

Three soil animal classes have been used as bioindicators: microfauna, mesofauna, and macrofauna. A selection of bioindicator and bioindex systems using these soil invertebrate groups to evaluate forest-soil quality (van Straalen, 2004) included

1. Soil microfauna (smaller than 0.02 mm): Comprised primarily free-living protozoa and Nematoda. In forest soils, the "nematode maturity index" (Bungers, 1990; Yeates and Bongers, 1999) uses nematodes classified on

a "colonizer–persister" scale to measure general response to stress (metals, acidification, eutrophication). The main problems associated with these indices are (a) systematic uncertainty, (b) spatial uncertainty due to the lack of definition of accurate models for the distribution of these organisms in the soil, (c) difficulty in achieving robust development indices without having a complete understanding of the systematic groups (species uncertainty), (d) the models and the data used that are processed in different countries (geographic uncertainty), and (e) the need for a reference site (control) in each study for a conclusive interpretation (Bongers, 1990; Wodarz et al., 1992; Ettema and Bongers, 1993; Korthals et al., 1996; Gupta and Yeats, 1997; Pankhurst et al., 1997).

2. Soil mesofauna (from 0.02 to 4 mm): Springtails (Acarida) are the most common soil arthropods, and other groups such as beetles (Carabidae), hymenoptera, and enchytraeids are present. Some of the most common biological indices are the "Acari/Collembola ratio," "oribatida/other Acari ratio," "biodiversity of oribatid Acari," and "Qualitá Biologica del Suolo (QBS)-ar and QBS Collembola" (Aoki et al., 1977; Bernini et al., 1995; van Straalen, 1997; Behan Pelletier, 1999; Knoepp et al., 2000; Parisi and Menta, 2008). Other mesofauna biological indices include the following:

   a. In forest soils, the "predatory mite maturity index" has been applied (Ruf, 1998) using mesostigmatid mites classified according to an r-K score to measure soil properties related to humus.

   b. The "arthropod acidity index" (Van Straalen and Verhoef, 1997; Van Straalen, 1998) using a classification of collembolans, oribatids, and isopods according to pH preference in order to estimate quantitative pH measures from the invertebrate community structure.

   c. The "oribatid mite life-history strategies" (Siepel, 1994, 1996) use classification of mites according to their reproductive and dispersal strategies to indicate the intensity of anthropogenic influence and successional events.

   d. The "life forms of Collembola" (Van Straalen et al., 1985; Faber, 1991) using a classification of collembolans according to morphological types reflecting their position in the soil profile to indicate profile buildup and ecological soil processes.

   e. The "dominance distribution of microarthropods" (Hagvar, 1994) uses lognormal distribution of numbers over species to obtain a general diagnostic of disturbance associated to heavy metals and acid rain in forest and grassland soils.

   f. The "biological index of soil quality (BQS)" (Parisi, 2001; Gardi, 2002) involves scoring systems that assign scores to groups of soil microarthropods to provide an indication of their biodiversity status.

Some of these methods are based on the general evaluation of microarthropods (Parisi, 2001), while others are based on the evaluation of a single taxon (Bernini et al., 1995; Iturrondobeitia et al., 1997; Paoletti, 1999; Paoletti and Hassal, 1999; Parisi, 2001).

The simplest approach is to select a single species as an indicator. Single species bioindicators will have reasonable specificity as long as the indicator species reacts in a specific way but will have a low resolution due to unavoidable fluctuations in density. Using single species could obscure interspecies relationships, while considering all the species in a system will not reveal effects on rare species, which may be more affected than abundant species (Cortet et al., 1999).

The application of these biodiversity or biological indicators is often limited by the difficulties in classifying microarthropods. The introduction of a simplified ecomorphological index, which does not require the classification of organisms to species level, allows these methodologies to be more widely applied:

3. Soil macrofauna (larger than 4 mm): Snails and slugs are very good bioindicators as they have a wide range, are easy to sample, and have a high tolerance and bioaccumulation capacity; earthworms are also a good bioindicator, as they migrate over short distances and are both resistant and sensitive. Other macrofauna biological indices are

    a. "Earthworm life-history strategies" have been used in forest soils (Bouché, 1977; Paoletti, 1999). Earthworms are classified according to the position they hold in the soil profile to measure aspects such as humus type, pH, and cultivation plowing.

    b. "Real model for earthworms" (Bouché, 1997) is based on an integrated database of various aspects related to the ecological and agronomic role of earthworms and has a very wide application.

    c. "Ant functional groups" (Andersen, 2002) are based on a classification of ants according to groups reflecting susceptibility to stress, with a wide application for evaluating nature restoration.

    d. "Diptera feeding groups" (Frouz, 1999) are based on a classification of dipteran larvae to measure the different types of organic materials in soils.

4.4.3 MICROBIOLOGICAL, BIOCHEMICAL, AND MOLECULAR INDICATORS TO ASSESS FOREST-SOIL QUALITY

While most of the chemical and physical property variables that are relevant to soil quality are well understood, measures of soil biological properties have so far been much more difficult to use as decision support tools for monitoring soil quality.

Descriptions of soil biological properties can range from single-parameter variables such as microbial biomass or respiration to multiparametric data describing biochemical profiles, measurements of enzyme activities, and molecular analyses of microbial communities. However, in contrast to the extreme complexity of microbial communities in the soil, the ideal soil microbiological and biochemical indicators for the determination of soil quality must be simple to measure, work equally well in all environments, and reliably reveal which problems exist and where.

It is unlikely that a sole ideal indicator can be defined with a single measure, due to the multitude of microbiological components and biochemical pathways. Therefore, an MDS is frequently applied (Carter et al., 1997). The basic indicators and the number of measures needed are still under discussion and depend on the aims of each investigation.

Bastida et al. (2008) indicate that given the complex nature of soils, it is important to select adequate indicators depending on the task; they also note the need to standardize indicator measurements. Further development of these tools should also help soil scientists to identify novel relationships and devise research to explore linkages between the biological, chemical, and physical properties of soils (Mele et al., 2008).

The star indicator for soil quality is SOM. In particular, since OM—or more specifically OC and the C cycle as a whole—can have an important effect on soil functioning, all the attributes linked to the soil C cycle are usually recommended as components in any MDS for soil-quality evaluation (Akeem, 2012). In addition, the visual output of the SOM analysis provides a rapid and intuitive means to examine covariance between variables, and with minimal training, it could be useful for assisting land managers to interpret multiparametric soil analyses.

According to a simplified scheme, SOM can be divided into two pools: nonliving and living. Nonliving SOM includes materials of different age and origin, which can be further divided into pools or fractions as a function of their turnover characteristics. For example, the humified fraction is more resistant to decay. The stability and longevity of this pool are a consequence of chemical structure and organomineral association. This pool of SOM influences different aspects of soil quality such as the fate of ionic and nonionic compounds, the increase in soil cation exchange capacity, and the long-term stability of microaggregates (Herrick and Wander, 1997). The interpretation guideline for the purpose of soil-quality assessment is that the higher the humified fraction of SOM, the higher its contribution to soil quality (Akeem, 2012).

The regulation of greenhouse gases is a largely overlooked function of soil quality in agroecosystems and can be calculated by using indicators such as gas flow or C sequestration (Liebig et al., 2001). This omission is surprising, especially in today's world where the soil is regarded as an important C sink and particularly since plants are CO_2 capturers, at a time when the greenhouse effect has become an issue of public concern (Bastida et al., 2008).

Regarding the living SOM pool, soil microbial communities are responsible for important physiological and metabolic processes of paramount interest for soil quality (Sessitsch et al., 2006). Soil microorganisms are mainly related to nutrient cycles, mineralization, humification, physical structure formation, degradation of contaminants, and soil fertility (Roldán et al., 1994; García et al., 2002).

Among all the SOM fractions, the most widely used indicator is microbial biomass C; another important microbial activity index for estimating soil quality is microbial respiration (Bastida et al., 2008). For this reason, national and international programs for monitoring soil quality currently include microbe biomass and respiration measurements but also extend to determining nitrogen mineralization, microbial diversity, and functional groups of soil fauna (Bloem, 2003). However, the structure, function, and relationships among soil microbial communities still represent a substantially uncharted territory that is intimately related to soil community and function.

From a biochemical soil approach, soil management practices can be monitored with other highly sensitive indicators such as enzymatic activity. The most widely used is dehydrogenase activity, an indicator of microbial activity linked to the oxidation of organic compounds and electron transport during microbial cell energy generation. Other commonly used enzymatic activities are related to C, P, N, and S cycles. However, only a few indices use enzymes for assessing soil quality, despite the fact that enzymes have been proposed as sensitive soil change indicators (Nannipieri et al., 1990; Bastida et al., 2008).

The first enzymatic indices may be the enzymatic activity number (EAN) proposed by Beck (1984), the biological index of fertility (BIF) of Stefanic et al. (1984), and the biochemical index of soil fertility (Koper and Piotrowska, 2003). As an example, Saviozzi et al. (2001) also used these indices and observed lower values for cultivated soils than for forest and native grasslands. Other examples were reported by Riffaldi et al. (2002) and indicated an increase in EAN and BIF in untilled management systems (natural grassland and orange groves) when compared to tilled systems in southeastern Sicily (Italy); Koper and Piotrowska (2003) established a biochemical soil fertility index for comparing the effect of organic and mineral fertilization. The biochemical index of soil fertility is

$$BISF = C_{org} + N_{total} + DH + P + PR + AM \qquad (4.1)$$

where

C_{org} is the OC content expressed as a percentage
N_{total} is the total nitrogen content expressed as a percentage
DH is the dehydrogenase activity, $cm^3 \ H_2 \ kg^{-1} \ 24 \ h^{-1}$
P is the phosphatase activity, μmol p-nitrophenylphosphate $g^{-1} \ h^{-1}$
PR is the protease activity, $\mu mol \ NH_4N \ kg^{-1} \ h^{-1}$
AM is the amylase activity, mg of decomposed starch h^{-1}

Until recently, one of the mean indices to take into account microbial diversity parameters was phospholipid fatty acids (PLFAs) devised by Puglisi et al. (2005), who established a soil alteration index. This methodology allows a quick and simple analysis and provides a profile of numerous fatty acids by gas chromatography, together with information on the abundance of microbial groups (fungi, bacteria Gram+ and Gram− and actinomycetes), but it provides little information on the alteration (and thus the quality or degradation) of a given soil, due to the scant number of parameters (only PLFAs) used, and one single indicator is not sufficient to evaluate the state of a soil.

However, PLFA as a sole technique is not sufficient to explain the processes and high-resolution microbial diversity of a soil. Subsequently, Puglisi et al. (2006) proposed three soil alteration indices, using various soil enzymatic activities to establish a soil degradation index for the assessment of agricultural practices, including crop density and the application of organic fertilizers in different locations:

1. The first index (AI 1) is expressed as a function of seven enzyme activities (arylsulfatase, β-glucosidase, phosphatase, urease, invertase, dehydrogenase, and phenoloxidase).
2. The second index (AI 2) uses β-glucosidase, phosphatase, urease, and invertase activities for its calculation.

3. The third index (AI 3) combines only three enzyme activities (β-glucosidase, phosphatase, and urease) and was established and validated from values obtained from the literature. This validation process could be considered as a valid option for endowing an index with greater spatial significance, although one problem is that the methods used must be identical, as minor methodological changes could lead to widely varying results.

Armas et al. (2007) determined a biological quality index for volcanic andisols and aridisols. This index presented the relationship among total C, different enzyme activities, and hot water-soluble C. The ratio between the predicted values for this model and the real total C values could be considered as a biological soil-quality index (Bastida et al., 2008):

$$\text{Total C} = -2.924 + 0.037 \times \text{extC} - 0.096 \times \text{cellulose}$$

$$+\, 0.081 \times \text{dehydrogenase} + 0.009 \times \text{respiration} \tag{4.2}$$

where
extC (hot water-soluble C) is expressed in g C m^{-1}
Cellulose is expressed in µmol glucose m^{-2} h^{-1}
Dehydrogenase is expressed in µmol INTF m^{-2}
Respiration is expressed in mg CO_2-C m^{-2} h^{-1}

More or less sensitivity is needed depending on the objectives. In fact, some microbial activity parameters may suffer from too much sensitivity, such as the case of ATP. The adenosine triphosphate molecule stores cellular energy and can suffer climate and temperature-dependent variations (Jorgensen and Raubuch, 2003; Bastida et al., 2006).

Some authors have proposed biological soil-quality indices for polluted soils affected by different contaminants. Bécaert et al. (2006) proposed a similar index, entitled the relative soil stability index (RSSI), to evaluate the capacity of a soil to recover after contamination as a function of enzymatic activity. In this case, they included an overlooked variable: time (Bastida et al., 2008).

New methodologies based on rapid and clean nucleic acids and protein extraction from soil open the door to the use of these molecular approaches for soil-quality assessment (Persoh et al., 2008). Although there is an increasing number of publications based on molecular biology, only a few indices that exploit molecular methodologies (genomic, transcriptomic, proteomic, and the so-called omics) have been applied in attempts to evaluate soil quality (Roldán et al., 1994; García et al., 2002).

However, the development of genomic, transcriptomic, or proteomic methodologies could have an importance for the evaluation of soil quality, not only from the standpoint of diversity, but also from a functional aspect. These methods can provide information on the role of specific microorganisms and their enzymes in key processes related to soil functionality (Bastida et al., 2012). These new techniques in molecular ecology therefore further the information concerning microbial diversity and function.

The analysis of metagenomic libraries offers the possibility of determining the gene content of noncultivable bacteria that could be the key drivers of processes related to soil quality, thus providing a better understanding of global microbial ecology (Schmeisser et al., 2007).

In the same vein, microbial diagnostic microarrays also represent a powerful tool for the parallel identification of many microorganisms (Sessitsch et al., 2006). Functional microarrays target genes that encode specific functions and can offer valuable information about functional soil characteristics. For example, functional gene arrays target genes that encode enzymes, which play a key role in various ecological processes such as nitrification, denitrification, nitrogen fixation, methane oxidation, sulfate reduction, and degradation of pollutants (Sessitsch et al., 2006).

Since the incorporation of the labeled substrate into microbial DNA or RNA implies that the populations are active under the tested conditions (Radajewski et al., 2003), the subsequent analysis of labeled biomarkers of subpopulations with stable-isotope probing (SIP) (DNA-SIP, RNA-SIP, PLFA-SIP) reveals related phylogenetic and functional information about the soil organisms that metabolize specific compounds (Neufeld et al., 2007).

Microarrays, real-time PCR, SIP linked to nucleic acids, and other molecular methods offer several advantages but also several limitations that have been widely developed in the review by Saleh-Lakha et al. (2005).

In summary, these methodologies attempt to analyze and study genes or transcripts in order to detect potential functions of microorganisms in the soil under different conditions. However, where does the soil microbial function reside?

Proteomics could provide new insights in this direction. Proteins provide more straightforward information about microbial activity in soils than real-time PCR functional genes or even their RNA transcripts (Wilmes and Bond, 2006; Benndorf et al., 2007).

Linking SIP with metaproteomic analysis could provide additional information, not only on proteins or enzymes that disappear due to adverse phenomena in the soil, but also on the precise role of proteins in specific soil processes and on the origin of these proteins. The use of SIP binding methods and metaproteomics could help define and quantify soil quality but in a more specific sense, taking into account the processes involving microorganisms and their proteins and their ecosystem function (Bastida et al., 2008).

4.4.4   Chemical Indicators to Assess Forest-Soil Quality

It is often difficult to clearly separate soil functions into chemical, physical, and biological processes due to the dynamic and interactive nature of these processes (Schoenholtz et al., 2000). This interconnection is particularly significant between chemical and biological indicators of soil quality. It is therefore necessary to identify a set of attributes that soils must possess in order to perform their functions and then translate these attributes into first- or second-level measurable soil properties or processes.

Consequently, there is rarely a one-to-one relationship between function and indicator; more likely, a given function is supported by a number of soil attributes, while any given soil property or process may be relevant to several soil attributes and/or soil functions simultaneously (Harris et al., 1996; Burger and Kelting, 1999; Schoenholtz et al., 2000). In this regard, Bastida et al. (2008) report that physical and chemical indicators have been extensively used for the design of soil-quality indices.

Although several soil chemical indicators are similar for agricultural and forest soils, there are nevertheless significant differences between agriculture and forestry as far as their use and assessment are concerned. As pointed out by Powers et al. (1998) and Schoenholtz et al. (2000), many analytical soil-testing methods frequently used in agriculture have proven to be only marginally useful in predicting forest growth.

The following is a summary of the mean soil chemical indicators used to characterize the soil quality in grassland and forest soils (Schoenholtz et al., 2000):

1. One of the oldest indicators that must be included is pH, as a basic indicator of soil quality and soil chemical "health." Of the chemical parameters of nutrient availability, pH has specific scoring functions that can be used for plant productivity and/or environmental components of soil quality, assessing pedotransfer functions for rooting depth and soil productivity attributes and assessing aggregate stability to evaluate a soil's resistance to erosion (Kiniry et al., 1983; Gale et al., 1991; Doran, 1994; Karlen and Stott, 1994; Larson and Pierce, 1994; Reganold and Palmer, 1995; Harris et al., 1996; Aune and Lal, 1997).

2. Soil OC status (Doran, 1994; Reganold and Palmer, 1995; Harris et al., 1996) is an important characteristic proposed as a first-order chemical indicator. It is also used as one of the chemical parameters of nutrient availability with specific scoring functions for plant productivity and/or environmental components of biological soil quality.

3. Regarding the indicators associated to nitrogen nutrient availability, the various soil N measures such as "total N," "organic N," and "mineral N" or "mineralizable N" (Doran, 1994; Reganold and Palmer, 1995; Powers et al., 1998) are soil chemical features that can be included as basic indicators of soil quality and have been proposed as good indices for the soil's nutrient-supplying capacity. Moreover, "extractable NH_4" and "NO_3-N" are chemical parameters of nutrient availability with specific scoring functions used for plant productivity and/or environmental components of soil quality (Harris et al., 1996).

4. Regarding the indicators associated to phosphorous nutrient availability such as "extractable P" (Burger et al., 1994; Reganold and Palmer, 1995) and "Bray P" (Harris et al., 1996; Aune and Lal, 1997), these include some nutrient-availability chemical parameters with specific scoring functions that can be used for plant productivity and/or environmental components of soil quality.

5. The CEC has also been suggested as a first-order chemical indicator and is calculated through the pedotransfer function using OC and clay content (Karlen and Stott, 1994; Larson and Pierce, 1994; Reganold and Palmer, 1995).

6. "Exchangeable K" and "extractable K, Ca, Mg" are soil chemical characteristics that can be included as basic indicators of soil quality, and the K
measure is one of the nutrient-availability chemical parameters with specific scoring functions to be used for plant productivity and/or environmental components of soil quality (Aune and Lal, 1997; Harris et al., 1996;
Reganold and Palmer, 1995).

7. "Salinity" could also be suggested as a first-order chemical indicator for its
inclusion as part of the MDS for agricultural and forest soils and as a basic
indicator of soil quality. It is used in the pedotransfer function for soil productivity attributes, to account for the salinity that reduces the productive
capacity of soils (Kiniry et al., 1983; Burger et al., 1994; Doran, 1994;
Karlen and Stott, 1994; Larson and Pierce, 1994).

Regarding the soil chemical indicators used in forest-soil-quality assessment, the productive indices (PI) should contain some measure of nutrient status, with the exact
chemical parameter depending on the external stressor or anthropogenic impact.
Possible critical soil chemical indicators may be OM, available N, soil P, and soil acidity (pH, base depletion, Al toxicity) (Henderson et al., 1990; Schoenholtz et al., 2000).

In connection with the soil biological indicators described, the inclusion of SOM
as a soil chemical indicator can measure several soil functions simultaneously
(Burger and Kelting, 1999), including OM decomposition and N mineralization as
functional components of soil productivity. Moreover, potentially mineralizable N
(anaerobic incubations) has also been proposed as a viable indicator of soil nutrient
supply based on the positive correlation with site index and foliar N and with total
organic C and N and its use as an index for microbial biomass (Powers et al., 1998).

4.4.5 PHYSICAL INDICATORS TO ASSESS FOREST-SOIL QUALITY

Regarding the soil's physical properties as indicators of forest-soil quality, productive forest soils have attributes that are, in part, a function of soil physical properties
and processes, as they (1) promote root growth; (2) accept, hold, and supply water; (3)
hold, supply, and recycle mineral nutrients; (4) promote optimum gas exchange; (5)
promote biological activity; and (6) accept, hold, and release C (Burger and Kelting,
1999). All of these factors will determine the extent to which each soil physical property or process is useful for measuring soil quality and monitoring the maintenance
of soil quality over time.

Some soil physical properties are static in time, and some are dynamic over
varying time scales. Some are resistant to change by forest management practices,
while some can be easily changed in positive and negative ways. If changed, some
properties and processes will recover at varying rates, while others are irreversible
(Schoenholtz et al., 2000). A summary of the mean soil physical properties used
to characterize soil quality in grassland and forest soils (Schoenholtz et al., 2000)
included the following:

1. "Soil texture" measures the percentage of sand, silt, and clay to evaluate the
soil's capacity for water and nutrient retention and transport (Doran, 1994).

2. "Soil depth" uses the measure of the soil's thickness (cm) to evaluate total nutrient, water, and oxygen availability (Larson and Pierce, 1991; Arshad and Coen, 1992; Doran, 1994).
3. "Soil bulk density" is a measure of core sampling density to assess root growth and water's rate of movement in the soil (Larson and Pierce, 1991; Arshad and Coen, 1992; Doran, 1994; Kay and Grant, 1996).
4. "Available water-holding capacity (WHC)" is used for the evaluation of water availability for plants and erosivity (Larson and Pierce, 1991; Arshad and Coen, 1992; Doran, 1994; Kay and Grant, 1996).
5. "Saturated hydraulic conductivity" uses the measure of water flow in a soil column to assess the water/air balance and soil hydrology regulation (Larson and Pierce, 1991; Arshad and Coen, 1992).
6. "Soil strength" measures the soil's resistance to penetration in order to evaluate root growth (Powers et al., 1998; Burger and Kelting, 1999).
7. "Porosity" uses the percentage of in-soil air volume to assess the water/air balance, water retention, and root growth (Powers et al., 1998).
8. "Aggregate stability and size distribution" uses a wet-sieving method to evaluate root growth and air/water balance (Arshad and Coen, 1992; Kay and Grant, 1996).
9. "Least limiting water range" uses water retention curves and the penetration resistance measure to assess water/air balance and root growth (Arshad and Coen, 1992; da Silva et al., 1994; Kay and Grant, 1996; Burger and Kelting, 1999).
10. "Leaching potential" uses a model to evaluate the transport, transformation, and attenuation of applied chemicals (Petach et al., 1991; Wagenet and Hutson, 1997).
11. "Erosion potential" uses models such as water erosion prediction project (WEPP) (Nearing et al., 1989) and site erosion potential (SEP) (Timlin et al., 1986) to assess the available soil, water, nutrient, root growth, and environmental concerns associated to soil erosion (Wagenet and Hutson, 1997).

4.5 MULTIPARAMETRIC INDICES TO ASSESS SOIL QUALITY

With the aim of developing practical measures for soil quality, integrative approaches are now being explored to sift out the interrelationships between several types of variables (Mele et al., 2008). In order to obtain a complete picture of soil quality, it is necessary to include different kinds of parameters (physical, chemical, and biological) (Frankenberger and Dick, 1983; Nannipieri et al., 1990; Dick, 1994; Gelsomino et al., 2006, Bastida et al., 2008). The reason behind the need to include a wider range of indicators is that some factors may affect some indicators but not others, and this may influence the accuracy of the picture of the soil quality.

The construction of soil-quality indices has been the subject of a large number of studies in the soil-science literature. However, the procedure currently used, which involves combining individual soil-quality scores into a soil-quality index, is the most controversial (Andrews et al., 2004), and no generally accepted criteria have been established for translating a set of individual soil attributes into simple soil-quality indices (Hailu and Chambers, 2012). The standard approach to generating an index to assess soil quality consists of the following three steps:

1. The first step is to choose an MDS of physical, chemical, and biological soil-quality variables (Larson and Pierce, 1991), based on either expert opinion or statistical data reduction methods such as PCA.
2. In the second step, the variables in the MDS are transformed into 0–1 scores, using either linear or nonlinear transformation functions.
3. In the final step, the indicator scores are combined into a soil-quality index using different integration techniques, including simple addition (Andrews and Caroll, 2001b), weighted addition (Harris et al., 1996), or decision support systems employing min-max objective functions (Yakowitz et al., 1993).

Both the second and the third steps are ad hoc. The transformation of individual soil-quality variables or indicators into 0–1 scores does not recognize that the impact of a soil-quality variable on outcomes (e.g., crop yield) might depend on the level of other soil variables or other production inputs.

4.5.1 MULTIPARAMETRIC INDICES INCLUDING PHYSICAL, CHEMICAL, AND BIOLOGICAL PARAMETERS

Bastida et al. (2008) proposed a description of the past, present, and future of soil-quality indices, which provides a very appropriate perspective for understanding the evolution of these important instruments and for selecting some of the most relevant. They suggest that the first multiparametric index for soil quality was probably established by Karlen et al. (1994a), who based this index on a framework established by Karlen and Stott (1994). The main drawback of their index is the fact that the weighting is subjective and does not depend on mathematical or statistical methods. The formula of the index uses selected soil parameters that are then weighted and integrated according to the following expression:

$$\text{Soil Quality} = q_{we}\,(\text{wt}) + q_{wma}\,(\text{wt}) + q_{rd}\,(\text{wt}) + q_{fqp}\,(\text{wt}) \qquad (4.3)$$

where
  q_{we} is the rating for the soil's ability to accommodate water entry
  q_{wma} is the rating for the soil's ability to facilitate water transfer and absorption
  q_{rd} is the rating for the soil's ability to resist degradation
  q_{fqp} is the rating for the soil's ability to sustain plant growth
  wt is the numerical weight for each soil function

Burguer and Kelting (1999) presented a soil-quality index for pinewoods using various soil functions and a method similar to that of Karlen et al. (1994a,b). The resulting index was suitable for the evaluation of forest-soil sustainability when subjected to different management systems. These authors proposed a boundary above which certain practices are sustainable and also discussed several spatial scale considerations of importance for obtaining a larger-scale functioning model, despite the fact that this factor has not been taken into account in many indices. The index was later applied by Kelting et al. (1999) to reflect the effect of different forest management practices. The information rendered was then used to help choose the most suitable index for sustainable forest-soil development in South Carolina (United States) (Bastida et al., 2008).

Andrews et al. (2002b) proposed a method that has been widely used by many authors (Sharma et al., 2005; Rezaei et al., 2006; Bastida et al., 2008). For the Andrews et al. (2002b) index, the indicators selected by factor analysis were bulk density, Zn, water-stable aggregates, pH, electrical conductivity, and SOM content; electrical conductivity and OM were the indicators assigned the highest importance in SQI. The equation proposed is

$$SQI = \sum_{i=1}^{n} 0.61 \times S_{SOMi} + 0.61 \times S_{EGi} + 0.16 \times S_{pHi} + 0.16 \times S_{WSAi}$$

$$+ 0.15 \times S_{zni} + 0.09 \times S_{BDi} \qquad (4.4)$$

where S is the score for the subscripted variable and the coefficients are the weighting factors derived from the PCA.

4.5.2 Physical–Chemical Indices to Assess Soil Quality

Foresters often integrate the analysis and knowledge of physical–chemical properties of soils to assess the capacity of sites to support productive forests, such as

1. The proposal by Pang et al. (2006) for the determination of an integrated fertility index (IFI) based on chemical, physical, and physical–chemical parameters for the study of soil-quality variations in Chinese forests.
2. Mohanty et al. (2007) applied multiple regressions and established a physical–chemical soil-quality index based on four parameters: (1) real density, (2) OM content, (3) resistance to root penetration, and (4) aggregate density.
3. Another example regarding future physical–chemical indices for soil quality is the use of spectroscopy as an inexpensive, noninvasive methodology for assessing soil quality. Shepherd and Walsh (2002) established a scheme for the development and use of soil spectral libraries for the rapid and non-destructive estimation of soil properties based on the analysis of diffuse reflectance spectroscopy and proposed r^2 regressions to validate several chemical and physical–chemical parameters. These authors concluded that the spectral library approach opens up new possibilities for soil-quality modeling.
4. Vagen et al. (2006) tested the potential of near-infrared spectroscopy as a tool for predicting and mapping soil physical–chemical properties in the highlands of Madagascar. Calibration models were developed for soil OC, total nitrogen, electrical conductivity, and clay contents in an area with a range of land uses and landscape forms.

5. Awiti et al. (2008) established a soil condition classification (good, average, poor) using infrared spectroscopy along a tropical-cropland chronosequence in sub-Saharan Africa. The study concludes that reflectance spectroscopy is rapid and offers the possibility for major efficiency and cost saving, allowing spectral case definition to define poor or degraded soils and enabling better targeting of management interventions.

4.5.3 Multiparametric Indices Based on Mathematical Models to Assess Soil Quality

Andrews et al. (2004) established a computer-based mathematical model (SMAF, soil management assessment framework). SMAF represents an additive nonlinear standardization tool for evaluating a soil's quality from its functions: (1) microorganism biodiversity and habitat, (2) filtration of contaminants, (3) nutrient cycling, (4) physical stability, (5) resistance to degradation and resilience, and (6) water relations (Bastida et al., 2008).

The SMAF tool was subsequently used by Cambardella et al. (2004) and Wienhold et al. (2006) and is based on food production and nutrient cycling as quality functions to objectively evaluate the effect of agricultural management practices on soil quality.

Erkossa et al. (2007) used the Andrews et al. (2004) methodology and various scoring functions in their attempt to evaluate soil quality in terms of different parameters. The purpose of this work was to compare the effects of four land preparation methods and their influence on soil quality: (1) broad bed and furrows, (2) ridge and furrow, (3) green manure, and (4) reduced tillage. However, they did not observe any significant differences in soil-quality values among the different treatments.

A trigonometric approach based on three subindices (nutritional, microbiological, and crop related) was used by Kang et al. (2005) to establish a sustainability index in soils under wheat amended with manures in Punjab (India), noting that the quality increased with amendment. Bastida et al. (2008) point out that this index used a wide variety of chemical, physical, and biological parameters.

Agricultural multiparametric quality indices outnumber nonagricultural ones. Trasar-Cepeda et al. (1998) chose a series of soils covered by a climax vegetation in Galicia (Spain) and established an equation to define total nitrogen from several microbial parameters (microbial biomass C, mineralized N, phosphatase, β-glucosidase, and urease). The resulting equation led them to report the presence of a balance between total N and several biochemical parameters. The equations are as follows:

$$\text{Total N} = (0.38 \times 10^{-3}) \text{ microbial biomass C} + (1.4 \times 10^{-3})$$
$$\text{mineralized N} + (13.6 \times 10^{-3}) \text{ phosphatase} + (8.9 \times 10^{-3})$$
$$- \text{glucosidase} + (1.6 \times 10^{-3} \times \text{urease}) \tag{4.5}$$

where
Total N is expressed as a percentage
Microbial biomass C and mineralized N are expressed in (mg kg^{-1})
The enzyme activities are expressed in μmol of liberated product g^{-1} h^{-1}

The model aims to approach the ideal state of soil quality in climax soils. However, in most of the developed world, these climax soils no longer exist. The validity and applicability of this ratio for soils contaminated by heavy metals, mine soils, and arable soils were demonstrated by Leirós et al. (1999).

Among the different statistical tools applied in the design of multiparametric indices for soil quality, an increasing number of studies have used artificial neural networks (ANNs) to probe complex data sets. As an example of how ANN can be used, we provide an example of the analysis of soils from two different regions of Southeast Australia using (1) Kohonen self-organizing maps, (2) data sets containing biochemical signatures of microbial communities determined by PLFA analysis, (3) genetic signatures obtained by terminal restriction fragment length polymorphisms (TRFLP), and (4) a range of single-parameter soil chemical, physical, and biological variables (Mele et al., 2008).

However, some of these approaches have limitations; for example, Gil-Sotres et al. (2003) criticized the work of De la Paz-Jiménez et al. (2002), who obtained an equation by multiple regression analysis relating OC to different enzymatic activities in La Paz (Argentina). The critique was based on the consideration that the combination of properties and experimental design was not adequate to design a soil-quality index for the purpose of evaluating the influence of soil use and management.

Regarding other multiparametric indices for soil quality based on statistical tools, Bastida et al. (2008) also highlight the design of Zornoza et al. (2007), which establishes two multilinear regression equation systems with the idea of evaluating soil environmental quality under natural vegetation with minimum human disturbance, working in semiarid conditions. The first established the affinity between nitrogen and different enzyme activities and physical–chemical indicators and was validated for mollisols, while the second defined soil organic C using similar indicators and was validated for entisols. The equations are as follows:

$$N_{Kjeldahl} = 0.448 \times P_{available} + 0.017 \times WHC + 0.410 \times (\text{phosphatase activity})$$
$$- 0.567 \times (\text{urease activity}) + 0.001 \times \text{microbial biomass C}$$
$$+ 0.410 \times \beta\text{-glucosidase} - 0.980 \tag{4.6}$$

$$SOC = 4.247 \times P_{available} + 8.183 \times \beta\text{-glucosidase} - 7.949 \times \text{urease} + 17.333 \tag{4.7}$$

where
$N_{Kjeldahl}$ is expressed in g kg^{-1}
$P_{available}$ is expressed in mg kg^{-1}
WHC is expressed in %
Phosphatase activity is expressed in μmol p-nitrophenylphosphate g^{-1} h^{-1}
Microbial biomass C is expressed in mg kg^{-1}
Urease activity is expressed in μmol NH^{+4} g^{-1} h^{-1}
β-glucosidase is expressed in μmol p-nitrophenylphosphate g^{-1} h^{-1}
SOC (soil organic carbon) is expressed in g kg^{-1}

Finally, Hailu and Chambers (2012) proposed a multiparametric index based on economic models to assess soil quality. They used the Luenberger productivity indicator, which is defined by differences in the values of the directional distance function. They introduced a distance function defined on the technology set, T, where

$$T = \{(x,y): \text{input} \times R_+^N \text{ can produce output } y \in R_+^M\} \tag{4.8}$$

and

$$\vec{D}(x,y;g_x,g_y) = \sup\{\beta : (x - \beta g_x, y + \beta g_y) \in T\} \tag{4.9}$$

is the directional distance function defined on T for the direction vector (g_x, g_y).

Defining this function for time period t, we obtain

$$\vec{D}^t(x^t,y^t;g_x,g_y) = \sup_\beta \{\beta : (x^t - \beta g_x, y^t + \beta g_y) \in T^t\} \tag{4.10}$$

Chambers and Pope (1996) and Chambers et al. (1996) previously defined the Luenberger productivity indicator and provided the formula.

This indicator is the arithmetic average of the change in productivity measured by the technology at time $t+1$ (the first two terms) and the change in productivity measured by the technology at time t (the last two terms).

4.6 PRACTICAL CASES: MODELS FOR THE INTEGRATION OF SOIL-QUALITY INDICATORS IN LANDSCAPE PLANNING PROJECTS

4.6.1 MODEL OF INTEGRATION OF QUALITATIVE AND QUANTITATIVE SOIL AND ENVIRONMENTAL INDICATORS: AN EXAMPLE OF THE ASSESSMENT OF THE ENVIRONMENTAL QUALITY OF A WHOLE COUNTRY (SPAIN)

4.6.1.1 Introduction

García-Montero et al. (2010a,b) proposed a geographic information systems (GIS) model using a qualitative approach to integrate SQI with other environmental indicators into a final map of environmental quality of Spain (1:500,000). This model was described in detail in an article published in the journal Environmental Impact Assessment Review. The map of the environmental quality of Spain was used in two strategic environmental assessments (SEAs) for two national transport infrastructure plans.

In this model, one vector—from a total of 102,240 different vectors with 12 components of environmental quality—was assigned to each one of the 50 million 1 ha grid squares (pixels) for Spain. Each vector was obtained by integrating qualitative and normalized values associated to four SQI and eight environmental indicators. The final classification of the 102,240 different vectors (based on the vector modulus) provided a raster map of Spain with five classes of territorial environmental quality. In this model, the GIS raster operations were able to deal with enormous amounts of information with no difficulty at all.

Finally, a check against the environmental quality map showed that these results had a good fit with reality (in relation to the scale and level of detail used). This checking process was based on a statistical analysis that showed that there was a very high frequency of 1 ha grid squares (pixels) with high environmental quality values associated to Spain's protected natural areas and that this was significantly greater than in the rest of the Spanish territory ($p < 0.0001$).

In summary, the GIS raster model developed to integrate SQI with other environmental indicators proved to be a simple and effective tool, which provided a useful environmental quality assessment for a large territory, based on panels of experts and on objective GIS calculations, and focused on SEA and landscape planning procedures.

4.6.1.2 Methodological Basis

The proposed model was used to compare territorial units (Ramos, 1979) and was combined with a multicriteria method to assess the quality of these territorial units (Norris and Farrar, 2001).

The map projections, GIS and mathematical software, and scale and level of detail used are described in García-Montero et al. (2010a,b). We analyzed enormous amounts of information, which made it impossible to apply vector-type GIS operations, and we thus chose to use a GIS raster-type model. The GIS models were based on logical operations of reclassifying and combining the raster layers, and the numerical vector operations were based on the vector module. Martínez-Falero and González (1995) and Tran et al. (2006) reported that the vector module is a suitable method for integrating multiple indicators into a single index for practical operating reasons in planning procedures.

We selected as the main valuation criteria "the conservation of biodiversity and the preservation of the environment," and we used Ramos' methodology (1979) and a GIS model proposed by Mancebo et al. (2005) and García-Montero et al. (2010b) (large territory integrated environmental (LATINO) model). These methodologies are based on models that compare the territorial units in relation to each other, based on the attributes or natural variables in each one.

The precision threshold was established at 100 m root-mean-square (RMS) (equivalent to a scale of 1:500,000).

Step 1: Valuation of SQI and environmental indicators (12 variables)

We looked for different evaluation criteria to assess SQI and environmental indicators, based on five existing Spanish digital maps on a national scale. Moreover, we consulted some panels of experts in order to obtain a complementary set of valuation qualities. Finally, we generated—either directly or by deduction—a set of 12 environmental quality indicators (including four SQI) representing 12 raster layers.

Regarding the four SQI, we assessed the productive capacity, biodiversity, naturalness, and uniqueness of the soils in Spain, using the information associated to the soil map (FAO, 2000), using the FAO's hierarchical classifications of soil taxonomy, and with a panel of five soil-science experts from Madrid.

We considered that with 12 variables, it was possible to assess the environmental quality of the Spanish territory. However, this model gave us an open system that allowed continuous incorporation of new quality scores.

Step 2: Objective assessment of territorial singularity

In order to safeguard biodiversity, we assessed the territorial singularity of the different categories or classes in the habitats, Corine Land Cover, landscape, and soil maps. This was done using an objective classification of their units calculated with the GIS. The following index of singularity was applied (Ramos, 1979; MMA, 2000):

$$S = Ln\left(\left(1 - \left(\frac{\text{Max} - x}{\text{Max} - \text{Min}}\right)\right) \times 100 + 1\right) \tag{4.11}$$

where
    S is the territorial singularity index
    Max is the ha of the map's largest category
    Min is the ha of the map's smallest category
    x is the ha of the map category being evaluated

Singularity was assessed on a logarithmic scale. This transformation made it possible to maximize the value of the categories with smaller areas and also to obtain a scale with fewer units. Thus, the category with the greatest surface area was awarded the lowest singularity value (0), and the category with the least surface area was awarded the maximum value (4.62). This continuous scale was then transformed into a discrete or qualitative scale of five classes, which were obtained by rounding each decimal value up to the next whole number. We thus obtained a higher singularity value for the least represented classes in the territory in order to safeguard biodiversity.

Step 3: Normalization of the 12 variables

The 12 variables were then normalized to avoid overlapping during their subsequent integration into the model. Normalization consisted of changing the original valuation scale for each variable, which was transformed into a common final continuous scale from 0 to 1 for all the variables. This was done by means of an equation applied to the original discrete or qualitative scales and another equation applied to the original continuous scales. The following formula was used to convert the discrete or qualitative scales into a continuous scale from 0 to 1:

$$Xn = \left(\frac{x - 0.5}{\text{Max}}\right) \tag{4.12}$$

The following equation was used to transform a continuous scale into another normalized continuous scale from 0 to 1:

$$Xn = \left(\frac{x - \text{Min}}{\text{Max} - \text{Min}}\right) \tag{4.13}$$

Step 4: Integration of the 12 variables into the model

The 12 normalized raster variables were integrated using GIS merging operations. Each pixel of 1 ha of territory was assigned a vector with the 12 natural variables valued. We obtained n vectors distributed among the 50 million 1 ha pixels in Spain.

The next step was to order the n vectors using the modulus or Euclidean distance, to assign a synthetic value of theoretical biodiversity and environmental quality. The vector was used to order the n vectors obtained based on their components, and

$$v = \sqrt{v_1^2 + \cdots + v_i^2 + \cdots + v_{12}^2} \quad i = 1 \ldots 12 \tag{4.14}$$

where

v is the vector modulus

i is a vector component

A total of 102,240 different vectors were obtained with 12 components, assigned to each of the 50 million 1 ha grid squares for Spain. Then the values obtained for each of the n Euclidean distances were normalized into five equivalent classes, corresponding to the five types of theoretical biodiversity and environmental quality (1 to 5). This normalized classification was obtained by applying the following formula:

$$\text{Biodiversity quality class} = ((V - V_{\min})/(V_{\max} - V_{\min})) * (5 + 0.5) \tag{4.15}$$

where

V is the vector modulus of each of the n vectors obtained

$V_{\min}$ is the minimum vector modulus obtained

$V_{\max}$ is the maximum vector modulus obtained

The very low biodiversity and environmental quality class is obtained when $0.5 \ll 1.5$

Low-quality class is obtained when $1.5 \ll 2.5$

Moderate-quality class is obtained when $2.5 \ll 3.5$

High-quality class is obtained when $3.5 \ll 4.5$

Very-high-quality class is obtained when $4.5 \ll 5.5$

García-Montero et al. (2010b) proposed 100 classes of environmental quality in the Spanish territory based on the LATINO model. However, in the described study, the use of five classes of biodiversity and environmental quality was sufficient for this classification, as it clearly distinguishes extreme cases of high and low environmental quality in a territory on a nation scale.

4.6.2 MODEL OF INTEGRATION OF QUANTITATIVE SOIL-QUALITY INDICATORS: AN EXAMPLE OF THE GENERATION OF A SOIL FERTILITY INDEX IN CARIBBEAN AGRICULTURAL AND FOREST AREAS

4.6.2.1 Introduction

Alexis et al. (2010) proposed a soil fertility index model based on a GIS tool, which integrated statistical multivariate methods and soil parameters, including heavy metal content, into models of land planning for agricultural and forestry development in a rural Caribbean area. This quantitative model was described in detail in an article published in the journal *Agroforestry Systems.*

In the Jaragua-Bahoruco-Enriquillo Biosphere Reserve, located on the southern border between the Dominican Republic and Haiti, there are depressed rural areas with soils with a high content of cadmium and other heavy metals that originate naturally in the geological substrate. Soil data and an inventory of coffee and kidney bean plantations were used to design a GIS tool to generate a soil fertility index in the study area.

This GIS tool was based on open-source raster models, using combination and reclassification operations based on the maps, geostatistical methods (kriging), statistical analyses external to the GIS, and a cartography limiting and excluding particular factors for crops (including heavy metal soil content). The GIS tool discriminated extreme soil fertility situations associated to sustainable agricultural and forest planning in contaminated rural areas of the Caribbean.

The soil and its fertility and contamination constitute highly significant elements for sustainable agricultural and forest planning processes. It is therefore advisable to incorporate fertility indices into agricultural and forest planning models. Vagen et al. (2006) state that when developing fertility indices, the challenge is generally to integrate a series of soil characteristics within a single-value index. Many studies apply a points system based on agronomic values of reference for a range of soil characteristics, and in some cases, certain soil characteristics are qualified with more weight than others (Karlen et al., 1998; Lal, 1998). However, one of the greatest problems posed by tropical soils is that there are no robust reference values and the valuation of the weight of soil characteristics or the assignment of weighting values is not submitted to any scientific criterion (Sánchez et al., 2003).

The region in the study has some extremely poor rural areas. Hernández et al. (2007) and Alexis (2008) described the characteristics of the study area. We selected the agroecosystems that were home to two of the most representative crops in the study area: coffee, which has considerable importance in the region's overall economy, and kidney beans (*Phaseolus vulgaris* L.), grown by families as a subsistence crop for their nutritional qualities.

4.6.2.2   Methodological Basis

The map projections, GIS and mathematical software, and scale and level of detail used are described in Alexis et al. (2010).

The soil maps and GIS models were based on the prior study of 80 samples of the surface soil layer (0–20 cm in depth) analyzed by Hernández et al. (2007), randomly selected within each of the ecosystems studied. These authors analyzed pH in water, OM, the main elements related to fertility (N, P_2O_5, K), and available levels of Ca, Mg, Mn, Fe, and Zn. Total levels of Cd were also determined. An inventory was made of the crops and geographic features in the study area. The precision threshold was established at 100 m RMS (RMS error), and the minimum cartographic unit or pixel selected was 1 ha. A pixel with a resolution of 100 m made it possible to generate a GIS soil tool of moderate accuracy, which would enable the knowledge acquired through a limited number of soil analyses to be integrated in the land planning procedures in real time.

One of the advantages of using GIS raster models is that GIS processes can be created using mathematical operations with numerical matrices, without the need to visualize each of the intermediate steps on the maps. Moreover, the scale of the raster layers always remains associated to the pixel size, so the GIS zoom display tool can be used without modifying the scale or distorting the level of detail in the maps. This means that the maps and models can be visualized and any corrections applied throughout the process, as described by García-Montero et al. (2008; 2010a,b). These authors propose GIS models based on successive combination and reclassifying operations applied to raster maps.

4.6.2.3 Model Steps and Logical Operations Applied

The statistical treatment associated to the GIS models was performed with the Statistica v.6 program (StatSoft, Inc., Tulsa, OK, 1999). When necessary, the variables used were transformed to comply with the requirements of the parametric statistical treatments. The normality of the data was checked using the Lilliefors and Kolmogorov–Smirnov tests, and the homogeneity of the variances was verified using the Levene test. The statistical treatments were integrated in the GIS raster models. The Excel spreadsheet program with dBase IV format was used as an intermediate step in the integration of the data sets.

The methodology included the use of GIS raster operations to integrate the soil study into a procedure, which classified the pixels relating to land with the greatest capacity for cultivation of coffee and kidney beans, and to distinguish them from pixels with a lesser capacity and from pixels relating to protected areas and/or with unacceptable levels of heavy metal content for agriculture.

The methodology was based on the procedures of Ramos (1979), MMA (2000), and García-Montero et al. (2008), which incorporate the notions of "limiting and excluding capacity factors" into land planning procedures. These methodologies use models that compare the land units according to the attributes or biotic and abiotic variables present in each one. These authors recommend that planning models should "dedicate their greatest efforts to the most significant problems." This philosophy should be part of both the environmental inventory stage and the design of the models. A GIS tool was devised to integrate the soil study into the kidney bean and coffee cultivation planning in the study area, using a model with five methodological stages, as described in the following paragraphs.

This first stage consisted of making a digital inventory of a representative sample of the kidney bean and coffee cultivation existing in the study area. This was done by georeferencing 32 bean cultivation sites covering a total of 311 ha and 44 coffee cultivation areas with a total of 413 ha.

The second part of the digital inventory consisted of applying geostatistical treatments with the ArcGIS program to generate themed soil maps. This was done by using the data analyzed from the 80 soil samples described and then applying a random selection procedure to obtain 40 soil samples. Ordinary kriging interpolation treatments were applied to each of the variables in these 40 soils in order to produce the raster layers corresponding to pH; OM; N; C/N; P_2O_5; available levels of K, Mg, Ca, Fe, Mn, and Zn; and total levels of Cd, following the procedures of Moncayo et al. (2006) and Rodríguez et al. (2006). The resulting raster maps were subsequently validated using the data from the remaining 40 soil samples that were not used in the kriging interpolation.

Several authors have demonstrated that statistical procedures applied to soil constitute a valuable tool in the planning procedures for certain crops, which are valid in particular study areas (Trangmar et al., 1985; Williams et al., 2008). The use of multivariate statistics can empirically simulate a probable cartography of fertility, which for practical operational reasons could be incorporated into the planning models.

The next stage of our model was to apply a multivariate statistical treatment to the data from the 80 soil samples described. A PCA was used to generate synthetic patterns of variation in soil variables, which could be potentially associated with the production of beans and coffee. With this in mind, variables more closely relating to fertility were used in the PCA: OM, N, P_2O_5, and available levels of K, Mg, Ca, Fe, Mn, and Zn. The principal factors or components obtained in the PCA were then examined, and one factor (PCx) was selected using the criterion of "greatest variance contained in the original matrix."

The synthetic variable PCx was incorporated into the GIS by means of the creation of a new raster layer PCx, which was considered the synthetic expression of the integrated variability of the raster layers of pH, OM, N, C/N, P_2O_5, and available levels of K, Mg, Ca, Fe, Mn, and Zn. This was done with the original layers for these variables, which were integrated into the GIS function "map calculator" using the mathematical equation, which groups the correlations between the original soil variables and the factor PCx. Thus, the value of the factor PCx was calculated for each pixel, based on the values presented by the soil variables in a pixel. With this procedure, we obtained a new raster layer PCx, which empirically simulates a fertility map in the study area.

The third stage of the model begins with the integration of the raster map for fertility PCx and the digital inventories of bean and coffee crops (the remaining classes in the vegetation and uses map were also integrated). The data obtained from the GIS integration of these raster maps has been used as numerical values for input in other statistical treatments. The objective was to estimate whether the soil fertility calculated for each pixel using the synthetic variable PCx was associated in any significant way with the 76 coffee and bean crops in the inventory. Thus, the planning models for coffee and beans (and other crops) incorporated the soil values PCx when they were statistically significant for these crops. The synthetic variable PCx for fertility will therefore have a predictive character in relation to soils in the corresponding land planning models.

The raster map for fertility (PCx) was then integrated with the digital inventory for beans (311 pixels or ha) and coffee (413 pixels or ha), as well as the remaining classes in the map of vegetation and uses, using the GIS procedures for combination and reclassification. The results obtained for each pixel were exported using Excel sheets in the dBase IV format, which were incorporated into the statistical program. Using ANOVA and post hoc Tukey tests, we studied the significant relationships between PCx and the crop inventory, vegetation types, and land uses. In cases where the ANOVA and the post hoc test indicated that there was a statistically significant relationship between the PCx value and the type of crop or use—for example, coffee crops—new GIS combination and reclassification operations were applied to create a new raster layer, which we called "map of soil compatibility for coffee cultivation."

In this new raster layer (compatibility of soil fertility with coffee cultivation), the value 2 was assigned to all the pixels in the territory with a PCx value between the average PCx value (calculated for the 413 coffee cultivation pixels) ± standard deviation (SD). The pixels with value 2 were called "soil compatible with coffee." The remaining pixels in the territory, with different PCx values from the average PCx range ± SD, were assigned the value 1 and were designated pixels with "soil not compatible with coffee."

Therefore, the predictive nature of the variable "soil compatibility with coffee" was due to the fact that the pixels with value 2 consisted of 66% of the values taken by PCx in the 413 coffee inventory pixels, corresponding to 66% of the most frequent fertility values for coffee, as its PCx values are around the average value of PCx established by the ANOVA and the post hoc test as a statistically significant value in the 413 coffee pixels. This compatibility model for soil fertility for coffee and bean cultivation could be repeated for each of the crops and vegetation classes it is desired to integrate into the planning process, provided that the ANOVA and post hoc test have assigned an average, statistically significant, PCx value for each class.

When the concepts of soil fertility and agrological capacity are applied in land planning processes, it is necessary to consider the appearance of "limiting factors," which are intrinsic to the territory, as proposed by Ramos (1979), Van Groenigen et al. (2000), and Rodríguez et al. (2006), who propose quantifying the different soil factors and limiting environmental characteristics in each study area in order to classify the sites according to their suitability for agricultural use.

Thus, in the fourth stage of the model, for each type of crop (beans or coffee), a set of raster maps of limiting factors was generated for each variable analyzed in the study area. These maps were made by using the reclassification of the raster maps for mean annual temperature, mean annual rainfall, slope of the terrain, and the following soil variables: pH, OM, N, P_2O_5, and available levels of K, Ca, Mg, Mn, Fe, and Zn. To establish the threshold limiting values for each variable, we used the bibliographical information available for bean and coffee crops in the Dominican Republic and comparable tropical countries. The threshold values relating to the behavior of each variable as a limiting factor for a crop were classified in each of the new raster maps, using the following valuables: value 1 for pixels designated "not apt" for cultivation, value 2 for pixels designated "apt" for cultivation, and value 3 for pixels designated "optimal" for cultivation.

The raster maps for limiting factors for each crop type were integrated with the corresponding soil compatibility maps obtained previously for this crop. The result was one raster layer for the agrological soil capacity for beans and another for coffee, in the studied area.

The final stage of the methodology consisted of the GIS integration of the agrological soil capacity maps for coffee and beans with new raster layers for various associated features or features that are "intrinsic" to the planning process. Within this global perspective, the fifth stage of this GIS tool involved a primary integration of the agrological soil capacity maps with two factors, which are intrinsic to the planning process: a raster layer that represents soil heavy metal content and another representing protected spaces in the study area. The integration of these two new elements was based on restrictive criteria that introduced excluding factors (intrinsic to the planning process) for beans and coffee. These layers were integrated using GIS outlining tools and combining and reclassification operations, which applied these excluding criteria.

The exclusion of areas for bean and coffee cultivation based on soil heavy metal content was exemplified with total Cd levels in the soil. We used the value of 3 mg/kg proposed by Cala et al. (1985) and Rodríguez et al. (2006) as the excluding criterion. This value agrees with the preliminary results obtained by Alexis (2008) in a series of bioassays on the impact of heavy metals on the roots and leaves of beans, coffee, sorghum (*Sorghum* sp.), and maize (*Zea mays* L.).

Areas for beans and coffee were then excluded by integrating the maps of protected natural spaces in the study area. The criterion for exclusion was the boundaries of the two national parks, which represent three quarters of Jaragua-Bahoruco-Enriquillo: the Jaragua National Park with 1651 km^2 and the Bahoruco National Park with 1125 km^2.

In this last stage, the soil study was integrated into the basic land planning process. The GIS tool for integrating the soil into the proposed plans was an open model that allowed the permanent incorporation of new environmental information on the soil (contaminants, soil processes, natural phenomena, etc.), as well as new information relating to other planning factors. In summary, the GIS tool developed discriminated extreme situations in sustainable agricultural and forest planning in contaminated rural areas of the Caribbean.

4.7   CONCLUSIONS

The process of sustainable forestry assessment requires evaluators to be reasonably informed about soil forest functions and SQI in order to increase their chances of achieving adequate joint decision making. Soil is one of the underlying matrices holding together many of the life-supporting processes and cycles on the planet.

The soil promotes water flow and retention, solute transport and retention, physical stability and support, retention and cycling of nutrients, buffering and filtering of potentially toxic materials, and maintenance of biodiversity and habitat. A healthy soil is capable of carrying out all the earlier functions. For this reason, soil quality is defined as the capacity of the soil to interact with the ecosystem in order to maintain biological productivity and environmental quality and promote animal and plant health.

Thus, the fundamental goals of protecting the quality, biodiversity, and productive capacity of soil for sustainable forest management can no longer be ignored. To support these goals, it is necessary to control or restrict agricultural and forestry activities that could reduce on-site agricultural and forest productivity and environmental quality in order to prevent soil degradation and restore its potential. Forest-soil-quality studies aimed at forest productivity are giving way to other kinds of studies based on forest ecological, edaphological, and physiological studies due to the increasing worldwide interest in sustainability.

Currently, there are different sets of SQI available for the purpose of identifying problem production areas, making realistic estimates of food and natural resource production, monitoring changes in sustainability and environmental quality in relation to forestry management, and assisting national and state or regional agencies to formulate and evaluate sustainable land-use policies. More research areas are starting to study soil quality due to the growing interest in this issue. However, a bibliographical review shows that there is insufficient incorporation of soil-quality concepts and soil indicators in forestry management, which could compromise the objectives of reaching widespread sustainable forest management.

The development of soil-quality indices should be sensitive to long-term alterations in soil management and climate but sufficiently robust not to vary as a consequence of short-term changes in weather conditions. They must correlate well to beneficial soil functions. Soil-quality indices have to be useful for understanding why a soil will or will not function as desired. They have to be comprehensible and useful to land managers, easy, and inexpensive to measure, and where possible, they should also be components of existing soil databases. For these reasons, we must design new procedures that integrate physical, chemical, biological, and other unforeseen elements into indicators and indices that can perform seamlessly in any given soil conditions worldwide in order to maintain all life functions and cycles constant on this planet. In this regard, the current knowledge of SQI has not reaped sufficient benefits from the wealth of tools available from numerical and quantitative models.

REFERENCES

Akeem, L. 2012. *Biodiversity Conservation in a Diverse World*. Rijeka, Croatia: Intech.

Alexis, S. 2008. Sustainable development strategies and the province and cross-border watershed of Pedernales (Dominican Republic-Haiti). PhD dissertation, Madrid, Spain: University of Alcalá.

Alexis, S., L.G. García-Montero, A. Hernández, A. García-Abril, and J. Pastor. 2010. Soil fertility and GIS raster models for tropical agroforestry planning in economically depressed and contaminated Caribbean areas (coffee and kidney bean plantations). *Agroforestry Systems* 79: 381–391.

Allievi, L., A. Marchesini, C. Salardi, V. Piano, and A. Ferrari. 1993. Plant quality and soil residual fertility six years after a compost treatment. *Bioresource Technology* 43: 85–89.

Andersen, A. 2002. Using ants as bioindicators in land management: Simplifying assessment of ant community responses. *Journal of Applied Ecology* 39: 8–17.

Andrews, S.S., D.L. Karlen, and C.A. Cambardella. 2004. The soil management assessment framework. *Soil Science Society of America Journal* 68: 1945–1962.

Andrews, S.S., D.L. Karlen, and J.P. Mitchell. 2002. A comparison of soil quality indexing methods for vegetable production systems in Northern California. *Agriculture, Ecosystems and Environment* 90: 25–45.

Andrews, S.S., J. Mitchell, R. Mancinelli, and D. Karlen. 2002. On-farm assessment of soil quality in California's Central Valley. *Agronomy Journal* 94: 12–23.

Andrews, S., J. Mitchell, R. Mancinelli, and D. Karlen. 2002b. On-farm assessment of soil quality in California's central valley. *Agronomy Journal* 94: 12–23.

Andrews, S.S. and C. Carroll. 2001a. Designing a soil quality assessment tool for sustainable agroecosystem management. *Ecological Applications* 11: 1573–1585.

Andrews, S.S. and C.R. Carroll. 2001b. Designing a decision tool for sustainable agroecosystem management: Soil quality assessment of a poultry litter management case study. *Ecological Applications* 11: 1573–1585.

ANPA. 2002. *Atlas of the Soil Indicators*. Rome, Italy: Agenzia Nationale per la Protezione dell'Ambiente.

Aoki, J. 1967. Microhabitats of oribatid mites on a forest floor. *Bulletin of the Natural Science Museum* 10: 133–138.

Armas, C., B. Santana, J. Mora, J. Notario, C. Arbelo, and A. Rodríguez-Rodríguez. 2007. A biological quality index for volcanic Andisols and Aridisols (Canary Islands, Spain): Variations related to the ecosystem development. *Science of the Total Environment* 378: 238–244.

Arshad, M.A. and A. Cohen. 1992. Characterization of soil quality: Physical and chemical criteria. *American Journal of Alternative Agriculture* 7: 25–31.

Aune, J. and R. Lal. 1997. Agricultural productivity in the tropics and critical limits of properties of Oxisols, Ultisols and Alfisols. *Tropical Agriculture* 74: 96–103.

Awiti, A.O., M.G. Walsh, K.D. Shepherd, and J. Kinyamario. 2008. Soil condition classification using infrared spectroscopy: A proposition for assessment of soil condition along a tropical forest-cropland chronosequence. *Geoderma* 143: 73–84.

Bachelier, G., G. Vannier, M. Pussard, M.B. Bouché, C. Jeanson, P. Boyer, Z. Massoud et al. 1971. *La vie dans les sols*. Paris, France: Gauthier-Villars.

Baker, G.H. 1998. Recognizing and responding to the influences of agriculture and other land-use practices on soil fauna in Australia. *Applied Soil Ecology* 9: 303–310.

Bardgett, R.D. and R. Cook. 1998. Functional aspects of soil animal diversity in agricultural grasslands. *Applied Soil Ecology* 10: 263–276.

Barth, H. and P. L'Heremite. 1987. *Scientific Basis for Soil Protection in the European Community*. London, U.K.: Elsevier.

Bastida, F., J.L. Moreno, T. Hernández, and C. García. 2006. Microbiological degradation index of soils in a semiarid climate. *Soil Biology and Biochemistry* 38: 3463–3473.

Bastida, F., A. Zsolnay, T. Hernández, and C. García. 2008. Past, present and future of soil quality indices: A biological perspective. *Geoderma* 147: 159–171.

Bautista-Cruz, A., R.F. del Castillo, J.D. Etchevers-Barra, M.C. Gutiérrez-Castorena, and A. Baez. 2012. Selection and interpretation of soil quality indicators for forest recovery after clearing of a tropical montane cloud forest in Mexico. *Forest Ecology and Management* 277: 74–80.

Bazzoffi, P., S. Pellegrini, A. Rocchini, M. Morandi, and O. Grasselli. 1998. The effect of urban refuse compost and different tractors tyres on soil physical properties, soil erosion and maize yield. *Soil and Tillage Research* 48: 275–286.

Beare, M., R. Parmelee, R. Hendrix, W. Cheng, D. Coleman, and D. Crossley. 1992. Microbial and faunal interactions and effects on litter nitrogen and decomposition in agroecosystems. *Ecological Monographs* 62: 569–591.

Bécaert, V., R. Samson, and L. Deschênes. 2006. Effect of 2, 4-D contamination on soil functional stability evaluated using the relative soil stability index (RSSI). *Chemosphere* 64: 1713–1721.

Beck, T.H. 1984. Methods and application of soil microbiological analysis at the Landensanstalt fur Bodenkultur und Pflanzenbau (LBB) for determination of some aspects of soil fertility. In *Proceedings of the Fifth Symposium on Soil Biology*, ed. M.P. Nemesís, pp. 13–20. Bucharest, Romania: Romanian National Society of Soil Science.

Behan Pelletier, V.M. 1999. Oribatid mite biodiversity in agroecosystems: Role for bioindication. *Agriculture, Ecosystems and Environment* 74: 411–423.

Benedetti, A. and O. Dilly. 2006. Approaches to defining, monitoring, evaluating, and managing soil quality. In *Microbiological Methods for Assessing Soil Quality*, eds. J. Bloem, D.W. Hopkins, and A. Benedetti, pp. 3–15. Oxfordshire, U.K.: CABI Publishing.

Benndorf, D., G.U. Balcke, H. Harms, and M. von Bergen. 2007. Functional metaproteome analysis of protein extracts from contaminated soil and groundwater. *The International Society for Microbial Ecology Journal* 1: 224–234.

Bernini, F., A. Avanzati, M. Baratti, and M. Migliorini. 1995. Oribatid mites (Acari Oribatida) of the Farma Valley (Southern Tuscany). Notulae Oribatologicae LXV. *Redia* LXXVIII: 45–129.

Bloem, J. 2003. Microbial indicators. *Trace Metals and Other Contaminants in the Environment* 6: 259–282.

Bloem, J., D. Hopkins, and A. Benedetti. 2003. *Microbiological Methods for Assessing Soil Quality.* Oxfordshire, U.K.: CABI Publishing.

Blum, W. 1993. Soil protection concept of the council of Europe and integrated soil research. In *Integrated Soil and Sediment Research: A Basis for Proper Protection,* eds. H.J.P. Eisjackers and T. Hammers, pp. 37–47. Dordrecht, the Netherlands: Kluwer Academic Publishers.

Bongers, T. 1990. The maturity index: An ecological measure of environmental disturbance based on nematode species composition. *Oecologia* 83: 14–19.

Bongers, T. 1999. The maturity index, the evolution of nematode life history traits, adaptive radiation and cp-scaling. *Plant and Soil* 212: 13–22.

Bouche, M. 1977. *Variability of Earthworm Populations—Tool for Agronomist.* Paris, France: Institut National de la Recherche Agronomique.

Bouché, M. B., F. Al-Addan, J. Cortez, R. Hammed, J. Heidet, G. Ferrière et al. 1997. Role of earthworms in the N cycle: A falsifiable assessment. *Soil Biology and Biochemistry* 29: 375–380.

Bowman, R., D. Nielsen, M. Vigil, and R. Aiken. 2000. Effects of sunflower on soil quality indicators and subsequent wheat yield. *Soil Science* 165: 516–522.

Brauns, A. 1955. Applied soil biology and plant protection. In *Soil Zoology,* ed. D.K. McE. Evan, pp. 231–240. London, U.K.: Butterworth.

Brejda, J., T. Moorman, D. Karlen, and T. Dao. 2000. Identification of regional soil quality factors and indicators. I. Central and southern high plains. *Soil Science Society of America Journal* 64: 2115–2124.

Bremer, E. and K. Ellert. 2004. Soil quality indicators: A review with implications for agricultural ecosystems in Alberta. Lethbridge, Alberta, Canada: Symbio Ag Consulting.

Brookes, P. 1995. The use of microbial parameters in monitoring soil pollution by heavy metals. *Biology and Fertility of Soils* 19: 269–279.

Brookes, P. 2001. The soil microbial biomass: concept, measurement and applications in soil ecosystem research. *Microbes and Environments* 16: 131–140.

Bruce, L., D. McCracken, G. Foster, and M. Aitken. 1997. The effects of cadmium and zinc-rich sewage sludge on epigeic Collembola populations. *Pedobiologia* 41: 167–172.

Bruce, L., D. McCracken, G. Foster, and M. Aitken. 1999. The effects of sewage sludge on grassland euedaphic and hemiedaphic collembolan populations. *Pedobiologia* 43: 209–220.

Brussard, P.F., J. Reed, and C.R. Tracy. 1998. Ecosystem management: What is it really? *Landscape and Urban Planning* 40: 9–20.

Burger, J.A., J.E. Johnson, J.A. Andrews, and J.L. Torbert. 1994. Measuring mine soil productivity for forests. In *International Land Reclamation and Mine Drainage Conference on Reclamation and Revegetation,* ed. United States Department of the Interior Bureau of Mines, pp. 48–56. Pittsburgh, PA: United States Department of the Interior Bureau of Mines.

Burger, J. and D. Kelting. 1999. Using soil quality indicators to assess forest stand management. *Forest Ecology and Management* 122: 155–156.

Cala, V., J. Rodríguez, and A. Guerra. 1985. Contaminación por metales pesados en suelos de la Vega de Aranjuez. (I) Pb, Cd, Cu, Zn, Ni, Cr. *Anales de Edafología Agrobiología* 14: 1595–1608.

Cambardella, C., T. Moorman, S. Andrews, and D. Karlen. 2004. Watershed-scale assessment of soil quality in the Loess hills of southwest Iowa. *Soil and Tillage Research* 78: 237–247.

Carter, M.R., E.G. Gregorich, D.W. Anderson, J.W. Doran, H.H. Janzen, and F.J. Pierce. 1997. Concepts of soil quality and their significance. *Developments in Soil Science* 25: 1–19.

Chaer, G. 2008. Response of soil microbial communities to physical and chemical disturbances: Implications for soil quality and land use sustainability. PhD dissertation, Corvallis, OR: Oregon State University.

Chambers, R.G. and R.D. Pope. 1996. Aggregate productivity measures. *American Journal of Agricultural Economics* 78: 1360–1365.

Chambers, R.G., R. Färe, and S. Grosskopf. 1996. Productivity growth in APEC countries. *Pacific Economic Review* 1: 181–190.

Chapin, F., O.E. Sala, and E. Huber-Sanwald. 2000. *Future Scenarios of Global Biodiversity.* New York: Springer-Verlag.

Cortet, J., D. Gillon, R. Joffre, J. Ourcival, and N. Poinsot-Balanguer. 2002. Effects of pesticides on organic matter recycling and microarthropods in a maize field: Use and discussion of the litterbag methodology. *European Journal of Soil Biology* 38: 261–265.

Cortet, J.A. Gomot-De Vauflery, N. Poinsot-Balaguer, L. Gomot, C. Texier, and D. Cluzeau. 1999. The use of invertebrate soil fauna in monitoring pollutant effects. *European Journal of Soil Biology* 35: 115–134.

Cuendet, G. and A. Ducommun. 1990. Peuplements lombriciens et activité de surface en relation avec les boues d'epuration et autres fumures. *Revue Suisse die Zoologie* 97: 851–869.

Curry, J.P. 1994. *Grassland Invertebrates. Ecology, Influence on Soil Fertility and Effects on Plant Growth.* London, U.K.: Chapman & Hall.

Da Silva, A., B. Kay, and E. Perfect. 1994. Characterization of the least limiting water range of soils. *Soil Science Society of America Journal* 58: 1775–1781.

De la Paz-Jiménez, M., A.M. De la Horra, L. Pruzzo, and R.M. 2002. Soil quality: A new index base microbiological and biochemical parameters. *Biology and Fertility of Soils* 35: 302–306.

Dick, R.P. 1994. Soil enzyme activities as indicators of soil quality. In: *Defining Soil Quality for a Sustainable Environment*, eds. J.W. Doran, D.C. Coleman, D.F. Bezdicek, and B.A. Stewart, pp. 107–124. Madison, WI: Soil Science Society of America.

Dindal, D.L. 1990. *Soil Biology Guide.* New York: John Wiley & Sons.

Doran, J.W. 1994. *Defining Soil Quality for Sustainable Environment.* Madison, WI: Soil Science Society of America.

Dylis, N. 1964. Principles of construction of a classification of forest biogeocoenoses. In: *Fundamentals of Forest Biogeocoenology*, eds. V.N. Sukachev and N.V. Dylis, pp. 572–589. Edinburgh, Scotland: Oliver and Boyd.

Ebel, A. and T. Davitashvili. 2007. *Air, Water and Soil Quality Modelling for Risk and Impact Assessment.* Dordrecht, the Netherlands: Springer.

EC European Commission. 2002. Decision No 1600/2002/EC of the European Parliament and of the Council of 22 July 2002 laying down the Sixth Community Environment Action Programme. In *Official Journal of the European Communities* L 242/1 (10.9.2002). Brussels, Belgium: European Parliament.

Erkossa, T., F. Itanna, and K. Stahr. 2007. Indexing soil quality: A new paradigm in soil science research. *Australian Journal of Soil Research* 45: 129–137.

Ettema, C. and T. Bongers. 1993. Characterization of nematode colonization and succession in disturbed soil using the maturity index. *Earth and Environmental Science* 16: 79–85.

European Commission. 2001. Biodiversity action plan for agriculture. http://eur-lex.europa.eu/LexUriServ/LexUriServ.do?uri=CELEX:52001DC0162%2803%29:EN:HTML

European Commission. 2004. Message from Malahide. http://ec.europa.eu/environment/nature/biodiversity/policy/pdf/malahide_message_final.pdf

European Commission. 2005. *Soil Atlas of Europe, European Soil Bureau Network.* Luxembourg, Europe: Office for Official Publications of the European Communities.

European Commission. 2006. Report from the commission to the European Parliament, the Council, the European Economic and Social Committee and the committee of the regions the implementation of the soil thematic strategy and ongoing activities. http://eur-lex.europa.eu/LexUriServ/LexUriServ.do?uri=CELEX:52012DC0046:EN:NOT

EU/ECE. 2003. United Nations Economic Commission for Europe–Kiev resolution on biodiversity. http://www.eea.europa.eu/data-and-maps/indicators/agriculture-area-under-management-practices/united-nations-economic-commission-for

Faber, H. 1991. Functional classification of soil fauna: A new approach. *Oikos* 62: 110–117.

FAO. 2000. World soil resources report 90. Rome, Italy: FAO.

FAO. 2009. World soil resources report 104. Rome, Italy: FAO.

Filip, Z. 1973. A healthy soil—foundation of a healthy environment. *Vesmir* 52: 291–293.

Filip, Z. 2002. International approach to assessing soil quality by ecologically-related biological properties. *Agriculture, Ecosystems and Environment* 88: 169–174.

Ford, D. 1983. What do we need to know about forest productivity and how can we measure it? In *IUFRO Symposium on Forest Site and Continuous Productivity*, eds. R. Ballard and S.P. Gessel, pp. 2–12. Washington, DC: USDA Forest Service.

Fox, C., E. Fonseca, J. Miller, and A. Tomlin. 1999. The influence of row position and selected soil attributes on Acarina and Collembola in no-till and conventional continuous corn on a clay loam soil. *Applied Soil Ecology* 13: 1–8.

Frankenberger, W. and W. Dick. 1983. Relationships between enzyme activities and microbial growth and activity indices in soil. *Soil Science Society of America Journal* 47: 945–951.

Frouz, J. 1999. Use of soil dwelling diptera (insecta, diptera) as bioindicators: A review of ecological requirements and response to disturbance. *Agriculture, Ecosystems and Environment* 74 (1): 167.

Gale, M., D. Grigal, and B. Harding. 1991. Soil productivity index: Predictions of site quality for white spruce plantations. *Soil Science Society of America Journal* 55: 1701–1708.

García, C., T. Hernández, A. Roldán, and A. Martín. 2002. Effect of plant cover decline on chemical and microbiological parameters under Mediterranean climate. *Soil Biology and Biochemistry* 34: 635–642.

García-Montero, L.G., E. López, A. Monzón, and I. Otero. 2010a. Environmental screening tools for assessment of infrastructure plans based on biodiversity preservation and global warming (PEIT, Spain). *Environmental Impact Assessment Review* 30: 158–168.

García-Montero, L.G., S. Mancebo Quintana, M.A. Casermeiro, I. Otero, and A. Monzón de Cáceres. 2010b. A GIS raster model for assessing the environmental quality of Spain focused on SEA and infrastructure planning procedures (LATINO model). In *Proceedings of Ninth Highway and Urban Environment Symposium*, eds. A. Rauch, G. Morrison, and A. Monzón, pp. 31–38. Dordrecht, the Netherlands: Springer.

García-Montero, L.G., I. Otero, S. Mancebo-Quintana, and M.A. Casermeiro. 2008. An environmental screening tool for assessment of land use plans covering large geographic areas. *Environmental Science and Policy* 11: 285–293.

Gardi, C., M. Tomaselli, V. Parisi, A. Petraglia, and C. Santini. 2002. Soil quality indicators and biodiversity in northern Italian permanent grasslands. *European Journal of Soil Biology* 38: 103–110.

Garrigues, E., M.S. Corson, D.A. Angers, H.M.G. van der Werf, and C. Walter. 2012. Soil quality in life cycle assessment: Towards development of an indicator. *Ecological Indicators* 18: 434–442.

Gelsomino, A., L. Badalucco, R. Ambrosoli, C. Crecchio, E. Puglisi, and S.M. Meli. 2006. Changes in chemicals and biological soil properties as induced by anthropogenic disturbance: A case study of an agricultural soil under recurrent flooding by wastewaters. *Soil Biology and Biochemistry* 38: 2069–2080.

Gilley, J., J. Doran, and B. Eggball. 2001. Tillage and fallow effects on selected soil quality characteristics of former conservation reserve program sites. *Journal of Soil and Water Conservation* 56: 126–132.

Gil-Sotres, F., M. Leirós de la Peña, and C. Trasar-Cepeda. 2003. A comment on the article by de la Paz-Jiménez et al. Soil quality: A new index based on microbiological and biochemical parameters. *Biology and Fertility of Soils* 37: 260.

Granatstein, D. and D. Bezdicek. 1992. The need for a soil quality index: Local and regional Perspectives. *American Journal of Alternative Agriculture* 17: 12–16.

Grime, J.P. 1997. Biodiversity and ecosystem function: The debate deepens. *Science* 277: 1260–1261.

Gupta, S.R. and Malik V. 1996. Soil ecology and sustainability. *Tropical Ecology* 37: 43–55.

Gupta, V.V. and G.W. Yeates. 1997. Soil microfauna as bioindicators of soil health. In *Biological Indicators of Soil Health*, eds. C. Pankhurst, B.M. Doube and V.V.S.R. Gupta, pp. 201–233. New York: CAB International.

Gzyl, J. 1999. Soil protection in Central and Eastern Europe. *Journal of Geochemical Exploration* 66: 333–337.

Hagvar, S., 1994. Log-normal distribution of dominance as an indicator of stressed soil micro-arthropod communities? *Acta Zoologica Fennica* 195: 71–80.

Hailu, A. and R.G. Chambers. 2012. A Luenberger soil-quality indicator. *Journal of Productivity Analysis* 38: 145–154.

Hansen, B., H.F. Alrøe, and E.S. Kristensen. 2001. Approaches to assess the environmental impact of organic farming with particular regard to Denmark. *Agriculture, Ecosystems and Environment* 83: 11–26.

Harris, R.F., D.L. Karlen, and D.J. Mulla. 1996. A conceptual framework for assessment and management of soil quality and health. In *Methods for Assessing Soil Quality*, eds. J.W. Doran and A.J. Jones, pp. 61–82. Madison, WI: Soil Science Society of America.

Henderson, T.L., M.F. Baumgardner, D.P. Franzmeier, D.E. Stott, and D.C. Coster. 1992. High dimensional reflectance analysis of soil organic matter. *Soil Science Society of America Journal* 56: 865–872.

Hendrix, P.F., R.W. Parmelee, D.A. Crossley, Jr., D.C. Coleman, E.P. Odum, and P. Groffman. 1986. Detritus food webs in conventional and no-tillage agro ecosystems. *Bioscience* 36: 374–380.

Hernández, A., S. Alexis, and J. Pastor. 2007. Soil degradation in the tropical forests of the Dominican Republic's Pedernales province in relation to heavy metal contents. *Science of the Total Environment* 378: 36–41.

Herrick, J.E. and M.M. Wander. 1997. Relationships between soil organic carbon and soil quality in cropped and rangeland soils: The importance of distribution, composition and soil biological activity. In *Soil Processes and the Carbon Cycle*, eds. R. Lal, J.M. Kimble, R.F. Follett and B.A. Stewart, pp. 405–425. Boca Raton, FL: CRC Press.

House, G. and M. Del Rosario. 1989. Influence of cover cropping and no-tillage practices on community composition of soil arthropods in a North Carolina agro ecosystem. *Environmental Entomology* 18: 302–307.

Howard, P.J.A. 1993. Soil protection and soil quality assessment in the EC. *Science of the Total Environment* 129: 219–239.

Iturrondobeitia, J., M. Saloña, J. Pereda, A. Caballero, and M. Andrés. 1997. Oribatid mites as an applied tool in studies on bioindication: A particular case. *Abhandlungen und Berichte des Naturkundemuseums Görlitz* 69: 85–96.

Jans Hammermeister, D.C. 1997. Evaluation of three simulation models used to describe plant residue decomposition in soil. *Ecological Modelling* 104: 1–13.

Jeffery, S., C. Gardi, A. Jones, L. Montanarella, L. Marmo, L. Miko, K. Ritz, G. Peres, J. Römbke, and W.H. van der Putten. 2010. *European Atlas of Soil Biodiversity*. Luxembourg, Europe: Publications Office of the European Union.

Joergensen, R.G. and M. Raubuch. 2003. Adenylates in the soil microbial biomass at different temperatures. *Soil Biology and Biochemistry* 35: 1063–1069.

Johnston, J.M. and D.A Crossley. 2002. Forest ecosystem recovery in the Southeast US: Soil ecology as an essential component of ecosystem management. *Forest Ecology and Management* 155: 187–203.

Joy, V. and P. Chakravorty. 1991. Impact of insecticides on non-target microarthropods fauna in agricultural soil. *Ecotoxicology and Environmental Safety* 22: 8–16.

Kang, G., V. Beri, B. Sidhu, and O. Rupela. 2005. A new index to assess soil quality and sustainability of wheat-based cropping systems. *Biology and Fertility of Soils* 41: 389–398.

Karlen, D. and D. Stott. 1994. A framework for evaluating physical and chemical indicators of soil quality. In *Defining Soil Quality for a Sustainable Environment*, eds. J.W. Doran, D.C. Coleman, D.F. Bezdicek, and B.A. Stewart, pp. 53–72. Madison, WI: Soil Science Society of America.

Karlen, D., S. Andrews, and D. Doran. 2001. Soil quality: Current concepts and applications. *Advances in Agronomy* 74: 1–40.

Karlen, D.L., C.A. Ditzler, and S.S. Andrews. 2003. Soil quality: Why and how? *Geoderma* 114: 145–156.

Karlen, D., J. Gardner, and M. Rosek. 1998. A soil quality framework for evaluating the impact of CRP. *Journal of Production Agriculture* 11: 56–60.

Karlen, D., M. Mausbach, J. Doran, R. Cline, R. Harris, and G. Schumann. 1997. Soil quality: A concept, definition and framework for evaluation. *Soil Science Society of America Journal* 61: 4–10.

Karlen, D., N. Wollenhaupt, D. Erbach, E. Berry, J. Swan, N. Eash, and J. Jordhal. 1994a. Crop residue effects on soil quality following 10-years of no-till corn. *Soil and Tillage Research* 31: 149–167.

Karlen, D., N. Wollenhaupt, D. Erbach, E. Berry, J. Swan, N. Eash, and J. Jordhal. 1994b. Long-term tillage effects on soil quality. *Soil and Tillage Research* 32: 313–327.

Kay, B. and C. Grant. 1996. Structural aspects of soil quality. In *Soil Quality is in the Hands of the Land Manager*, eds. R.J., MacEwan and M.R. Carter, pp. 37–41. Victoria, BC, Canada: University of Ballarat.

Kelting, D., J. Burguer, S. Patterson, M. Aust, M. Miwa, and C. Trettin. 1999. Soil quality assessment in domesticated forests—a southern pine example. *Forest Ecology and Management* 122: 167–185.

Kettler, T., D. Lyon, J. Doran, W. Powers, and W. Stroup. 2000. Soil quality assessment after weed-control tillage in a no-till wheat-fallow cropping system. *Soil Science Society of America Journal* 64: 339–346.

Killham, K. 1994. *Soil Ecology*. Cambridge, U.K.: University Press.

Kiniry, L., C. Scrivner, and M. Keener. 1983. *A Soil Productivity Index Based upon Predicted Water Depletion and Root Growth*. Columbia, MO: University of Missouri.

Knoepp, J., D. Coleman, D. Crossley, and J. Clark. 2000. Biological indices of soil quality: An ecosystem case study of their use. *Forest Ecology and Management* 138: 357–368.

Koper, J. and A. Piotrowska. 2003. Application of biochemical index to define soil fertility depending on varied organic and mineral fertilization. *Electronic Journal of Polish Agricultural Universities* 6: 6.

Korthals, G.W., A. van de Ende, H. Van Megen, T.M. Lexmond, J.E. Kammenga, and T. Bongers. 1996. Short-term effects of cadmium, copper, nickel and zinc on soil nematodes from different feeding and life-history strategy groups. *Applied Soil Ecology* 4: 107–117.

Kovda, V.A. 1975. *Biogeochemical Cycles in Nature and their Disturbance Caused by Humans*. Moscow, Russia: Nauka.

Lal, R. 1998. Soil erosion impact on agronomic productivity and environment quality. *Critical Reviews in Plant Sciences* 17: 319–464.

Larsen, K., F. Purrington, S. Brewer, and D. Taylor. 1986. Influence of sewage sludge and fertilizer on the ground beetle (Coleoptera: Carabidae) fauna of an old-field community. *Environmental Entomology* 25: 452–459.

Larson, W. and F. Pierce. 1991. Conservation and enhancement of soil quality. In *Evaluation for Sustainable Land Management in the Developing World*, ed. International Board for Soil Research and Management, pp. 175–203. Thailand, Bangkok: IBSRM.

Larson, W. and F. Pierce. 1994. The dynamics of soil quality as a measure of sustainable management. *Soil Science Society of America* 35: 37–51.

Lawton, J.H. and R.M. May. 1995. *Extinction Rates*. Oxford, U.K.: Oxford University Press.

Leirós, M., C. Trasar-Cepeda, F. García-Fernández, and F. Gil-Sotres. 1999. Defining the validity of a biochemical index of soil quality. *Biology and Fertility of Soils* 30: 140–146.

Li, Y., M. Lindstrom, J. Zhang, and J. Yang. 2001. Spatial variability of soil erosion and soil quality on hillslopes in the Chinese Loess Plateau. *Acta Geologica Hispanica* 35: 261–70.

Liebig, J. and J. Doran. 1999. Impact of organic production practices on soil quality indicators. *Journal of Environmental Quality* 28: 1601–1609.

Liebig, M., G. Varvel, and J. Doran. 2001. A simple performance-based index for assessing multiple agroecosystem functions. *Agronomy Journal* 93: 313–318.

Lübben, B. 1989. Influence of sewage sludge and heavy metals on the abundance of Collembola on two agricultural soils. In *Third International Seminar on Apterygota*, ed. R. Dallai, pp. 419–428. Siena, Tuscany, Italy: Università di Siena.

Mäder, P., A. Fließbach, D. Dubois, L. Gunst, P. Fried, and U. Niggli. 2002. Soil fertility and biodiversity in organic farming. *Science* 296: 1694–1697.

Maharning, A.R., A.A.S. Mills, and S.M. Adl. 2008. Soil community changes during secondary succession to naturalized grasslands. *Applied Soil Ecology* 41: 137–147.

Mancebo, S., L.G. García-Montero, M.A. Casermeiro, I. Otero, A. Esplugas, and M. Navarra. 2005. Modelo preliminar de la calidad natural de España 1:500.000. In *Proceedings of III Congreso Nacional de Evaluación de Impacto Ambiental*, eds. M.A. Casermeiro, L. Desdentado, M. Díaz, A.P. Espluga, L.G. García-Montero, D.E. Nelly, J. Puig, and I. Sobrini, pp. 205–236. Pamplona, Navarre, Spain: Asociación Española de Evaluación de Impacto Ambiental.

Mansourian, S., D. Vallauri, and N. Dudley. 2005. *Forest Restoration in Landscapes: Beyond Planting Trees*. New York: Springer.

Martínez-Falero, E. and S. González. 1995. *Quantitative Techniques in Landscape Planning*. Boca Raton, FL: CRC Lewis Publishers.

Mele, P., M. Pauline, and D. Crowley. 2008. Application of self-organizing maps for assessing soil biological quality. *Agriculture, Ecosystems and Environment* 126: 139–152.

Menta, C. 2012. Soil fauna diversity-function, soil degradation, biological indices, soil restoration. In *Biodiversity Conservation in a Diverse World*, ed. G.A. Lameed, pp. 59–94. Rijeka, Croatia, Kvarner Bay: InTech.

Miller, R.E., J.D. McIver, S.W. Howes, and W.B. Gaeuman. 2010. *Assessment of Soil Disturbance in Forests of the Interior Columbia Basin: A Critique*. Portland, OR: U.S. Department of Agriculture, Forest Service.

Minnesota Forest Resources Council. 1999. *Sustaining Minnesota Forest Resources: Voluntary Site-Level. Forest Management Guidelines for Landowners, Loggers, and Resources Managers*. St. Paul, MN: Minnesota Forest Resources Council.

MMA Ministerio de Medio Ambiente. 2000. *Guía para la elaboración de estudios del medio físico. Contenido y metodología*. Madrid, Spain: Ministerio de Medio Ambiente.

Mohanty, M., D. Painuli, A. Misra, and P. Ghosh. 2007. Soil quality effects of tillage under rice-wheat cropping on a Vertisol in India. *Soil and Tillage Research* 92: 243–250.

Moncayo, F., A. Hincapié, J. Betancur, and L. Tafur. 2006. Spatial variability of chemical and physical properties of a sandy typic udivitrands in the Colombian Central Andean zone. *Revista de la Facultad de Ciencias Médicas (Córdoba, Argentina)* 59: 3217–3235.

Nannipieri, P., S. Grego, and B. Ceccanti. 1990. Ecological significance of the biological activity in soils. In *Soil Biochemistry*, eds. J.M. Bollag and G. Stotzky, pp. 293–355. New York: Marcel Dekker.

Nearing, M., G. Foster, L. Lane, and S. Finckner. 1989. A process-based soil erosion model for USDA-Water Erosion Prediction Project technology. *Transactions of the American Society of Agricultural Engineers* 32: 1587–1593.

Neufeld, J., M. Wagner, and J.C. Murrell. 2007. Who eats what, where and when? Isotope labelling experiments are coming of age. *International Society for Microbial Ecology Journal* 1: 103–110.

Norris, W. and D.A. Farrar. 2011. Method for the natural quality evaluation of Central Hardwood forests in the Upper Midwest, USA. *Natural Areas Journal* 21: 313–323.

Page-Dumroesea, D., M. Jurgensen, W. Elliota, T. Ricea, J. Nesserc, T. Collinsd, and R. Meurisse. 2000. Soil quality standards and guidelines for forest sustainability in northwestern North America. *Forest Ecology and Management* 138: 445–462.

Pang, X., W. Bao, and Y. Zhang. 2006. Evaluation of soil fertility under different *Cupressus chengiana* forests using multivariate approach. *Pedosphere* 16: 602–615.

Pankhurst, C. 1997. Biodiversity of soil organisms as an indicator of soil health. In *Biological Indicators of Soil Health*, eds. C.E. Pankhurst, B.M. Doube, and V.V.S.R. Gupta, pp. 297–324. Wallingford, U.K.: CABI Publishing.

Pankhurst, C., B. Doube, and V. Gupta. 1997. Biological indicators of soil health: Synthesis. In *Biological Indicators of Soil Health*, eds. C.E. Pankhurst, B.M. Doube, V.V.S.R. Gupta, pp. 419–435. Wallingford, U.K.: CABI Publishing.

Paoletti, M.G. and M. Hassall. 1999. Woodlice (Isopoda: Oniscidae): their potential for assessing sustainability and use as bioindicators. *Agriculture, Ecosystems and Environment* 74: 157–165.

Parisi, V. 2001. The biological soil quality, a method based on microarthropods. *Acta Naturalia de L'Ateneo Parmense* 37: 97–106.

Parisi, V. and C. Menta. 2008. Microarthropods of the soil: Convergence phenomena and evaluation of soil quality using QBS-ar and QBS-c. *Fresenius Environmental Bulletin* 17: 1170–1174.

Persoh, D., S. Theuerl, F. Buscot, and G. Rambold. 2008. Towards a universally method for quantitative extraction of high-purity nucleic acids from soil. *Journal of Microbiological Methods* 75: 19–24.

Petach, M.C., R.J. Wagenet, and S.D. deGloria. 1991. Regional water flow and pesticide leaching using simulations with spatially distributed data. *Geoderma* 48: 245–269.

Pfotzer, G. and C. Schuler. 1997. Effects of different compost amendments on soil biotic and faunal feeding activity in an organic farming system. *Biological Agriculture and Horticulture* 15: 177–183.

Pierce, F.J. and W.E. Larson. 1993. Developing criteria to evaluate sustainable land management. In *Proceedings of the Eighth International Soil Management Workshop: Utilization of Soil Survey Information for Sustainable Land Use*, ed. J.M. Kimble, pp. 7–14. Washington, DC: USDA, Soil Conservation Service, National Soil Survey Center.

Pinamonti, F., G. Stringari, F. Gasperi, and G. Zorzi. 1997. The use of compost: Its effects on heavy metal level in soil and plant. *Resource, Conservation and Recycling* 21: 129–143.

Powers, R., A. Tiarks, and J.R. Boyle. 1998. Assessing soil quality: Practicable standards for sustainable forest productivity in the United States. In *The Contribution of Soil Science to the Development and Implementation of Criteria and Indicators of Sustainable Forest Management*, eds. E.A. Davidson, M.B. Adams, and K. Ramakrishna, pp. 53–80. Madison, WI: Soil Science Society of America.

Puglisi, E., A. Del Re, M. Rao, and L. Gianfreda. 2006. Development and validation of numerical indices integrating enzyme activities of soils. *Soil Biology and Biochemistry* 38: 1673–1681.

Puglisi, E., M. Nicelli, E. Capri, M. Trevisan, and A. Del Re. 2005. A soil alteration index based on phospholipid fatty acids. *Chemosphere* 61: 1548–1557.

Radajewski, S., I. McDonald, and J. Murrell. 2003. Stable-isotope probing of nucleic acids: A window to the function of uncultured microorganisms. *Current Opinion in Biotechnology* 14: 296–302.

Ramachandran, P.K., V.D. Nair, B. Mohan, and J.M. Showalter. 2010. Carbon sequestration in agroforestry systems. *Advances in Agronomy* 108: 237–307.

Ramos, A. 1979. *Planificación física y ecología. Modelos y métodos*. Madrid, Spain: E.M.E.S.A.

Reganold, J. and A. Palmer. 1995. Significance of gravimetric versus volumetric measurements of soil quality under biodynamic, conventional, and continuous grass management. *Journal of Soil and Water Conservation* 50: 298–305.

Rezaei, S., R. Gilke, and S. Andrews. 2006. A minimum data set for assessing soil quality in rangelands. *Geoderma* 136: 229–234.

Riffaldi, R., A. Saviozzi, R. Levi-Minzi, and R. Cardelli. 2002. Biochemical properties of a Mediterranean soil as affected by long-term crop management systems. *Soil and Tillage Research* 67: 109–114.

Rodríguez, J., M. López, and J. Grau. 2006. Heavy metals contents in agricultural top soils in the Ebro basin (Spain). Application of the multivariate geostatistical methods to study spatial variations. *Environmental Pollution* 144: 1001–1012.

Roldán, A., F. García-Orenes, and J. Albaladejo. 1994. Microbial populations in the rhizosphere of the *Brachypodium retusum* and their relationship with stable aggregates in a semiarid soil of southeastern Spain. *Arid Land Research and Management* 8: 105–114.

Ruf, A. 1998. A maturity index for predatory soil mites (Mesostigmata: Gamasina) as indicator of environmental impacts of pollution on forest soils. *Applied Soil Ecology* 9: 447–452.

Ryan, M. 1999. Is an enhanced soil biological community, relative to conventional neighbours, a consistent feature of alternative (organic and biodynamic) agricultural systems? *Biological Agriculture and Horticulture* 17: 131–144.

Saleh-Lakha, S., M. Miller, R. Campbell, K. Schneider, P. Elahimanesh, M. Hart, and J.T. Trevors. 2005. Microbial gene expression in soil: Methods, applications and challenges. *Journal of Microbiological Methods* 63: 1–19.

Sanchez, P.A., C.A. Palm, and S.W. Buol. 2003. Fertility capability soil classification: a tool to help assess soil quality in the tropics. *Geoderma* 114: 157–185.

Saviozzi, A., R. Levi-Minzi, R. Cardelli, and R. Riffaldi. 2001. A comparison of soil quality in adjacent cultivated, forest and native grassland soils. *Plant and Soil* 233: 251–259.

Schmeisser, C., H. Steele, and W.R. Streit. 2007. Metagenomics, biotechnology with nonculturable microbes. *Applied Microbiology and Biotechnology* 75: 955–962.

Schoenholtz, S., H. Van Miegroetb, and J. Burger. 2000. A review of chemical and physical properties as indicators of forest soil quality: Challenges and opportunities. *Forest Ecology and Management* 138: 335–356.

Secretariat of the Convention on Biological Diversity. 2000. http://www.cbd.int/secretariat/

Sessitsch, A., E. Hackl, P. Wenzl, A. Kilian, T. Koostic, N. Stralis-Pavese, B. Tankouo-Sandjong, and L. Bodrossy. 2006. Diagnostic microbial microarrays in soil ecology. *New Phytologist* 171: 719–736.

Seybold, C.A., M.J. Mausbach, D.L. Karlen, and H.H. Rogers. 1998. Quantification of soil quality. In *Soil Processes and the Carbon Cycle*, eds. R. Lal, J.M. Kimble, R.F. Follet, and B.A. Stewart, pp. 387–403. Boca Raton, FL: CRC Press.

Shannon, C.E. and W. Weaver. 1949. *The Mathematical Theory of Communication*. Urbana, IL: University of Illinois Press.

Sharma, K., U. Mandal, K. Srnivas, K. Vittal, B. Mandal, J. Grace, and V. Ramesh. 2005. Long-term soil management effects on crop yields and soil quality in a dry-land Alfisol. *Soil and Tillage Research* 83: 246–259.

Shepherd, K. and M. Walsh. 2002. Development of reflectance spectral libraries for characterization of soil properties. *Soil Science Society of America Journal* 66: 988–998.

Siepel, H. and C. van de Bund. 1988. The influence of management practices on the microarthropod community of grassland. *Pedobiologia* 31: 339–354.

Siepel, H. 1994. Structure and function of soil microarthropod communities. PhD dissertation, Wageningen, the Netherlands: University of Wageningen.

Siepel, H. 1996. Biodiversity of soil microarthropods: The filtering of species. *Biodiversity and Conservation* 5: 251–260.

Six, J., E. Elliott, and K. Paustian. 2000. Soil structure and soil organic matter. II. A normalized stability index and the effect of mineralogy. *Soil Science Society of America Journal* 64: 1042–1049.

Stefanic, F., G. Ellade, and J. Chirnageanu. 1984. Researches concerning a biological index of soil fertility. In *Fifth Symposium on Soil Biology*, eds. M.P. Nemes, S.P. Kiss, C. Papacostea, M. Stefanic, and M. Rusan, pp. 35–45. Bucharest, Romania: Romanian National Society of Soil Science.

Thomson Reuters. 2011. http://thomsonreuters.com

Timlin, D.J., R.B. Bryant, V.A. Snyder, and R.J. Wagenet. 1986. Modeling corn grain yield in relation to soil erosion using a water budget approach. *Soil Science Society of America Journal* 50: 718–723.

Tran, L., R. O'Neill, and E. Smith. 2006. A generalized distance measure for integrating multiple environmental assessment indicators. *Landscape Ecology* 21: 469–476.

Trangmar, B.B., R. S. Yost, and G. Uehara. 1985. Application of geostatistics to spatial studies of soil properties. *Advances in Agronomy* 38: 45–94.

Tranvik, L., G. Bengtsson, and S. Rundgren. 1993. Relative abundance and resistance traits of two Collembola species under metal stress. *Journal of Applied Ecology* 30: 43–52.

Trasar-Cepeda, C., C. Leirós, F. Gil-Sotres, and S. Seoane. 1998. Towards a biochemical quality index for soils: An expression relating several biological and biochemical properties. *Biology and Fertility of Soils* 26: 100–106.

Turbé, A., A. De Toni, P. Benito, P. Lavelle, N. Ruiz, W.H. Van der Putten, E. Labouze, and S. Mudgal. 2010. Soil biodiversity: Functions, threats and tools for policy makers. Bio Intelligence Service, IRD, and NIOO report for the European Commission. European Commission.

Turco, R., A. Kennedy, and M. Jawson. 1994. Microbial indicators of soil quality. In *Defining Soil Quality for a Sustainable Environment*, eds. J.W. Doran, D.C. Coleman, D.F. Bezdicek, and B.A. Stewart, pp. 73–90. Madison, WI: Soils Science Society of America.

United Nations, 1992. National biodiversity strategies and action plans. http://www.cbd.int/nbsap

Vagen, T., K. Shepherd, and M. Walsh. 2006. Sensing landscape level change in soil fertility following deforestation and conversion in the highlands of Madagascar using Vis-NIR spectroscopy. *Geoderma* 133: 281–294.

Van Bruggen, A.H.C. and A.M. Semenov. 2000. In search of biological indicators for soil health and disease suppression. *Applied Soil Ecology* 15: 13–24.

Van Gestel, C.A.M. and N.M. van Straalen. 1994. Ecotoxicological test systems for terrestrial invertebrates. In *Ecotoxicology of Soil Organisms*, eds. M.H. Donker, H. Eijsackers, and F. Heimbach, pp. 205–229. Lansing, MI: Lewis Publishers.

Van Groenigen, M., M. Gandah, and J. Bruma. 2000. Soil sampling strategies for precision agricultural research Ander Sahelian conditions. *Soil Science Society of America Journal* 64: 1674–1680.

Van Straalen, N.M., T.B.A. Burghouts, and M.J. Doornhof. 1985. Dynamics of heavy metals in populations of Collembola in a contaminated pine forest soil. In *Proceedings of International Conference on Heavy Metals in the Environment*, ed. T.D. Lekkas, pp. 613–615. Edinburgh, Ireland: CEP Consultants Ltd.

Van Straalen, N.M. and D.A. Krivolutsky. 1996. *Bioindicator Systems for Soil Pollution.* Dordrecht, the Netherlands: Kluwer Academic Publishers.

Van Straalen, N. 1997. Community structure of soil arthropods. In *Biological Indicators of Soil Health*, eds. C.E. Pankhurst, B.M. Doube, and V.V.S.R. Gupta, pp. 235–264. Wallingford, U.K.: CABI Publishing.

Van Straalen, N.M. and H.A. Verhoef. 1997. The development of a bioindicators system for soil acidity based on arthropod pH preferences. *Journal of Applied Ecology* 34: 217–232.

Van Straalen, N. 1998. Evaluation of bioindicator systems derived from soil arthropod communities. *Applied Soil Ecology* 9: 429–437.

Van Straalen, N. 2004. Chapter 6—The use of soil invertebrates in ecological surveys of contaminated soils. *Developments in Soil Science* 29: 159–195.

Wagenet, R. J. and J.L. Hutson. 1997. Soil quality and its dependence on dynamic physical processes. *Journal of Environmental Quality* 26: 41–48.

Wallwork, J. 1970. *Ecology of Soil Animals.* London, U.K.: McGraw-Hill.

Warkentin, B. 1995. The changing concept of soil quality. *Journal of Soil Water Conservation* 50: 226–228.

Wells, C. 1984. Nutrient balances and high productivity. In *Southeastern Forest Experiment Station, Research Paper*, pp. 439–443. Asheville, NC: USDA Forest Service.

Wienhold, B., J.L. Pikul, Jr., M.A. Liebig, M.M. Mikha, G.E. Varvel, J.W. Doran, and S.S. Andrews. 2006. Cropping system effects on soil quality in the Great Plains: Synthesis from a regional project. *Renewable Agriculture and Food Systems* 21: 49–59.

Wilmes, P. and O. Bond. 2006. Metaproteomics: Studying functional genes expression in microbial ecosystems. *Trends in Microbiology* 14: 92–97.

Williams, M., C. Ryan, R. Rees, E. Sambane, J. Fernando, and J. Grace. 2008. Carbon sequestration and biodiversity of re-growing miombo woodlands in Mozambique. *Forest Ecology and Management* 254: 145–155.

Wodarz, D., E. Aescht, and W. Fissner. 1992. A weighted coenotic index (WCI): Description and application to soil animal assemblages. *Biology and Fertility of Soils* 14: 5–13.

Yakowitz, D., J. Stone, L. Lane, P. Heilman, J. Masterson, J. Abolt, and B. Imam. 1993. A decision support system for evaluating the effect of alternative farm management systems on water quality and economics. *Water Science and Technology* 28: 47–54.

Yeates, G.W. and T. Bongers. 1999. Nematode diversity in agro ecosystems. *Agriculture, Ecosystems and Environment* 74: 113–135.

Zanella, A., M. Englisch, B. Jabiol, K. Katzensteiner, R. de Waal, H. Hager, B. van Delft et al. 2006. Towards a common humus form classification, a first European approach: Few generic top-soil references as functional units. In *18th World Congress of Soil Science*, Philadelphia, PA. http://a-c-s.confex.com/crops/wc2006/techprogram/P16690.HTM (accessed on October 2012).

Zornoza, R., J. Mataix-Solera, C. Guerrero, V. Arcenegui, F. García-Orenes, J. Mataix-Beneyto, and A. Morugán. 2007. Evaluation of soil quality using multiple lineal regression based on physical, chemical and biochemical properties. *Science of Total Environment* 378: 233–237.

5 Functionality Indicators for Sustainable Management

José Antonio Manzanera, Susana Martín-Fernández, and Antonio García-Abril

CONTENTS

5.1   SUMMARY

Environmental indicators are quantitative assessments providing information on the state of conservation and health of the environment. Some environmental indicators have a wide range, for instance, greenhouse gas emissions, protection areas, biodiversity indices, proportion of forest land area, net annual forest growing stock volume, and average condition of health and vigor. All of them help to characterize the current status and track or predict significant changes in the ecosystem.

In contrast, narrow-range indicators are those based on the ecophysiological state of the vegetation. These indicators are of a technical and scientific nature and are aimed at a more limited and specialized audience. Nevertheless, they can be adapted to a wider-ranging audience for managerial purposes, for instance, in policy evaluation. Most of the time, functionality indicators may be used as an indicator system, rather than individually, as a collection of indicators viewed as a whole to provide a better assessment of the integrity and health of the ecosystem. In this chapter, we focus our interest on the functioning of plants as an important part of the ecosystem.

Forest functionality indicators are an especially complex area, as the estimation of something as simple as biomass demands a large number of measurements, estimations, and models. This scenario is aggravated by the substantial influence of environmental factors and climate change. Nevertheless, certain indices have proven to be efficient estimators of vegetation biomass and net primary and ecosystem production, such as leaf area index (LAI), light use efficiency, and extinction coefficient of radiation in a forest canopy.

A different approach is the use of models for predicting the ecosystem functionality. There are two main types of models. The first is based on growth processes governed by physiological and environmental variables that regulate the ecosystem. Growth process models cover a wide range of complexity, from detailed studies of environmental and physiological variables at the plant level to simple models for large-scale and long-term general applications. The second type is based on empirical data that analyze the balance of inputs and outputs. Examples of the latter are the gap models of population dynamics and classic forest models, which calculate growing stock volume in a stand from diameter at breast height, tree height, stand density, tree age, site index, and/or other forest variables. Due to the empirical nature of these models, their utility is restricted to climate, soil, and other environmental conditions similar to those of the site where the model was developed, and they cannot thus be applied to large-scale scenarios.

The indicators of interest for our particular case are specifically known as ecophysiological indicators, an example of which is presented in this chapter. This case study considers the strategic role of riparian ecosystems for the surface protection of floodplains. Riparian vegetation exhibits high levels of biodiversity, nutrient cycling, and productivity and provides specialized ecological functions, such as improving water quality. Furthermore, riparian ecosystems may constitute greenbelts and ecological corridors to connect natural areas and ecosystems that would otherwise be isolated by human pressure. In many semiarid environments of Mediterranean ecosystems, white poplar (*Populus alba* L.) is the dominant riparian tree and has been used to recover degraded areas, together with other native species such as ash (*Fraxinus angustifolia* Vahl.) and hawthorn (*Crataegus monogyna* Jacq.). Gas-exchange patterns were investigated and compared to those of natural stands. In the restoration zones, the planted white poplars had higher rates of net assimilation and water use efficiency (WUE) than the mature trees in the natural stand. Significant differences in physiological performance between species were also found. The net assimilation and transpiration rates of white poplar were higher than those for ash and hawthorn. White poplar also showed higher levels of stomatal conductance and behaved as a colonizing, water-consuming species with a more active gas-exchange and ecophysiological adaptation than the other species used for restoration purposes. Nevertheless, ash and hawthorn play a complementary role for the purposes of biodiversity.

5.2   INTRODUCTION: STATE OF THE ART ON ECOPHYSIOLOGICAL INDICATORS

Ecophysiological indicators are measures with varying degrees of complexity of various important plant functions such as photosynthesis, water balance, and nutrient status, which can reveal what is happening in a particular ecosystem. Indeed, the functioning of plants is an important part of many ecosystems and often plays a key role in the energy, water, and biogeochemical cycles of ecosystems such as forests and prairies, in addition to significantly contributing to other indicators like environmental and biological indicators. The ecological role of living organisms such as plants can be assessed with quantitative measurements or with statistics over time. These types of indicators are of the type known as sustainable development indicators, which track sustainability with respect to the environment (Smeets and Weterings 1999).

Some wide-ranging environmental indicators related to plant physiology include the well-known greenhouse gas emissions measured in CO_2 equivalent mass forest area for protection purposes, and biodiversity at the specific, intraspecific, and ecosystem level (Government of Canada 2010). Various widely used forest indicators are also related, including proportion of forest land area, net annual growing stock volume, and average crown condition with regard to health and vigor.

Ecological indicators are required to characterize the functioning of a complex ecosystem by means of direct measurements. This is a difficult task, mainly due to issues of size and scale, complexity of interactions, cost, lack of methodological standards on an international scale, lack of past reference levels for long-term monitoring, etc. However, ecological indicators and especially functionality or ecophysiological indicators can be used to identify major ecosystem stress and should therefore be retained. Bioindicator species may be a special case, when their function can be used to determine environmental integrity. Biological indicators can also consist of a measure, an index of measures, or a model that characterizes an ecosystem or one of its critical components, for instance, the dominant tree species in a forest. Therefore, their primary use is to characterize the current status and track or predict significant changes in the ecosystem.

Ecophysiological indicators are of a technical and scientific nature, which means that they are geared to a more limited and specialized audience. Nevertheless, they can be adapted to a wider-ranging audience for managerial purposes, for instance, in policy evaluation. Most of the time, functionality indicators may be used as an indicator system, rather than individually, as a collection of indicators viewed as a whole to provide a better assessment of the environment, integrity, and health of a forest.

Forest functionality indicators are an especially complex area, as the estimation of something as simple as biomass demands a large number of measurements, estimations, and models and is approximate and inaccurate. One approach is the estimation of net primary productivity (NPP). It is known that the NPP of a forest ecosystem is determined by environmental factors subjected to climate change, such as shortwave energy available for photosynthesis, temperature, water availability in the soil, and other factors determining the plant water content, which influence plant development. More specifically, water cycle plays a limiting role in the forest ecosystem. This is especially acute in the Mediterranean and other arid or semiarid climates. This circumstance reduces forest growth and productivity, as a dry soil or a high evaporative demand of the atmosphere may cause a negative water balance in the vegetation. An adequate quantification of the water balance in such circumstances is a priority for forest management within the framework of climate change.

As an essential research tool, ecophysiological indicators in the form of indices and models have been developed to further the knowledge of the functional phenomena implicated in a forest ecosystem. These models hypothesize the ecosystem function and can explain and predict responses to changes in the variables in the model, thus clearly constituting a perfect management tool.

5.3   VEGETATION INDICES

Various indices have been developed in view of the difficulties in accurately assessing environmental indicators such as ecosystem biomass. One of these is the LAI, which is defined as the projected surface area of plant leaves divided by the soil area occupied by the plant, that is, its vertical projection. LAI is directly related to NPP by introducing the concept of leaf efficiency in the conversion of photosynthetically active radiation (PAR) into biomass:

$$\text{LAI} = \frac{\text{NPP}}{\varepsilon(h) * c} \tag{5.1}$$

where
   $\varepsilon(h)$ is the efficiency of leaves in the conversion of PAR into biomass
   c is a constant that depends on the cover type

An obvious difficulty is the measurement of LAI over wide forest areas. The remote sensing approach attempts to resolve this obstacle in a practical way. For instance, Nemani and Running (1989) found a log–linear relationship between LAI and the normalized difference vegetation index (NDVI), by which the normalized ratio of near-infrared to red reflectivity calculated by NDVI permitted an acceptable estimation of LAI.

Additionally, NDVI has been used as an indicator of chlorophyll content. However, the relationship was not linear and the estimation was only successful for low chlorophyll contents. NDVI has had more useful applications in the estimation of absorbed PAR (APAR) by plants. In this case, the relationship between APAR and NDVI appeared to be linear and was used in the scattering by arbitrarily inclined leaves (SAIL) model (Goward and Huemmrich 1992).

Another variable of interest is photosynthetic radiation use efficiency (PRUE), the rate of CO_2 absorption per quantum of energy used in photosynthesis. This variable can successfully be estimated by another physiological index, the photochemical reflectance index (PRI), which is defined as

$$PRI = \frac{R531 - R570}{R531 + R570} \qquad (5.2)$$

where
  R531 and R570 are the reflectivities of the vegetation cover at 531 and 570 nm, respectively (Gamon et al. 1992)

The reflectivities at these particular wavelengths are characteristic of xanthophylls present in the leaves, which are indicators of PRUE and responses to stress. The PRI can be estimated by means of hyperspectral remote sensing.

Apart from other very important uses, PRUE has been used in a model for NPP estimation:

$$NPP = PRUE*[f(APAR)*f(\text{soil moisture})*f(VPD)*f(T)] \qquad (5.3)$$

where
  VPD is the vapor pressure deficit in the atmosphere
  T is the temperature

When there is no limitation due to water or nutrient shortage, NPP is linearly related to APAR, PRUE being the slope of the model (Monteith 1972, 1977; Rosati et al. 2004).

Another useful ecophysiological indicator is the extinction coefficient of radiation (k) in a forest canopy, which is defined following the Beer–Lambert law as

$$k = LAI^{-1} * Ln\left(\frac{PPFD_o}{PPFD_{LAI}}\right) \qquad (5.4)$$

where
  $PPFD_o$ is the photosynthetic photon flux density on top of the forest canopy
  $PPFD_{LAI}$ is the density under the forest canopy

Although Beer–Lambert law was first applied to turbid solutions, it also adequately describes the phenomenon of light absorption by foliage (Norman 1979, Monteith and Unsworth 1990, Landsberg and Gower 1997).

Another index that can be estimated by remote sensing is the water index (WI), which is the ratio between the reflectivities at 900 nm and at 970 nm, respectively. This index reveals changes in the relative water content of plants, water potential, and other physiological variables such as stomatal conductance. Among the multiple applications of WI, it is used to estimate forest fire risk (Gao 1996, Garcia et al. 2008).

5.4 MODELS OF SYSTEM FUNCTIONALITY

There are two main types of models. The first is based on growth processes governed by physiological and environmental variables that regulate the ecosystem, and the second is based on empirical data analyzing the balance of inputs and outputs. Examples of the latter are the gap models of population dynamics and classic forest models, which calculate growing stock volume in a stand from tree height, diameter at breast height, stand density, tree age, site index, and/or other forest variables. Due to the empirical nature of these models, their utility is restricted to climate, soil, and other conditions similar to those of the site where the model was developed, and they cannot thus be applied to large-scale scenarios.

Growth process models, on the other hand, are based on ecosystem functionality, to which vegetation physiology offers a major contribution. Growth process models cover a wide range of complexity, from detailed studies of environmental and physiological variables at the plant level to simple models for large-scale and long-term general applications. Among others, the model MAESTRO (Wang and Jarvis 1990) is based on the assessment of net carbon assimilation, transpiration, and forest cover structure. The model BIOMASS (McMurtrie et al. 1990) has been successfully used for pine stand studies based on net assimilation, stomatal conductance, soil moisture, and precipitation. Other models have focused on the analysis of the biogeochemical cycles of the ecosystem. The forest ecosystem version is the model FOREST-BGC (Running and Coughlan 1988), which studies carbon, nitrogen, and water cycles as the main factors in forest development. This model has been used to estimate the aboveground NPP of conifers. The model PnET (Aber and Federer 1992) estimates net ecosystem productivity (NEP) in temperate broad-leaved forests. In summary, there is a large range of models to cover most situations, which provide a good explanation of the main factors governing growth responses to changing environmental factors (Landsberg and Gower 1997).

The environmental and ecophysiological variables feeding these growth process models can be either measured directly or, when this is not possible, estimated from other variables. For instance, net carbon assimilation of autotrophic plants is principally governed by the photosynthetic photon flux density (PPFD) reaching the plant and can be more easily estimated on a large-scale than net assimilation; stomatal conductance may be estimated from the vapor pressure gradient (VPG) from the plant mesophyll to the atmosphere, etc. Therefore, most physiological variables can be estimated from a set of meteorological or environmental variables, which are easier to measure or calculate, given the species, type of population, and behavior. We may hypothesize that growth process models can be used on medium to large scales and based on environmental data for managerial purposes at the stand or forest level. These data would determine the functioning of the models and their translation into ecophysiological processes of biological productivity, from the individual to the ecosystem scale.

Data gathering has been conducted using traditional methods of field data campaigns in ecosystems, floras, forests, etc. In particular, the high complexity of forest ecosystems has received special attention, and their behavior in the context of climate change is today a priority. However, classic forest inventories are insufficient to provide the amount of new data necessary for modeling, and the expense is prohibitive on such a large scale.

However, new technologies have been developed, which may help to remove these obstacles. Since the 1960s, remote sensing has delivered new applications of enormous interest for environmental studies. This new science is based on the observation of the Earth's surface from sensors installed on satellites or other platforms that capture relevant environmental information at regular intervals from across broad extensions and at a moderate cost (Chuvieco and Congalton 1989, Arroyo et al. 2008). The regular revisit period of satellite imagery at a territorial scale makes remote sensors the ideal tool for this purpose. In spite of all these advantages, the main drawback of this technology is the need to calibrate the data through field validation. Therefore, accurate estimations of the main physiological variables are necessary for the correct adjustment of the models.

5.5 CASE STUDY: INDICATORS FOR THE ECOPHYSIOLOGICAL COMPETENCE OF WOODY SPECIES FOR RIPARIAN ECOSYSTEM RESTORATION

5.5.1 INTRODUCTION

The conservation of riparian forests is an important goal, as floodplain vegetation plays a strategic role both in the surface protection of river banks from floods and as an environment, which shelters high levels of biodiversity and fulfills specialized ecological functions such as improving water quality (Glenz 2005). Furthermore, they may constitute greenbelts and ecological corridors to connect natural areas and ecosystems that would otherwise be isolated by human pressure. Riparian trees and shrubs maintain high carbon assimilation and growth rates under conditions that cause drought-induced growth reduction in upland trees (Foster 1992, Hart and Disalvo 2005). Nevertheless, in many semiarid Mediterranean-climate regions, riparian ecosystems receive irregular annual precipitations and may suffer from stream flow variations, leading some riparian species to be possibly more water stressed in these conditions than those from mesic ecosystems (Smith et al. 1998). In cottonwood (*Populus fremontii* Wats.), transpiration occurred at the highest rates with maximum temperature and vapor pressure deficit. In this species, transpiration decreased in parallel with radiation input on cloudy days, while willow (*Salix gooddingii* Ball) behaved differently (Schaeffer et al. 2000). These results suggest that the relationship of riparian tree physiology at the leaf level with environmental factors such as the light and VPG may affect their ecological role in floodplain areas. Leaf-level gas-exchange parameters vary among species and can therefore be used as indicators of the response to changes in the riparian ecosystem and as predictors of plant behavior in restoration activities.

5.5.2 EVALUATION METHODOLOGY

The study was conducted in a protected space, the floodplain of the Henares river (Madrid, Spain), located at 40° 02′ N, 3° 36′ W and at 588 m elevation.

The protected area was established as a measure to recover the site, in which over 26 ha had been cleared of all vegetation for the installation of quarries. Several years later, the restoration activities conducted in these disturbed areas involved the establishment of plantations with native species, for example, white poplar (*P. alba* L.), narrow-leaved ash (*F. angustifolia* Vahl.), and hawthorn (*C. monogyna* Jacq.). The latter is a fast-growing thorny deciduous native shrub, tolerant of wet soils and frequently present along water streams. Natural riparian forests had been preserved close to these restored areas, formed by mature white poplar and elm (*Ulmus minor* Mill.) groves. Some patches of natural regeneration dominated by young white poplars were also present. The soil is alluvial and well drained, with sandy–gravel subsoil derived from deposits from the nearby river. Meteorological data were recorded from an existing local weather station located in the vicinity of the study site. Mean annual precipitation was 800 mm and mean annual temperature 14°C for the 2-year period 2002–2003, with a broad range of daily temperatures of between −8.9°C and 39.7°C during the same period (Manzanera and Martínez-Chacón 2007).

The plantations were established in 1994 and 1999. The study area was divided into three zones: a zone of natural vegetation (called zone A), consisting of alternating groves of both mature and young white poplars accompanied by elms, and two plantation zones, one planted in 1994 (9 years old during the study, zone B) and a 4 year old zone planted in 1999 (zone C). In each zone, a 25 m * 25 m square plot was randomly installed, and 10 plants of each species and type were selected within each plot. Plant establishment after planting was guaranteed by supplemental water to each tree by drip irrigation for 4 h every 2 weeks during the summer months for the first 3 years after planting.

Gas-exchange parameters were measured in natural light with a portable LCI (ADC Bioscientific Ltd.) gas analyzer and ranged from 26 to 2640 μmol m^{-2} s^{-1} during measurements. These measurements were performed under a clear sky on four fully developed leaves, each located at the apex of four lateral branches, oriented south, north, west, and east in the crown of each plant. We registered gas-exchange variables, for example, net assimilation rate (A, μmol CO_2 m^{-2} s^{-1}), stomatal conductance to water vapor (g_s, mmol water vapor m^{-2} s^{-1}), transpiration rate (E, mmol water vapor m^{-2} s^{-1}), and PPFD (μmol m^{-2} s^{-1}). WUE (the ratio of A–E), intercellular CO_2 concentration (Ci, μmol mol^{-1}), and leaf to air water VPG (kPa) were calculated using the recorded data. Physiological measurements were initiated 9 years after the first plantation was established, during the period from August 2002 to August 2003.

5.5.2.1   Poplar and Elm in the Natural Riparian Forest

The first set of data was recorded as a preliminary study in the natural riparian forest during summer, from August to October 2002, and the physiological parameters of both mature and young poplars and elms were measured at midday.

5.5.2.2   Gas Exchange from Natural versus Planted White Poplar during Summer

Mature poplar trees from the natural zone (A) and poplar plants from both plantation zones (B and C) were compared through gas-exchange measurement from June to

August 2003, by sampling 10 mature and 10 young poplars every 2 h, in 4 periods: from 8 to 10 a.m., from 10 to 12 a.m., from 12 a.m. to 2 p.m., and from 2 to 4 p.m., solar time.

5.5.2.3 Comparison of Gas-Exchange Characteristics among Species in Plantations during the Summer Period

In both plantations (zones B and C), white poplar, ash, and hawthorn were compared by sampling 10 plants per species per zone along the same four daily periods as the former experiment, from June to August 2003.

5.5.2.4 Statistical Analysis

For all mean comparisons, multifactorial ANOVA (Statgraphics Plus 5.1) was used with PPFD as a covariate to adjust gas-exchange characteristics to the same PPFD. In natural conditions, this parameter ranged from 26 to 2640 $\mu mol\ m^{-2}\ s^{-1}$. The multiple range test of the least significant difference (LSD) at the 0.05 level was used to discriminate means.

After statistical analysis, leaf measurements were grouped into two sunlit (south and east) and two shaded (west and north) leaf records per plant. Responses of leaf gas exchange were also related to environmental variables (VPG and PPFD) by boundary-line analysis, and regression models were fitted to the higher part of the data (Chambers et al. 1985). The model of best fit was selected and the adjusted R-squared statistic was calculated for each model. Assimilation vs. PPFD curves of the planted species in June and July 2003 and of the natural mature white poplars and elms from August to October 2002 were fitted to the rectangular hyperbola model for canopy carbon assimilation (Landsberg and Gower 1997):

$$A = \frac{\Phi c * PPFD * A_{max}}{\Phi c * PPFD + A_{max}} \tag{5.5}$$

where A_{max} is the photon-saturated assimilation rate and the apparent maximum quantum efficiency (Φc) is given by the initial slope of the A vs. PPFD curve. Conductance to water vapor vs. VPG curves of the planted species in June and July 2003 and of the natural mature white poplars and elms from August to October 2002 was fitted to a polynomial regression model.

5.5.3 RESULTS

5.5.3.1 Poplar and Elm in the Natural Riparian Forest

In the natural stand, young white poplar plants showed a higher net assimilation rate than the mature trees in August (Figure 5.1a). This result was associated with differences in the PPFD received, which were higher in the young white poplar plants than in the mature trees and elms (Figure 5.1f). Elms assimilated significantly less CO_2 than poplars during the whole period, although no statistical differences in intercellular CO_2 concentration were found in comparison with mature poplars

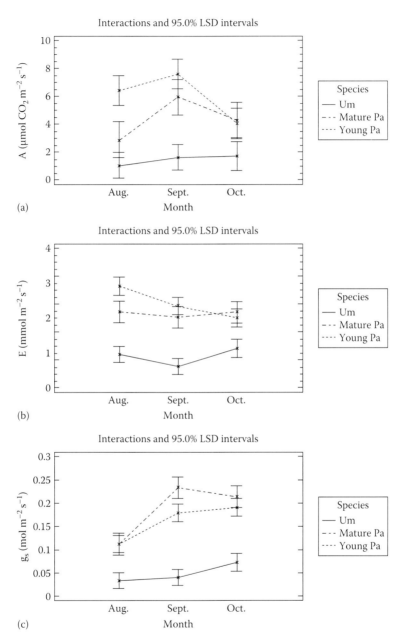

FIGURE 5.1　(a) Net assimilation rate (A, μmol CO_2 m^{-2} s^{-1}), (b) transpiration (E, mmol m^{-2} s^{-1}), (c) conductance to water vapor (g_s, mol m^{-2} s^{-1}), (d) intercellular CO_2 concentration (Ci, μmol mol^{-1}), (e) WUE, and (f) PPFD (μmol m^{-2} s^{-1}) of three plant types: mature trees of *P. alba* (mature Pa), young regenerated plants of the same species (young Pa), and mature *U. minor* (Um) in the natural stand of the Henares floodplain during the period August–October 2002.

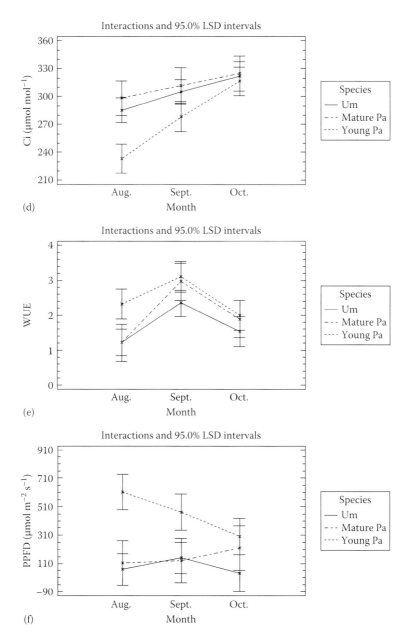

FIGURE 5.1 (continued)   (a) Net assimilation rate (A, $\mu mol\ CO_2\ m^{-2}\ s^{-1}$), (b) transpiration (E, $mmol\ m^{-2}\ s^{-1}$), (c) conductance to water vapor (g_s, $mol\ m^{-2}\ s^{-1}$), (d) intercellular CO_2 concentration (Ci, $\mu mol\ mol^{-1}$), (e) WUE, and (f) PPFD ($\mu mol\ m^{-2}\ s^{-1}$) of three plant types: mature trees of *P. alba* (mature Pa), young regenerated plants of the same species (young Pa), and mature *U. minor* (Um) in the natural stand of the Henares floodplain during the period August–October 2002.

TABLE 5.1

Quantum Efficiency of CO_2 Assimilation (Φc, Parameter Estimate $\pm$ Asymptotic Standard Error), Light-Saturated Assimilation (Amax, μmol m^{-2} s^{-1}, Parameter Estimate $\pm$ Asymptotic Standard Error), and Adjusted R-Squared Statistic (%) of the Regression Model of the Fitted A vs. PPFD Curves (Equation 5.5) for White Poplar (*P. alba*) and Elm (*U. minor*) in the Natural Stand of the Henares Floodplain Protected Area

Species	Φc	Amax	Adj. R^2
P. alba	0.10 ± 0.06	13.62 ± 4.54	69.79
U. minor	0.05 ± 0.01	8.43 ± 0.82	92.34

(Figure 5.1d). Assimilation was higher in September than in August and October. Gas exchange was higher in sunlit than in shaded leaves. These differences were statistically significant in September for all species, but not in October, when leaf activity lowers prior to senescence.

Transpiration and g_s also were higher in poplars than in elms (Figure 5.1b and c). Transpiration reached the maximum in August and the minimum for the period in September, probably due to accumulated drought stress (Figure 5.1b). In contrast, g_s increased in September and October. The intercellular CO_2 concentration of young poplars was significantly lower than in both mature poplars and elms, but there were no statistical differences between the latter (Figure 5.1d). Young poplars were more efficient in the use of water in August than the mature poplars and elms (Figure 5.1e).

The fitted rectangular hyperbola model for canopy carbon assimilation (Equation 5.5; Landsberg and Gower 1997) explained a high percentage of variability in both poplars and elms (between 69.79% and 92.34%, Table 5.1). Figure 5.2a shows typical A vs. PPFD boundary-line curves for both species. White poplar showed the highest estimated Φc and Amax and elm had the lowest estimated values of both parameters (Table 5.1). Boundary-line analysis also showed a good association between g_s and VPG (Figure 5.2b), with R^2 values between 70.05% and 83.49% for the polynomial regression models of both species (Table 5.2).

5.5.3.2 Gas Exchange of Natural versus Planted White Poplar during Summer

A typical summer pattern of daily variation in the gas exchange of white poplar is shown in Figure 5.3. No data were recorded at dawn as there were no sunlit leaves, or they were out of reach. Net assimilation of the sunlit leaves reached the maximum early in the morning and progressively descended, showing a midday depression. The same behavior was observed for transpiration, conductance to water

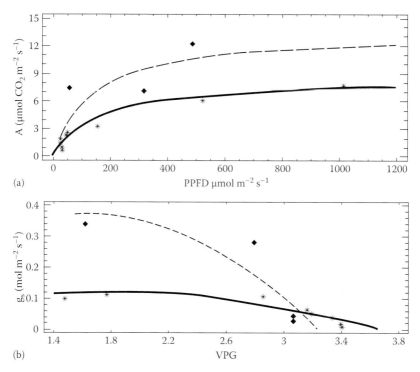

FIGURE 5.2   (a) Net assimilation rate (A, μmol CO_2 m^{-2} s^{-1}) vs. PPFD (μmol m^{-2} s^{-1}) and (b) conductance to water vapor (g$_s$, mol m^{-2} s^{-1}) vs. VPG (kPa) curves of *P. alba* (thin dotted line, ·) and *U. minor* (thick solid line,*) in the natural stand.

TABLE 5.2

Polynomial Regression Models Fitted between Water Vapor Conductance (g$_s$, mmol m^{-2} s^{-1}) and Water VPG (kPa) and Adjusted R-Squared Statistic (%) of White Poplar (*P. alba*) and Elm (*U. minor*) in the Natural Stand of the Henares Floodplain Protected Area

Species	Model	Adj. R^2 (%)
P. alba	$g_s = 0.460173\text{VPG} - 0.142108\text{VPG}^2$	70.05
U. minor	$g_s = 0.12698\text{VPG} - 0.0348974\text{VPG}^2$	83.49

---

vapor, and WUE. As expected, in all cases, sunlit leaves had higher assimilation rates than shaded leaves (p < 0.0001; Figure 5.3a). In contrast, the transpiration rate was highest at midday, and the differences between light and shadow leaves were greater (Figure 5.3b). The conductance to water vapor progressively declined from the maximum at 8–10 a.m. (Figure 5.3c).

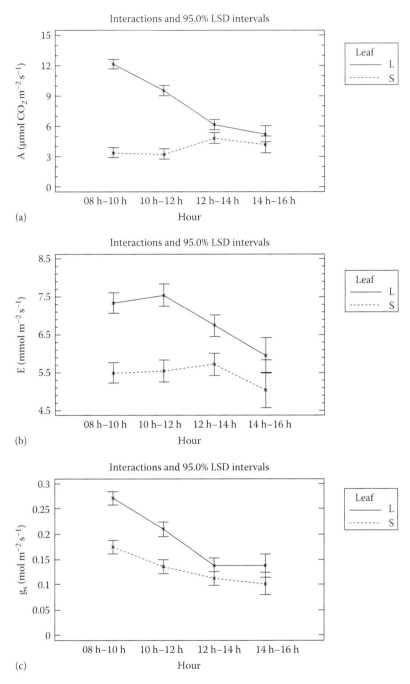

FIGURE 5.3 Daily changes in (a) net assimilation rate (A, μmol CO_2 m^{-2} s^{-1}), (b) transpiration (E, mmol m^{-2} s^{-1}), (c) conductance to water vapor (g_s, mol m^{-2} s^{-1}), and (d) PPFD (μmol m^{-2} s^{-1}) of white poplar leaves exposed to light (L) or shadow (S) during June–August 2003.

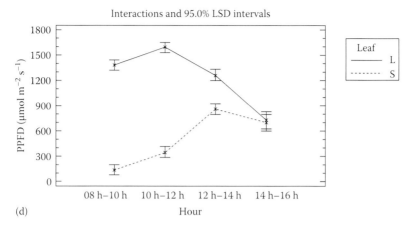

(d)

FIGURE 5.3 (continued)   Daily changes in (a) net assimilation rate (A, μmol CO_2 m^{-2} s^{-1}), (b) transpiration (E, mmol m^{-2} s^{-1}), (c) conductance to water vapor (g$_s$, mol m^{-2} s^{-1}), and (d) PPFD (μmol m^{-2} s^{-1}) of white poplar leaves exposed to light (L) or shadow (S) during June–August 2003.

5.5.3.3   Comparison of Gas-Exchange Characteristics among Species in Plantations during the Summer Period

The survival rate of the plantations was high for all the species used for the restoration of the Henares floodplain: 97.8% for ash, 91.3% for white poplar, and 87% for hawthorn. White poplar showed a higher growth in diameter and height than ash and hawthorn (Figure 5.4). White poplar plants in the zone C (1999 plantation) reached similar sizes to those of ash plants established in 1994 (zone B). Differences were found in net assimilation among the three species.

The maximum net assimilation rate was obtained in July for all species. The fitted rectangular hyperbola model for canopy carbon assimilation (Equation 5.5; Landsberg and Gower 1997) explained a high percentage of variability in all three species (R^2 values between 83.08% and 89.41%; Figure 5.5). When A vs. PPFD boundary-line curves for all three species were fitted, ash showed the highest estimated Amax and Φc, and hawthorn had the lowest estimated values of both parameters (Figure 5.6).

5.5.4   Discussion

The results obtained in our study confirm the key role of PPFD, that is, light, as a major environmental variable controlling the ecophysiological performance of plants. Indeed in the natural stand, young white poplar plants showed higher rates of net assimilation, transpiration, and WUE than the adult trees, associated with greater availability of light in the regeneration openings. These results agree with those obtained by Wittig et al. (2005) in white poplar and other *Populus* species, in which canopy closure caused a decline in light availability and the subsequent reduction in the carbon assimilation rate and gross primary production. Elms had

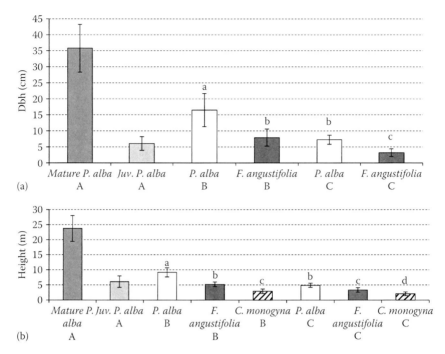

FIGURE 5.4 (a) Mean diameter at breast height (1.3 m; dbh, cm ± standard deviation) and (b) mean tree height (m, ± standard deviation) of white poplar (*P. alba*), ash (*F. angustifolia*), and hawthorn (*C. monogyna*) in the natural stand (zone A), the 1994 plantation (zone B), and the 1999 plantation (zone C) for the restoration of the Henares floodplain. Values with the same letter are not significantly different at the 0.05 level (LSD test).

a lower carbon assimilation rate, transpiration, stomatal conductance, and apparent maximum quantum efficiency than white poplar trees. Wittig et al. (2005) found in white poplar and other *Populus* species that canopy closure caused the decline of light availability and a subsequent reduction in the carbon assimilation rate and gross primary production. Stomatal conductance and photosynthetic rates have long been recognized to be positively associated (Kozlowski and Pallardy 1997, Wang et al. 2000, Peña-Rojas et al. 2004). Similar results have also been observed in *Populus tremuloides* (Noormets et al. 2001), in *P. fremontii* (Horton et al. 2001a), and in the floodplain tree *Acer negundo* (Foster 1992), in which PPFD was the primary factor influencing net carbon assimilation. The higher light availability in the restoration zones was probably due to the lower stand density of the planted white poplars, which had higher rates of assimilation and WUE than the mature trees in the natural stand.

The other main environmental variable, VPG, also showed a major influence on the ecophysiological performance of most woody plants. Specifically, high values of VPG in this study were a strongly significant factor influencing g_s and carbon assimilation. Horton et al. (2001b) observed that the vapor pressure deficit limited both net carbon assimilation and g_s in the species *P. fremontii*, which is adapted to

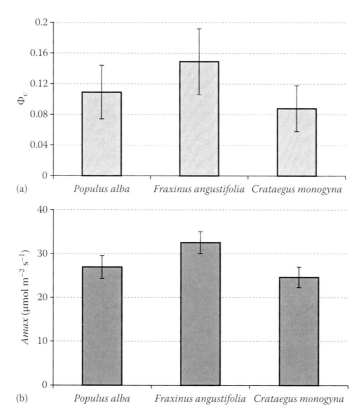

FIGURE 5.5 (a) Quantum efficiency of CO_2 assimilation (Φc, parameter estimate ± asymptotic standard error) and (b) light-saturated assimilation (Amax, $\mu mol\ m^{-2}\ s^{-1}$, parameter estimate ± asymptotic standard error) of the fitted A vs. PPFD curves for the species used in the restoration of the Henares floodplain.

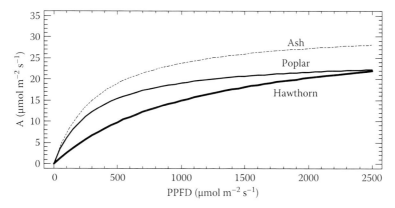

FIGURE 5.6 Net assimilation rate (A, $\mu mol\ CO_2\ m^{-2}\ s^{-1}$) vs. PPFD ($\mu mol\ m^{-2}\ s^{-1}$) curve of white poplar (*P. alba*, thin solid line), narrow-leaved ash (*F. angustifolia*, dotted line), and hawthorn (*C. monogyna*, thick solid line) plantations.

warm and dry climates. Disalvo and Hart (2002) reported that the increment in the relative basal area of *Populus trichocarpa* was negatively correlated with vapor pressure deficit. In box elder, a negative association was also found between A and g_s and vapor pressure deficit (Kolb et al. 1997). White poplars from both plantations and natural stands showed the highest rates of transpiration, net assimilation, and WUE at the beginning of the day and decreased later on in the day, in parallel with stomatal conductance. This midday depression has been described in eastern cottonwood (*Populus deltoides*), in which a high vapor pressure deficit made a more significant contribution than a high PPFD to the reduction in g_s and A (Pathre et al. 1998). A midday depression of photosynthesis has also been observed in white poplar leaves as a consequence of stomatal closure (Barta and Loreto 2006). Mature trees from the natural stand had greater conductance to water vapor than juvenile plants, implying that WUE was lower in mature than in young trees. This difference might be due to the fact that the mature trees had deeper roots and easier access to phreatic water for longer periods. In Douglas-fir riparian forests, age was the main factor influencing water use at the stand level, with young stands showing higher transpiration rates than mature stands (Moore et al. 2004). These results are in agreement with those obtained from g_s–VPG boundary-line analysis in white poplar (Manzanera and Martinez-Chacón 2007).

The net assimilation and transpiration rates of planted white poplar were higher than those of ash and hawthorn in association with greater conductance rates. The same can be said in comparison with elms in the natural stand. *Populus* shows higher levels of stomatal conductance than other woody plant genera with the same PPFD and vapor pressure deficit conditions (Will and Teskey 1997). A narrow cavitation safety margin and a tight stomatal regulation of transpiration have been observed in other *Populus* species (Sparks and Black 1999). This more active gas-exchange rate in poplar implies that this tree behaves as a water consumer, maximizing carbon assimilation and growth (Hetherington and Woodward 2003) but with a lower WUE than ash. This agrees with the boundary-line analysis of the g_s–VPG curves (Figure 5.2b), which shows that white poplar has a higher g_s than elm at low VPG, but the steeper slope at high VPG indicates a greater sensitivity to this parameter.

The photosynthetic capacity of ash was also measured by Kazda et al. (2000) in Central Europe, where lower estimated Amax values were found (16.67 μmol m^{-2} s^{-1}) than in our study (32.54 μmol m^{-2} s^{-1}); our higher value was probably due to a greater availability of light. Hawthorn had a lower assimilation rate than poplar and ash in the same zones. These differences in carbon uptake among species can influence ecosystem processes such as decomposition and nutrient cycling (Fischer et al. 2004).

We have therefore observed significant differences in physiological performance between species. White poplar behaved as a water-consuming, fast-growing species with a more sensitive gas-exchange dynamic than the other species used in the restoration project. The ecophysiological adaptation and tolerance of the young regenerating poplars are probably due to the development of a deep root system and to their ability to tap groundwater reserves from lower phreatic levels, according to the model of drought-avoiding, water-spending plants, thereby maintaining photosynthetically active leaves during longer periods than the species with a higher WUE.

Ash had a lower growth in diameter and height than white poplar, in spite of having higher Amax, probably due to its shorter leaf duration and to other limiting factors of the net assimilation rate.

We conclude by proposing the use of white poplar for the rapid restoration of riparian vegetation in semiarid Mediterranean environments. Ash and hawthorn can also play a role as accompanying species for the purposes of enhancing biodiversity. These findings should be taken into consideration by environmental managers for the establishment of specific objectives, including the conservation of all the important native species present in Mediterranean riparian ecosystems.

REFERENCES

Aber, J.D. and C.A. Federer. 1992. A generalized, lumped-parameter model of photosynthesis, evapotranspiration and net primary production in temperate and boreal forest ecosystems. *Oecologia* 92: 378–386.

Arroyo, L.A., C. Pascual, and J.A. Manzanera. 2008. Fire models and methods to map fuel types: The role of remote sensing. *Forest Ecology and Management* 256: 1239–1252.

Barta, C. and F. Loreto. 2006. The relationship between the Methyl-Erythritol phosphate pathway leading to emission of volatile isoprenoids and abscisic acid content in leaves. *Plant Physiology* 141: 1676–1683.

Chambers, J.L., T.M. Hinckley, G.S. Cox et al. 1985. Boundary-line analysis and models of leaf conductance for four oak-hickory forest species. *Forest Science* 31(2): 437–450.

Chuvieco E. and R.G. Congalton. 1989. Application of remote-sensing and geographic information-systems to forest fire hazard mapping. *Remote Sensing of Environment* 29(2): 147–159.

Disalvo A.C. and S.C. Hart. 2002. Climatic and stream-flow controls on tree growth in a Western montane riparian forest. *Environmental Management* 30 (5): 678–691.

Fischer, D.G., S.C. Hart, T.G. Whitham et al. 2004. Ecosystem implications of genetic variation in water-use of a dominant riparian tree. *Oecologia* 139: 288–297.

Foster, J.R. 1992. Photosynthesis and water relations of the floodplain tree, boxelder (*Acer negundo* L.). *Tree Physiology* 11: 133–149.

Gamon, J., J. Penuelas, and C.B. Field. 1992. A narrow-waveband spectral index that tracks diurnal changes in photosynthetic efficiency. *Remote Sensing of Environment* 41: 35–44.

Gao, B.C. 1996. NDWI. A normalized difference water index for remote sensing of vegetation liquid water from space. *Remote Sensing of Environment*, 58: 257–266.

García, M., E. Chuvieco, H. Nieto, and I. Aguado. 2008. Combining AVHRR and meteorological data for estimating live fuel moisture content. *Remote Sensing of Environment* 112: 3618–3627.

Glenz, C. 2005. Process-based, spatially-explicit modelling of riparian forest dynamics in Central Europe—Tool for decision-making in river restoration. PhD thesis, Faculty of Natural Environment, Polytechnic School of Lausanne, Switzerland.

Government of Canada, 2010. *The National Inventory Report 1990–2008: Greenhouse Gas Sources and Sinks in Canada*. Environment Canada. 582 p.

Goward, S.N. and K.F. Huemmrich. 1992. Vegetation canopy PAR absorptance and the normalized difference vegetation index: An assessment using the SAIL model. *Remote Sensing of the Environment* 39: 119–140.

Hart, S.C. and A.C. Disalvo. 2005. Net primary productivity of a western montane riparian forest: potential influence of stream flow diversion. *Madroño* 52(2): 79–90.

Hetherington, A.M. and F.I. Woodward. 2003. The role of stomata in sensing and driving environmental change. *Nature* 424: 901–908.

Horton, J.L., T.E. Kolb, and S.C. Hart. 2001a. Responses of riparian trees to interannual variation in ground water depth in a semi-arid river basin. *Plant Cell and Environment* 24: 293–304.

Horton, J.L., T.E. Kolb, and S.C. Hart. 2001b. Leaf gas exchange characteristics differ among Sonoran Desert riparian tree species. *Tree Physiology* 21: 233–241.

Kazda, M., J. Salzer, and I. Reiter. 2000. Photosynthetic capacity in relation to nitrogen in the canopy of a *Quercus robur*, *Fraxinus angustifolia* and *Tilia cordata* flood plain forest. *Tree Physiology* 20: 1029–1037.

Kolb, T.E., S.C. Hart, and R. Amundson. 1997. Boxelder water sources and physiology at perennial and ephemeral stream sites in Arizona. *Tree Physiology* 17: 151–160.

Kozlowski, T.T. and S.G. Pallardy. 1997. *Physiology of Woody Plants*. 2nd edn. Academic Press, San Diego, CA, 411 p.

Landsberg, J.J. and S.T. Gower. 1997. *Applications of Physiological Ecology to Forest Management*. Academic Press, San Diego, CA, 354 p.

Manzanera, J.A. and M.F. Martínez-Chacón. 2007. Ecophysiological competence of *Populus alba* L., *Fraxinus angustifolia* Vahl. and *Crataegus monogyna* Jacq. used in plantations for the recovery of riparian vegetation. *Environmental Management* 40(6): 902–912. DOI 10.1007/s00267-007-9016-z

McMurtrie, R.E., D.A. Rook, and F.M. Kellinher. 1990. Modelling the yield of *Pinus radiata* on a site limited by water and nutrition. *Forest Ecology and Management* 30: 381–413.

Monteith, J.L. 1972. Solar radiation and productivity in tropical ecosystems. *Journal of Applied Ecology* 9: 747–766.

Monteith, J.L. 1977. Climate and the efficiency of crop production in Britain. *Philosophical Transaction of the Royal Society of London, Series B* 281: 277–294.

Monteith, J.L. and M.H. Unsworth. 1990. *Principles of Environmental Physics*, 2nd edn. Arnold, London, U.K.

Moore, G.W., B.J. Bond, J.A. Jones et al. 2004. Structural and compositional controls on transpiration in 40- and 450-year-old riparian forests in western Oregon, USA. *Tree Physiology* 24: 481–491.

Nemani, R.R. and S.W. Running. 1989. Estimation of regional surface resistance to evapotranspiration from NDVI and Thermal-IR AVHRR data. *Journal of Applied Meteorology* 28: 276–284.

Noormets, A., A. Sober, E.J. Pell et al. 2001. Stomatal and non-stomatal limitation to photosynthesis in two trembling aspen (*Populus tremuloides* Michx.) clones exposed to elevated CO2 and/or O3. *Plant Cell and Environment* 24: 327–336.

Norman, J.M. 1979. Modeling the complete crop canopy. In *Modification of the Aerial Environment of Plants*, ed. B.J. Barfield and J.F. Gerber, pp. 249–277. American Society of Agricultural Engineering, St. Joseph, MI.

Pathre, U., A.K. Sinha, P.A. Shirke, and P.V. Sane. 1998. Factors determining the midday depression of photosynthesis in trees under monsoon climate. *Trees* 12: 472–481.

Peña-Rojas, K., X. Aranda, and I. Fleck. 2004. Stomatal limitation to CO$_2$ assimilation and down-regulation of photosynthesis in *Quercus ilex* resprouts in response to slowly imposed drought. *Tree Physiology* 24: 813–822.

Rosati, A., S.G. Metcalf, and B.D. Lampinen. 2004. A simple method to estimate photosynthetic radiation use efficiency of canopies. *Annals of Botany* 93: 567–574, doi:10.1093/aob/mch081.

Running, S.W. and J.C. Coughlan. 1988. A general model of forest ecosystem processes for regional applications. I. Hydrologic balance, canopy gas exchange and primary production processes. *Ecological Modelling* 42: 125–154.

Schaeffer, S.M., D.G. Williams, and D.C. Goodrich. 2000. Transpiration of cottonwood/willow forest estimated from sap flux. *Agricultural and Forest Meteorology* 105: 257–270.

Smeets, E. and R. Weterings. 1999. Environmental indicators: Typology and overview. Technical report No 25. European Environment Agency, Copenhagen, Denmark, 19 p.

Smith, S.D., D.A. Devitt, A. Sala, et al. 1998. Water relations of riparian plants from warm desert regions. *Wetlands* 18: 687–696.

Sparks, J.P. and R.A. Black. 1999. Regulation of water loss in populations of *Populus trichocarpa*: The role of stomatal control in preventing xylem cavitation. *Tree Physiology* 19: 453–459.

Wang, X., P.S. Curtis, K.S. Pregitzer, and D.R. Zak. 2000. Genotypic variation in physiological and growth responses of *Populus tremuloides* to elevated CO_2 concentration. *Tree Physiology* 20: 1019–1028.

Wang, Y.P. and P.G. Jarvis. 1990. Description and validation of an array model—MAESTRO. *Agriculture and Forest Meteorology* 51: 257–280.

Will, R.E. and R.O. Teskey. 1997. Effect of irradiance and vapour pressure deficit on stomatal response to CO_2 enrichment of four tree species. *Journal of Experimental Botany* 48(317): 2095–2102.

Wittig, V.E., C.J. Bernacchi, X.G. Zhu et al. 2005. Gross primary production is stimulated for three *Populus* species grown under free-air CO_2 enrichment from planting through canopy closure. *Global Change Biology* (2005)11: 1–13.

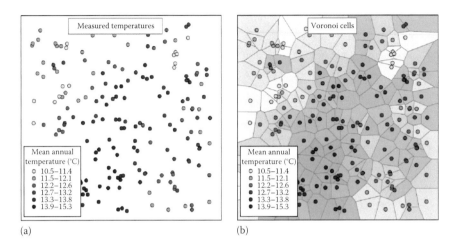

FIGURE 2.2 Voronoi or Thiessen interpolation algorithm. (a) Measured temperatures; (b) Voronoi cells.

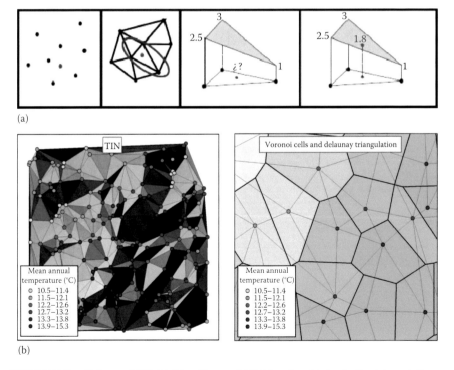

FIGURE 2.3 Delaunay TIN. (a) Algorithm and (b) triangles obtained after interpolation.

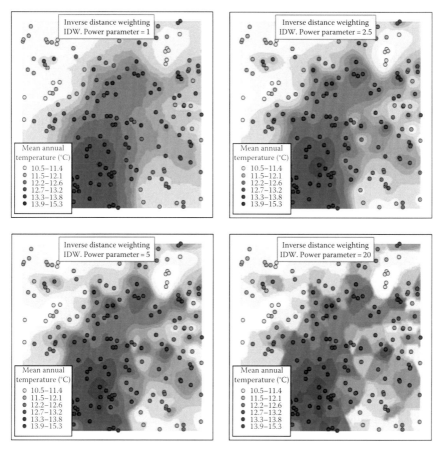

FIGURE 2.5 Inverse distance weighted (IDW) algorithm with increasing values for power parameter a.

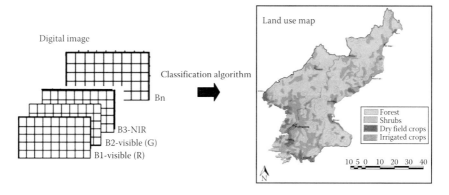

FIGURE 2.18 Classification of digital image process.

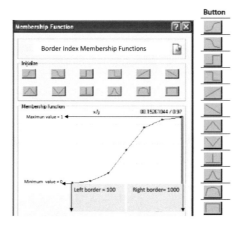

FIGURE 2.22 Types of membership functions.

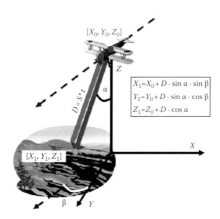

$$X_1 = X_0 + D \cdot \sin \alpha \cdot \sin \beta$$
$$Y_1 = Y_0 + D \cdot \sin \alpha \cdot \cos \beta$$
$$Z_1 = Z_0 + D \cdot \cos \alpha$$

FIGURE 2.23 LiDAR systems provide the distance between a sensor and a target surface based on the time between the emission of a pulse and the detection of a reflected return. D=distance of target surface; S=speed of light; t=time recorded by the lidar sensor. GNSS receivers and INSs allow the source of the return signal to be located in three dimensions.

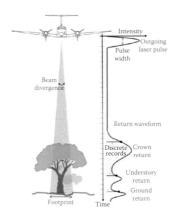

FIGURE 2.24 Differences between waveform recording and discrete-return lidar devices. At the left is the intersection of the laser illumination area (footprint). In the center, the hypothetical return signal of a waveform recording sensor. To the right, the signal recorded by discrete-return lidar sensors. (From Fernández-Díaz, J.C., *Imag. Notes*, 26(2), 31, 2011.)

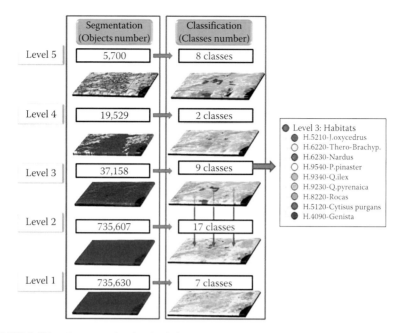

FIGURE 2.25 Segmentation levels (left) and their corresponding classifications (right).

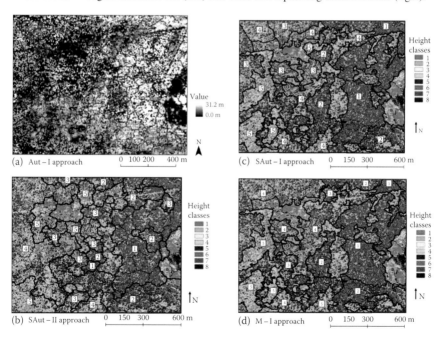

FIGURE 2.27 Results for the four proposed forest structure characterization approaches: (a) Aut-I approach, (b) SAut-II approach, and (c) SAut-I approach. The numbers inside the polygons indicate the forest structure type (1, 2, 3, 4, or 5) to which each polygon was assigned. Polygons with no number indicate that they were not forest stands and were not included in the forest structure classification. Dashed line indicates the automated segmented (a) and (b) and manually delineated polygons (c) and (d).

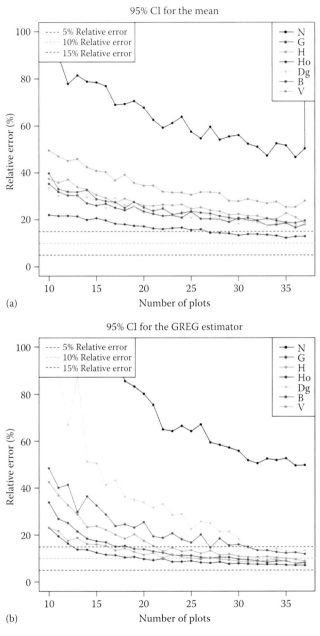

FIGURE 2.28 (a) 95% relative error for SSS sampling (sample mean). (b) 95% relative error for DSS sampling (GREG estimator).

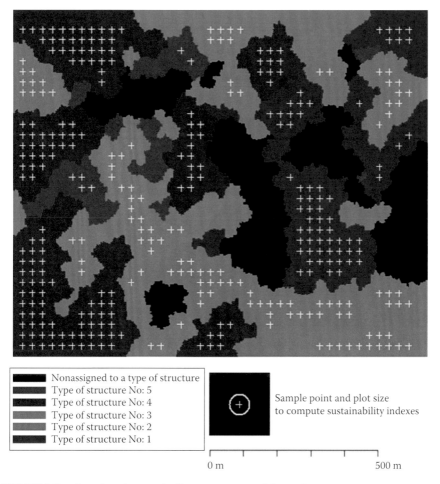

	Nonassigned to a type of structure
	Type of structure No: 5
	Type of structure No: 4
	Type of structure No: 3
	Type of structure No: 2
	Type of structure No: 1

Sample point and plot size
to compute sustainability indexes

0 m   500 m

FIGURE 3.9   Sample points on the forest structures of the study area.

FIGURE 6.1   Land uses in the Mediterranean. Agriculture, pastures, and forestry, preserving high biodiversity of flora and fauna. Jerez de la Frontera (Cadiz, Spain). (Photo: Antonio García-Abril.)

FIGURE 6.2 Agroforestry landscape in Abancay, Peru. (Photo: Antonio García-Abril.)

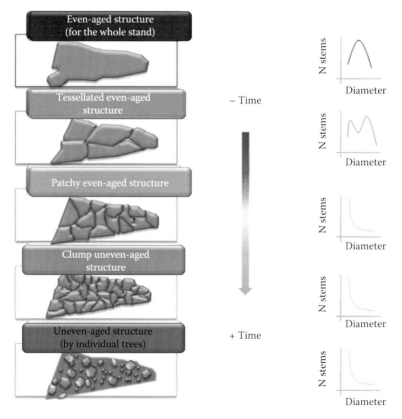

FIGURE 6.3 Increase in complexity of forest structure from a homogeneous, regular structure. Natural succession under a regime of moderate perturbations. (From Velarde, M.D. et al., *Integración paisajística de las repoblaciones forestales.*, *Serie Técnica de Medio Ambiente*, Dirección General de Medio Ambiente, Consejería de Medio Ambiente, Vivienda y Ordenación del Territorio de la Comunidad de Madrid, Madrid, Spain, 2013.)

FIGURE 6.4   Virgin beech forest in the Karpathos range (Romania). (Photo: Antonio García-Abril.)

FIGURE 6.5   Old-growth forest. H.J. Andrews Experimental Forest (Oregon). (Photo: Antonio García-Abril.)

FIGURE 6.6   Crown fire in Cascades Central Range (Metolius area, OR). (Photo: Francisco Mauro.)

FIGURE 6.13 Clear-cutting, plowing, and sowing contiguous to previously regenerated areas. Tierra de Pinares (Soria, Spain). Landscape impact is negative. Adverse landscape effects also increase with seed-tree clear-cutting and if the soil is plowed and artificially planted. (Photo: Antonio García-Abril.)

FIGURE 6.16 Continuous-cover physiognomy of a shelterwood-managed forest (120 year rotation and 60 year RP). Pinar de Valsaín (Segovia, Spain). (Photo: Antonio García-Abril.)

FIGURE 6.17 Area of mature trees beneath a uniform and group shelterwood-managed area in Pinar de Valsaín (Segovia, Spain). (Photo: Antonio García-Abril.)

FIGURE 6.18 Successful regeneration area in a group shelterwood system in Pinar de Valsaín (Segovia, Spain). In this case, if the RP increases and trees are retained, an irregular forest structure can be achieved. (Photo: Antonio García-Abril.)

FIGURE 6.19 Mature 110-year-old trees in an even-aged forest managed by a shelterwood system in Pinar de Valsaín (Segovia, Spain). (Photo: Antonio García-Abril.)

FIGURE 6.22 Regeneration gaps in an uneven-aged beech forest. (Photo: Antonio García-Abril.)

Landscape visual criteria for *exterior landscape*: broad-scale approach		
1. Avoid fragmentation looking for connection among relevant ecosystem.	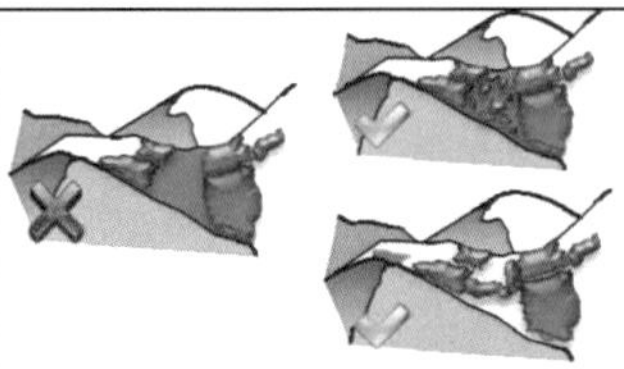	This criterion indicates that connectivity is not only important for ecological reasons but promotes landscapes preferred by the public.
2. Increase biodiversity a. Of species and ecosystems b. Of landscape elements.	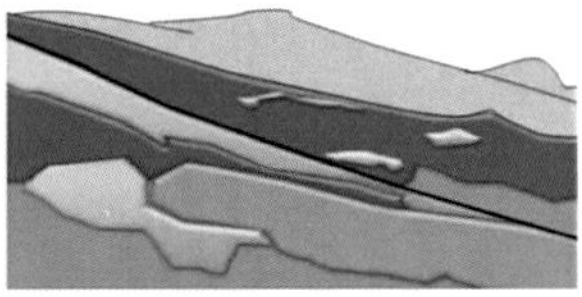	Stands of mixed species tend to be more stable against biotic and abiotic damages than monospecific ones. Also multilayered stands offer high level of diversity. Besides, forest landscapes with aesthetic complexity are preferred.
3. Soften margins of forest and new plantations.	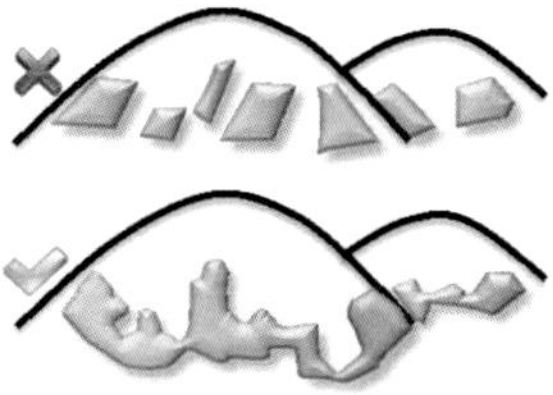	Create or maintain wavy edges with indentations improve visual diversity and introduce irregularity to straight forest edges. This also regards the importance of edges for the connectivity.
4. Adapt infrastructures, equipment and other artificial elements to forest environment.	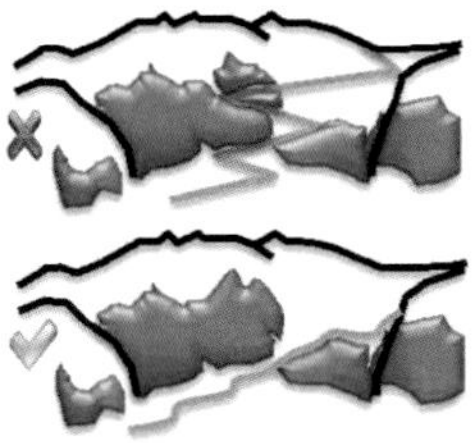	This criterion involves visual fragility and aesthetic values as well as the effects of anthropogenic disturbances affecting biodiversity conservation.
5. Work at a landscape scale in order to forest integration of management activities.		Match interventions to landscape scale refers to give importance to the perception of relative and absolute sizes. Furthermore, management areas are intended to have independent structure and functioning.

FIGURE 6.23 Landscape criteria for exterior landscape: broad-scale forest management approach.

Landscape visual criteria for *interior landscape*: small-scale approach	
1. Protect riversides and shores.	
2. Pay attention to the singular function of forest edges.	
3. Increase ecosystem and species diversity.	
4. Preserve large old trees, large fallen trees and trees of different species.	
5. Integrate structures and squipment into the forest landscape.	
6. Do not disturb the *genius loci* (spirit of the place).	

FIGURE 6.24 Landscape criteria for interior landscape: small-scale forest management approach.

Pair-wise comparison:

(a) Selection of a set of meaningful alternatives:

$$\Omega = \{ \text{Alternative 1}, \text{Alternative 2}, \text{Alternative 3}, \text{Alternative 4}, \ldots \}$$

(b) Figure out all feasible pairs of alternatives in Ω.

(c) For every pair in (b), ask for the decision maker preference between the alternatives within the pair: which of the following alternatives do you consider to be more preferred?

Alternative 4 — Alternative 1

| Alternative 4 |
| Alternative 1 |
| They are equally preferred |
| I do not know |

(d) Register the answers of the decision-maker.

1: Row alternative preferred 2: Column alternative preferred 3: Both alternatives are indifferent 4: No preference stated by the decision maker	...	Alternative $i-1$	Alternative i	Alternative $i+1$	...
...	...	...	...	...	...
Alternative i-1		3	3	1	...
Alternative i		3	3	4	...
Alternative i+1		2	4	3	...
...		...	...	...	...

FIGURE 7.1 Steps in a pair-wise comparison process.

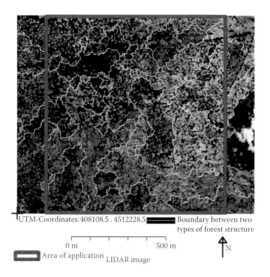

UTM-Coordinates: 408108.5 . 4512228.5 ▬▬▬ Boundary between two
types of forest structure

0 m ⊢——⊢——⊢——⊢——⊢ 500 m

↑ N

▭ Area of application LIDAR image

FIGURE 7.6   LiDAR image of the DCHM of the area selected for the case of application.

Location: A   Location: B

Which of these locations do
you consider more sustainable?   A   B   No
response

FIGURE 7.11   Example of data entry screen for pair-wise comparison of sustainability of
each evaluator and information provided for the comparison: (a) real image of the compared
points.

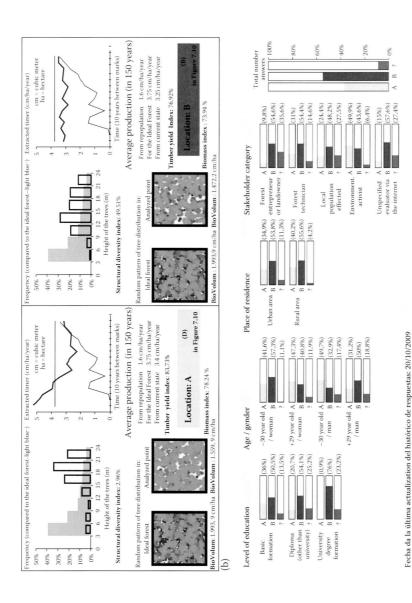

FIGURE 7.11 Example of data entry screen for pair-wise comparison of sustainability of each evaluator and information provided for the comparison: (b) information about the indices of sustainability on each point to compare; (c) statistics of previous answers provided by other evaluators.

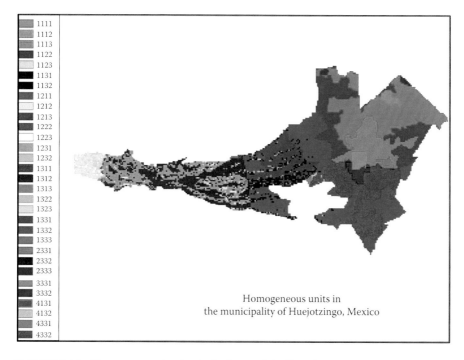

FIGURE 8.10 Homogeneous land units in the municipality of Huejotzingo, Mexico.

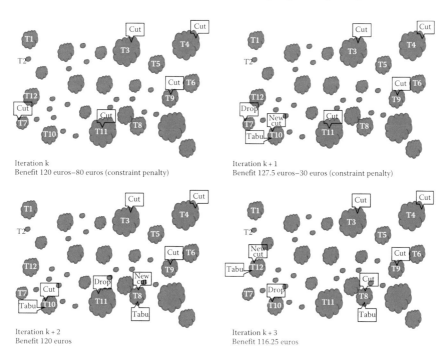

FIGURE 8.18 Iteration k through $k+3$ of the TS process. Crown projection of the Scots pine trees in the study area. Labeled trees to be cut or in the tabu list.

6 Landscape Indicators for Sustainable Forest Management

Antonio García-Abril, M. Victoria Núñez,
M. Angeles Grande, M. Dolores Velarde,
Particia Martínez-Obispo, and
Roberto Rodríguez-Solano

CONTENTS

6.1 INTRODUCTION

6.1.1 Chapter Content

This chapter presents a number of landscape indicators that are currently used in sustainable forest management (SFM) and describes the trends and efforts to integrate landscape into SFM. Most of the examples and discussions presented here have been taken from the temperate and boreal regions, although the main international initiatives on criteria and indicators (C&I) for SFM are also analyzed.

The chapter is divided into three sections. The first is an introductory review on the state of the art of landscape indicators and their relationship with SFM.

Section 6.2 focuses on the visual and ecological landscape. The section describes the landscape and stand characteristics linked to biodiversity (as the key element for determining ecological functions) and shows examples of man-made landscapes that successfully integrate high biodiversity, production, and landscape beauty. It also describes the landscape changes, which take place under natural dynamics, and the role of successional steps in mature forests. Although mature forests are the key element for conservation of biodiversity, they have practically disappeared all over the world. There is also an explanation of the meaning of landscape in different forest management methods and the possibilities of increasing biodiversity. This section identifies the most valuable characteristics of landscape for forest users in order to integrate visual landscape preferences and significant features of biodiversity conservation. Finally, 5 principles and 11 criteria are proposed for landscape integration into forest management on large and small scales. All these principles and criteria link visual preferences with biodiversity conservation and guide further analysis of landscape indicators.

The third section includes a bibliographic review of visual and ecological indicators for integrating landscape into SFM. Finally, this section identifies an extensive common ground for visual and ecological landscape indicators. The conclusions of this chapter are shown in an epilogue.

6.1.2 General Scope

Forests and wild vegetation areas are an essential part of the landscape. However, the landscape itself is progressively becoming an important and essential factor in forest management, which includes landscape as an objective or as a condition for other activities.

The visual landscape is the area of territory, which we can see with our eyes and perceive with our senses. The way it is perceived determines the characterization and assessment of the aesthetic value of landscape and its meaning.

Human perception of landscape changes under the influence of economic, social, aesthetic, ethical, and cognitive factors. Perception thus entails physical and psychological elements (Bell 2001). This perception is mainly visual but also involves other senses (Aguiló et al. 1995). The visual expression of the landscape affects people in different ways and can influence aesthetic appreciation, health, and/or welfare (Ode et al. 2009).

The ecological landscape is the area of land containing the pattern that affects and is affected by a process of interest (in our case the forest management) (e.g., Wu 2006, 2008). This definition includes the landscape patterns, the interaction among its elements, and how these patterns and interactions change over time. The ecological landscape consists of its external manifestation, or *fenosystem*, and the hidden ecological relationships, or *cryptosystem* (González Bernáldez 1981; Bell 2001).

From the ecological point of view, the landscape has certain basic elements whose distribution and form influence plant and animal populations and their dynamics in order to maintain stability, regulate their expansion, and avoid the fragmentation and disappearance of their habitats (Velarde et al. 2013).

The distribution of a landscape's elements defines the spatial configuration, structure, or pattern characterizing each scene. This structure or pattern has properties relating to the functioning of the landscape: the movement and flow of animals, plants, water, wind, materials, and energy through the structure (Forman 1995; Dramstad et al. 1996; Velarde et al. 2013).

Fry et al. (2009) state that both the visual and ecological characters are dependent on the landscape structure, and they can, therefore, also share the same theoretical basis for landscape assessment.

Research focused on assessing landscape from the visual point of view usually values the common landscape attributes perceived by a number of observers (Hull and Stewart 1992; Peron et al. 1998; Bell 2001; Dramstad et al. 2006; Tveit et al. 2006) by means of questionnaires. However, it should also be understood that the landscape has an intrinsic value of its own, which is difficult to separate from its ecological value and, therefore, less dependent on fashions, attitudes, or circumstantial criteria (Ramos 1993).

Recognizing and understanding the relationship between ecology and visual appearance (perception and aesthetics) at the conceptual level are of great importance for the planning and management of the landscape. The identification of a possible common theoretical basis, as well as useful indicators for both ecological and visual aspects of the landscape, may enable an integrated approach to landscape assessment and monitoring programs and also provide the basis for the development of new tools for the analysis of changes in the landscape (Tveit 2009).

The consideration of the landscape in forest management has been developed and integrated progressively over the last four decades, serving as an instrument to improve forest management in a process that continues to enhance visual quality and the conservation of ecosystems.

Forest planners, designers, and managers must incorporate visual landscape management into their plans as part of SFM. The last 30 years have seen numerous developments in visual management and design processes and techniques. Other developments in SFM, such as ecosystem management and the need for more public participation in forestry planning, have also influenced the direction of forest management (Bell 2001).

The current perception of visual landscape conservation is a consequence of the conservation of the ecological landscape. Visual landscape design cannot ignore the ecological requirements of the environment, but ecological conditions alone are not enough; it also needs to integrate activities into the surrounding landscape (Velarde and Ruíz 2007).

Landscape perception by people implies that public opinion is a necessary component of management. Therefore, it is necessary to provide information and public participation. The information presented for analysis in participatory processes must be intelligible and as visually realistic as possible.

The last 15 years have brought new findings, which contribute to understanding the ecological landscape (for reviews, see Bauhus et al. 2009; Wirth et al. 2009). Today, landscape is considered as a complex reality, which is home to both species and people, a reality shaped by processes and ecological relationships, activities, and interactions, resulting in a shifting mosaic that changes over time and space. The spatiotemporal landscape composed of patterns of different land uses and natural vegetation must contain a sufficient representation of the communities of species to ensure their conservation. This is a priority of landscape management, which contributes to sustainable development. In this pattern, mature phases of succession represented by complex forests are of paramount importance. Their preservation and creation have also become a fundamental objective of landscape management in the future (e.g., Kuuluvainen 2009).

Landscape indicators are currently high on the policy agenda and are increasingly used in the assessment of different landscape values. Efficient indicators can help evaluate whether or not changes are proceeding in a desirable direction (Fry et al. 2009).

The development of ecological landscape indicators has been a very active area of research and has resulted in a wide range of measures and indicators, based on the principles of ecological landscape. For the visual appearance of landscape, however, this conceptual base is often absent, thus hindering the progress in the development of indicators (Ode and Fry 2006; Fry et al. 2009).

Apart from the integration of multiple landscape assessment principles, there is no standard method for the evaluation of landscape (Gulinck et al. 2001). Sustainable management of forest landscapes has not yet been completely defined or achieved. Procedures are currently under development and revision, and research results are incorporating new information relevant to forest sustainability. For this reason, the growing number of indicators for the assessment of SFM is still an ongoing process. This chapter seeks to integrate the visual and ecological landscape, since the convergence of the two approaches is possible and many of the discrepancies can be reconciled.

6.1.3 MAIN TRENDS IN THE CONSIDERATION OF THE LANDSCAPE IN FOREST MANAGEMENT

The formal incorporation of the landscape in forest management in developed countries has been driven mainly by the negative effects of certain forestry activities, which have resulted in social protests, triggered primarily by the loss of scenic landscapes, the visual effects of clear-cutting, disturbing afforestation, and the destruction of old-growth forests.

The visual landscape was first incorporated in forest management in the United States in the late 1960s, in the form of constraints to adapt or preserve the landscape quality.

Ecological landscape conservation began to be included in forest planning in the 1990s for several reasons: the progress in the knowledge of forest ecosystems, the environmental damage caused by certain forest practices, the declining importance of wood as an economic resource in the global economy, greater social awareness of the need to conserve species and ecosystems, and the recognition of sustainable development as the guiding principle of human activities on landscapes.

As a consequence of the previously mentioned trends, the importance of landscape in all aspects of forest planning and management has been promoted in SFM, through international efforts, such as the Rio–Helsinki process, the requirements of certification, and an international movement favoring more natural forest management (Núñez et al. 2010). However, the integration of visual and ecological landscape concepts in forest planning is still an ongoing process subject to discussion and reflection. New knowledge is incorporated in a process of continuous improvement through the analysis of experiments and measures taken directly from real forests.

Key points of discussion are whether the visual and ecological aspects of landscape coincide, whether there is an association between environmental quality and visual quality, and whether the measures for the conservation and improvement of visual quality are useful for achieving ecological sustainability, and vice versa (Gobster 2001, Gobster et al. 2007; Sheppard 2001).

6.1.3.1 Europe

Landscape integration in forest management has been improved in the last three decades, thanks to more in-depth research into landscape ecology and several regulations concerning environmental impact assessment and specific forest management policies. The most important reference at the European level is the *European Landscape Convention* in Florence (2000) agreed by the European Council. This convention aims to protect, manage, and plan all landscapes in Europe. The convention emphasizes the role of perception when it states that "landscape" means an area, *as perceived by people*, whose character is the result of the action and interaction of natural and/or human factors. This convention comes on the tail of others dealing with cultural heritage conservation and nature conservation.

For example, the Ministerial Conference on the Protection of Forest in Europe (MCPFE) proposed forest operational guidelines for the Natura 2000 biodiversity network, including the landscape. "*Management practices should, where appropriate, promote diversity of structures, horizontal and vertical, e.g. multiage-stands, and species diversity, e.g. mixed stands. Where appropriate, the practices seek to maintain or restore landscape diversity*" (MCPFE 2003).

The European Landscape Convention has promoted some changes such as encouraging the study of landscape from an integrated approach, where visual, cultural, and social qualities are included along with ecological functions. The importance and challenges for this approach have been highlighted by several researchers (Fry 2001; Opdam et al. 2001; Tress et al. 2001, 2005; Wissen et al. 2008).

Germany has a long history of research and applications in the area of visual landscape, dating back to the late nineteenth century (von Salisch 1885). Strong demand for recreational areas gave rise to planning aimed at controlling both the number of visitors and their impacts on forests. Furthermore, landscape conservation was considered a source of social wealth, comfort, and recreation for the urban population (Ammer and Pröbstl 1991).

German authors point out some of the major problems of afforestation—sometimes isolated in small plots and inside sharply geometric limits, with power lines and other infrastructures—which are all generally a result of poor or nonexistent planning. Some of the German forests serve as illustrative examples of good design and proper planning and also provide guidelines on the various species of trees and shrubs with different ecological and visual qualities that can be used. The current concept of forest landscape, therefore, differs from the earlier view, in that it is no longer defined by personal opinions but is the result of scientific research, with special emphasis on empirical–social studies (Ammer and Pröbstl 1991).

In Germany, more than 150 visual landscape assessment methods have been developed and described, mainly based on expert ratings. Still, further analysis on landscape perception is being made available, as it is considered that single expert ratings are insufficient to properly understand and assess landscape scenery (Stölb 2005; Gruehn and Roth 2010; Roth and Gruehn 2010).

France, through the work of *Institut national de recherche en sciences et technologies pour l'environnement et l'agriculture*—formerly known as CEMAGREF—has played a special role, along with Switzerland, in the proposal for new forms of silvicultural management that are more in line with natural processes and landscape (CEMAGREF 1981).

In Scandinavia, the work of Dramstad, Ode, Tveit, and Fry (Dramstad et al. 2006; Tveit et al. 2006; Ode et al. 2008, 2009; Fry et al. 2009) has contributed to identifying nine key concepts reflecting the dominant aspects of the visual landscape as presented in visual guidelines and the research literature (Tveit et al. 2006). Fry et al. (2009) analyzed the correspondence between ecological indicators and visual indicators in order to explore whether there is a common ground in both concept and operation.

In southern Europe, the landscape planning research group (School of Forestry in Madrid, Spain), led by A. Ramos, focused their research on the viewshed, developing metrics and models for assessing visual quality and fragility of landscape, in general and forest landscape in particular (Ramos 1979; Escribano et al. 1987; Aguiló et al. 1995; Aguiló and Iglesias 1995; Martínez-Falero and González-Alonso 1995). Another line of this group, supplementary to the aforementioned, is the integration of afforestation in landscape (Ramos et al. 1986; González-Alonso et al. 1989; Velarde et al. 2013).

In Britain, Lucas (1991), Bell (1993, 1999, 2001, 2003, 2011), and the Forestry Commission (1988, 1989, 1990, 1991, 1992, 1994, 2003) were pioneers in the development of design guidelines for afforestation and forest treatments. Several guides for the conservation of ecosystems and forest landscapes have been published for use as management tools by owners, managers, and other agents involved in the forestry sector. These guidelines first establish general design principles and evaluation criteria for the landscape.

The principles of forest landscape design are deeply rooted within the British Forestry Commission, which is mindful of the importance of maintaining the natural qualities of the landscape, as forests must meet the needs of society in its multiple aspects. These principles are implemented by carefully defined techniques for managing different types of forests and topography. For example, in some areas where the forest landscape is not of high quality, it is advisable to make an appropriate assessment of its features prior to reforestation. If this does not represent an improvement in the landscape, the commission considers that it should not be reforested (Forestry Commission 1994).

The close-to-nature silviculture movement has promoted a close-to-nature forestry management, also known as near-natural forest management or continuous-cover forestry (CCF). This movement has brought about changes in European forest management by promoting conversion from clear-cut managed homogeneous forest to managed forests with a mixed structure and composition, containing large trees and excluding clear-cuts (de Turckheim 1992, 1993, 1999; Duchiron 1994; Schutz 1997; García-Abril 1999; Mason et al. 1999; Mason and Kerr 2004; Bruciamacchie and de Turckheim 2005; Pukkala and von Gadow 2012).

6.1.3.2   North America

In the United States, the integration of landscape in forest management took place in two stages. The first, in the late 1960s, incorporated the visual landscape concept driven by the reaction to far-reaching changes in highly attractive landscapes caused by felling and logging systems and was facilitated by the emergence of legislation on environmental impact. The second stage is the inclusion of the concept of ecological landscape that took place in the 1990s.

The visual landscape concept responds to aesthetic considerations that must be protected and enhanced, whereas the set of operative principles is part of visual resource management programs (USDA Forest Service 1974) such as the USDA Forest Service Scenery Management Systems (SMS) (USDA Forest Service 1995). These programs have been adopted in various parts of the world such as British Columbia and Australia (Sheppard 2001; Bell and Apostol 2008).

According to Gobster (2001), this first consideration of the visual landscape is revolutionary in that it incorporates the perception of nature for its protection in forest management and in landscape planning, in general. It is based on formal aspects and expert assessment that take into account the visual preferences of the population by comparing photographs relating to a particular moment. These assessments helped to protect the most scenic areas from timber harvesting and provided guidelines for forest managers on how to mitigate visual impacts elsewhere by leaving vegetation screens along roadsides and undulating the edges of clear-cuts (Gobster 2001).

They monitored the landscape changes in terms of the scale of alteration seen from a particular place with a particular view and focused forestry practices on hiding the view from the observer. These activities were aimed at the common goal of preserving the appearance of the landscape as it is. The landscape quality objectives were the preservation and maintenance of the forest. The main problem was to separate the visual resources from others, such as ecological resources (Bell 2001).

This view of the landscape—based on formal aspects and focusing on the characteristics of the most scenic landscapes and using static landscape information—has attracted considerable criticism, although it has served to mitigate landscape impacts and the hostility of the population toward forestry on public lands (Sheppard 2001).

The second step in the integration of landscape in forest management was the incorporation of the ecological landscape, as a result of disputes over the felling of old-growth forests in the Pacific Northwest during the 1970s and 1980s.

Research in old-growth forests has highlighted their great complexity, with specific species under threat—and requiring hundreds or thousands of years for their creation—following disturbances, usually from crown fires. Logging endangered these forests and the species inhabiting them (e.g., Spies and Duncan 2009).

The northern spotted owl (*Strix occidentalis caurina*) became the symbol of the need to safeguard old-growth forests, which have become the hallmark for the change in forest management based on ecosystem management in North America and elsewhere.

Knowledge of the structure and functioning of forest ecosystems is the key to proper forest management. The type and extent of natural disturbances are critical for this forest structure and development, shaping the landscape over long time scales.

Ecosystem management takes account of the complexity of ecosystems, as well as the complexity of social and economic relations in forest areas (Kohm and Franklin 1997). Ecosystem management recognizes the complexity of each place and the requirements of individuals at each site. It also integrates biological, social, and technical knowledge and facilitates social participation and shared decision making. The attitude of ecosystem management is one of humility and caution, due to the appreciation of the complexity of ecosystems and the limitations of our knowledge. The paradigm of silvicultural actions to reduce the negative effects of clear-cuts was based on the necessity to retain trees, wildlife trees, deadwood, and vegetation corridors for habitat connectivity (Kimmins 2001).

The need for a visual landscape management that considers ecological connections and conservation is highlighted by the SMS (USDA Forest Service 1995), which attempts to integrate the scenic aesthetic with ecosystem management principles, albeit more by explicitly allowing trade-offs than by aligning the fundamentals of two sets of values (Sheppard 2001). A series of plans and regulations have been generated as a result of the process of including landscape in forest management.

The objectives of the *National Environmental Policy Act* of 1969 include ensuring an aesthetically pleasing environment, preserving the natural appearance of the historical legacy (thus, forests), and maintaining diversity. The *Forests and Rangeland Renewable Resources Planning Act* of 1974 recognizes the vital importance of renewable resources in forests for social and economic welfare, as well as the need for long-term planning. Among the aspects of this planning, the Act includes

- The inventory of intangible assets such as landscape, understood as "naturalness"
- A forest management consistent with the aesthetic resources
- The evaluation of management benefits in relation to environmental quality factors, such as natural and aesthetic values

As an amendment to the *Forests and Rangeland Renewable Resources Planning Act* of 1974, the *National Forest Management Act* of 1976 added the following aspects of the landscape, considered in terms of protection:

- Planning should include the potential impacts of forestry activities, including landscape.
- Cuttings should be consistent with the protection of aesthetic resources.

The *SMS*, for the inventory and analysis of the aesthetic values of National Forest lands (USDA Forest Service 1995), evolved from and replaced the *Visual Management System (VMS)* of 1974.

The *SMS* increases the role of stakeholders throughout the inventory and planning process and interacts with the basic concepts and terminology of ecosystem management. It also provides for improved integration of aesthetics with other biological, physical, and social/cultural resources in the planning process.

The Northwest Forest Plan of 1994 heralded a new stage in forest management and applied the principles and knowledge of ecosystem management to 9.7 million hectares of federal lands (Forest Service and Bureau of Land Management) in the three states of the Pacific Northwest, limited by the area of the northern spotted owl. The plan was aimed at protecting and restoring old-growth forests, watershed, and streamside conditions in order to maintain stable populations of animals and plants. In addition to protecting native biodiversity, the plan sought to obtain a predictable and sustainable level of timber sales from federal forests (Spies and Duncan 2009).

The Act to Save America's Forests (1996) aims to strengthen the protection of biodiversity and to suppress felling in old-growth forests and other protected areas within their national forests and seeks to restore their original biodiversity through active and passive management actions. It also considers the aesthetic and ecological value of biological resources and, more specifically, of the flora.

This Act calls for selective logging systems and states the following reason—among others—related to landscape ecology: *"Selective logging maintains the structure and function of the natural forest, works on behalf of the natural processes inherent to the forest and allows the development of natural processes of succession towards old-growth forests."*

The Omnibus Public Land Management Act of 2009, which includes the forest landscape restoration plan, aims to promote the restoration of priority forest areas based on landscape ecology.

The *Bureau of Land Management's National Landscape Conservation System (NLCS)* was created in the year 2000, and the National Landscape Conservation System Act was signed into law in 2009.

The National Landscape Conservation System aims to protect, preserve, and restore significant landscapes due to their cultural, ecological, and scientific values. National monuments, national scenic trails, national wild and scenic rivers, and national wilderness preservation systems are included in this network.

6.1.3.3   Visual Landscape in International Initiatives for Sustainable Forest Management

There are several monitoring schemes around the world that focus on SFM: the so-called nine international processes for SFM that evolved into sets of C&I (see Chapter 3). These C&I consider visual landscape from different points of view.

Within Europe, the Pan-European indicators for SFM include the landscape pattern indicator, which is directly related to the landscape. However, there are other landscape features indirectly related to these indicators, for example, indicators for age structure and/or diameter distribution, tree species composition, regeneration, naturalness, deadwood, or cultural and spiritual values. These can be used as surrogates of landscape indicators for SFM but must be adapted for a proper interpretation of the landscape in SFM.

The indicators in the Montreal Process for SFM include landscape values according to the criteria of cultural, social, and spiritual needs and values. More specific indicators of forest landscape are not explicit. However, similar surrogates to those mentioned earlier are within the C&I of this process, which also focus attention on public participation (e.g., indicator 7.1.c provides opportunities for public participation in public policy and decision making related to forests and public access to information).

The African Timber Organization and the International Tropical Timber Organization (ITTO) processes identified 28 criteria and 60 indicators for SFM. Neither of these processes directly developed forest landscape indicators. Both consider as important characteristics the maintenance of permanent forest, the state of the forest, and the planning of forest activities but do not propose particular measures of forest landscape values. Nevertheless, participation of stakeholders is one of their main proposals.

The other processes, Dry Forest in Asia, Dry-Zone Africa, Lepaterique, and Tarapoto, propose similar criteria to the processes already mentioned. Forest landscape indicators are, therefore, required to report on the state of forests.

6.1.4   DISCREPANCY BETWEEN VISUAL AND ECOLOGICAL LANDSCAPE VALUES

The visual and ecological landscape explicitly belongs to forest management. We will now discuss whether the two views are divergent or may overlap at least partially. "Forest scientists and resource managers seem to be divided between those who see a strong association between ecological health and visual quality, and those who do not…. This second view is held by those, especially among the forest sciences fraternity, who see sustainability as too complex to be directly related to visual landscape indicators, or to be assessed by a visual analysis approach" (Sheppard et al. 2001).

There is an ongoing debate in the scientific community as to whether ecological SFM also has aesthetic benefits (e.g., Gobster 1999; Daniel 2001; Sheppard and Harshaw 2001; Haider and Hunt 2002; Gobster et al. 2007; Velarde and Ruíz 2007). On the one hand, efforts have been made at the conceptual level to identify the theoretical common ground and indicators for both aspects of landscape (Tveit et al. 2006; Fry et al. 2009), and on the other hand, there is a practical search for common management criteria, which can fulfill both ecological and aesthetic objectives, including the ways in which aesthetics and ecology may have either complementary or contradictory implications for a landscape (Gobster et al. 2007; Velarde and Ruíz 2007; Velarde et al. 2013).

In general, we can accept the hypothesis that there is coincidence between aesthetic appreciation and a healthy and sustainable ecosystem. That is, aesthetically speaking, sustainable landscapes are preferred (Sheppard et al. 2001; Tindall 2001).

Discrepancies can often be resolved through education and public information as to the meaning, function, and temporal variation of landscapes. Knowledge, experience, and learning play an important role in landscape appreciation. A greater emphasis on understanding the landscape and methods of measurement will provide opportunities for people to appreciate sustainable landscapes, which could lead to an expansion in the idea of landscape beauty (Gobster 2001).

Policies for landscape planning, landscape design, and management activities can be used to achieve ecological and aesthetic objectives (Gobster et al. 2007).

More in tune with the new forestry or ecosystem management is the idea that aesthetic appreciation should be informed by ecological knowledge, so that what is good ecologically also looks good to us (Sheppard 2001). The second revolution in the aesthetic assessment of forest landscapes is aesthetic ecology (Gobster 2001).

However, a possible problem is trying to teach preconceived or partial models of ecological landscape where our knowledge is limited and where nature may evolve unpredictably.

Current developments in the findings and knowledge related to the structure, functioning, and dynamics of forest ecosystem suggest that landscape forestry models and images are not definitive; they cannot, therefore, be interpreted as equivalent to "Nature's voice." This would also be a human interpretation.

6.1.5 PUBLIC PARTICIPATION AND MODELING

Forest management has come under scrutiny from society in many ways, due to not only its visual but also its ecological impact. There is a duty to explain forest management and facilitate public participation by providing intelligible information that is as realistic as possible and reflecting temporal variations in the short, medium, and long terms.

The first interactive tools for planning, design, and visualization of planning include realistic visualizations (Sheppard et al. 2001) and spatiotemporal map models (Luymes 2001).

The visualization methods are not only realistic final rendering tools but also interactive tools for the planning and design of the management plan (Sheppard et al. 2001). Both of these should be careful, objective, and rigorous (Luymes 2001).

In public participation processes, it is necessary to identify the public, stakeholders, and community. There is a toolbox for public participation that can be used for different levels such as informing, consulting, involving, and working in partnership (Bell and Apostol 2008).

The need for detailed forest parameters, wildlife diversity, and other biophysical information has increased markedly in the last 20 years, driven in large part by the demand for information for modeling forest ecosystems and SFM.

In the past two decades, forest models have benefited from ongoing improvements in technology and data availability (Mladenoff 2004). These forest management systems have been developed by research institutes and commercial companies in user-friendly applications that can be adapted to a wide range of project sizes.

Three-dimensional visualizations of forest landscapes are quantitative ecological information-based techniques that can be used to visualize forest structure, dynamics, landscape transformations, and regional plans (Wang et al. 2006).

Broadly defined, forest landscape simulation models (FLSMs) are computer programs for projecting landscape change over time. FLSMs can also be used to test hypotheses about the interactions among processes and patterns across forested landscapes (Scheller and Mladenoff 2007) and as a model that predicts changes in the spatial characteristics (distribution, shape, abundance, etc.) of model objects (He 2008). Modeling requires making many choices between extent and resolution, precision and generality, accuracy and meaningful prediction, and parameterization and validation (Mladenoff 2004).

FSLMs vary widely in their algorithms, complexity, and input requirements. Computer languages now allow tremendous flexibility for joining diverse numerical or computation methods within a single model or modeling system (Woodbury et al. 2002). Forest simulators encompass a wide range of models dealing with different ecological processes and operating across a large range of spatial and temporal scales (Dietze and Latimer 2011). Most of the simulations are made using algorithms or equations selected by the user, either from the software applications or the literature review, or calculated by the user.

The main difference among FSLMs is whether the community itself is static or dynamic. For a static community, tree species composition and associated characteristics are defined a priori and do not evolve during the simulation, although the spatial community will change over time (Scheller and Mladenoff 2007). From an ecological perspective, FSLMs can be classified based on spatial interactions (inclusive or exclusive), ecosystem processes (inclusive or not), and community dynamics (static or dynamic). These models can incorporate spatiotemporal processes such as natural disturbances and human influences. It is very important to know the changes in the forest landscape pattern under anthropogenic interference for the planning and management of forest landscape and the sustainable use of forest resources (Wang et al. 2006).

Data sources with a high accuracy and resolution are essential to develop reliable landscape visualizations. These data may include land cover maps, digital elevation models (DEMs), tree images, tree diameter (not always available), tree heights, stand densities, and species composition.

The drawback of FSLMs is the uncertainty and stochasticity of ecological processes, which is transferred to the quantification of parameters, model sensitivity, and spatial resolution (He 2008).

The direct outputs of a forest landscape model are the spatiotemporal patterns of the forest study area. This makes it possible to compare scenarios, anticipate results, and facilitate management decisions—for example, spatial modeling framework (Landscape Management Policy Simulator, LAMPS)—for forest landscape planning (Bettinger et al. 2005; Nonaka and Spies 2005; Thompson et al. 2006; Johnson et al. 2007; Spies et al. 2007a,b).

6.2 VISUAL LANDSCAPE AND ECOLOGICAL LANDSCAPE

6.2.1 BIODIVERSITY AS A REPRESENTATIVE ELEMENT OF ECOLOGICAL FUNCTION

Most citizens taking part in a participatory forest management process have not conceptualized the similarities and differences between the visual and ecological landscape (subjective perception and ecological function). In general, they are unaware that the scientific community has accepted certain results as scientific knowledge. Clearly, even though these results may be known, each individual has his or her own tastes, which differ from the scientists' results. In the interests of fairness, the available scientific information is explained to the evaluators before they take decisions. As we shall see in the following, biodiversity is a key element in this information. The ecological landscape is also contained within the broad concept of biodiversity. For this reason, the variables, elements, and processes that take place in the ecological landscape are directly related to biodiversity and its consequences.

6.2.1.1 Biodiversity: An Encompassing Concept

Biodiversity* is defined as the totality of genes, species, and ecosystems in a region (Global Biodiversity Strategy 1992), including genetic diversity, taxonomic or species diversity, and ecological diversity (Di Castri and Younés 1996).

The maintenance of biological richness requires not only the conservation of the communities of genes, species, and ecosystems but also the relationships among them (the ecological processes) in time, space, and at all scales of observation (from the organism to the ecosystem) over landscape and territories.

It is not known how many species there are in the world, nor even the role many of them play or their relationships for the maintenance of life on the planet and, therefore, for human life. The loss of species (either through ignorance or due to productive or recreational activities) is irreversible, and if only from a utilitarian point of view, mankind should do its utmost to avoid it.

Besides human activities, biodiversity means interrelations and interrelations mean complexity. Complexity in turn involves resilience and stability (Margaleff 1993), and thus biodiversity is a sign of quality whenever it applies. Unfortunately, complex developed ecosystem stages have been mostly destroyed by anthropogenic causes and induced degradation processes. The remaining species obtained over time scales of hundreds or thousands of years in mature stages of succession must be protected in order to conserve biodiversity.

Landscape simplification and elimination of complex stages occur in very populated areas but are progressing rapidly around the world, spreading to places where disturbance was rare until the twentieth century, such as the primary tropical forests and boreal forests (Millennium Ecosystem Assessment 2005). All over the world, biodiversity is threatened by human activities and by the disappearance of complex and unchanged ecosystems.

The conservation of the planet's biodiversity requires a representative ecosystem integrity to be maintained, but there is a worldwide decline in wholly undisturbed areas, sometimes due to unbridled economic exploitation, sometimes due to poverty and subsistence needs.

Preserving biodiversity means maintaining the conditions for the existence of a sufficient representation of the communities of species, at least at the regional level, and ensuring that these communities are distributed in a spatial mosaic that can guarantee their existence indefinitely in a dynamic equilibrium.

As biodiversity includes the processes and fluxes of ecosystems, its maintenance also involves preserving healthy ecological conditions and ecosystem functioning. Biodiversity represents the natural or ecological function that must be maintained as an indispensable goal of sustainable development. It is generally accepted that sustainable development is the necessary approach for managing natural resources and human activities, as it meets the needs of the present without compromising the ability of future generations to meet their own needs (World Commission on Environment and Development 1987).

---

* Initially, biodiversity was related to species richness and the relative abundance of each species (Margaleff 2002). Currently, its meaning has been extended to ecosystems and landscape.

The aim of achieving sustainable development rests on the nonnegotiable condition of adapting to natural processes. Any activity that seeks to persist over time must have an ecological basis and be environmentally consistent and compatible. Thus, the idea of maintaining a continuous, long-term productive activity entails preserving the equilibrium and functioning of nature. The overall maintenance of the ecological functioning of natural resources must become a target and guide for sustainable development, landscape planning, forest planning, and forest management. Sustainable management needs to include ecosystem-based approaches and management adapted to natural processes and the socioeconomic context, grounded in research and learning.

The real challenge and the key to biodiversity conservation is to integrate it into all human activities and land uses, from traditional forestry through forest plantations, agriculture, industry, transport systems, cities, and in the gardens of every private house. Social awareness is an essential commitment for everyone and requires no major effort or cost, merely the use of a few simple good practices in planning and design and in routine activities, including the domestic garden. This is a necessary step for creating a global network of biodiversity.

Landscape patterns for the conservation of biodiversity may be different. The same objective can certainly be achieved through several alternative approaches to the ecological landscape, perhaps not all created and maintained with the same effort and aesthetic appearance. More research and knowledge are necessary, but one thing is clear from the concept of sustainable development: we must not make decisions that cause irreversible or long-lasting impacts. Caution is paramount and allows nature to develop solutions from the repository of ecosystem information and changing environmental conditions. Paradoxically, this is also an economic approach, as there is no need to expend money and effort when nature can do it by itself.

6.2.1.2 Biodiversity and Landscape Diversity at Different Scales

This section shows that complex and heterogeneous landscape structures are recognized to be more biodiverse than homogeneous ones and are more effective in preserving species and communities. Let us analyze some of these structures.

6.2.1.2.1 Agroforestry Systems and Historic Agrarian Landscapes

It is a fact that humankind has transformed the Earth. The existence of agricultural and historical landscapes, which remained largely unchanged for centuries and with a high biodiversity and level of attractiveness, indicates an adaptation to natural processes and the concurrence of aesthetic and sustainable landscapes. These landscapes formed a lasting bond between current and future production and were seen as an inheritance. Long-term survival depended on the proper management of the land.

Thus, for example, in Europe, the traditional models of agricultural production have created a landscape of great spatial heterogeneity linked to major ecological diversity (de Miguel and Gómez-Sal 2002).

The coexistence of different land uses and vegetation patches fosters habitat richness and allows the cohabitation of groups of species occupying different niches, resulting in greater overall diversity (e.g., Atauri and de Lucio 2001; Díaz-Pineda and Schmidtz 2003; de Lucio et al. 2003; Farina 2006; de Zavala et al. 2008).

Hedges, borders, and small forests in croplands provide an agricultural landscape with a network structure and play an essential role in the conservation of biodiversity in agricultural areas. All these elements are also defenses against wind erosion and pest control (Bennet 1998; Burel and Baudry 2001).

However, things have changed. Between the 1940s and 1990s, many reticular mosaic landscapes, hedges, and borders were destroyed in Europe due to land consolidation. These landscapes had a regulatory effect even over long distances, even decreasing winds at a distance of dozens of kilometers (Dajoz 2006).

Furthermore, the current lack of connection in Europe between agricultural production and food supply is interrupting the transfer of family-owned land from one generation to another. Short-term decision making, technological capabilities, landscape transformation, and administrative measures (grants, subsidies, etc.) do not take account of the fact that sustained agricultural activities endanger the existence of the landscape and its adaptation to nature.

There are other examples of good practices: the wooded pasture known as the *dehesa* is a well-known model of sustainable silvopastoral management in the Mediterranean regions of Spain and Portugal. These are forests transformed by agricultural and forestry uses to obtain an optimal system resulting in a patchwork landscape comprising original vegetation on the steepest and highest areas; scrub on degraded or abandoned pastures and croplands; scattered trees on poor, moderately sloping soils (a physiognomy associated with the typical landscape of trees and pastures); grazing areas; and farming on fertile soils (García-Abril et al. 1989) (Figure 6.1).

The diversity of the *dehesa* is equivalent to that of forest mosaics of vegetation types in different successional stages, so that it conserves natural richness and is

FIGURE 6.1   (**See color insert.**) Land uses in the Mediterranean. Agriculture, pastures, and forestry, preserving high biodiversity of flora and fauna. Jerez de la Frontera (Cadiz, Spain). (Photo: Antonio García-Abril.)

compatible with various human activities (Díaz Pineda and Schmidtz 2003). The products are multiple: pasture and livestock production from goats, cows, sheep, pigs, and horses, which graze on pastures and also eat the shrubs and acorns of *Quercus spp.* Other products obtained besides charcoal and cork include honey and fungi. Trees are essential for providing shelter for livestock in winter and summer and prolong the period of vegetative activity in their shade. Game species are another resource, and the existence of numerous protected species and attractive landscapes encourages ecotourism (García-Abril et al. 1989; San Miguel 1994; Montero et al. 2000).

The habitat of many protected Mediterranean carnivores is linked to the *dehesa* and to similar Mediterranean forest transformations, which combine dense and scattered trees or shrub vegetation and open spaces with grass and sometimes crops. They provide a variety of prey, especially the rabbit (*Oryctolagus cuniculus*). The rabbit is an essential part of the diet of these carnivores and is the key species for maintaining fauna richness (Delibes-Mateos et al. 2007; Delibes-Mateos and Gálvez-Bravo 2009). Forty-eight species of vertebrates prey on the rabbit in the *dehesa* landscape, to a greater or lesser extent (Delibes-Mateos and Hiraldo 1981).There are some species that are completely dependent on it.

Open areas with grass are necessary for the conservation of fauna diversity. The most effective way to conserve these areas is still through extensive livestock management. The *dehesa* system maintains a higher diversity than the evergreen Mediterranean forest, with a predominance of *Quercus spp.*

Fortunately, silvopastoral systems can be found around the world, some of which are entirely equivalent in their objectives and functioning to the *dehesa* (e.g., Harvey et al. 2003, 2005). In temperate regions, some woodland types are also silvopastoral systems.

More generally, agroforestry systems are integrated production systems with trees, croplands, and/or livestock, which maintain a high diversity and a complementary spatial and temporal sequence of products that exploit the resources to their maximum advantage (Nair 1993) (Figure 6.2). These systems have been created by farmers through the method of trial and error. They combine diversified production and biodiversity and have an ecological basis. There are currently various successful models that researchers seek to understand, improve, and replicate (Kumar and Nair 2006).

The example of agroforestry systems and other ecological studies is calling into question the concept of intensive agriculture. "During the last 30 years, the positive benefits of agroforestry to the producer and the environment have been increasingly recognized. Combining trees and crops in spatial or temporal arrangements has been shown to improve food and nutritional security and mitigate environmental degradation, offering a sustainable alternative to monoculture production.

As the plethora of benefits of agroforestry are realized, modern land-use systems are evolving towards a more sustainable and holistic approach to land management" (Nair and Ramachandran 2007).

The combination of multispecies systems reveals potential advantages in agriculture: Productivity is improved, there is greater control of pests and diseases, and ecological and economic benefits are provided (Malezieux et al. 2009; Mediene et al. 2012; Ratnadass et al. 2012).

FIGURE 6.2 **(See color insert.)** Agroforestry landscape in Abancay, Peru. (Photo: Antonio García-Abril.)

New strategies incorporating ecological knowledge gained from the observation of natural ecosystems are an alternative for the design of "ecologically intensive" agroecosystems (Malezieux 2012). These systems are indeed both ecological and productive. Designing ecologically intensive agroecosystems calls for an in-depth knowledge of biological regulations in ecosystems and for the integration of the traditional agricultural knowledge held by local farmers.

Biodiverse agroforestry and silvopastoral systems have a mosaic and often layered structure over wide areas, where the natural disturbance regime has been replaced by human use, creating areas with different succession stages and with different productions. It is a system that has been modified and maintained by man but conserves many elements of the natural vegetation. Natural disturbances are controlled, and regeneration phases are recreated. In this case, biodiversity and naturalness (understood as ecosystems with a natural disturbance regime) do not correspond, and biodiversity may be very high in some managed systems.

6.2.1.2.2 Forest and Forest Landscapes

The key to biodiversity is spatial heterogeneity both at the landscape scale and within the forest structure.

The presence of all serial states created by the natural disturbance regime allows the presence of all their species.

The disturbance regime creates a pattern of patches with a different composition, size, and shape, as well as corridors that connect them and that have their own edge effect equivalent to spatial and temporal discontinuities (Crow and Gustafson 1997). The matrix landscape element is the most dominant one and has a significant effect on connectivity and movement of organisms (Lindenmayer and Franklin 2002). The combination of patches, corridors, and edges gives rise to a mosaic (Forman 1995; Dramstad et al. 1996), and the degree of heterogeneity depends on the combination of these elements.

On the downside of sustainability, fragmentation implies loss of habitats, isolation, and the division of natural habitats into small scattered patches (Dramstad et al. 1996). Thus, strategies based on managing the landscape pattern are fundamental for the improvement of biodiversity and the resilience of ecosystems in productive landscapes (Fischer et al. 2006): a minor perturbation can create a patch equivalent to a forest harvest or a windthrow. On the other hand, extensive and intense perturbations may create a landscape similar to a deforestation action or a fire.

There are many examples that link spatial heterogeneity and biodiversity. Typical forest species have different needs throughout their life cycle and are linked to heterogeneous landscapes at different scales, as is the case of the capercaillie (*Tetrao urogallus*) (Pollo et al. 2005; Bañuelos et al. 2008). At the landscape scale, vegetation patches are needed at different successional stages, in addition to favorable habitats interconnected by over 100 ha (Suchant and Baritz 2000). Regarding the forest structure, the capercaillie requires spatial heterogeneity and irregularity in the diameter distribution of trees, with a mosaic of vertically varied structures, depending on the different periods of its life cycle, age, and sex. However, the main factor is the diversity of horizontal structures (Pollo et al. 2005). The most favorable habitat is a forest mosaic with clumps of different age, structure, and species composition (Pollo et al. 2005).

In forest areas, vertical and horizontal structure heterogeneity relates to high species diversity due to different contiguous habitats inhabited by numerous species of flora, thanks to the microecological diversity. These habitats are also used by various animals of different sizes and capacities of movement (e.g., Camprodón 2001; Smith and Smith 2001).

A complex vertical structure is composed of the different layers—or small groups—of similar trees or shrub sizes. This type of complexity allows the presence of a higher number of birds than broad homogeneous even-aged structures (e.g., Smith and Smith 2001). In general, richness of species increases with structural heterogeneity (spatial stages of forest development and tree species diversity) (e.g., Gil-Tena et al. 2007).

The minimum size of an independent or autonomous forest unit—from the forest dynamic point of view—is the size that warrants independence from the exterior structures and functions.

A preserved area is independent if it always contains the forest stage of development on which a certain species of flora or fauna depends exclusively (e.g., a saprophytic mushroom or a woodpecker, which exclusively lives on mature trees, will only be present if these niches exist). Small preserved areas cannot contain all the stages of forest development, nor the species linked to them. In this case, the preserved area would be dependent on external niches for the survival of the specialized taxon.

Therefore, preserved areas are intended simultaneously to involve all stages of development so that a forest can achieve independence. The following are some examples of minimum forest areas, which allow the presence of all stages of forest development and the habitat of small vertebrates, as cited in the scientific literature:

- A minimum size of 30 ha was assessed for monospecific beech forests (Fagus sylvatica) in Central Europe (Korpel 1982, 1995).
- Seventy ha for Picea abies (Korpel 1982) in the same site (Korpel 1982, 1995).
- Birds and small mammals adapted to the shade conditions of forests (those living inside the forest and need a minimum area to be distant from edges and external generalist predators) require a minimum of 100 ha (e.g., Tellería and Santos 2001; Santos et al. 2002; Dajoz 2006).

Regarding these assessments, the minimum area would be 400 ha for the both habitat. To conserve forest habitats, a network of large and small forest cores is needed in addition to corridors that facilitate connectivity.

Old-growth forests are mature phases of forest development. They present large trees and deadwood, often with a multilayered structure to which specialized organisms have been adapting for a period of centuries. Standing dead trees or fallen stumps and wood debris are essential for forest biodiversity conservation. They complete the life cycle of trees and provide substrate, shelter, feeding, and reproduction for many species dependent on them. Deadwood is more abundant in natural forests than in productive forests. Thresholds for the minimum amount of deadwood were assessed based on research into old-growth forests in boreal and temperate regions (e.g., Nilsson et al. 2002; Christensen et al. 2005; Müller and Bütler 2010).*

Dry and decayed trees, as well as fallen or standing dead trees, are often perceived as being out of place in an aesthetic landscape. Most people are users of forest parks and gardens or managed forests where these kinds of trees were considered the result of disease or mismanagement until recent years. At least two situations must be distinguished. The first entails general processes of decay and destruction throughout the forest area. This corresponds to the initial fragmentation phase of an even-aged forest that has not been harvested or altered by events such as wind, snow, plagues, or diseases that partially disturb the forest. The second relates to processes of decay and destruction in certain forest locations or in small areas. This refers to endogenous dynamics of heterogeneity at the local scale, for example, the

---

* There are many well-known examples such as the following:
- Woodpeckers use dead and decayed trees for nesting. They open cavities of different shapes and sizes depending on their type of beak. These cavities are later occupied by insectivorous or frugivorous birds, owls, bats, bees, and other insects (e.g., Otto 1998; Humphrey et al. 2002; Vallauri et al. 2002; Camprodón et al. 2007).
- Cavities are often found in trees at their physiological end. Thus, the percentage of cavities due to woodpeckers in forests with this kind of tree is not of great significance. They are more important in managed forests (Remm and Lohmus 2011).
- The northern spotted owl (*S. occidentalis caurina*), a typical forest species, lives in old-growth forests with a multilayered structure and is over 150–200 years old. It nests in tree cavities above 75 cm in diameter (e.g., Agee 1997; McComb et al. 2002).

forest develops through cyclical phases of growth, maturity, decay, and destruction in localized areas of different sizes.

6.2.1.3 Long-Term Forest Dynamics and Landscape

In forest ecosystems, perturbations may be diverse and range from far-reaching destructive processes such as fire to smaller events such as windthrow, avalanches, or clumps of old/decayed trees.

The destruction of large areas leads to the development of even-aged tree populations. Subsequently, succession promotes differentiation due to the environment and the species temperament and strategies. When destruction does not occur, the forest structure becomes a shift of homogeneous units at different stages of development and composition. If this situation lasts for long periods, the forest structure becomes uneven aged, generally with mixed species, in dynamic equilibrium, fluctuating around a tractor point (stable stage) (Otto 1998) (Figure 6.3). This type of scenery may be found in the surviving temperate virgin forests of Europe (e.g., Korpel

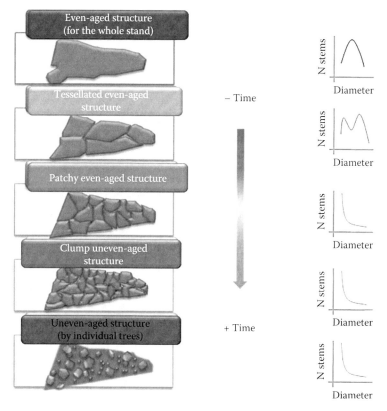

FIGURE 6.3   (See color insert.) Increase in complexity of forest structure from a homogeneous, regular structure. Natural succession under a regime of moderate perturbations. (From Velarde, M.D. et al., *Integración paisajística de las repoblaciones forestales.*, *Serie Técnica de Medio Ambiente*, Dirección General de Medio Ambiente, Consejería de Medio Ambiente, Vivienda y Ordenación del Territorio de la Comunidad de Madrid, Madrid, Spain, 2013.)

1982, 1987; Koper et al. 2009; Trotsiuk et al. 2012), in boreal forests (Angelstam and Kuuluvainen 2004; Shorohova et al. 2009; Keneeshaw et al. 2011; Kuuluvainen and Aakala 2011), and in primary tropical forests.

Virgin, natural, or primary tropical forests in Europe under minor perturbation regimes show irregular structures comprising small homogeneous clumps, often less than 1 ha. Large trees at their physiological end are also present.

Another characteristic of old-growth forests and virgin forests is the high volume of permanent biomass, as revealed by studies in broad-leaved forests in Karpathos (eastern Europe) (e.g., Korpel 1982; Trotsiuk et al. 2012). In Slovakian natural forests, total variation of biomass per hectare between the phase of large (maturity) and small accumulation (end of destructive phase) is 30% of the total biomass. This means that the volume never falls below 70% of the maximum volume (Korpel 1982). Other studies also reveal significant quantities of live trees remaining in the destructive and regenerative phases (e.g., Trotsiuk et al. 2012).

In afforestations and forests that were clear-cut hundreds of years ago, the process of fragmentation and spatial variability can also be observed (e.g., Koop 1987; Otto 1998; von Oheimb et al. 2005).

Shade-intolerant or slightly shade-tolerant species reach their physiological limit and create disperse regeneration areas, which promote old-growth forest characteristics. The stand age for *Pseudotsuga menziesii* is 150–200 years (Agee 1997; Spies 1997) and 200–300 years for *Pinus sylvestris* (e.g., Kuuluvainen and Aakala 2011). As time evolves, other shade- and semishade-tolerant species appear. Thus, the development of successional stages involves shade-tolerant species, with a high structural diversity, but may present fewer species or even become a monospecific forest, as can be seen in some types of beech forests (Korpel 1982, 1995).

In temperate and boreal forests, a mosaic landscape develops as the fragmentation processes of shade-intolerant species advance. The mosaic consists of broad patches with decayed and fallen trees and coarse wood debris, which represents an important characteristic of old-growth forests. In addition, many regeneration areas can be seen in clumps with low tree density and gaps in the canopy caused by dead trees. Other features of old-growth forests are the presence of a multilayer canopy, cavities, a wide range of tree sizes, and large trees that have reached their physiological end and that accumulate the most forest biomass (Bauhus et al. 2009) (Figure 6.4).

These early old-growth forest stages, which include shade-intolerant species and gap dynamics, develop into shade-tolerant species and other related species phases. Regeneration will spread over most of the forest area, although it may involve the presence of gaps when parent trees disappear from the upper canopy layer. The characteristic disturbance regime in old-growth structures is gap dynamics (Angelstam and Kuuluvainen 2004; Shorohova et al. 2009).

The multilayered foliage canopy is distinctive of many old-growth types (Spies 1997).

Partial perturbations of greater significance than gaps and that occur over long periods, such as snowstorm, windstorm, pests, and diseases, promote succession with multiple variables (Otto 1998) and a landscape mosaic comprising mature stages, depending on the local environmental conditions. In this case, a dynamic

FIGURE 6.4 **(See color insert.)** Virgin beech forest in the Karpathos range (Romania). (Photo: Antonio García-Abril.)

cohort is the characteristic disturbance regime (Angelstam and Kuuluvainen 2004; Shorohova et al. 2009).

Perturbations can also be far reaching and intense such as fire. This can be observed in coniferous forests in the western United States and in the boreal forests of areas in the interior of Eurasia and North America, where they evolve with even-aged dynamics. Large-scale stand-replacing disturbances initiate succession and allow forests to regenerate over large areas simultaneously (Angelstam and Kuuluvainen 2004; Shorohova et al. 2009). These major perturbation dynamics may last for hundreds or thousands of years, showing different phases of old growth, with large trees, coarse wood debris, snags, logs, and regeneration in extensive areas of the undercanopy.

Different phases of forest development can be distinguished: Oliver and Larson (1990) assessed four, Spies and Franklin (1996) considered six, Spies (1997) suggested seven, Franklin et al. (2002) proposed eight, and Angelstam and Kuuluvainen (2004) characterize six.

According to these authors, the old-growth phase may start after 100–250 years for different species (Franklin et al 2002).

For *P. menziesii* (Douglas fir), the early transition growth phase (Spies 1997) corresponds to vertical diversification (Franklin et al. 2002), after the stand is

FIGURE 6.5   (See color insert.) Old-growth forest. H.J. Andrews Experimental Forest (Oregon). (Photo: Antonio García-Abril.)

aged 250–350 years. It is characterized by the remaining trees of Douglas fir, which occupy a scattered layer of emergent trees, above western hemlock (*Tsuga heterophylla*) and other shade-tolerant species evolved from gaps in the original forest of Douglas fir. This phase also shows relatively large amounts of woody debris from fallen dead trees of the original cohort (Figure 6.5).

In the coastal coniferous forests of the northwest United States, these phases can be very stable, and there is a low likelihood of fire, so the development cycle may last for more than 500 years. Precipitation facilitates decomposition of woody material, which lowers fire risk (Agee 1997).

In the less humid coniferous forests of the northwest United States, the rate of decomposition is slower and facilitates accumulation of deadwood, wood debris in different phases, decadent old-growth stages, and pests. All these may ease crown forest fire due to dry lightning storms (Agee 1997) (Figure 6.6).

Keneeshaw et al. (2011) state that in the boreal forests of North America and Eurasia, there are more dynamics than only major forest fires. In central Siberia and North America, where there are long dry periods and continental conditions, major fires are more frequent than in the coastal regions with shade-tolerant species on both continents. In Eurasia, surface fires are more frequent than crown fires. In North America, large-scale disturbances include spruce budworms in the eastern forest, in addition to crown fires.

In Fennoscandia,* natural fire frequency is lower than was supposed; "The surface fire interval in upland dry *P. sylvestris*-dominated forests, was on the order of 150–250 years, when human impact was low" (Pitkänen et al. 2002; Keneeshaw et al. 2011).

---

* Fennoscandia is Finland and Scandinavian countries.

FIGURE 6.6 (See color insert.) Crown fire in Cascades Central Range (Metolius area, OR). (Photo: Francisco Mauro.)

In mesic sites dominated by *P. abies*, it can be hundreds, even thousands, of years. Some places with similar characteristics have not been burned since the last glaciations, more than 10,000 years ago. However, human activities have increased the number and frequency of fires (Keneeshaw et al. 2011).

Old-growth forests were predominant among boreal forests before the advent of significant human activities, especially in Eurasia where they evolved through small-scale disturbances.

In addition to gap dynamics, disturbances other than fire may promote conditions for even-aged stands of extensive areas due to drought, plagues, or large blowdowns (Shorohova et al. 2009; Kuuluvainen and Aakala 2011).

Biodiversity at the landscape scale contains species linked to all successional stages. In forest areas, if large-scale perturbation frequency is low such as in the case of fire, most of the area will be old growth at different successional stages. Gap dynamics from early stages and shade-intolerant species is followed by phases with shade-tolerant species. Therefore, to preserve biodiversity, it is necessary to maintain the different successional stages described.

Current old-growth forests are essential for conserving the organisms that inhabit their communities and can migrate to settle on developing mature forests. This reason alone is sufficient to warrant extending protection to all current old-growth forests (Perry and Amaranthus 1997). Thus, the aim of forest management based on landscape and ecological criteria is to promote complex forest systems.

6.2.1.4   Forest-Fire Dynamics and Man-Modified Forest Landscapes

Human activities interact with landscape by modify structures, which may increase or decrease the effects of perturbations. In Europe, fire means disaster. In the Mediterranean region, they are very frequent and extensive and may arrest the processes toward more complex forests. Its occurrence is even more widespread due to arson and the current homogeneous forest landscapes.

The significance of fire in the Mediterranean region has for centuries been distorted by human actions.

Several examples show that extensive crown fires were sporadic and limited in the past. For instance, *Pinus nigra* old forests in central and southern Spain, whose seeds do not sprout after fire (Tapias and Gil-Sánchez 2005), have been described by foresters since the 1850s. Those forests included irregular stands and trees of hundreds, or even thousands, of years old in inhabited zones with dry storms (Gómez Manzaneque et al. 2005; Tíscar 2005a,b).

Since the 1950s, arson has become more frequent in the Iberian Peninsula. As a consequence, the area of the pyrophyte *Pinus pinaster* has increased because fires have eliminated all shade- and semishade-tolerant species.

The Yellowstone crown fire of 1988 was facilitated by the homogeneous forest structure comprising extensive mature stages and the accumulation of wood and was extinguished thanks to the great efficiency of firefighting, surveillance, and prevention actions implemented since the nineteenth century. In this landscape, crown fires are the natural perturbation driving forest dynamics. An accumulation of young stand, old-growth and destruction phases over wide areas can lead to large-scale perturbations (Otto 1998). In Yellowstone, natural fires promoted spatial heterogeneity, composed of completely and partially burned areas in addition to unburned areas. This process created a mosaic of patches in different successional stages and with similar features to original forests (Turner et al. 2003).

In the boreal and temperate forest landscapes of North America, many natural forests were transformed into plantations and commercial forests and managed like even-aged forests with rotations of less than 100 years and a clear-cut system. The regeneration system was often seedling or plantation. These forests have very dense young phases—the thinning phase and also the phase known as the stem exclusion phase (Kuuluvainen 2009)—which are particularly vulnerable to crown fires. Moreover, higher frequency of fires due to human activities is well documented, and major fires have come to the attention of the public.

In certain types of forest, wide patches of mature and healthy forest with a complex structure act as barriers against crown fires and preserve forest structure (Perry and Amaranthus 1997).

> "Once some threshold proportion of the landscape becomes fragmented and permeated by flammable young forests or grasses, the potential exists for a self-reinforcing cycle of catastrophic fires– an absorbing landscape crosses a threshold and becomes a magnifying one." (Perry and Amaranthus 1997)

Forest fires are natural perturbations in continental, boreal, or Mediterranean regions with dry periods or with areas of deadwood accumulation, but they are not the only or the main perturbation that drives forest dynamics.

6.2.1.5   Clear-Cutting: A Contrast with Forest Fires

Commercial forestry treatments based on short rotations and clear-cutting do not correspond to forest landscapes subject to natural dynamics in many boreal forest areas (Angelstam and Kuuluvainen 2004; Kuuluvainen 2009; Shorohova et al. 2009; Keneeshaw et al. 2011) (see Section 6.2.1.3).

Clear-cuttings have been justified as comprising a management that is similar to fire perturbation, to which species—especially shade-intolerant species—are adapted.

These species show certain plasticity to regenerating undercanopy conditions and with gradual gaps in young phase canopies (Tappeiner et al. 1997; Velarde et al. 2013). Extensive regeneration gaps are not inherent to these species; this is simply a management option.

The clear-cutting system differs from natural fires, even in the case of crown fires. Traditional clear-cutting removes all valuable commercial wood while destroying the undergrowth during logging. Natural fires conserve structural and functional features, which increase the spatial heterogeneity of nutrients and humidity; they can also preserve propagules and seed sources to reforest the burned area. Some structural features remain: unburned areas, partially burned areas, live trees, snags, and logs. These legacies are capable of recovering the stages prior to perturbation in the long run (e.g., Halpern and Spies 1995; Franklin et al. 1997; Kuuluvainen 2009). These are the main reasons why clear-cuts are not the ecological equivalent of natural disturbance (Perry and Amaranthus 1997).

In natural forests, clear-cuts followed by short rotation period (RTP) cuttings reduce the populations of many species and completely remove shade-tolerant species. Exclusion due to dense phases entirely suppresses species that may appear during the regeneration or establishment phase (Spies 1997).

In Fennoscandia, the structural homogeneity caused by clear-cutting as the dominant method of harvesting, and the growing of even-aged stands, entails another dramatic landscape change: the sharp decrease in old forest, old trees, and deadwood (Kuuluvainen 2009). "If forest management practices continue to drastically change ecosystem structures from those that occur naturally, the continued decline in diversity and local species extinctions also seems inevitable in the future" (Kuuluvainen 2009).

To reduce the effects of clear-cutting in natural forests, forestry management systems were developed based on the retention of structural features. These systems can be used in intensive silviculture (e.g., Halpern and Spies 1995; Franklin et al. 1997; Bauhus et al. 2009; Kuuluvainen 2009; Gustafsson et al. 2010, 2012).

6.2.2   CHALLENGES AND TRENDS IN SUSTAINABLE FOREST LANDSCAPE MANAGEMENT

Forest activities have effects on landscape. Thus, the incorporation of the concepts and knowledge from ecology, landscape ecology, landscape design, and landscape planning into forest management has been of great importance in the last 50 years. Forest management produces its own visual appearance and has different consequences on the structure and functioning of forest landscapes. Each management system has clear visual and ecological effects. The following is an overview of trends that have emerged in forest management and their evolution.

For about 20 years, the prevalence of concepts of ecological landscape has been shaping the new boundaries of forest management. Ecosystem management, close-to-nature forestry or CCF, and the conservation and recovery of old-growth forests are innovations and milestones in the process of improving SFM.

6.2.2.1 Forestry Trends

In the last 50 years, the concept of forestry, its objectives, and its methodologies have undergone a considerable change, although these processes are still ongoing due to advances in forest ecosystem functioning, increased knowledge, and new priorities.

Sustainable timber yield, soil and water conservation, wildlife conservation, and recreational and leisure activities were the traditional objectives of forestry and of the different forest and forest landscape management, although timber production was the traditional and most frequent use.

In the 1990s, the introduction of the concept of SFM (as part of sustainable development) guided forest management and was developed in international initiatives for SFM application and standardization.*

New findings in forest ecology and the impacts of forest activities have had a significant impact on changes in forest management. Society's growing interest in conservation and recreation has also played a part, in addition to the declining importance of timber and the traditional forest sector in the gross domestic product.

Social aspects of forest management have gained in importance, due to the involvement of residents and users in the actions undertaken in forests.

Three main types of forest management can be considered based on silviculture systems (Schutz 1991, 1997): monofunctional silviculture aimed at timber production and characterized by the use of intensive silviculture, multifunctional silviculture with a range of objectives and the production of marketable and nonmarketable values (e.g., recreation, water and soil protection, conservation of species and ecosystems), ecological silviculture or ecosystem silviculture aimed mainly at ecosystem conservation, especially through species, habitats, or biotopes. The Natura 2000 network in Europe is a good example of the latter.

Ecological silviculture considers timber production as a subordinate objective or a secondary production derived from conservation management.

Historically, in Europe, monofunctional silviculture has been applied for more than 200 years and is still in use. Multifunctional silviculture has been in use for over

---

* The term sustainable (forestry) development has its origins in the Rio Conference in 1992 and has occupied much of the international debate on forests. Resolution H1 of the Second Ministerial Conference on the Protection of Forests in Europe (Helsinki Conference, 1993) defined the term sustainable forest management as *the stewardship and use of forests and forest lands in a way, and at a rate, that maintains their biodiversity, productivity, regeneration capacity, vitality, and their potential to fulfill, now and in the future, relevant ecological, economic, and social functions, at local, national, and global levels, and that does not cause damage to other ecosystems.*

This definition of SFM has established the guiding concepts of forestry since the 1990s, as a basis for current and future production.

100 years but became a consolidated and popular approach in the 1970s. Finally, ecological silviculture emerged in the 1980s but was applied after the 1990s.

This European evolution has its parallels in the United States, where new forestry and ecosystem management responded to intensive silviculture in the 1980s and 1990s, respectively.

Integration of the visual landscape into forest management occurred in the 1960s and is today consolidated, and there are methodologies available for its assessment (see Section 6.1.3).

Ecological landscape (linked to visual effects) is also currently in the process of integration, as increasing information on structure, ecosystem functions, and effects of forest treatments and activities becomes available. All these systems involve the management of complexity.

Different terms have been used to describe the inclusion of ecological landscape and scenic values into traditional forestry, such as "landscape forestry" (Boyce 1995). Visible stewardship (Sheppard 2001) insists on care and attention in visible forestry actions. Visible stewardship takes agrarian man-modified landscapes as its guide, due to their respect for nature and their acceptance by the public, in contrast with certain forest management practices.

Kimmins (2002) considers that sustainable forestry today is a social forestry with an ecologically based, multivalue ecosystem management. This social forestry involves new paradigms such as ecosystem management and adaptive management. From an ecological perspective, it is based on respect for (Kimmins 2002)

- Ecological diversity
- Biological diversity
- Sustainability
- The ecological role of disturbance

6.2.2.2 Ecosystem Management: The Ecosystem Approach and Adaptive Management

"The adoption of ecosystem management as a guiding philosophy for 21st-century forestry represents a move from simplified to complex conceptions of ecological and organizational systems" (Kohm and Franklin 1997). Kohm and Franklin (1997) and Spies (1997) emphasize prudence and humility as essential attitudes for forestry for the twenty-first century, due to the uncertainty of our decisions and our insufficient knowledge of ecological complexity. For Spies (1997), prudence justifies the assumption of an adaptive ecosystem approach: "Given our imperfect knowledge of forest stand, structure and function and poor understanding of forest management effects on biodiversity and long-term ecosystem function, it is uncertain how well we can sustain ecosystem values while providing commodity resources."

Kohm and Franklin (1997) highlight our insufficient knowledge, which is based on revisable hypotheses and the results of research and experiments. A number of advances in forest composition, structure, and function have evolved in the last decades and are expected in the future. "From this we are reminded of the very tentative state of our current knowledge and the iterative nature of learning.

We begin, finally, to appreciate that each management prescription is a working hypothesis whose outcome is not entirely predictable. And, hopefully, we adopt humility as a basic attitude in all approaches to forests–whether as scientists, advocates, managers or policy makers. Adaptive management is the only logical approach under the circumstances of uncertainty and the continued accumulation of knowledge. Management must be designed to enhance the learning process and provide for systematic feedback from monitoring and research to practice" (Kohm and Franklin 1997).

Since these words were written, events have confirmed the accuracy of this prediction, as notable contributions have been made to forest dynamics in the last 15 years, revealing that no definitive models can be proposed for long-term decisions from this developing science. Nature has many faces, and assuming solutions attributed to Nature's voice is merely a human interpretation.

Forestry deals with complex systems in which the variability and heterogeneity of processes and structures are key elements in the overall forest dynamics and resilience. They cannot be controlled or simplified to produce goods and services without long-term consequences on the environment and a likely reduction in their ability to change and adapt (Messier and Puettmann 2011).

Adaptive management seeks sensible solutions under conditions of risk and uncertainty in a complex social and biophysical context. It is built on learning, collaboration, and integrated management. The key to the process is to learn from policy outcomes, from actions already taken, and from knowledge derived from research and practical experience (Stankey et al. 2005).

Models and modeling are fundamental tools that can be used to represent important elements of the system in space and time, to simulate the complexity of stands and landscapes, and to incorporate environmental changes in order to generate possible scenarios.

The new models and tools must integrate complexity and self-organization with their multiple relationships and take into account the unexpected results of the dynamics. These models will help society to learn how to make use of the natural capacity of systems to guide them in the right direction and produce the necessary goods and services (Messier and Puettmann 2011), "but this requires more than new tools, it implies a totally new way of looking at the forest and forestry" (Messier and Puettmann 2011). It requires above all prudence to adapt to nature and to changing conditions, it does not cause irreversible problems, and it facilitates different practical solutions to obtain a desired result. "Instead, creativity in thoughts and diversity in practices are needed in designing new forest management policies for the future" (Messier and Puettmann 2011).

6.2.2.3   Close-to-Nature Forestry

Prudence, humility, continuous learning, and adaptation to nature to obtain complex productive forests similar to the mature stages of forest succession are part of what is known as close-to-nature silviculture, close-to-nature forestry, or CCF (Bruciamacchie and de Turckheim 2005; Pukkala and von Gadow 2012). This is a movement in European management that has its roots in the nineteenth century and emerged for reasons similar to those that led to its revival in the 1980s and 1990s.

In the nineteenth century, intensive forestry conifer plantations were established with a regime of sowing or planting and harvesting, in much the same way as an agricultural crop, but with the primary goal of rebuilding European forests degraded by logging, grazing, fire, and the devastation dating from the beginning of the industrial revolution. In the late nineteenth century, the susceptibility of these forests to pests and diseases, windstorms, snowstorms, and forest fire—in contrast to enduring forests—was observed. A key moment in the history of forestry occurred around 1880, a period which saw the birth of the idea of a natural or close-to-nature forestry management. The idea was proposed by K. Gayer (1822–1907) in response to intensive forestry and its proven consequences of ecological and economic instability. He basically advocated an individualized treatment of each forest adapted to its natural characteristics. This principle is analogous to one of the conclusions of ecosystem management: the development of site-specific knowledge (Kohm and Franklin 1997). Natural silviculture was also based on the principles of reliance on natural processes, natural regeneration, and obtaining uneven and mixed forests. In the late nineteenth century, irregular silviculture using the single-tree selection system was also articulated through the method of *jardinage*.

The ideas of natural silviculture, permanent forest, continuous forest, and continuous cover continued to spread until the Second World War. After this conflict, the world was caught up in a sense of euphoria and an immense confidence in man's abilities. Technical capacity was regarded as a means of controlling and guiding nature, merely by applying the appropriate techniques and investments. This was a time when the forest was understood as a factory, the era of industrial forestry, the forest as a crop, and the profitable forest. Plantations, clear-cut, and even-aged forests were the main tools of silviculture. Profitability was understood as an extractive activity (e.g., mining). Thus, natural forests created after thousands of years were clear-cut and replaced by monospecific plantations with a short rotation. Profitability was calculated simply by minimizing costs and maximizing timber yield, regardless of the cyclical process involving stages with low income and expenses, which would be recovered decades later. This is far removed from the system used to manage a renewable resource.

As in the nineteenth century, the differing response of forests to natural disasters (even-aged stands compared to mixed uneven-aged or heterogeneous forest with natural regeneration) gave cause for reflection. Repeated windstorms in Central and North Atlantic Europe, pests and diseases, snowstorms, and acid rain tested the stability and resilience of forests managed with different types of silvicultural methods. Structurally, diverse forests absorb disturbance better and conserve elements that promote rapid recovery. Moreover, in different parts of Europe, permanent forest management remained in the public administration and in the hands of various private owners and family properties with more than a century of experience. These properties managed their production to obtain high-quality large-volume wood by means of microecological variability and dynamic forest processes to avoid the costs of regeneration, pruning, and thinning. Why spend money on what nature can do for free?

Close-to-nature forestry or CCF was not initially considered as a goal in itself but as a means to achieve an optimal economic benefit. Forests were intended for timber production. The reflection prompted by natural disasters should also be seen in this light and gave rise to very creative solutions (Jacobsen 2001). High-quality wood and lower costs produce a better economic balance in forests with continuous-cover management than in even-aged-managed stands with major regeneration, clearing and pruning costs, and lower wood value (de Turckheim 1993; Schutz 1997).

Laiho et al. (2011) conclude for Finnish forests of *P. sylvestris* and *P. abies* that "uneven-aged management is more profitable than even-aged rotation forestry (RF), especially with high discount rates. Uneven-aged management seems to be superior to current even-aged RF also with respect to environmental and multifunctional aspects, such as carbon sequestration, bilberry yield, structural diversity and scenic values."

Another recent source of knowledge to support close-to-nature forestry is the dynamics of virgin forests in Europe, in which large-volume trees provided continued stability to the whole and there is a huge quantity of regeneration, of which only a very small number is needed to replace the upper canopy trees.

In close-to-nature silviculture, the focus on individual trees or small groups of trees and the use of microecological variability has led to a high structural diversity and mixed stands. In this type of management, deadwood and minority understory species are also retained. Even the minority species are safeguarded from cuttings in order to conserve biodiversity. The biodiversity of species and structures is regarded as a necessary condition to ensure a sustained production. Treatments take account of the cooperation—rather than competition—relationships between the trees and include high thinning in order to promote objective trees and slowly reduce density in the surroundings. These cuttings also facilitate crown development and offer tree stability due to crowns that are suited to an upper-tree position. Low thinning is not used.

Prudence advises frequent and nonirreversible actions, with extraction volumes below 80 m^3 and rotations of 5–15 years. Extraction is more expensive and complicated than with clear-cuts, but there is extensive experience in forests with large volumes and complex topography, such as in Switzerland and Slovenia. These are forests with a diverse species composition and structure, which were also created based on the wise principle of not eliminating anything that is unknown and does not pose a problem for the whole forest or trees focusing on production.

We distinguish a principal tree category and a complementary tree category. The producer tree category includes dominant trees, which confer stability on the forest structure, are a source of regeneration, and concentrate the profit value. These are the producer trees. A second group entails fast-growing trees, which grow freely up to the dominant canopy. The third group includes viable regenerated trees, which can be promoted by future clearing. The principal tree category holds half the total volume; the other half comes from the complementary tree category. It protects the soil and the producer tree category. The complementary stand contributes to natural pruning and also includes minority species and decayed or dead trees. A tree from the complementary tree category is crucial for diversity. Only those trees that disturb the development of trees in the producer tree category or are susceptible to diseases or plagues should be cut.

European temperate forests under this type of management are diverse, stable, productive, and beautiful when compared to even-aged forestry. Nature is also to some degree involved in this form of management and especially in family forests where the forest is an extension of the home.

In conventional forestry, with even-aged management and short regeneration periods (RP), control and geometric order are fundamental. The implicit maxim of this management is: control is good.

In close-to-nature forestry, the number of degrees of freedom is greater due to the diversity of the elements, and there is a high frequency of cuttings, which allows a quick return to the same situation. These factors allow a wide margin for creativity and freedom, under the responsibility of maintaining the same previous degrees of freedom for successors. Forests must improve or remain in the same initial condition. The maxim for this type of management is: freedom is better.

In Europe, there are several associations that serve as a focus for people and institutions interested in this type of management. The two main associations are the German ANW "Arbeitsgemeinschaft Naturgemässe Waldwirtschaft" (working group for close-to-nature forestry) and Pro Silva Europe (European federation of foresters advocating forest management based on natural processes), consisting of researchers, teachers, owners, technicians, and foresters. Almost all the German Länder were inspired by the ideas of the ANW (Bruciamacchie and de Turckheim 2005).

Close-to-nature forestry is an example of sustainable management, where the economic returns and ecological diversity find common ground.

In the past, biodiversity was considered as a means to an end. Today—especially in public forests—it has become a goal, and timber production is now a secondary objective. In this context, cuttings are ecosystem management tools.

This system of forestry management has prompted the transformation of many even-aged stands into uneven-aged stands and of coppice forest or coppice with standards into irregular high forest. Empirical knowledge is vast; there are numerous practical experiences that can be studied, which have been largely overlooked by research and teaching.

6.2.2.4 Maintenance and Creation of Old-Growth Forest as a Forestry Objective

Forests are the most altered and destroyed vegetation type. Old-growth forests correspond to late forest development successional stages and are most at risk, since hundreds of years are required for an old-growth stand to make up after disturbance. They can remain, under gap dynamics, for hundreds or thousands of years.

Certain organisms require old-growth forests for their continued existence, either because they live exclusively on these forests or because they need them at different times of their life cycle.

These are essential features of the landscape, in that they represent the entire space-time dynamic and structural and functional complexity associated with it. Old-growth forests are needed for the maintenance of biodiversity and structural diversity at the stand level, as well as the diversity of ecosystems and their successional stages throughout the landscape (e.g., Bauhus et al. 2009; Kuuluvainen 2009).

"The challenge to management is to find the mix of stand and landscape practices that meets biological and social objectives" (Spies 1997). The conservation, restoration, and creation of these forests are goals of management and should be included in forest management priorities (Bauhus et al. 2009; Kuuluvainen 2009).

It is generally accepted that there is a lack of knowledge of the methods to be used while maintaining the characteristics and creation of these forests (Franklin et al. 1997; Bauhus et al. 2009; Kuuluvainen 2009; Schütz et al. 2012).

For Messier and Puettmann (2011) and Puettmann et al. (2011), silviculture is based on "control and command," searching for a single steady or cyclical state focused on efficiency, control, and predictability. In contrast, the forest as a complex adaptive system (CAS) focuses on attributes of persistence, adaptability, and variability.

This opinion identifies all types of silviculture with intensive regular silviculture—the so-called traditional silviculture, the silviculture of "control and command."

Experiences on uneven-aged forest management and the methods and experiences of close-to-nature silviculture are a good starting point for generating suitable silvicultural treatments. Some of the objections regarding the capacity of traditional silviculture to create the characteristics of old-growth forests were resolved by close-to-nature silviculture, such as the conservation of minority species, deadwood, vertical and horizontal heterogeneity, multi-canopy layers, stable crowns, advanced regeneration, and undercanopy species.

The conservation of biodiversity is an essential requirement in all forest management activities, including productive forests. Bauhus et al. (2009) propose the introduction of varying degrees of old-growth characteristics in managed forests, instead of differentiating landscape into old growth and regrowth. This active management for old growthness is currently being applied (e.g., Tappeiner et al. 1997; Carey 2009; Kuuluvainen 2009).

The importance of matrix elements in landscape structure for biodiversity conservation (Lindenmayer and Franklin 2002; Fischer et al. 2006) highlights the concept of integrating biodiversity conservation into forest management in their entirety, instead of only applying it to reserves or certain types of forests.

6.2.3   Spatiotemporal Dynamics in Managed Forests for Timber

Forest management affects landscape through actions aimed at timber and nontimber forest products, maintenance of biodiversity conservation, protection of water and soil, health and vitality of ecosystems, social practices, and recreation. Wood is the most traditional and widespread forest production and focuses on cutting certain trees. Cutting models are also linked to regeneration and the future structure of forests. Thus, different methods of forest management and silviculture give rise to different landscapes, which can differently fulfill the functions of conservation, protection, production, recreation, or visual appeal.

In this section, we analyze the main silvicultural methods from the landscape point of view and relate them to the main processes of natural dynamics. We also analyze the meaning and consequences of the spatiotemporal dynamics of silvicultural methods specified in a management plan.

6.2.3.1   Main Silvicultural Systems and Similarities with Main Disturbance Regimes

Silvicultural systems are classified by regeneration cutting types. They focus on tree regeneration—either at the end of the RTP or periodically—depending on a certain cutting diameter limit. Regeneration can be natural or artificial. Regeneration cuttings are also the most productive, since they involve the largest trees. They model forest structure and can influence the size and distribution of trees and forest species composition. Cuttings are not just a harvesting or regenerative system; they can be tools for ecosystem management.

The time of harvest is crucial, as it shapes the future forest structure. A forest management plan or working plan area divides the forest into sequential management units and thus provides a complete and successive system for future forest development. Moreover, the forest landscape is related to forest structure.

6.2.3.1.1　Basic Forest Structures

Diversity of spatial structures defines forest landscape at the *exterior* landscape scale, while the structure in a particular place defines internal landscape. There are two basic structures, even-aged and uneven-aged, which correspond to two silvicultural models: even-aged high forest and uneven-aged high forest (Schutz 1991) (Figure 6.7).

The vertical and horizontal structure of even-aged stands is homogeneous. In the vertical structure, there is essentially only one canopy level. Competition between trees is horizontal, at the crown level. Trees compete for space. In uneven-aged stands, crowns stratify and the horizontal profile is irregular, as there is alternation of trees and groups of trees of different diameter, height, and development. Competition among trees is diffused, or rather the trees are ordered and subordinated by access to light.

The structure in horizontal or vertical projection is heterogeneous. The horizontal structure includes single trees or various groups, while the vertical profile is a broken line.

This broken line and the absence of a large homogeneous upper canopy involve special conditions of regeneration. It is usually agreed that the irregular structure is one in which trees grow and regenerate under the shade or influence of mature adjacent trees in the upper canopy, and this occurs if the diameter of a regeneration gap is approximately less than twice the dominant height of the adjacent trees (Smith et al. 1997). In natural or managed forests, ecounits (Oldeman 1983, 1990) are similar to this size and become a mosaic structure.

Semi-even-aged stands entail two or three age classes or a mosaic of stratified trees, normally bistratified. From an even-aged structure for the whole stand to an uneven-aged structure in terms of individual trees, complexity is gradually increased by expanding the boundaries among the groups and decreasing the patch size of the mosaic. The texture of the matrix becomes thin, and it is more difficult to localize the groups as the process of irregularity advances.

6.2.3.1.2　Main Disturbance Regimes and Silvicultural Systems

Here, we consider three main disturbance regimes, their linked structures, and the silvicultural systems that resemble them.

Silvicultural systems mimic natural disturbances, but all commercial wood is removed, and there is an evolution in the regeneration and rejuvenation of stands. The natural disturbance regimes and associated dynamics we present here come

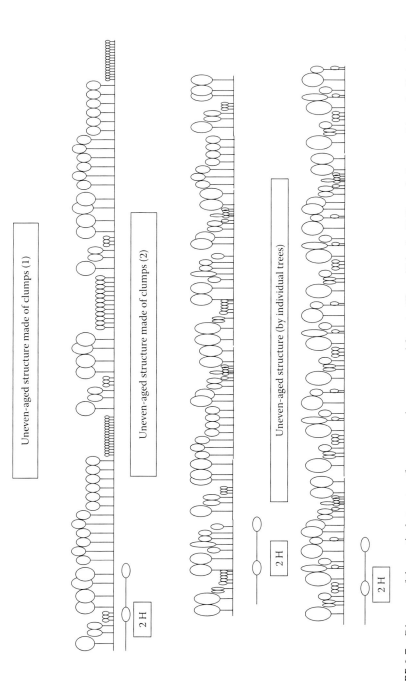

FIGURE 6.7 Diagram of the vertical structure for even- and uneven-aged forests. (From Velarde, M.D. et al., *Integración paisajística de las repoblaciones forestales., Serie Técnica de Medio Ambiente*, Dirección General de Medio Ambiente, Consejería de Medio Ambiente, Vivienda y Ordenación del Territorio de la Comunidad de Madrid, Madrid, Spain, 2013.)

from the findings in boreal and temperate forests by Angelstam and Kuuluvainen (2004) and Shorohova et al. (2009):

1. *Severe stand-replacing disturbance*, with succession dynamics: Crown fire perturbation triggers the regeneration process for the whole forest. Even-aged forests evolve composed of trees with a range of ages of less than 40 years. The diameter and age distributions are unimodal and near normal: The succession advances depending on fire recurrence. Thus, after internal processes encourage irregularity, gap dynamics may occur before the next crown fire, hundreds of years later.
2. *Partial disturbances* with cohort dynamics. The partial death of adult trees to a varying extent determines different structural subtypes: two canopy layers comprising adult trees and regenerated trees. The tree diameter and age distributions are bi- or plurimodal. This is a temporary situation that develops into a regime of gap dynamics and even-aged structure if there are no more partial or severe disturbances. It is originated by partial disturbances such as windstorms, snowstorms, surface fires, pests, and diseases.
3. *Endogenous disturbances* with gap dynamics: All ages and diameter distributions are represented in small areas or through individual trees. These distributions are "negative exponential" or "reverse J-shaped." Gaps may come from the death of individual trees or groups of trees by senescence, winds, or diseases.

Shorohova et al. (2009) state that "all-aged stands are gradually formed over several hundreds of years of endogenous succession with gap dynamics. This endogenous succession can be interrupted by stand-replacing or partial disturbances leading to successions with even-aged or cohort-structured stands."

We consider the following silvicultural systems for high forest (Matthews 1991): clear-cutting, shelterwood system, and selection system. We discuss the associated perturbations, the differences between them, and their associated dynamics; the structures obtained allow the classification of subtypes.

6.2.3.1.2.1 Even-Aged Cutting Systems   *Clear-cuttings* are the widespread cuttings that give rise to even-aged forests. All trees are removed from the stand (usually from monospecific stands). They can be classified into

- Clear-cuttings, clear-felling: Removal of all trees from the stand.
- Seed-tree clear-cutting: Reserve trees or parent trees remain after the first cut. They can be isolated, in groups or in lines.

These are similar to *severe stand-replacing disturbance*. Differences with fire perturbations were discussed in Section 6.2.1.5. They can be summarized as follows: removal of nearly all structural elements that could serve as refuge and propagules capable for regeneration.

Shelterwood systems gradually remove trees over a period of time equal to or greater than the duration of an artificial age class (normally 20–40 years) or longer periods, depending on the regeneration development. This system maintains even-aged structures with periods of two or more strata. The two main subtypes are the *uniform shelterwood system* and the *group shelterwood system.* The disturbance is partial and the remaining trees are maintained until death in natural process. However, in the managed stand, trees from the upper canopy remain until regeneration develops, then trees are cut. Forest structure is regular, with a bi- or plurimodal diameter and age distribution, while crop trees remain. The main difference with clear-cuttings is the continuous covering of soil by crop trees and regenerated trees. Regeneration and crop trees coincide in space and time.

The irregular shelterwood system is applied to create or maintain uneven-aged forests. This type is included in the following group.

6.2.3.1.2.2   *Uneven-Aged Cutting Systems*   *Tree selection systems* are selective cuttings that generate and conserve uneven-aged stands. All management units are constantly in the process of regeneration. The soil is always protected by trees. This is equivalent to endogenous disturbances with gap dynamics in small areas caused by the death of one or more trees or small groups of trees for episodes of wind, snow, disease, or plague.

To achieve and maintain uneven-aged forests, the three principal cutting types are *single-tree selection* (removal of one or several trees, *jardinage*), *group selection system* (removal of groups of trees, similar to clear-cuttings in very small areas), and *irregular shelterwood system* (shelterwood system applied to very small areas). In this case, all development stages are present in the stand. In contrast to gap dynamics, trees are removed before their physiological end, even before their maximum size is reached. Sometimes undercanopy and secondary species may be controlled.

From the point of view of promoting structural diversity, clear-cutting, the shelterwood system, and selective cuttings inevitably involve the management unit.

Each of the cutting systems improves high or low structural diversity depending on the size of the management unit. The smaller the regeneration areas, the more similar to minor disturbance dynamics and complex structures of the final succession stages (Figure 6.8).

To achieve high structural diversity, it is advisable to replace clear-cuttings with a shelterwood system and the shelterwood system with a selection system (Pommerening and Murphy 2004).

Habitat and biodiversity goals can best be described in terms of forest stand structure and species composition. Thus, many—but not all—components of stand structure are affected or can be created by silvicultural practices (Tappeiner et al. 1997).

Each structure type can be recognized by a vertical profile (crown line) that can serve as a surrogate for heterogeneity and visual attractiveness (Figure 6.9).

6.2.3.1.3   *Retention Forestry*

Retention forestry is "an approach to forest management based on the long-term retention of structures and organisms, such as live and dead trees and small areas of intact forest, at the time of harvest. The aim is to achieve a level of continuity in forest structure, composition, and complexity that promotes biodiversity and sustains ecological functions at different spatial scales" (Gustafsson et al. 2012).

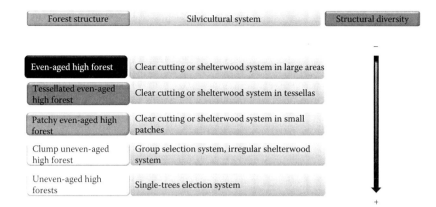

FIGURE 6.8 Increase in structural diversity by silvicultural systems and size of management unit. (From Velarde, M.D. et al., *Integración paisajística de las repoblaciones forestales., Serie Técnica de Medio Ambiente*, Dirección General de Medio Ambiente, Consejería de Medio Ambiente, Vivienda y Ordenación del Territorio de la Comunidad de Madrid, Madrid, Spain, 2013.)

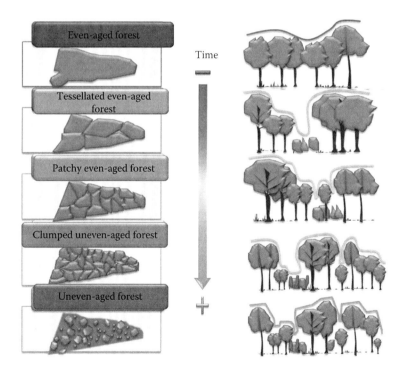

FIGURE 6.9 Different stand or compartment structures. Broken crown line corresponds to spatial heterogeneity and high vertical stratification and to more mature successional stages. (From Velarde, M.D. et al., *Integración paisajística de las repoblaciones forestales., Serie Técnica de Medio Ambiente*, Dirección General de Medio Ambiente, Consejería de Medio Ambiente, Vivienda y ordenación del Territorio de la Comunidad de Madrid, Madrid, Spain, 2013.)

This concept and its application emerged to reduce the impact of clear-cutting in natural forests.

The variable retention harvest system is based on the retention of structural forest elements at the time of harvest, at least until the next rotation, for three main purposes (Franklin et al. 1997):

1. "Lifeboating" species and processes immediately after logging and before forest cover is reestablished.
2. "Enriching" reestablished forest stands with structural features that would otherwise be absent.
3. "Enhancing connectivity" in the managed landscape.

Other authors add more objectives (Gustafsson et al. 2012) such as ecosystem services and productivity, public acceptance of forest harvesting, continuity in key habitat elements and processes, minimizing the off-site impact of harvesting, and improving the aesthetics of harvested forests.

Retention may include individual trees, snags, logs, or small patches of forest at the time of harvest, at least until the next RTP (Franklin et al. 1997). *Disaggregated group retention* and *dispersed retention* indicate different spatial distributions of retained structures (Gustafsson et al. 2012) (Figure 6.10).

The percentage of surface or biomass retained can vary widely depending on the characteristic of the forest and the local context. The minimum area or minimum volume to be retained for achieving target ecological effects is 5%–10% (Gustafsson et al. 2012).

FIGURE 6.10 Variable retention near Squamish (British Columbia, Canada). (Photo: Rubén Valbuena.)

The remaining structural elements are "legacies" from the previous forest and perform similar functions to fire "legacies" in contrast with traditional clear-cuttings (Halpern and Spies 1995; Franklin et al. 1997; Perry and Amaranthus 1997).

The positive effects of variable retention on carbon storage and the conservation of several taxonomy groups have been proven, and there are a number of experiments under way to assess the long-term effects (e.g., Franklin et al. 1997; Bauhus 2009; Gustafsson et al. 2012). Variable retention, in many cases, allows diversity to be maintained and restored (Bauhus et al. 2009).

The retained elements suffer sudden exposure, and many trees and copses die in the years after the cut. Exposure to wind and sun, heat oscillations, and frost are among the reasons for their decay and death, which is also sometimes caused by pests and diseases. Very few isolated trees and small groups come through to the next rotation. Other problems related to future timber yield come from retained trees, which compete with the regeneration and can generate diseases or plagues (Bauhus et al. 2009).

Another drawback is the exposure of small structural elements to predator species among the species in the forest interior (Kimmins 2001).

Variable retention is successfully applied in different ecosystems all over the world (Gustafsson et al. 2012). As time evolves, we can weigh the numerous implications of this concept in order to incorporate it into traditional silvicultural systems as a tool to improve diversity and incorporate old-growth forest elements.

6.2.3.2 Old-Growthness Features to Be Included in Silvicultural Systems

Forest management in many regions is linked to even-aged, monospecific stands with short rotations and often to plantations after clear-cutting. "The first prerequisite, is to modify the current silvicultural approach based on regulation and homogenization of stand and landscape structures to one that fosters natural ecosystem complexity" (Kuuluvainen 2009).

However, even in forests managed using shelterwood and selection systems, structural complexity may be greatly reduced, and the features corresponding to mature stages may be very low, as will the biodiversity associated with these stages.

Bauhus (2009) has made a comprehensive review of the concept of old growthness and of the state of the art in the current knowledge and experiences of its implementation at stand level. Old growthness is an approach that promotes or maintains the structural attributes of old-growth forests. It is also an objective that has emerged in forest management to increase biodiversity (Bauhus et al. 2009).

Actions to maintain the existing old-growth forests* have focused on

- The creation of reserves
- The use of "variable retention" procedures in natural forests (e.g., Franklin et al. 1997; Gustafsson et al. 2012)

---

* An old-growth forest is a forest in the later stages of development, characterized by the presence of old trees and structural diversity (Spies and Duncan 2009). Strictly speaking, it is limited to the stages of the forest where the pioneer species (shade intolerant) have disappeared.

- Promotion of actions in regrowth and secondary forest to incorporate old-growth attributes (e.g., Kuuluvainen et al. 2002; Bauhus et al. 2009; Gustafsson et al. 2010)
- Development of specific silvicultural operational methods (e.g., Kuuluvainen 2002, 2009)

In any case, active management is preferable to inhibition in order to achieve structural objectives and make the results more predictable (Keeton 2006). Nowadays, active management or active restoration to achieve old growthness is already a field of experimentation and practical application, and the results must be included in practical traditional silviculture and the forest management plan. The more experience and knowledge there is of old-growth forests in the silviculture of regrowth and secondary forests, the better they can be transferred to natural forests to promote old-growth silviculture.

As forests managed for timber objectives—and even some in wilderness areas—often do not contain the attributes of late successional stages, it is necessary to incorporate these attributes in order to improve biodiversity at the landscape level.

Despite the definition in footnote under this section, to apply these concepts to timber forests, we must bring forward the concept of old-growth forest and consider old-growth forests to be those stages of development in long-lived pioneer species when part of the trees has come to their physiological limit and begins an endogenous process of renewal.

It is, therefore, necessary to incorporate the characteristics of old-growth forests into the management of shade-intolerant species: the restoration practices are directed primarily to increasing the structural complexity of managed forest for timber.

We will consider the following elements of old growth from Bauhus et al. (2009) and Franklin et al. (2002):

- Large trees, some at the end of their physiological life span
- Wood debris from dead and decayed trees (including coarse wood debris)
- Spatial variability of tree sizes
- Size of homogeneous regeneration units
- Presence of several species
- Presence of advanced regeneration
- Stable crowns

The concept of variable retention must be used to apply old growthness in various aspects. "In this context the term retention implies that an attribute that would be removed under conventional management is deliberately retained for conservation purposes" (Bauhus 2009).

The prescriptions that can be established in a management plan will include the structural attributes of old-growth forests. These prescriptions may involve some additional costs, but the increase in biodiversity also has consequences on the improvement of stability and resilience to different types of disturbance and brings environmental and economic changes.

In natural spaces and on public property, this type of management offers substantial possibilities for application as it has further added values such as watershed protection, increased permanent carbon storage, and the improvement of landscape and recreational possibilities (Carey 2009).

Multiple scales must be considered for an improvement in biodiversity, from the stand to the landscape pattern. They must all represent the biodiversity contained in the different successional stages. Old growthness requires a management approach based on the goals of maintaining continuous forest cover and structural heterogeneity of forests over much of the landscape area (Kuuluvainen 2009).

The next section will analyze the temporal dynamics in three fundamental models of forestry organized in a forest management plan. For each one, we will discuss the means of including measures to incorporate characteristics of old-growth forests.

6.2.3.3 Spatiotemporal Stand Physiognomies for Different Silvicultural Methods in a Managed Forest for Timber

Different units are considered for the purposes of forest management. Forests are divided into working circle units, each with a particular cutting system, RTP, or diameter limit. There are also compartments and inventory units for locating cuttings, in addition to other activities. These units can be assembled into blocks and subdivided into stands. Stands are homogeneously developed groups or with homogeneous quality, and their size may vary.

Working circle units: blocks, compartments, and stands, are regular if the distribution of the number of stems is unimodal, and all trees belong to the same generation. They are known as semiregular when two generations are involved and the diameter distribution is bimodal or plurimodal. They are irregular when three or more generations coincide. These are properly multidiametric and irregular (irregular and uneven aged). Irregularity arises from shade influence on regeneration and young groups. However, an increasing number of homogeneous groups in a certain area lead to discontinuities and boundaries, promoting habitat diversity and opportunities for species.

In plantations and in regrowth and secondary forests, it is necessary to consider the target forest to be achieved, since the visual and ecological landscape is modeled in the forest management plan.

The following sections explain the spatial and temporal sequence of the forest structure (physiognomy) within a management area or working circle, depending on the management system.

6.2.3.3.1 Model for Landscape Evolution of Regular Stands under the Clear-Cutting System

Even-aged-managed forests present regular structures, largely repeated in cutting units or blocks. Trees in each block have a similar height, and there is one canopy layer formed by dominant and codominant trees.

For a 120 year rotation with a RP of 20 years, forests can be divided into six blocks that can include minor units such as compartments or stands. In Figure 6.11, the blocks match the compartments. We have depicted them all together, but they can be scattered throughout the forest. Each part corresponds to a particular stage of

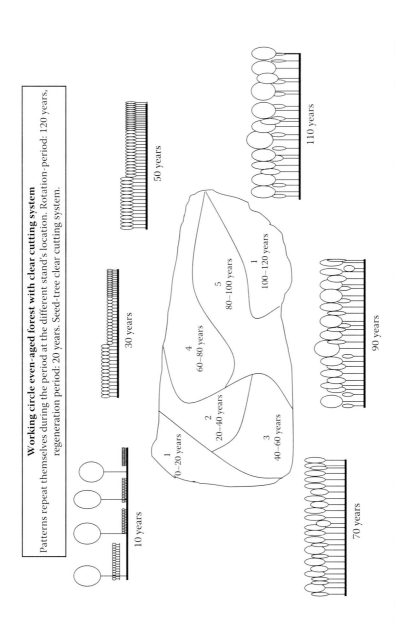

FIGURE 6.11 Spatial variation of stand structure for a whole even-aged-managed forest at a certain development stage. (From Velarde, M.D. et al., *Integración paisajística de las repoblaciones forestales.*, *Serie Técnica de Medio Ambiente*, Dirección General de Medio Ambiente, Consejería de Medio Ambiente, Vivienda y Ordenación del Territorio de la Comunidad de Madrid, Madrid, Spain, 2013.)

development. This is a cyclical development; thus, every part evolves into the next stage through growth and cuttings.

Figure 6.12 shows the evolution of the vertical structure and biomass of a block or compartment.

At all stages, almost all the biomass is in the top layer of the dominant and codominant trees. The trees dominated below the canopy have low vitality and reduced dimensions, since they are of the same age as the dominant trees. The landscape of the mature phase near the time of harvest, with columnar trees and sparse understory, is usually highly valued as there are no obstructions to the view and the height of the trees produces a gratifying sense of monumentality. Nevertheless, we must keep in mind that for this landscape to exist, other landscapes in 80% of the forest area in our example will have densely homogeneous structures with small trees that are not very suitable for recreational uses and are not particularly attractive (Figures 6.11 and 6.12).

Studies on visual preferences based on photographs may be incorrect if all development stages are not considered. If the columnar phase is preferred, the associated dense young phases should be included in the final value.

Negative visual impacts increase for clear-cuttings followed by plowing and artificial sowing (Figure 6.13).

Several procedures can be implemented to incorporate old-growth characteristics into managed forests. If the RTP increases (Curtis 1997), larger diameters can be achieved and a greater area of soil can be preserved. For instance, if the soil is bare for 10 years, this means that for a RTP of 50 years, 20% of the forest area is exposed, while for a 100 year RTP, only 10% of the area is exposed. From the recreational point of view, if areas with trees over 80 years are preferred, only 20% of the forest area would be available for RTPs of 100 years.

Large trees can also come from thinning aimed at increasing the diameter of dominant and codominant trees. This method also increases the wood value. Thinning from above is adequate for this purpose, whereas thinning from below is useless.

Dead and decadent trees must also be conserved except when they can spread diseases or plagues. Thinning from above also allows the decrease of the competitive-exclusion phase and the establishment of undergrowth species. In addition, the resulting well-shaped crowns promote tree stability against wind or snow.

Variable retention management can also be applied to trees in order to achieve large tree size and deadwood. Moreover, if small clumps are retained, they can provide for flora and fauna species.

Gaps can be created during thinning activities to serve as advanced regeneration cores and recovery mechanisms against major perturbations.

Another problem associated with clear-cuttings is CO_2 losses, since felling and plowing cause soil carbon mineralization and transfer to the atmosphere, whereas small residues are burned in situ, used as fuel, or mineralized. The timber has a short shelf life as paper, wood chips, etc., and ends in the atmosphere.

CCF has been used in numerous forests in Britain to increase the biodiversity and stability of plantations, mostly nonnative species with short RTPs, harvested by clear-cuttings and subsequent planting (Yorke 1998; Mason and Kerr 2004; Davies et al. 2008; Kerr et al. 2010; Davies and Kerr 2011).

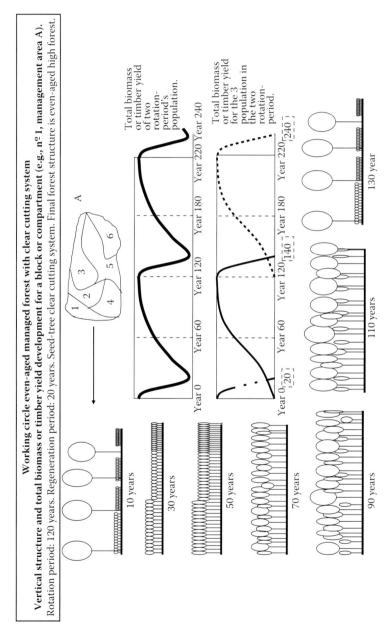

FIGURE 6.12 Cyclical development of vertical structure and biomass of a block or compartment within an even-aged-managed forest. (From Velarde, M.D. et al., *Integración paisajística de las repoblaciones forestales.*, *Serie Técnica de Medio Ambiente*, Dirección General de Medio Ambiente, Consejería de Medio Ambiente, Vivienda y Ordenación del Territorio de la Comunidad de Madrid, Madrid, Spain, 2013.)

FIGURE 6.13   **(See color insert.)** Clear-cutting, plowing, and sowing contiguous to pre-viously regenerated areas. Tierra de Pinares (Soria, Spain). Landscape impact is negative. Adverse landscape effects also increase with seed-tree clear-cutting and if the soil is plowed and artificially planted. (Photo: Antonio García-Abril.)

CCF in Britain, in its simplest version simply excludes clear-cuts and uses shelter-wood or selection systems. A clear-cut is an opening greater than 0.25 ha. It is applied to exotic or native species and natural regeneration and also includes planting.

The objectives of CCF are to increase use of natural regeneration, to maintain a continuous canopy of tree species, to diversify the forest structure, and in many cases to obtain large trees.

An example of diversifying the forest structure consists of reducing the regu-larization of monospecific coniferous even-aged forests and obtaining multispecific uneven-aged conifer forests by increasing the RP.

The system is accomplished by gradually opening up gaps and planting them with species that are to be included in the final forest. All the species can be exotic; in this case, structural heterogeneity is obtained as a result. Additional benefits can be achieved through natural regeneration, native species, large trees, and deadwood.

6.2.3.3.2   Model for Landscape Evolution of Regular
Stands under the Shelterwood System

A working circle or management area of a shelterwood-managed forest presents regular and stratified structures, which may be repeated. Different canopy layers can be distinguished in some areas.

Figure 6.14 shows the spatial variation of a management area under the shelter-wood system at a particular time.

Regeneration extends in three blocks, which occupy half the total area (TA) during half the period. During regeneration, the cuttings for two populations coincide: one

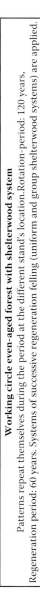

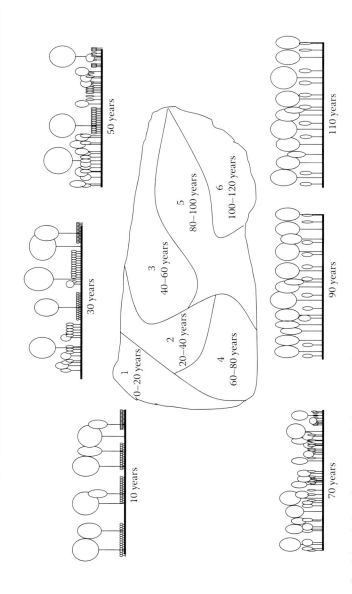

FIGURE 6.14   Spatial variation of a working circle structure for a shelterwood-managed forest at a certain stage of development. (From Velarde, M.D. et al., *Integración paisajística de las repoblaciones forestales.. Serie Técnica de Medio Ambiente*, Dirección General de Medio Ambiente, Consejería de Medio Ambiente, Vivienda y Ordenación del Territorio de la Comunidad de Madrid, Madrid, Spain, 2013.)

composed by adult trees and the other by young trees. This produces a stratified appearance with large or small trees depending on the phase of development. Regeneration is through small clumps while harvesting. The populations originated contain trees with age differences of up to 60 years in the example (Figures 6.14 and 6.15).

After the cutting of all adult trees, groups of different heights persist for a time until a typical regular structure emerges. The upper canopy comprises dominant and codominant trees. The dominated trees below the canopy are very slim and have low vitality, since they are of the same age as the trees in the upper canopy.

The mature phase landscape near harvest time is similar to an even-aged stand with columnar trees and sparse undergrowth (Figure 6.19).

In this case, structural diversity is higher than in even-aged stands over half the forest area. This can be more clearly seen as the management units decrease in size and spread throughout the whole forest.

Figure 6.15 shows the cyclical development of the vertical structure and biomass of a block or compartment within a shelterwood-managed forest, and Figure 6.16 shows an example of continuous cover achieved with this system.

This system has advantages over clear-cutting systems for structural heterogeneity and tree size, as the parent trees remain during the RP (20–80 years). In addition, several canopy layers are present over half the TA (Figures 6.17 and 6.18).

As in the previous case, to obtain larger trees, the rotation and RP must be lengthened. From the recreational point of view, if the most attractive part corresponds to the stratified area with trees aged over 80 years, two-thirds of the entire management area may have an attractive landscape.

To obtain large specimens among the dominant and codominant trees, appropriate clearings can be applied, and the RTP can be increased. The most suitable thinning is from above, as opposed the common practice of thinning from below.

Thinning must maintain dead and decayed trees that are not detrimental to the stand's health.

Thinning from above facilitates the competitive-exclusion phase and the establishment of undergrowth species. It also promotes good crown shapes, which offer stability against wind and snow, especially during the regeneration phase. Regeneration comes through dense small groups under parent trees. Due to the high tree density, thinning and pruning develop naturally.

Some trees can be retained until the next RTP in order to achieve large dimensions and deadwood. Groups of retained trees can fulfill the same function and favor species of flora and fauna.

Regeneration over almost two-thirds of the area constitutes a reserve for recovery in the event of a serious disturbance.

Figures 6.17 through 6.19 show the different aspects of an adult and a regenerating stand in a forest managed by the shelterwood system.

As in the case of clear-cuttings, the problem of the mineralization of organic matter does not arise since the soil is always covered with trees. This system is aimed at shade-intolerant and semishade-tolerant species but has also been applied to shade-tolerant species. Irregular structures may develop when the RP equals rotation. This structural heterogeneity varies between forests obtained under even-aged management by clear-cutting or under uneven-aged management by selection cuts.

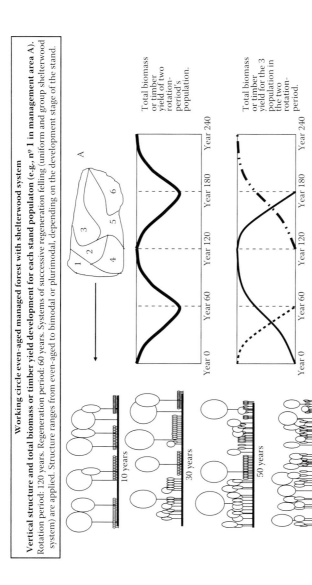

FIGURE 6.15 Evolution of vertical structure and biomass in a shelterwood-managed forest. (From Velarde, M.D. et al., *Integración paisajística de las repoblaciones forestales., Serie Técnica de Medio Ambiente*, Dirección General de Medio Ambiente, Consejería de Medio Ambiente, Vivienda y Ordenación del Territorio de la Comunidad de Madrid, Madrid, Spain, 2013.)

FIGURE 6.16 **(See color insert.)** Continuous-cover physiognomy of a shelterwood-managed forest (120 year rotation and 60 year RP). Pinar de Valsaín (Segovia, Spain). (Photo: Antonio García-Abril.)

FIGURE 6.17 **(See color insert.)** Area of mature trees beneath a uniform and group shelterwood-managed area in Pinar de Valsaín (Segovia, Spain). (Photo: Antonio García-Abril.)

FIGURE 6.18 (See color insert.) Successful regeneration area in a group shelterwood system in Pinar de Valsaín (Segovia, Spain). In this case, if the RP increases and trees are retained, an irregular forest structure can be achieved. (Photo: Antonio García-Abril.)

FIGURE 6.19 (See color insert.) Mature 110-year-old trees in an even-aged forest managed by a shelterwood system in Pinar de Valsaín (Segovia, Spain). (Photo: Antonio García-Abril.)

6.2.3.3.3   Model for Landscape Evolution of Irregular Stands

An uneven-aged-managed forest involves regular and layered structures in ecounits (Oldeman 1983, 1990) with a very small area throughout all the stands. There is stratum differentiation in most of the forest area.

The following example illustrates spatial variation for the management unit at a certain time within an uneven-aged forest.

Strictly speaking, the RTP does not exist. We can assume a cutting diameter limit for wood removal and to promote gaps and regeneration. The method of implementing the management plan is to calculate the annual allowable cut and to divide the forest into stands to match the years of intervention. The RTP between cuts is termed the transition period in this case. Annual cuts in each unit must not exceed 80 m³/ha in order to avoid adverse effects. The transition period lasts between 5 and 15 years, depending on the forest growth. In this system, regeneration, crop harvest, final cuttings, and thinning occur simultaneously.

Figure 6.20 shows a forest divided into 10 management units, corresponding to a transition period of 10 years. Each one can also contain several minor units or stands. The structure depicted for each management unit is arbitrary, as each one is composed of a wide variety of irregular structures.

Figure 6.21 represents the cyclical evolution of an uneven-aged management unit. We assumed a 15 year transition period. The cyclical variation of biomass is similar to that of virgin forests under small perturbations. Most of the standing biomass remains, and the forest appears timeless even if a part of the wood is removed.

The continued presence of a significant part of the standing biomass produces a shady environment and encourages shade-tolerant species of flora and fauna (Figure 6.22). This function is also available in shelterwood-managed forests to a lesser extent.

Typically, we associate irregular structures with shade-tolerant species. The reason is that we identify irregular structures with uneven-aged structures due to individual trees. Shade-intolerant species (e.g., pines and other coniferous trees) can also be managed as uneven aged, as irregularity can be achieved through groups or clumps. Cuttings can involve the removal of individual trees, clear-cuttings of several trees, or using the shelterwood system in small areas. These promote a complex structure and landscape. The appeal of an uneven-aged-managed forest encompasses areas of irregularity due to individual trees, small homogeneous areas, or semiregular layered areas.

Uneven-aged management promotes more diverse forest structures than other management types. Continuous cover and the removal of the annual allowable cut can emulate the gap dynamics of old-growth forests. However, other attributes of old forests may be absent (Bauhus et al. 2009). Some other drawbacks for the application of this system to old-growth forests have been assessed (Halpern and Spies 1995; Franklin et al. 1997).

Although structural diversity is important for old-growth forests, other aspects must also be achieved. The cutting diameter limit may reduce the presence of large trees. Decadent or dead trees may be removed, and the network of patches may be highly regular or with only a few species. The variable retention procedure can also be used

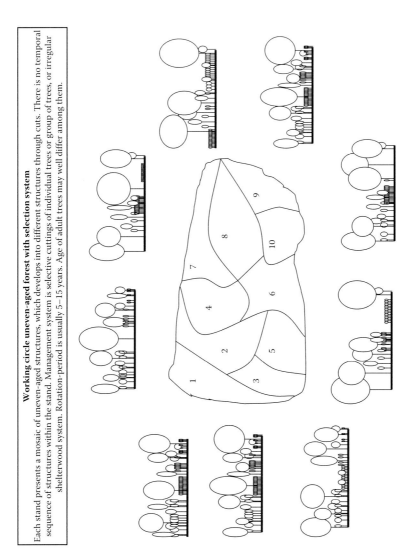

Working circle uneven-aged forest with selection system

Each stand presents a mosaic of uneven-aged structures, which develops into different structures through cuts. There is no temporal sequence of structures within the stand. Management system is selective cuttings of individual trees or group of trees, or irregular shelterwood system. Rotation-period is usually 5–15 years. Age of adult trees may well differ among them.

FIGURE 6.20 Spatial variation of a working circle structure for a selection-system-managed forest. (From Velarde, M.D. et al., *Integración paisajística de las repoblaciones forestales., Serie Técnica de Medio Ambiente*, Dirección General de Medio Ambiente, Consejería de Medio Ambiente, Vivienda y Ordenación del Territorio de la Comunidad de Madrid, Madrid, Spain, 2013.)

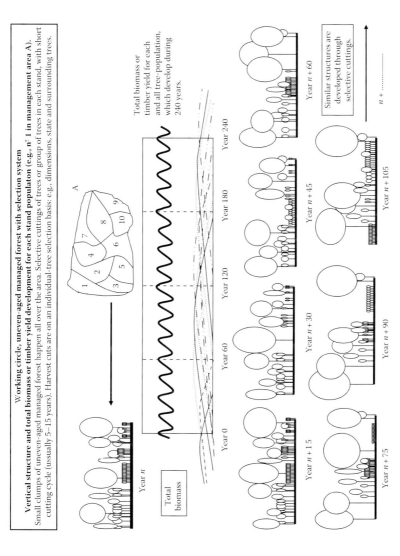

FIGURE 6.21 Vertical structure and biomass evolution in an uneven-aged-managed area. (From Velarde, M.D. et al., *Integración paisajística de las repoblaciones forestales., Serie Técnica de Medio Ambiente*, Dirección General de Medio Ambiente, Consejería de Medio Ambiente, Vivienda y Ordenación del Territorio de la Comunidad de Madrid, Madrid, Spain, 2013.)

FIGURE 6.22 (See color insert.) Regeneration gaps in an uneven-aged beech forest. (Photo: Antonio García-Abril.)

to promote certain trees to achieve their physiological limit and preserve some clumps from intervention. A better procedure may involve a very high cutting diameter limit, the retention of some trees until death, and the promotion of species diversity.

Uneven-aged management methods are well understood (Bruciamacchie and de Turckheim 2005). The transformation of an even-aged forest into an uneven-aged forest and its continued maintenance may be a product of close-to-nature management. The main principles for uneven-aged forest management are the following (de Turckheim 1992):

- Always keep the maximum possible biomass compatible with regeneration over most of the area. This increases the stability of the system.
- Individual management of trees, according to their function in the ecosystem. Age is never asked. The concepts of period, age class, balanced age distribution, age of maturity, and cuttings for improvement and for regeneration are not used. Periodic growth is more interesting than average growth.
- Wide areas of natural regeneration under large trees over a long period. The growth of young trees in a semishade environment is highly beneficial to promoting vertical stems, with clearing and pruning occurring through competition among trees in dense zones. Thus, actions to artificially reduce tree density are avoided.
- Cuttings involve trees at their cutting limit diameter, with a high quality and shape. They are chosen for their high wood value. These gaps allow

regeneration. Further clearings and selections maintain the elite trees in the upper canopy layer.

- Trees with diseases will be removed while conserving dead trees for their ecological worth, despite their low marketable value.
- The forest structure is modeled by cuttings.

The irregular structure observed in forests that have been under the close-to-nature system for decades is a consequence—not the objective—of the management regime (de Turckheim 1999).

Individual tree management gives a continuous-cover forest result, whose appearance is static over a large scale but shows high heterogeneity at the small scale. It consists of a mosaic of varied clumps, small habitats, and biotopes. Standing volume tends to be stable over time and from one zone to another. The volume of tall trees is high, and regeneration spreads throughout large areas. Multifunctionality is guaranteed for the whole area, and thanks to the high stability of the system, functional continuity is strongly guaranteed (de Turckheim 1999).

Generally, timber production goals are regarded as the priority in even-aged-managed forestry, while protection and conservation are more closely associated with uneven-aged management. However, it has been demonstrated that total timber production for the same environment and for both management types is very similar. Nevertheless, the volume of wood with large dimensions is higher in uneven-aged- than in even-aged-managed forests. This makes uneven-aged management financially more advantageous. The overall economic balance between regular and irregular equivalent forests favors the irregular forest, not only because of the higher wood value but also due to the lower cost of pruning, thinning, and clearings (de Turckheim 1993; Schutz 1997; Bruciamacchie and de Turckheim 2005; Laiho et al. 2011).

6.2.4 VISUAL FOREST LANDSCAPE AND PUBLIC PREFERENCES

The field of forest preference research has contributed much to our understanding of the impacts of management interventions on the aesthetic and recreational values of forests (Edwards et al. 2012) and particularly to discovering whether people prefer managed or unmanaged forests (Ribe 1989; Tyrvainen et al. 2003).

To determine how public opinion perceives the (visual) landscape effects of forest management, we evaluated their preferences for the visual expression of different forest management practices, that is, what the public may perceive directly, rather than the practices themselves (e.g., the public perceives the lower density of trees as a visual expression of clearings). Identifying different preferences, as a specification of preferences in landscape studies, can be of assistance in a multifunctional management that takes into account the productive, environmental, aesthetic, and heritage value of the forests (Purcell et al. 1994; Tarrant and Cordell 2002; Gruehn and Roth 2010).

In recent years, forests have become favorite places for recreation and outdoor activities in European urban society, and social demand for activities related to forest areas is growing rapidly (de Lucio and Múgica 1994; Harshaw et al. 2006; Ode and Fry 2006; Oku and Fukamachi 2006). Often, as the demand for the multiple use of the forest increases, forest management becomes a source of conflict (Parsons 1995; Tyrvainen et al. 2003). For example, it is well known that clear-cutting is one of the least popular silvicultural treatments with the public (Ribe 1989; Silvennoinen et al. 2001, 2002; Tahvanainen et al. 2001; Ribe 2005, 2006), although it is often preferred by owners and forest managers for financial reasons and ease of management. Some treatments may be necessary for forest maintenance but less suitable for delivering improved landscape quality. Educational initiatives may serve as a useful tool for raising awareness of the value of forest landscape elements of which the public may be unaware. Such is the case of dead trees, either standing or fallen, which are essential for the proper functioning of forest dynamics but whose presence can be interpreted by the public as negative for the landscape (Gundersen and Frivold 2011).

The users' preferences can be assessed according to their responses to different types of vegetation and landscape elements (Ulrich 1986; Kaplan and Kaplan 1989). Several studies have linked forest management to visual preferences (Williams and Cary 2002; Ribe 2005), and others have analyzed visual preferences in order to minimize the impact of forest management in recreational use (Karjalainen and Komulainen 1999; Silvennoinen et al. 2001, 2002; Tahvanainen et al. 2001; Gruehn and Roth 2010).

The question of whether people can perceive and appreciate ecology as a part of their perception of the landscape has not yet been clearly answered (Gobster et al. 2007). It has been suggested that this appreciation is enhanced by knowledge (Fudge 2001; Matthews 2002) and also by the fact of being an inhabitant of such places, as these subjects appreciate landscapes more intimately and less visually—more significantly than visually—than expert approaches assume (Dakin 2003). Some authors also consider the time factor as a variable in forest aesthetics (Ribe 1989; Silvennoinen et al. 2002).

The history of landscape quality assessment has involved a contest between expert- and perception-based approaches, paralleling a long-standing debate in the philosophy of aesthetics (Daniel 2001). A recent study by Roth and Gruhen (2010) shows that at the average group level, the differences between lay people's and experts' judgments are of no practical relevance. On the other hand, they point out that the results of their study have also shown that no single experts' judgment can replace broad empirical data on landscape perception, which agrees with the European Landscape Convention that defines landscape as "an area, as perceived by people," and stresses the importance of participatory procedures in landscape planning.

Attempts to integrate recreational values in forest planning are relatively common (Graefe and Vaske 1987; Zhou and Liebhold 1995; Silvennoinen et al. 2002). An example is the efforts made to include design in the VMS of the U.S. Forest Service, leading to the zoning of activities to reduce visual impacts (Bell 2001). However, there have been very few studies in Mediterranean areas, and these have been mostly centered on preferences on agricultural or agroforestry landscapes (Gomez-Limon and de Lucio 1999; Ruíz Sánchez and Cañas Guerrero 2001; Arriaza et al. 2004; Sayadi et al. 2005). Consequently, knowing how different types of forest management—or lack of management—affect scenic values is an element that should be included in forest policy and management, particularly in places where landscape and recreational value are the primary functions.

6.2.4.1 Summary of the Results of Forest Landscape Preference Studies

6.2.4.1.1 *Close to Nature*

The fact that public appreciation of a visual landscape relies on the creation of natural shapes and perceived naturalness is considered a key factor in landscape aesthetics (Purcell et al. 1994; Gobster 1999; Tveit et al. 2006). This is an aspect that visitors consider of great importance. It is, therefore, desirable to conceal all evidence of management as far as possible, to avoid straight-lined afforestation and geometric shapes in the design of edges, and it is also recommended to hide timber harvesting and forest roads from the sight of visitors (Ammer and Pröbstl 1991; Forestry Commission 1994; Velarde and Ruíz 2007).

At the same time, intact forests are surprisingly not preferred by the public, since the presence of many dead trees and the lack of a certain order are considered unattractive (Ammer and Pröbstl 1991). Stewardship is generally believed to have a positive impact on visual preferences (Nassauer 1995; Tveit et al. 2006). Sheppard (2001) developed an aesthetic theory of care (visual stewardship) that combined the ecology and aesthetics of forest landscapes. The consensus appears to be that a low level of management intensity is most highly valued but a degree of intervention is preferred to "tidy up" the forest landscape (Edwards et al. 2012).

6.2.4.1.2 *Mix of Young and Adult Trees, High Proportion of Old Trees, and Retention of Old Trees*

There is a preference for stands resulting from selective cutting, with a stratified or irregular structure, which contains groups of trees from different generations. Visitors consider this structure more similar to the natural state of the forest and more harmonious on the whole.

The larger and older trees are the most attractive and noticeable to the public. Young stands may be more enjoyable if some of the larger specimens are retained. Tree size appears to be the quality with the most important and common link to recreational value, where larger trees are preferred (Ammer and Pröbstl 1991; Gundersen and Frivold 2011; Edwards et al. 2012).

Some studies make recommendations for forest management with regard to maintaining a certain ratio of the timber material, which is cut and which remains on the ground (Brown and Daniel 1986; Ribe 2005, 2006). Maintaining a light density that produces a sensation of openness rather than closure endows the interior landscape with high quality.

6.2.4.1.3 *Mix of Conifers and Broadleaves*

Although there are some species in Germany and Finland (Silvennoinen et al. 2001) that are preferred on their own merits (spruce, oak, birch, and, to a lesser extent, beech), generally speaking, different species are associated with each place as being the most suitable. According to the results of the European scale study carried out by

Edwards et al. (2012), tree species is of relatively little importance, and, on balance, broadleaves are marginally preferable to conifers, while mixed stands are marginally preferable to monocultures. Of the factors mentioned earlier, proximity to nature and proportion of very old trees are the ones given greater importance. Another aspect that is being increasingly identified as important is the health of the forest, so that the value for recreation and landscape decreases in woods where there is a noticeable decline in the forest's health.

6.2.4.1.4   Presence of Water

Water is seen as a key element shaping human preferences (Litton 1972; Kaplan and Kaplan 1989). Lake shorelines are particularly sensitive areas from the aesthetic point of view, and natural river vegetation also represents an attractive element of contrast in the visual scene (Lucas 1991; Haider and Hunt 2002).

6.2.4.1.5   Presence of Dead Trees, Either Standing or on the Ground

Regarding the presence of coarse woody debris, the common ground between ecological and aesthetic values is not directly evident, at least if aesthetics are considered from the point of view of public preference (Velarde et al. 2005). The presence of dead or decaying trees is perceived as negative by the public and considered a symptom of disease and lack of care and management or else held responsible for hindering movement or creating a sense of lack of stewardship in the forest. However, it appears that from both the aesthetic and the ecological points of view, the presence of a few large logs is preferable to numerous small ones. This divergence can also be removed by the appreciation of the ecological integrity and health of the forest, considering that knowledge, experience, and education play an important role in the assessment (Gobster 2001; Gundersen and Frivold 2011).

These findings provide managers and policy makers with evidence to support the long-term retention of forest stands and the conversion of intensively managed forests to CCF and other low-impact silvicultural systems, in which recreation is an important management goal (Edwards et al. 2012).

The following is a summarized table (Table 6.1) that highlights some criteria and conclusions of European and American authors.

6.2.5   Integrating Features for Visual Landscape and Biodiversity Conservation

6.2.5.1   Synthesis of Principles for Landscape Integration of Forest Management Activities

Now the relationship between biodiversity conservation and forest landscape has been explored, and after considering the preferences of the public and the rules governing forest landscape, this section will now proceed to analyze this point in greater detail.

The proposal that follows is the result of an extensive review of the literature conducted on the previously mentioned topics (biodiversity conservation, forest landscapes, and public preferences) combined with the identification of good practices. Field visits were made to various kinds of forests, and interviews were conducted with forest managers as described in Núñez et al. (2010). We first present five

TABLE 6.1

Summary of Some Conclusions from European and American Authors on Analysis of Visual Preference in Forestry

Reference	Place	Variables Considered in the Analysis of Visual Preferences	Comments
Brown (1984)	Fort Collins, Colorado, United States	1. Herbaceous understory vegetation 2. Areas with a lower planting density 3. Little intense felling	Concludes that there is greater scenic beauty when there is plant cover and that scenic quality decreases with both the geometric boundaries that occur in felling and the poor-quality woody debris left after the cuts
Brown and Daniel (1986)	Fort Collins, Colorado, United States	1. Increased vegetation 2. Color	Continuation of the previous work, introducing the color variable in photos and slides used for the survey
Ribe (1989)	United States	Surveys from 15 photos have been measured: 1. Tree density 2. Land cover 3. Felling–thinning 4. Effect of time	Study focused on the basis of forest timber production Shows greater scenic beauty with sparse tree retention and with felling done to create irregular stands
Tahvanainen et al. (2001)	Isla Ruissalo (Southwest Finland archipelago)	1. Thinning type 2. Woody debris 3. Natural state or degree of naturalization 4. Cultural landscape with traditional management	Establishes an assessment of scenic beauty and recreational capacity applied to respondents in rural suburban areas and the city center In rural areas, the highest value is assigned to removal of woody debris and traditional forest management, while in the city center, the highest values are for the natural state.
Haider and Hunt (2002)	Lake Ontario (Canada)	1. Presence of water 2. Tree size 3. Presence of dead trees 4. Shrubs 5. Density 6. Amount of timber	There are two possibilities of aesthetic appreciation of the scenic beauty of coastal forests: 1. Based on psychophysical models 2. Based on an ecological classification of vegetation types (aesthetic variations in vegetation accounted for 50%) The presence of dead trees may explain 54% of the aesthetic variations.

(continued)

TABLE 6.1 (continued)
Summary of Some Conclusions from European and American Authors on Analysis of Visual Preference in Forestry

Reference	Place	Variables Considered in the Analysis of Visual Preferences	Comments
Silvennoinen et al. (2002)	Finland	1. Felling type 2. Time after felling 3. Differences in preferences between forest owners and the rest of the population	Surveys based on photographs • The effect of felling is negative, while the effect of clearing is positive. • Visible crop residues. • Natural regeneration with seed trees is positive (almost as positive as the valuation with mature trees). • Women, young people, urban public, and those who are not forest owners or have no relation to forestry assessed clearing less positively and viewed felling as more harmful.
Tarrant and Cordell (2002)	United States	Aimed at obtaining maximum value from the forest (recreational, environmental, and productive) based on 1. Timber production 2. Contribution of fresh air 3. Scenic beauty 4. Cultural heritage	Variation with forest ownership (private or public) and category of value (utilitarian, subsistence, spiritual, or aesthetic) Women, young people, and white people rated the utilitarian forest lower and were more inclined toward environmental values. The highest spiritual values (i.e., heritage conservation) were associated to a stronger pro-environmentalist attitude. Highest values for the elderly corresponded to quality of life, which in private forests was associated to a more consistent environmental attitude.
Ribe (2006)	United States	1. Ratio between old and young trees 2. Socially acceptable forestry	40%–50% retention of green trees is optimal in terms of social acceptance of silvicultural practices. 25% retention decreases social acceptance. There is a totally negative opinion of clear-cuts and all fellings with 15% retention.
Kaltenborn and Bjerke (2002)	Røros, southern Norway	Preferences for local landscapes and environmental value orientations • Presence of water • Cultural landscapes • Traditional farm environments • Modern agricultural practices	The highest preference was expressed for wildland scenes containing water, followed by cultural landscapes and traditional farm environments. Landscapes with elements of modern agricultural practices were the least preferred category. Significant positive correlations were found between environmental value orientation and preferences.

Tyrvainen et al. (2003)	Helsinki	• Whether aesthetic and ecological values can be combined in the management of urban forests. • Covered the main conflict situations in urban forest management: thinning, understory management, the leaving of dead snags, and decaying groundwood	Results show that the majority of residents in Helsinki prefer managed forests and that the preferences are however closely connected to the background characteristics of the respondents.
Silvennoinen et al. (2001)	Finland	• Quantitative models for forest stand level landscape preferences	The priority of a stand increases with mean tree height, skewness of the height distribution, and volume of large pines and birches. The priority decreases with an increasing number of trees per hectare.
Thompson et al. (2005)	Central belt of Scotland	• Explores who uses woodlands near their homes, why they visit, what benefits they believe they obtain, and what makes the difference	• Proximity of woodlands is important (within 5 km of town). • Freedom from rubbish is the physical quality people care about most. • Directional signs, good information panels, variety of trees, and tidiness of appearance • Most people feel at peace in a woodland. • People asked what the countryside meant to them said the following: trees (most often mentioned), quiet, fields, hills, peace, animals.
Nasar and Li (2004)		• Water and a key aspect of water: its reflection	• Individuals gave the most favorable ratings to the scene with reflective water, suggesting the potential desirability of reflective ponds as a design element.
Haider and Hunt (2002)		• Forested shorelines	• Tree size, tree mortality, conifer shrubs, tree density, amount of hardwood, and slope explained 60.2% of the variance in scenic beauty between the study sites.

general principles of landscape integration into SFM, which transversely guide the criteria to be proposed later. These principles are as follows:

1. Multiple-scale approach to forest management actions
2. Seeking to match aesthetic and ecological criteria
3. Taking account of public preferences
4. Seeking to simulate nature when taking forest management decisions
5. Masking unavoidable negative landscape impacts

6.2.5.1.1 Multiple-Scale Approach to Forest Management Actions

There is a need when carrying out forestry activities to consider a broad scale—the landscape scale—and to integrate different scales of work, moving progressively from a region, district, or basin to smaller systems such as a mountain, reforestation, or a recreational area. This can be done from the existing land-use plans and then by continuing to move in closer to the landscape: first from a distance or from the outside—what we have termed the exterior landscape (margins, species and colors, skylines, etc.)—to the interior landscape, with greater detail (roads, clearings, borders, banks, etc.). Thus, the quality and fragility of the management area can be previously determined and integrated into the management criteria.

6.2.5.1.2 Seeking to Match Aesthetic and Ecological Criteria

The visual landscape cannot ignore ecological requirements, that is, the spatial configuration of the landscape is related to the existence of species, habitats, and biotopes, and vice versa, thereby creating or designing a landscape structure with ecological and aesthetic implications. From the aesthetic point of view, the success of a landscape design depends more than any other factor on the creation of natural forms (Ammer and Pröbstl 1991; Forestry Commission 1994).

6.2.5.1.3 Taking Account of Public Preferences

Today, public participation is considered to be an important element of forest planning. SFM attempts to respond to the various interests expressed by society—which frequently reflect opposing visions of the relationship between man and the natural environment—by orientating management toward multifunctionality and multiple uses of forest resources (Martins and Borges 2007; Cantiani 2012).

6.2.5.1.4 Seeking to Simulate Nature When Taking Forest Management Decisions

Seeking to simulate nature when taking forest management decisions means that forests should be considered and managed as an ecosystem, for example, understanding the importance of disturbances for natural forest dynamics (Trotsiuk et al. 2012). As a result, ecological principles deliver valuable forest landscapes (Ramos 1993).

6.2.5.1.5 Masking Unavoidable Negative Landscape Impacts

Forest management involves a series of actions that inevitably impact the landscape. Such is the case of the removal of trees, fire wall or certain infrastructure development, silvicultural treatments, or regeneration treatments. For actions in which an impact on the landscape is unavoidable, the extent and intensity of the impact must be hidden or diminished.

6.2.5.2   Landscape Design Criteria Linking Visual Preferences and Biodiversity

Research on visual preferences, elements, and patterns of landscape characteristics has evolved linkages between visual and ecological landscape, highlighting a common ground for both approaches. As noted in the first principle, when carrying out forestry activities, there is a need to consider a broader scale—the landscape scale—and to integrate various different scales. The reason for starting at the exterior landscape scale is to bring unity to the whole, as this is the essential purpose of landscape design (Forestry Commission 1994). Once this exterior landscape has been analyzed, we need to move in closer to the target area and study the landscape from the inside or the interior landscape (Núñez et al. 2010).

After these principles, the criteria for landscape design have been provided for each of these two landscape dimensions. They are summarized in Figures 6.23 and 6.24.

The following criteria have been selected for a forest management that satisfies both visual preferences and biodiversity conservation:

1. Landscape criteria for the *exterior landscape*: broad-scale approach to forest management
   A. Avoiding fragmentation by seeking connection between relevant ecosystems
   B. Increasing biodiversity
      a. Of species and ecosystems
      b. Of landscape elements
   C. Softening margins of forest and new plantations
   D. Adapting infrastructures, equipment, and other artificial elements to the forest environment
   E. Working at a landscape scale in order to foster integration of management activities
2. Landscape criteria for the *interior landscape*: small-scale approach to forest management
   A. Protecting riversides and shores
   B. Paying attention to the singular function of forest edges
   C. Increasing ecosystem and species diversity
   D. Preserving large old trees, large fallen trees, and trees of different species
   E. Integrating structures and equipment into the forest landscape
   F. Not disturbing the *genius loci* (spirit of the place)

6.3   LANDSCAPE INDICATORS FOR THE ASSESSMENT OF SUSTAINABLE FOREST MANAGEMENT

A large number of indicators have been developed in the area of ecological landscape that are applicable to forest landscape management—particularly for the application of SFM systems—whereas in the area of visual and aesthetic

Landscape visual criteria for *exterior landscape*: broad-scale approach		
1. Avoid fragmentation looking for connection among relevant ecosystem.	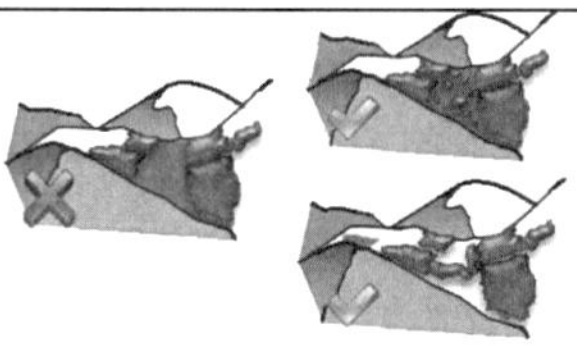	This criterion indicates that connectivity is not only important for ecological reasons but promotes landscapes preferred by the public.
2. Increase biodiversity a. Of species and ecosystems b. Of landscape elements.	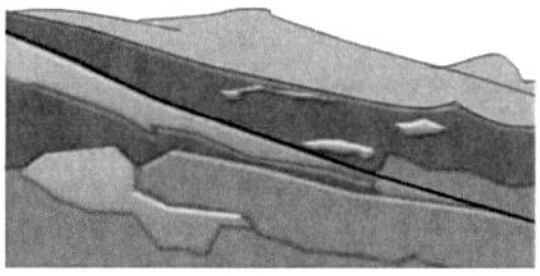	Stands of mixed species tend to be more stable against biotic and abiotic damages than monospecific ones. Also multilayered stands offer high level of diversity. Besides, forest landscapes with aesthetic complexity are preferred.
3. Soften margins of forest and new plantations.	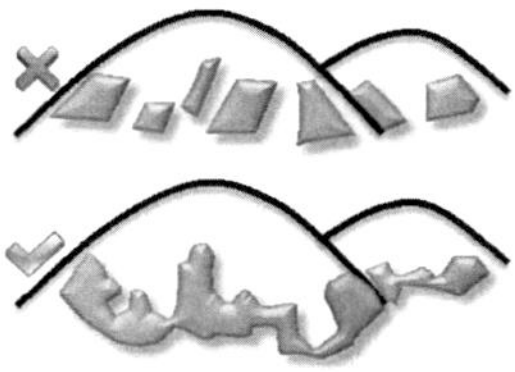	Create or maintain wavy edges with indentations improve visual diversity and introduce irregularity to straight forest edges. This also regards the importance of edges for the connectivity.
4. Adapt infrastructures, equipment and other artificial elements to forest environment.	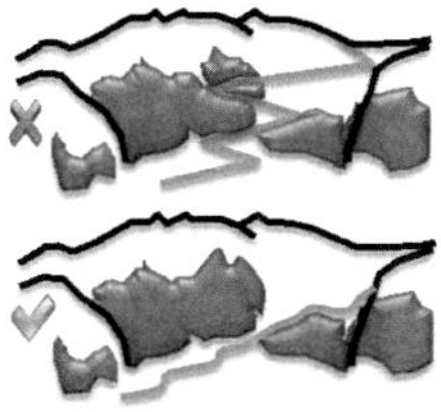	This criterion involves visual fragility and aesthetic values as well as the effects of anthropogenic disturbances affecting biodiversity conservation.
5. Work at a landscape scale in order to forest integration of management activities.		Match interventions to landscape scale refers to give importance to the perception of relative and absolute sizes. Furthermore, management areas are intended to have independent structure and functioning.

FIGURE 6.23 **(See color insert.)** Landscape criteria for exterior landscape: broad-scale forest management approach.

landscapes, the number of indicators proposed is much lower. Landscape ecology has been a very active area of research and has resulted in the development of a wide range of indicators, measurements, and landscape indices based on landscape ecological principles. For the visual aspects of the landscape, however, this conceptual basis is often lacking and hinders progress in the development of indicators (Ode and Fry 2006; Fry et al. 2009).

Landscape visual criteria for *interior landscape*: small-scale approach		
1. Protect riversides and shores.		Structure and energy of rivers depend to some extent on the materials from the forest, e.g. logs form small pools, are essential for many fishes and deliver nutrients. The presence of trees maintain certain conditions of light and temperature in water, necessary for the survival of some species of aquatic life, while minimizing the incidence of erosive effects. Natual rivers are usually of high aesthetic level and fragility.
2. Pay attention to the singular function of forest edges.		Edges are important for the maintenance of ecotones. Changes in their length and width may affect several species of fauna. From the visual point of view, their natural appearance leads to more appreciated landscapes.
3. Increase ecosystem and species diversity.		The presence of small habitats can promote visual and ecological diversity. From the aesthetic point of view they can break the monotony of the landscape, and from the ecological viewpoint, they allow coexistence of species adapted to different light conditions, soil, vegetation, etc.
4. Preserve large old trees, large fallen trees and trees of different species.		Large and dead trees are essential for biodiversity, completing the life cycle of trees and providing shelter, food or breeding to a large number of species. From the aesthetic point of view, the public also prefers forests with several generations of trees, and the presence of large individual trees or groups of trees.
5. Integrate structures and squipment into the forest landscape.		The visual significance of this criterion is based on the visual fragility and aesthetic value of the forest landscape. The more structures and equipment are integrated in the environment, the lower the ecological impact.
6. Do not disturb the *genius loci* (spirit of the place).		Management actions should conserve this spirit that is unique to that particular places, which represents a value and an important incentive for a good design, and that must be preserved.

FIGURE 6.24 **(See color insert.)** Landscape criteria for interior landscape: small-scale forest management approach.

This section describes the different sources of the landscape indicators used to assess the landscape. There are numerous indicators from several disciplines and SFM standards; most of those from SFM initiatives refer to the ecological landscape and to biological conservation. We will first discuss biodiversity or ecological indicators.

The authors propose several indices within the framework of landscape planning, landscape ecology, and forest management. These indices can be integrated with other already existing indices.

We will aim to highlight the common ground for both kinds of indicators, as many ecological indicators may also be useful for visual aspects, and we will then link the indicators identified in the literature with the conceptual proposal for forest landscape management criteria previously described in Section 6.2.5.

6.3.1   Diverse Sources of Indicators

6.3.1.1   Landscape Ecology

Ecology in general—and landscape ecology in particular—has given rise to numerous indicators on aspects such as biodiversity, spatial heterogeneity, spatial pattern, connectivity, and edge effect (Forman 1995; Turner et al. 2003; Farina 2006; Miller et al. 2006).

In this subsection, we propose several indices derived from Section 6.2.2, where we highlighted some landscape patterns such as agroforestry and silvopastoral systems, which accomplish high diversity, complex structure, and varied and sustained production. Another concept we analyzed was the size of the patches required for forest species conservation.

6.3.1.1.1   Agroforestry System

   a. PAFA, Proportion of agroforestry system area relative to TA

$$PAFA = \frac{AFA}{TA} \qquad\qquad (6.1)$$

  where
     PAFA is the proportion of agroforestry system area
     AFA is the agroforestry area
     TA is the total area

   b. PSPA, Proportion of silvopastoral system area relative to TA

$$PSPA = \frac{SPA}{TA} \qquad\qquad (6.2)$$

  where
     PSPA is the proportion of silvopastoral system area
     SPA is the silvopastoral system area
     TA is the total area

   c. Spatial metrics: Metrics from landscape ecology can be used (e.g., Farina 2006) for agroforestry and silvopastoral system landscapes. We propose the following:

     • Proportion of forest area relative to TA
     • Patch density
     • Border length of forest area

6.3.1.1.2 *Available Forest Area for Species Conservation (AFC)*

1. ACM, Available conservation minimum area for key and endangered species
   a. NACM, Number of patches with area bigger than ACM
      i. DACM, Density of patches with area bigger than ACM relative to TA (areas for comparison can be assessed by species and taxonomic groups)

$$DACM = \frac{NACM}{TA} \qquad (6.3)$$

   where
   > DACM is the density of patches with area bigger than ACM
   > NACM is the number of patches with area bigger than ACM
   > TA is the total area
   > ACM is the available conservation minimum area for key and endangered species

   b. SACM, TA of patches with area bigger than ACM
      i. PACM, Proportion of area of patches bigger than ACM relative to TA (areas for comparison can be assessed by species and taxonomic groups)

$$PACM = \frac{SACM}{TA} \qquad (6.4)$$

   where
   > PACM is the proportion of area of patches bigger than ACM
   > SACM is the total area of patches with area bigger than ACM
   > TA is the total area

   ACM is the available conservation minimum area for key and endangered species

2. AHF, Available area for habitat forest species
   Thresholds range from 100 to 400 ha. For each one, we calculate the number and density of patches, total, and proportion of areas.
   a. N400 or N100, number of patches with area bigger than 400 and 100 ha, respectively
      i. DN400, Density of patches with area bigger than 400 ha relative to TA

$$DN400 = \frac{N400}{TA} \qquad (6.5)$$

   where
   > DN400 is the density of patches with area bigger than 400 ha
   > N400 is the number of patches with area bigger than 400 ha
   > TA is the total area

ii. DN100, Density of patches with area bigger than 100 ha relative to TA

$$DN100 = \frac{N400}{TA} \qquad (6.6)$$

where
DN100 is the density of patches with area bigger than 100 ha
N100 is the number of patches with area bigger than 100 ha
TA is the total area

b. A400 or A100, TA of patches with area bigger than 400 and 100 ha, respectively

   i. PA400, Proportion of area of patches bigger than 400 ha to TA

$$PA400 = \frac{A400}{TA} \qquad (6.7)$$

where
PA400 is the proportion of area of patches bigger than 400 ha.
A400 is the total area of patches with area bigger than 400 ha
TA is the total area

ii. PA100, Proportion of area of patches bigger than 100 ha to TA

$$PA100 = \frac{A100}{TA} \qquad (6.8)$$

where
PA100 is the proportion of area of patches bigger than 100 ha
A100 is the total area of patches with area bigger than 100 ha
TA is the total area

6.3.1.2 Landscape Planning

Landscape planning incorporates the visual landscape as a crucial element of planning, considering it in terms of visual quality and visual fragility or sensitivity and using spatial indices of land use and landscape structure (Ramos 1979; Aguiló et al. 1995; Aguiló and Iglesias 1995). The key concept used in landscape planning is the viewshed (Aguiló and Iglesias 1995). Absolute viewshed (AV) is the area seen from a certain viewpoint or view zone, with a certain range. Relative viewshed (RV) is the percentage of the visible area related to the maximum visible area, which is calculated as the area of a circle centered at the observation point with a radius of a desired visual range (R) (Equation 6.9):

$$RV = 100 \cdot \frac{AV}{\pi R^2} \qquad (6.9)$$

Diverse metrics and ranges for viewshed have been applied to assess landscape quality and fragility in landscape planning and environmental impact assessment (Aguiló and Iglesias 1995; Aguiló et al. 1995).

Another important concept is visual accessibility, which refers to the possibility of a landscape unit being more or less viewed by observers. Factors that influence accessibility are observation distance, the position of the observer, contrast, backlight or atmospheric effects, elements enhancing or obscuring the observed area, diversity, color, etc. (Aguiló and Iglesias 1995).

6.3.1.3 Proposal for Forest Management

We should point out that the meaning, importance, and measurement of indicators must be established at various spatial and temporal scales: (1) regional landscape, (2) landscape, (3) forest, (4) management area, (5) compartment, and (6) stand. The time scale may be hundreds or even thousands of years if forest succession stages require this time to be attained.

Several of the following indicators in this section have significance for visual quality and forest planning, in addition to attributes related to old-growth forests. Indicators are based on measurements and data that may be available in current timber inventories and forest management plans. Indicators related to deadwood, cavities, and tree hollows are old-growth forest attributes, which are included for their ecological importance. These indicators have been proposed as indicators of biodiversity and SFM by different authors (Lassauce et al. 2011) and in international initiatives for SFM.

Deadwood and cavities are currently included in forest inventories as ecological items. Several methodologies have been developed to measure them (Woodall and Monleón 2007; du Cros and Lopez 2009; Rondeux et al. 2012).

Both deadwood and cavities must be considered at the scale of the forest management unit and forest stand. Also, dimensions of decaying trees, snags, fallen trees, stumps, wood debris, and trees with cavities and hollows must be distinguished by species.

Here, we present our proposal for several landscape ecological indicators related to forest management, which also have an aesthetic dimension.

6.3.1.3.1 *Difference of Tree Distribution from a Reference Distribution (DRD)*

DRD refers to the difference between current tree distribution (CTD) and *tree diameter reference distribution (TDRD)* or with *tree height reference distribution (THRD)*. It assesses the variability of tree sizes and the existence of balanced tree distributions in all stages of development at the stand level, given a sustainable timber yield. It also indicates spatial heterogeneity at the compartment or stand scale. At the forest level, it must be calculated by aggregating the results from individual stands. In the case of small private forests, information on trees in all development phases may not be available, and assessment at the landscape scale may be complex. Reference distributions must be developed by forest types and geographic regions and for each site index. The difference in CTD with either TDRD or THRD can be calculated by means of the *Kolmogorov–Smirnov* test.

SILVANET software (Martínez-Falero et al. 2010) provides a computational procedure for these assessments. In Chapter 3, we describe the methodology to assess THRD from yield tables of *P. sylvestris* in central Spain.

6.3.1.3.2 Proportion of Rotation Period Relative to a Reference Rotation Period

RTP relates to large diameter sizes. Moreover, long RTP promotes high levels of wood in the forest. Proportion of rotation period (PRTP) indicator (Equation 6.10) must be related to reference rotation period (RRTP) and must also be assessed by species and region at the management area or working circle scale. Average results are aimed at broader scales. In even-aged-managed forests planned for old-growthness transformation, this indicator may help to assess the period:

$$PRTP = \frac{RTP}{PRTP} \tag{6.10}$$

where
   PRTP is the proportion of rotation period
   RTP is the rotation period
   RRTP is the reference rotation period

6.3.1.3.3 Proportion of Large Tree Volume Relative to Total Volume (PLTV)

PLTV indicates the degree of accumulation of biomass in large-sized trees of Equation 6.11:

$$PLTV = \frac{LTV}{TV} \tag{6.11}$$

where
   PLTV is the proportion of large tree volume
   LTV is the large tree volume
   TV is the total volume

If the management plan defines a cutting diameter limit, it must distinguish species, geographic region, and site index in order to guarantee young trees and regeneration within the forest structure. Nevertheless, diameters over 60–80 cm are the usual thresholds for large diameters.

This indicator can be assessed at the scale of management area, compartment, or stand. For a global assessment, the landscape scale is proposed.

6.3.1.3.4 Proportion of Stratified Canopy Area and Fine-Sized Structure over Total Forest Area (PSA)

PSA is a structural diversity indicator, which can be assessed at all scale levels (Equation 6.12).

The assessment of PSA requires timber data from the management plan (see Section 6.2.3.3 and Figures 6.12, 6.15, and 6.21). Moreover, RTP and RP data are also necessary, since the proportion of ages is linked to the size of the area:

$$PSA = \frac{2RP}{\left(RTP + RP\right)} \tag{6.12}$$

where
PSA is the proportion of stratified canopy area and fine-sized structure
RTP is the rotation period
RP is the regeneration period

- In the case of even-aged forests with strict clear-cutting and subsequent plantation and a 60 year RTP, RP = 0 and PSA = 0.
- In the case of uneven-aged forests with seed-tree clear-cutting, a 10 year RP, and a 100 year RTP, PSA = 20/100 = 0.2.
- In a forest managed by the shelterwood system, with an RTP of 120 years and an RP of 60 years, PSA = 120/180 = 0.66.
- In the case of uneven-aged forest, all the forest management area is in continuous regeneration. Therefore, RP = RTP, and PSA = 1.

This is an indicator of the stratified canopy area or an area with small clumps. This indicator assumes that merely a start of regeneration or some disperse trees may indicate a stratified structure. For a better representation of a stratified structure, ranges of areas should be smaller for all silvicultural models.

6.3.1.3.5 Proportion of Clear-Cutting System Area Relative to Total Forest Area

The clear-cutting system has been widespread in many regions and has been a source of controversy for decades. In principle, its use should be restricted for visual reasons and also to protect the soil, water, and ecosystem.

This indicator can be applied at different scales and relates to preserved zones or trees. At the landscape and forest scale, it assesses the proportion of clear-cut area with respect to other treatments and preserved or retained areas (Equation 6.13):

$$PCC = \frac{CC}{TFA} \tag{6.13}$$

where
PCC is the proportion of clear-cutting system area relative to the total forest area
CC is the area treated by clear-cutting system
TFA is the total forest area

6.3.1.3.6 Proportion of Structural Retention Relative to Total Forest Area

The term retention implies that an attribute that would be removed under conventional management is deliberately retained for conservation purposes (Bauhus et al. 2009).

This system is successful for reducing the effects of clear-cutting in natural forests. It also provides an effective means of restoring old-growth attributes in regrowth and secondary forests.

It assesses the proportion of retained area (or volume) relative to the total forest area (or volume), since retention can be applied by areas or by individual or disperse trees (Equations 6.14 and 6.15):

$$PSRA = \frac{RA}{TFA} \tag{6.14}$$

$$PSRV = \frac{SRV}{TFV} \tag{6.15}$$

where

PSRA is the proportion of structural retention area relative to the total forest area

RA is the area of retention

TFA is the total forest area

PSRV is the proportion of structural retention volume relative to the total forest volume

SRV is the volume of retention

TFV is the total forest volume

It can be applied from the stand to forest level.

6.3.1.4   International Initiatives for Sustainable Forest Management

International initiatives for SFM have developed their own checklist for C&I during the last decades. These have evolved from working groups based on both research and experience (see Section 6.1.3).

Although the ITTO, the European Union, and the Montreal Process for temperate and boreal forests outside Europe have refined C&I for SFM over the past two decades to the point that there is now a substantial consensus (McDonald and Lane 2004), landscape is not explicitly considered in any of these processes.

While international standards have important effects on the definition of SFM in participating countries, they have been criticized for the lack of any indicators that are directly related to social aspects—especially scenic beauty—and for the predominance of economic and ecological indicators (Lim 2012).

Nevertheless, several criteria in the international SFM standards have indicators that are implicitly associated with landscape from the ecological, aesthetic, and visual point of view (Table 6.2). Most of these were intended for biodiversity but also have implications for visual landscape in terms of connectivity of patches, importance of edges, impact of infrastructures, protection of riversides and shores, and the spirit of the place, among others.

In Table 6.5, meaningful indicators for landscape from SFM standards have also been assigned to the interior and exterior landscape criteria as defined in Section 6.2.5. In some cases, the selected indicators can be related to more than one criterion. Therefore, results in the last two columns may exceed the number of SFM indicators related to landscape criteria in each international process.

The Pan-European Process is the process with the highest proportion of indicators related to landscape, FSC results are similar. Dry Asia and Dry Africa

TABLE 6.2

Indicators Related to Visual Landscape within SFM Standards

International Process	Total Number of SFM Indicators	No. of SFM Indicators Related to Landscape Criteria	% of SFM Indicators Related to Landscape Criteria	No. of Times That "Exterior Landscape" Criteria[a] Are Considered in SFM Indicators Related to Landscape	No. of Times That "Interior Landscape" Criteria (*) Are Considered in SFM Indicators Related to Landscape
Dry Africa	47	10	21.28	10	5
ITTO	66	10	15.15	14	5
Lepaterique	53	10	18.87	12	6
Montreal	67	8	11.94	7	7
Oam	60	10	16.67	6	9
Near East	65	11	16.92	17	3
Pan-European	35	14	40.00	18	10
Dry Asia	49	13	26.53	19	7
Tarapoto	57	7	12.28	7	2
FSC	345	29	8.41	47	26
PEFC	35	14	40.00	18	10

[a] Interior and exterior landscape criteria as defined in Section 6.2.5.

also have high percentages. However, these are currently scarcely implemented. For the rest of the processes, the landscape is poorly represented, with less than 20%.

6.3.1.5 Indicators from Landscape Character Assessment

While ecological values have been on the policy agenda for a long time, visual quality has received less attention, for example, in Europe, at least until the European Landscape Convention was launched in the year 2000.

One of the main sources of visual indicators is the assessment of landscape character, which has been developed as a tool for a landscape description that includes the experience of landscape and which could form a useful basis for the subsequent evaluation of landscape visual quality in a management or policy setting (Ode et al. 2008). Landscape character is defined as a distinct, recognizable, and consistent pattern of elements in the landscape that makes one landscape different from another, rather than evaluating what makes it better or worse (Swanwick 2002). The nature of these indicators varies greatly, some having strong links to landscape aesthetic theory, visual disturbance, or perceived natural vegetation, whereas others have been borrowed and applied directly from landscape ecology (Ode et al. 2008).

A broad review of the literature covering papers on landscape aesthetics, visual concepts, and landscape preferences (Tveit et al. 2006) resulted in the

TABLE 6.3
Visual Landscape Concepts Identified by Tveit et al. (2006) and Ode et al. (2008), Meaning of Related Indicators, and Application to Forest Landscapes

Visual Landscape Concept	Meaning of Indicators	Application to Forest Landscapes
Complexity refers to the diversity and richness of landscape elements and features and the interspersion of patterns in the landscape.	Indicators describe the complexity of landscape with regard to both content and spatial configurations.	Forest landscapes are visually valued by the number and spatial organization of landscape elements such as number of forest layers, presence of different species, and spatial patterns.
Coherence relates to the unity of a scene, the degree of repeating patterns of color and texture, as well as the correspondence between land-use and natural conditions. Coherence is a factor for predicting preference within information processing theory and refers to a more immediate understanding and readability of our environment (Kaplan and Kaplan, 1989).	The indicators of coherence focus on correspondence with expected natural conditions, fragmentation, repetition of pattern across the landscape, presence of water, etc.	*Idem*
Disturbance refers to the lack of contextual fit and coherence in a landscape.	It relates to the presence, extent, and visual impact of disturbing elements.	Large clear-cutting areas, burned or unhealthy forest areas, infrastructures.
Stewardship refers to the sense of order and care present in the landscape, reflecting an active and careful management.	Level of management for vegetation (level of abandonment, presence of weeds, management detail, etc.) and status and conditions of man-made structures (farm buildings, fences, etc.)	In forest landscapes, stewardship relates to signs of the type and conditions of management and succession stage.
Imageability reflects the ability of a landscape to create a strong visual image in the observer, thereby making it distinguishable and memorable. Imageability can be a product of the totality of a landscape or its elements.	It focuses on spectacular, unique, and iconic elements and their visibility.	Large trees, waterfalls, viewpoints, historical elements, etc.
Visual scale describes landscape perceptual units in relation to their size, shape, diversity, and the degree of openness in the landscape.	Proportion of open land, viewshed size, depth of view, and obstruction of view have been suggested for the assessment of visual scale indicators.	*Idem*

TABLE 6.3 (continued)

Visual Landscape Concepts Identified by Tveit et al. (2006) and Ode et al. (2008), Meaning of Related Indicators, and Application to Forest Landscapes

Visual Landscape Concept	Meaning of Indicators	Application to Forest Landscapes
Naturalness describes the perceived closeness to a preconceived natural state.	Indicators focus on the quality of the current vegetation in relation to its perceived naturalness, as well as the pattern in the landscape, perceived as natural or not.	Water in the landscape and a close-to-nature appearance are often used as indications of naturalness.
Historicity describes the degree of historical continuity and richness present in the landscape.	Historical continuity is reflected by the visual presence of a different era.	Historical forest landscapes and landscape heritage.
Ephemera refers to landscape changes related to season or weather.	Season-linked activities (events taking place in relation to the season), landscape attributes with seasonal change and weather characteristics.	Within forest environments, this character considers seasonal visual variations in vegetation, the extent and frequency of changes, and the presence of water with seasonal change.

identification of nine key visual concepts supported by different theories explaining people's experience of landscape and their landscape preferences. Each concept focuses on different aspects of the landscape that are important for visual quality. Measurable indicators, including both suggested and empirically tested ones, were later linked to each of these concepts (Ode et al. 2008; Fry et al. 2009). These nine key concepts, closely linked to visual landscape characterization, are described in Table 6.3.

6.3.2 VISUAL AND ECOLOGICAL INDICATORS LINKED TO FOREST LANDSCAPE MANAGEMENT CRITERIA

For many landscape values, the search for indicators has been data driven rather than theory driven, allowing us to neglect the essential aspects of what the indicators are intended to indicate (Ode et al. 2008). It is, therefore, crucial to be aware of which indicators are useful and to have a solid theoretical base for their application.

Section 6.2.5 presented a proposal for forest management criteria that combined landscape aesthetic and biodiversity conservation goals in both existing and new forest plantations. This proposal, which is intended to be simple and easy to understand, is the result of a thorough review of the literature, combined with the contribution of various forest managers with field experience. Two dimensions of forest landscapes are considered, known as the *exterior landscape* and the *interior landscape*, and within each one, a series of *forest landscape criteria* have been considered.

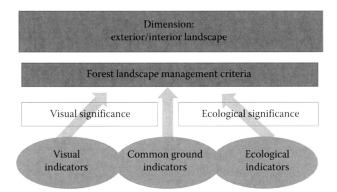

FIGURE 6.25 Conceptual structure of the proposal for a common ground of visual and ecological landscape indicators.

The conceptual structure of the proposal described in Figure 6.25 is based on the methodology presented by Fry et al. (2009) exploring the common ground for visual and ecological landscape indicators.

The following tables present both visual and ecological indicators from various sources, including some proposed by the authors, aimed at contributing to describe the extent to which proposed forest landscape management criteria are being successfully implemented (Tables 6.4 and 6.5).

From these tables, we can conclude that visual and ecological landscape indicators share a broad common ground. For criterion 1 of exterior landscape, this common ground reveals the visual and ecological importance of landscape connectivity. There is a high proportion of indicators related to this criterion as well as to criterion 2, since biodiversity of visual elements, species, and ecosystems has been primary concerns in recent decades.

Indicators for the third criteria of exterior landscape appear to be more specific, and only a small number were identified. All were classified in the common ground group for their ecological importance and visual meaning. Indicators related to infrastructures and equipment (criterion 4) mainly evaluate visual aspects. Their ecological impact tends to be evaluated by specific studies depending on the activity. The fifth criterion considers the importance of public participation and acceptance of management and planning and provides a common ground that can be easily improved and adapted to local conditions (Tables 6.4 and 6.5).

From Tables 6.4 and 6.5, we can conclude that the majority of indicators were assigned to the common ground group. Protecting rivers and shores (criterion 1), connectivity and aesthetic value of forest edges, actions aimed at biodiversity, and integration of structures directly concern both visual and ecological aspects at the interior landscape scale. The common group of indicators for criterion 4 comprises the presence of large trees or seasonal changes in natural vegetation, whereas the presence of dead trees refers to ecological indicators that are visually valued through the presence of spectacular unique elements. Finally, the "genius loci" is best described visually or with indicators from the common ground.

TABLE 6.4

Visual and Ecological Forest Landscape Indicators for the Exterior Landscape Dimension

Exterior Forest Landscape

1. Avoid fragmentation, seeking connection among relevant ecosystems

Visual significance: Fragmentation affects visual landscape complexity, coherence, and naturalness (Tveit et al. 2006; Ode et al. 2008; Fry et al. 2009).

Ecological significance: Fragmentation leads to the disappearance of fragile fauna and increases fire risk. The number of species and their abundance increase with fragment size and conversely decrease rapidly with a decrease in the size and number of patches. The presence of ecological corridors fosters connectivity for the species.

Visual Indicators	Common Ground	Ecological Indicators
• Form of vegetation (Palmer 2004)	• PCC. Proportion of clear-cutting system area (authors' proposal)	• Presence of ecological corridors (Fry et al. 2009)
	• Number of patches in the landscape, average size of patches, density of patches (Llausas and Nogue 2012)	• Intact riparian vegetation corridors (Sheppard and Harshaw 2001)
	• Heterogeneity, fragmentation, and connectivity (Diaz-Varela et al. 2009)	• ACM area for key and endangered species. NACM, DACM, SACM, PACM
	• Fragmentation index (Taylor 2002; Palmer 2004)	
	• Distance indices, neighborhood, and connectivity (Hilty et al. 2006; Vila et al. 2006)	• AHF species. DN400, DN100, A400, A100

(continued)

TABLE 6.4 (continued)
Visual and Ecological Forest Landscape Indicators for the Exterior Landscape Dimension

Exterior Forest Landscape

2. Increase diversity (both diversity of species and ecosystems and diversity of landscape elements)

Visual significance: A landscape with mixed elements of different sizes and shapes is more attractive than homogeneous landscapes Diversity affects visual landscape complexity, coherence, naturalness, and ephemera (Tveit et al. 2006; Ode et al. 2008; Fry et al. 2009).

Ecological significance: Heterogeneous landscapes or complex structures are recognized as being more biodiverse than homogeneous ones, and they are also more effective in preserving species and communities. The coexistence of different vegetation and land uses produces a richer habitat and allows the cohabitation of groups of species occupying different niches, resulting in greater overall diversity (Atauri and de Lucio 2001; Díaz-Pineda and Schmidz 2003; de Lucio et al. 2003; Farina 2006; de Zavala et al. 2008). The composition and structure of the landscape affect the organisms occupying the habitats.

Visual Indicators
- Aggregation (de la Fuente de Val et al. 2006)
- Density elements (Gulinck et al. 2001; Schüpbach 2002; de la Fuente de Val et al. 2006)
- Repeating patterns in the landscape (Kaplan and Kaplan 1989; Pearson 2002)
- Degree of contrast between landscape elements (Hands and Brown 2002; Arriaza et al. 2004)

Common Ground
- PCC, Proportion of clear-cutting system area (authors' proposal)
- PAFA (authors' proposal)
- PSPA (authors' proposal)
- Change in the form of landscape elements (Weinstoerffer and Girardin 2000; Gulinck et al. 2001; Palmer 2004; de la Fuente de Val et al. 2006)
- Heterogeneity (Fjellstad et al. 2001; Dramstad et al. 2006)
- Diversity of landscape attributes (Germino et al. 2001; Gulinck et al. 2001; Palmer 2004; de la Fuente de Val et al. 2006)
- Change the size of the elements (Gulinck and Wagendorp 2002; Palmer 2004; de la Fuente de Val et al. 2006)

Ecological Indicators
- ACM area for key and endangered species (number of patches, density, area, proportion to TA) (authors' proposal)
- AHF species bigger than 400 and 100 ha (number of patches, density, area, proportion to TA) (authors' proposal)
- PSA relative to the total forest area (authors' proposal)
- DRD. Differences from a reference tree distribution (authors' proposal)
- PLTV (authors' proposal)
- SILVANET indicator of *structural diversity* (Martínez-Falero et al. 2010)
- Type of habitat patches connected (Fry et al. 2009)
- Signs of natural regeneration in forest restoration (Sheppard 2001)
- Tree species composition (Oxbrough et al. 2012)
- Percentage of natural vegetation (Schüpbach 2002; Arriaza et al. 2004; Palmer 2004; Ayad 2005)
- Level of vegetation succession (Schüpbach 2002; Palmer 2004)

- Structure and land-use spatial indices (Aguiló et al. 1995)

- Form of vegetation (Gulinck and Wagendorp 2002; Palmer 2004; de la Fuente de Val et al. 2006)
- Spatial organization of vegetation (Kuiper 2000)

- Mixed forests (Oxbrough et al. 2012)
- Proportion of landscape with a long continuity of vegetation (Jessel 2006)
- Proportion of landscape with traditional land use (Gulinck et al. 2001; Jessel 2006)
- Attributes classified as disturbance (Gulinck and Wagendorp 2002; Arriaza et al. 2004)
- Percentage of area affected by natural and anthropogenic disturbances (Fry et al. 2009)
- Shannon diversity index (Vila et al. 2006)

3. Soften margins of forests and new plantations

Visual significance: The prominence and fragility of the upper margins of forests reach a maximum in the skyline, as they are seen from most sites (Forestry Commission 1994). It is, therefore, necessary to be particularly careful with thinning and other forest management activities carried out in these areas.

Shape of forest margins affects visual landscape naturalness, stewardship and complexity (Ode and Fry 2006; Tveit et al. 2006; Fry et al. 2009)

Ecological significance: Changes in the forest edges affect the diversity of species present in the territory (Velarde and Ruiz 2007).

Visual Indicators

Ecological Indicators

Common Ground

- Edge density (Germino et al. 2001; Palmer 2004)
- Shape-linear characteristics (Darlington 2002; Fairclough et al. 2002)
- Area and/or percentage of forestland where silviculture is designed to protect scenery (e.g., thinning, fire prevention, and insect control) (Lim 2012)

(continued)

TABLE 6.4 (continued)

Visual and Ecological Forest Landscape Indicators for the Exterior Landscape Dimension

Exterior Forest Landscape

4. Adapt infrastructures, equipment, and other artificial elements to forest environment

Visual significance: Take into account visual fragility and aesthetic value when planning and designing the construction of new roads, forest roads, and firebreaks as well as in the maintenance of the existing ones.

Artificial elements affect visual landscape naturalness, coherence, stewardship, and disturbance (Ode and Fry 2006; Tveit et al. 2006; Fry et al. 2009).

Ecological significance: These structures have been considered as anthropogenic disturbances affecting biodiversity conservation. The more they are integrated in the environment, the lower the ecological impact.

Visual Indicators	Common Ground	Ecological Indicators
• Area visually affected by the disturbance (Gulinck et al. 2001; Duckert et al. 2009; Thomas et al. 2012) • Distance to the disturbing elements (Fry et al. 2009) • Area and/or percentage of forestland where silviculture is designed to protect scenery (e.g., thinning, fire prevention, and insect control) (Lim 2012)	• PCC, Proportion of clear-cutting system area (authors' proposal) • Attributes classified as disturbances (Gulinck and Wagendorp 2002; Arriaza et al. 2004) • Presence of roads and/or tracks (Sheppard and Harshaw 2001)	

5. Work at the landscape scale in order to foster integration of management activities

Visual significance: Idem Management activities are to be integrated in the environment, respecting the texture of the landscape.

This criteria affects visual scale and complexity (Ode and Fry 2006; Tveit et al. 2006; Fry et al. 2009).

Ecological significance: Management activities (new afforestation, thinning, infrastructure development, etc.) must conform to the scale of the surrounding landscape so as to preserve and promote the ecological functions of the forest.

Visual Indicators

Common Ground

Ecological Indicators

- Area and/or percentage of land conversion from nonforest to forest to protect scenery (e.g., afforestation) (Meitner et al. 2006)

- Success in meeting approved visual quality objectives (VQOs), public acceptance of visual impacts in visually sensitive areas outside established VQOs (e.g., provincial parks and special use areas) (Meitner et al. 2006)

- Success in demonstrating SFM to the public through enhanced visual treatments and by providing information along public access routes (Meitner et al. 2006)

- PCC, Proportion of clear-cutting system area (authors' proposal)

- Proportion of clear areas (Weinstoerffer and Girardin 2000; Palmer 2004)

- Spatial indexing structure and land use (Aguiló et al. 1995)

- Area and/or percentage of forestland used for timber yield that is protecting scenery by adopting alternative harvesting techniques to clear-cutting (e.g., CCF, selective cutting, and regeneration before cutting) (Lim 2012)

- PSA (authors' proposal)

- Proportion of landscape with a long continuity of vegetation (Jessel 2006)

- Percentage of area affected by natural and anthropogenic disturbances (Fry et al. 2009)

(continued)

TABLE 6.4 (continued)

Visual and Ecological Forest Landscape Indicators for the Exterior Landscape Dimension

Exterior Forest Landscape	Common Ground
Visual Indicators	• Area and/or percentage of forestland where silviculture is designed to protect scenery (e.g., thinning, fire prevention, and insect control) (Lim 2012)
• Depth of visual range (Germino et al. 2001; Gulinck et al. 2001)	
• Area, surface density, and variability indices (Vila et al. 2006)	
• Size of viewshed (Germino et al. 2001; Gulinck et al. 2001; de la Fuente de Val et al. 2006; Weitkamp et al. 2012)	
• Density of obstructing objects (Weinstoerffer and Girardin 2000; Weitkamp et al. 2012)	
• Degree of visual penetration of vegetation (Weinstoerffer and Girardin 2000)	
• Area and/or percentage of visually sensitive forestland or scenic forestland being protected (Lim 2012)	

TABLE 6.5
Visual and Ecological Forest Landscape Indicators for the Interior Landscape Dimension

Interior Forest Landscape

1. Protect riversides and shores

Visual significance: These areas are of particular visual sensitivity, so care must be taken over the preservation or enhancement of the landscape and its visual quality (Forestry Commission 2003). The forest–water interaction represents a very attractive element of contrast from the visual point of view (Velarde and Ruiz 2007). Landscape change in these areas may affect naturalness, complexity, coherence, imageability, and historicity (Ode and Fry 2006; Tveit et al. 2006; Fry et al. 2009).

Ecological significance: The banks of rivers, streams, lakes, and reservoirs must be maintained with natural vegetation to protect water quality, minimize erosion, and provide shelter and refuge for aquatic life.

Visual Indicators	Ecological Indicators
Common Ground	• Presence of water (Van Mansvelt and Kuiper 1999; Kuiper 2000)
	• Proportion of water in the landscape (Palmer 2004)
	• Presence of riparian vegetation (Sheppard 2001)

2. Pay attention to the singular function of forest edges

Visual significance: From the visual perspective, edges should be adapted to the geometry of the ground. The edges are of great value to increase visual diversity (Bell 2001).

Forest edge management may affect naturalness, coherence, and complexity (Ode and Fry 2006; Tveit et al. 2006; Fry et al. 2009).

Ecological significance: Increased patch edge associated to the geometry of fragmented landscapes modifies the microclimate created by forest density and favors the invasion of generalist species (Velarde et al. 2013).

(*continued*)

TABLE 6.5 (continued)
Visual and Ecological Forest Landscape Indicators for the Interior Landscape Dimension

Interior Forest Landscape

Visual Indicators	Common Ground	Ecological Indicators
• Shape-linear characteristics (Darlington 2002)	• PSR, Proportion of structural retention (authors' proposal) • Edge permeability (Fry et al. 2009) • Edge density (Germino et al. 2001; Palmer 2004)	• Ecotone, indices, and interior habitat (Vila et al. 2006)

3. Increase ecosystem and species diversity

Visual significance: The presence of some clearings increases the amount of edge and can promote visual diversity. They break the monotony of the landscape and are often used as playgrounds for the public. The contrast between grasslands and forests is aesthetically attractive (Velarde et al. 2013). Diversity may affect visual scale, imageability, complexity, historicity, and naturalness (Ode and Fry 2006; Tveit et al. 2006; Fry et al. 2009).

Ecological significance: Diversity allows coexistence of species adapted to different light, soil, and vegetation conditions. It also prevents fire disturbance.

Visual Indicators	Common Ground	Ecological Indicators
	• PSA relative to the total forest area (authors' proposal) • DRD, Differences from a reference tree distribution (authors' proposal) • PLTV (authors' proposal)	• PRTP relative to an RRTP (authors' proposal)

(continued)

- PSR (authors' proposal)
- Presence of clearings, small tree patches, meadows (Sheppard 2001)
- Proportion of clearings (Weinstoerffer and Girardin 2000; Palmer 2004)
- Diameter distribution (authors' proposal)
- Proportion of large trees (authors' proposal)
- Percentage of multispecies–multilayer forest area over the whole forest area (authors' proposal)

4. Preserve large old trees, large fallen trees, and trees of different species

Ecological significance: Large trees are helpful to prevent forest disturbance, as they may be a source of invertebrate predators for adjacent young stands.

Both fallen or standing dead trees are essential for the proper functioning of forest dynamics (Velarde and Ruiz 2007).

Visual significance: Forest stands containing several generations of trees as well as the presence of unique individual trees or groups of trees, particularly in open spaces, are usually preferred by the public (Velarde et al. 2013).

There is a conflict between ecology and aesthetics in the case of the presence of deadwood. Pristine forests are not preferred by the public, since the presence of many dead trees and the lack of order are considered unattractive (Ammer and Pröbstl 1991).

Presence of large trees may affect historicity, naturalness, complexity, coherence, stewardship, and imageability (Ode and Fry 2006; Tveit et al. 2006; Fry et al. 2009).

TABLE 6.5 (continued)
Visual and Ecological Forest Landscape Indicators for the Interior Landscape Dimension

Interior Forest Landscape

Visual Indicators
- Presence of historical elements (Fry et al. 2009)
- Presence of spectacular unique elements (Fry et al. 2009)

Common Ground
- Large trees in open spaces (Sheppard 2001)
- Seasonal changes in natural vegetation (Ahas et al. 2005)
- DRD, Differences from a reference tree distribution (authors' proposal)
- PLTV (authors' proposal)
- PSR (authors' proposal)

Ecological Indicators
- PRTP relative to an RRTP (authors' proposal)
- Signs of natural regeneration in forest restoration (Sheppard and Harshaw 2001)
- Volume of deadwood (Lassauce et al. 2011)
- Presence of old wood and trees (Fry et al. 2009)

5. Integrate structures and equipment into the forest landscape

Visual significance: From the point of view of public preference, it must be ensured that visitors do not perceive the evidence of forest management, such as road construction and stacked wood (Velarde and Ruíz 2007).

Stewardship, disturbance, and naturalness are the landscape character dimensions potentially affected (Ode and Fry 2006; Tveit et al. 2006; Fry et al. 2009).

Ecological significance: Infrastructures are regarded as anthropogenic disturbances affecting the environment.

Visual Indicators

Common Ground
- Area affected by the disturbance (Gulinck et al. 2001)
- Presence of roads and/or tracks (Sheppard and Harshaw 2001)
- Disturbing attributes (Gulinck et al. 2001; Arriaza et al. 2004)
- Distance to the disturbing elements (Fry et al. 2009)

Ecological Indicators

6. Do not disturb the genius loci (spirit of the place)

Visual significance: The unique charm of the place or *genius loci* refers to the presence of singularities or features of the terrain that give it a special value. When meeting the design principles, the spirit that is unique to that particular place must remain, as it represents an important value and stimulus (Forestry Commission 1994).

Historicity and imageability are the related landscape character dimensions (Ode and Fry 2006; Tveit et al. 2006; Fry et al. 2009).

Visual Indicators

• Presence of historical elements (Jessel 2006)
• Density of cultural elements (Van Mansvelt and Kuiper 1999)

Ecological significance

Common Ground

• PAFA, AFS (authors' proposal)
• PSPA, SPS (authors' proposal)
• Proportion of landscape with ancient vegetation (Jessel 2006)
• Proportion of landscape with traditional use (Gulinck et al. 2001; Jessel 2006)

Ecological Indicators

6.4 EPILOGUE

The concepts and methods for landscape and sustainable forestry—as well as our knowledge of forest ecosystems and the evaluation of the results of our actions—are undergoing a constant process of development and improvement. Indeed, our knowledge is incomplete and our vision is partial.

For this reason, there is no complete and accepted framework of landscape indicators; however, valid approaches can be made to avoid irreversible damage to forest ecosystems, which are the source of goods and services.

The appreciation of the visual attractiveness of a landscape addresses questions about its meaning and the adaptation of human activities. There are several productive and sustainable landscapes that are maintained by humans and their activities and are valued for their beauty.

Many features that reflect visual attractiveness correspond to elements that are useful or necessary to conserve biodiversity and adapt our activities to nature. Minor discrepancies can occur in some cases, but they are easily corrected with public information.

Through science and research, new meanings are being discovered, and explanations provided for the structure, function, and dynamic relationships found in nature and the ecological landscape. However, scientific knowledge and subsequent models are no guarantee of making the right decision. Furthermore, the models are a simplification of reality, and they require review and improvement.

Nature contains many pathways and multiple variables. Landscapes and forests are complex systems in which humans live and intervene. The multiple possibilities that exist in nature are specified by human management decisions and resolved within a complex social, economic, and political system. Ecosystems and landscapes are modified by humans, and their effect penetrates into the most remote and isolated sites. In this sense, the before-man nature model is unattainable. Man in nature is the realistic model. Two goals to guide forest landscape management, with consequences for human survival and the integrity of the surrounding nature, emerge in the context of sustainable development:

- To leave a better world to our descendants and a greater power of decision than we ourselves have had
- To create and maintain landscapes with great resilience to change and to sustain viable and sufficient populations of all organisms

To achieve these objectives, the approach is to lean on natural processes through ecosystem management. By taking advantage of natural dynamics and biodiversity, it will be possible to achieve activities and landscapes which—while not exactly natural—are close to nature or nearly natural. Integrating nature in all our activities is a priority for the coming decades.

Achieving this goal is a challenge for society and for the whole range of disciplines and professions and should produce results in the areas of social, economic, administrative decision, and regulatory policy.

High biodiversity, visual attractiveness, and sustained economic production have been achieved in some landscape systems. These examples demonstrate that sustainable development is possible and that moreover there are procedures and approaches available to ensure it.

REFERENCES

Agee, J. K. 1997. Fire management for the 21st century. In: *Creating a Forestry for the 21st Century: The Science of Ecosystem Management*, eds. Kohm K. A. and J. F. Franklin. Washington, DC: Island press, pp. 191–202.

Aguiló, M. et al. 1995. *Guía para la elaboración de estudios del medio físico: contenido y metodología*. Madrid, Spain: Ministerio de Medio Ambiente.

Aguiló, M. and E. Iglesias. 1995. Landscape inventory. In: *Quantitative Techniques in Landscape Planning*, eds. E. Martínez-Falero and S. González-Alonso. New York: Lewis, pp. 47–85.

Ahas, R., A. Aasa, S. Silm, and J. Roosaare. 2005. Seasonal indicators and seasons of Estonian landscapes. *Landscape Research* 30:173–191.

Ammer, U. and U. Pröbstl. 1991. *Freizeit und Natur. Probleme und Lösungsmöglichkeiten einer ökologisch verträglichen Freizaitnutzung (Tiempo Libre y Naturaleza. Problemas y soluciones de un uso ecológico y sostenible del recreo)*. Hamburg, Germany: Verlag Paul Parey.

Angelstam, P. and T. Kuuluvainen. 2004. Boreal forest disturbance regimes, successional dynamics and landscape structures—A European perspective. *Ecological Bulletins* 51:117–136.

Arriaza, M., J. F. Canas-Ortega, J. A. Canas-Madueno, and P. Ruiz-Aviles. 2004. Assessing the visual quality of rural landscapes. *Landscape and Urban Planning* 69(1):115–125.

Atauri, J. A. and J. V. de Lucio. 2001. The role of landscape structure in species richness distribution of birds, amphibians, reptiles and lepidopterans in Mediterranean landscapes. *Landscape Ecology* 16(2):147–159.

Ayad, Y. A. 2005. Remote sensing and GIS in modeling visual landscape change: a case study of the northwestern and coast of Egypt. *Landscape and Urban Planning* 73(4):307–325.

Bañuelos, M. J., M. Quevedo, and J. R. Obeso. 2008. Habitat partitioning in endangered Cantabrian capercaillie *Tetrao urogallus cantabricus*. *Journal of Ornithology* 149:245–252.

Bauhus, J., K. Puettmann, and C. Messier. 2009. Silviculture for old-growth attributes. *Forest Ecology and Management* 258(4):525–537.

Bell, S. 1993. *Elements of Visual Design in the Landscape*. London, U.K.: E & F N Spon.

Bell, S. 1999. *Forest Practice Guide: Forest Design Planning, a Guide to Good Practice*. Edinburgh, Scotland: Forestry Commission.

Bell, S. 2001. Landscape pattern, perception and visualisation in the visual management of forests. *Landscape and Urban Planning* 54(1–4):201–211.

Bell, S. (ed.). 2003. *The Potential for Applied Landscape Ecology to Forest Design Planning*. Edinburgh, Scotland: Forestry Commission.

Bell, S. 2011. *Forest and Landscape Guidelines*. Edinburgh, Scotland: Forestry Commission.

Bell, S. and D. Apostol. 2008. *Designing Sustainable Forest Landscapes*. London, U.K.: Taylor & Francis.

Bennet, A. F. 1998. *Linkages in the Landscape: The Role of Corridors and Connectivity in Wildlife Conservation*. Gland, (Suiza) and Cambridge: IUCN.

Bettinger, P., M. Lennette, K. N. Johnson, and T. A. Spies. 2005. A hierarchical spatial framework for forest landscape planning. *Ecological Modelling* 182(1):25–48.

Boyce, S. G. 1995. *Landscape Forestry*. New York: John Wiley & Sons.

Brown, T. C. 1984. The concept of value in resource allocation. *Landscape and Economy* 60:231–246.

Brown, T. C. and T. C. Daniel. 1986. Predicting scenic beauty of timber stands. *Forest Science* 32(2):471–487.

Bruciamacchie, M. and B. de Turckheim. 2005. *La futaie irrégulière. Théorie et pratique de la sylviculture irrégulière, continue et proche de la nature.* Aix-en-Provençe, France: Edisud.

Burel, F. and J. Baudry. 2001. *Ecología del paisaje. Conceptos, métodos y aplicaciones.* Madrid, Spain: Mundi Prensa.

Camprodón, J. 2001. Tratamientos forestales y conservación de la fauna vertebrada. In: *Conservación de la biodiversidad y gestión forestal. Su aplicación a la fauna vertebrada,* eds. J. Camprodón and E. Plana. Barcelona, Spain: Centre tecnologic forestal de Cataunya i Edicions de la Universitat de Barcelona, pp. 119–189.

Camprodón, J., D. Campión, R. Martínez- Vidal, A. Onrubia, H. Robles, J. L. Romero et al. 2007. Estatus, selección del hábitat y conservación de los pícidos ibéricos. In: *Conservación de la biodiversidad, fauna vertebrada y gestión forestal,* eds. J. Camprodon and E. Plana. Barcelona, Spain: Publicacions i Edicions de la Universitat de Barcelona, pp. 391–434.

Cantiani, M. G. 2012. Forest planning and public participation: A possible methodological approach. *iForest—Biogeosciences and Forestry* 5(2):72–82.

Carey, A. B. 2009. Maintaining biodiversity in managed forest. In: *Old Growth in a New World: A Pacific Northwest Icon Reexamined,* eds. T. A. Spies and S. L. Duncan. Washington, DC: Island press, pp. 58–69.

CEMAGREF. 1981. *Reboisament et paysage.* Paris, France: Forêt-loisirs et équipements de plein air, nouvelle série n°4.

Christensen, M., K. Hahn, E. P. Mountford, P. Odor, T. Standovar, D. Rozenbergar et al. 2005. Dead wood in European beech (Fagus sylvatica) forest reserves. *Forest Ecology and Management* 210(1–3):267–282.

Crow, T. R. and E. J. Gustafson. 1997. Ecosystem management: Managing natural resources in time and space. In: *Creating a Forestry for the 21st Century: The Science of Ecosystem Management,* eds. K. A. Kohm and J. F. Franklin. Washington, DC: Island Press, pp. 215–228.

Curtis, R. O. 1997. The role of extended rotations. In: *Creating a Forestry for the 21st Century: The Science of Ecosystem Management,* eds. K. A. Kohm and J. K. Franklin. Washington, DC: Island Press, pp. 165–170.

Dajoz, R. 2006. *Précis d'écologie.* Paris, France: Dunod.

Dakin, S. 2003. There's more to landscape than meets the eye: Towards inclusive landscape assessment in resource and environmental management. *Canadian Geographer/Le Géographe Canadien* 47(2):185–200.

Daniel, T. C. 2001. Aesthetic preference and ecological sustainability. In: *Forest and Lanscapes: Linking Ecology, Sustainability and Aesthetics,* eds. S. R. J. Sheppard and H. W. Harshaw. Cambridge, U.K.: CABI Publishing, pp. 15–30.

Darlington, J. 2002. Mapping lancashire's historic landscape: The lancashire historic landscape characterisation programme In: *Europe's Cultural Landscape: Archaeologists and the Management of Change,* eds. G. Fairclough and S. Rippon. Exeter, England: Short Run Press, pp. 97–105.

Davies, O., J. Haufe, and A. Pommerening. 2008. *Silvicultural Principles of Continuous Cover Forestry: A Guide to Best Practice.* Bangor, U.K.: Bangor University, p. 111.

Davies, O. and G. Kerr. 2011. The cost and revenues of transformation to continuous cover forestry: Modelling silvicultural options with Sitka spruce. Forest research, Forestry Commission.

de la Fuente de Val, G., J. A. Atauri, and J. V. de Lucio. 2006. Relationship between landscape visual attributes and spatial pattern indices: A test study in Mediterranean-climate landscapes. *Landscape and Urban Planning* 77(4):393–407.

de Lucio, J. V., J. A. Atauri, P. Sastre, and C. Martínez. 2003. Conectividad y redes de espacios naturales protegidos. Del modelo teórico a la visión práctica de la gestión. In: *Conectividad ambiental: las áreas protegidas en la cuenca mediterránea*, eds. R. García Mora. Sevilla, Spain: Consejería de Medio Ambiente. Junta de Andalucía, pp. 29–53.

de Lucio, J. V. and M. Múgica. 1994. Landscape preferences and behaviour of visitors to Spanish national parks. *Landscape and Urban Planning* 29(3):145–160.

de Miguel, J. M. and A. Gómez-Sal. 2002. Diversidad y funcionalidad de los paisajes agrarios tradicionales. In: *La diversidad biológica en España*, eds. F. Díaz Pineda, J. M. de Miguel, and M. A. Casado. Madrid, Spain: Prentice Hall, pp. 273–284.

de Turckheim, B. 1992. Pour une sylviculture proche de la nature. *Forêts de France* 350:14–20.

de Turckheim, B. 1993. Bases economiques de la sylviculture proche de la nature. Paper read at 1° Congrés européen de PRO SILVA, 21–24, juin 1993, at Besançon.

de Turckheim, B. 1999. Planification et contrôle en futaie irrégulière et continue. *Revue forestière française* (n° spécial 1999: L'aménagement forestier hier, aujourd'hui, demain), pp. 76–86.

de Zavala, M. A., R. Zamora, F. Pulido, J. A. Blanco, B. Imbert, T. Marañón et al. 2008. Nuevas perspectivas en la conservación, restauración y gestión sostenible del bosque mediterráneo. In: *Ecología del bosque mediterráneo en un mundo cambiante*, eds. F. Valladares. Madrid, Spain: Ministerio de Medio Ambiente. Organismo autónomo de Parques Nacionales, pp. 511–532.

Delibes-Mateos, M. and L. Gálvez-Bravo. 2009. El papel del conejo como especie clave multifuncional en el ecosistema mediterráneo de la Península Ibérica. *Ecosistemas* 18(3):14–25.

Delibes-Mateos, M. and F. Hiraldo. 1981. The rabbit as pray in the Iberian Mediterranean ecosystem. In: *Proceedings of the World Lagomorph Conference*, eds. K. Myers and C. D. MacInnes. Guelph, Ontario, Canada: University of Guelph, pp. 614–622.

Delibes-Mateos, M., S. M. Redpatha, E. Angulo, P. Ferrerasa, and R. Villafuertea. 2007. Rabbits as a keystone species in southern Europe. *Biological Conservation* 137:149–156.

Di Castri, F. and T. Younés. 1996. Introduction: Biodiversity, the emergence of a new scientific field. It's perspectives and constraints. In: *Biodiversity, Science and Development: Towards a New Partnership*, eds. F. Di Castri and T. Younés. Cambridge, U.K.: CAB International. IUBS, pp. 3–16.

Díaz-Pineda, F. and F. M. Schmidtz. 2003. Tramas espaciales del paisaje. Conceptos, aplicabilidad y temas urgentes para la planificación territorial. In: *Conectividad ambiental: las áreas protegidas en la Cuenca Mediterránea*, eds. M. R. García Mora. Sevilla, Spain: Consejería de Medio Ambiente Junta de Andalucía, pp. 9–28.

Diaz-Varela, E. R., M. F. Marey-Perez, A. Rigueiro-Rodriguez, and P. Alvarez-Alvarez. 2009. Landscape metrics for characterization of forest landscapes in a sustainable management framework: Potential application and prevention of misuse. *Annals of Forest Science* 66(3):301–311.

Díaz Pineda, F. and F. M. Schmidtz. 2003. Tramas espaciales del paisaje. Conceptos, aplicabilidad y temas urgentes para la planificación territorial. In: *Conectividad ambiental: las áreas protegidas en la Cuenca Mediterránea*, eds. M. R. García Mora and (Coord.). Sevilla, Spain: Consejería de Medio Ambiente Junta de Andalucía, pp. 9–28.

Dietze, M. C. and A. M. Latimer. 2011. Forest simulators. In: *Sourcebook of Theoretical Ecology*, eds. A. M. Hastings and L. Gross. Berkeley, CA: University of California Press, pp. 307–316.

Dramstad, W. E., J. D. Olson, and R. T. T. Forman. 1996. *Landscape Ecology, Principles in Landscape and Land—Use Planning*. Washington, DC: Harvard University Graduate School of Design, Island Press & the American Society of Landscapes Architects.

Dramstad, W. E., M. Sundli Tveit, W. J. Fjellstad, and G. L. A. Fry. 2006. Relationships between visual landscape preferences and map-based indicators of landscape structure. *Landscape and Urban Planning* 78(4):465–474.

du Cros, R. T. and S. Lopez. 2009. Preliminary study on the assessment of deadwood volume by the French national forest inventory. *Annals of Forest Science* 66(3):302.

Duchiron, M. S. 1994. *Gestion de futaies irregulieres et melangées*, eds. M.-S. Duchiron. Nancy, France: Marie-Stella Duchiron.

Duckert, D. R., D. M. Morris, D. Deugo, S. Duckett, and S. McPherson. 2009. Developing site disturbance standards in Ontario: Linking science to forest policy within an adaptive management framework. *Canadian Journal of Soil Science* 89(1):13–23.

Edwards, D. M., M. Jay, F. S. Jensen, B. Lucas, M. Marzano, C. Montagne et al. 2012. Public preferences across Europe for different forest stand types as sites for recreation. *Ecology and Society* 17(1):27–38.

Escribano, R., M. Frutos, E. Iglesias, E. Mata, and I. Torrecilla. 1987. *El Paisaje*. Madrid, Spain: Ministerio de Obras Publicas y Transportes. Secretaría de Estado para las Políticas del Agua y el Medio Ambiente.

Fairclough, G., G. Lambrick, and D. Hopkins. 2002. Historic landscape characterisation in England and a Hampshire case study. In: *Europe's Cultural Landscape: Archaeologists and the Management of Change*, eds. G. F. S. Rippon. Exeter, U.K.: Short Run Press, pp. 69–83

Farina, A. 2006. *Principles and Methods in Landscape Ecology: Toward a Science of Landscape*. Dordrecht, the Netherlands: Springer.

Fischer, J., D. B. Lindenmayer, and A. D. Manning. 2006. Biodiversity, ecosystem function, and resilience: ten guiding principles for commodity production landscapes. *Frontiers in Ecology and the Environment* 2(4):80–86.

Fjellstad, W. J., W. E. Dramstad, G. H. Strand, and G. L. A. Fry. 2001. Heterogeneity as a measure of spatial pattern for monitoring agricultural landscapes. *Norwegian Journal of Geography* 55:71–76.

Forestry Commission. 1988. *Forests and Water Guidelines*. Londres, U.K.: Forestry Commission.

Forestry Commission. 1989. *Native Pinewoods: Grants and Guidelines*. Londres, U.K.: Forestry Commission.

Forestry Commission. 1990. *Forest Nature Conservation. Guidelines*. Londres, U.K.: Forestry Commission.

Forestry Commission. 1991. *Community Woodland Design. Guidelines*. Londres, U.K.: Forestry Commission.

Forestry Commission. 1992. *Lowland Landscape Design. Guidelines*. Londres, U.K.: Forestry Commission.

Forestry Commission. 1994. *Forest Landscape Design. Guidelines*. Londres, U.K.: Forestry Commission.

Forestry Commission. 2003. *Forests and Water Guidelines*, 4th edn. Londres, U.K.: Forestry Commission.

Forman, T. T. 1995. *Land Mosaics*. Cambridge, U.K.: Cambridge University Press.

Franklin, J. F., R. B. Berg, D. A. Thorburgh, and J. C. Tappeiner. 1997. Alternative silvicultural approaches to timber harvesting: Variable retention harvest system. In: *Creating a Forestry for the 21st Century: The Science of Ecosystem Management*, eds. K. A. Kohm and J. F. Franklin. Washington, DC: Island Press, pp. 111–140.

Franklin, J. F., T. A. Spies, R. Van Pelt, A. B. Carey, D. A. Thornburgh, D. R. Berg et al. 2002. Disturbances and structural development of natural forest ecosystems with silvicultural implications, using Douglas-fir forests as an example. *Forest Ecology and Management* 155(1–3):399–423.

Fry, G. L. A. 2001. Multifunctional landscapes—Towards transdisciplinary research. *Landscape and Urban Planning* 57(3–4):159–168.

Fry, G., M. S. Tveit, A. Ode, and M. D. Velarde. 2009. The ecology of visual landscapes: Exploring the conceptual common ground of visual and ecological landscape indicators. *Ecological Indicators* 9(5):933–947.

Fudge, R. S. 2001. Imagination and the science-based aesthetic appreciation of unscenic nature. *The Journal of Aesthetics and Art Criticism* 59(3):275–285.

García-Abril, A. 1999. La gestión próxima a la naturaleza. La restauración de la armonía hombre-naturaleza. In: *Homenaje a D. Ángel Ramos Fernández (1926–1998)*, ed. Real Academia de Ciencias Exactas Físicas y Naturales. Madrid, Spain: Academia de Ingeniería y E. T. S. I. de Montes, pp. 1287–1306.

García-Abril, A., L. Yoldi, and J. L. Canga. 1989. Las repoblaciones forestales. In: *El libro rojo de los bosques españoles*, eds. ADENA. Madrid, Spain: ADENA-WWF-España, pp. 237–276.

Germino, M. J., W. A. Reiners, B. J. Blasko, D. McLeod, and C. T. Bastian. 2001. Estimating visual properties of Rocky Mountain landscapes using GIS. *Landscape and Urban Planning* 53(1–4):71–83.

Gil-Tena, A., S. Saura, and L. Brotons. 2007. Effects of forest composition and structure on bird species richness in a Mediterranean context: Implications for forest ecosystem management. *Forest Ecology and Management* 242(2–3):470–476.

Global Biodiversity Strategy. 1992. *Global Biodiversity Strategy: Guidelines for Action to Save, Study and Use Earth's Biotic Wealth Sustainably and Equitably*. Washington, DC: WRI, IUCN, UNEP.

Gobster, P. H. 2001. Visions of nature: Conflict and compatibility in urban park restoration. *Landscape and Urban Planning* 56(1–2):35–51.

Gobster, P. H. 1999. An ecological aesthetic for forest landscape management. *Landscape Journal* 18:54–64.

Gobster, P. H., J. I. Nassauer, T. C. Daniel, and G. Fry. 2007. The shared landscape: What does aesthetics have to do with ecology? *Landscape Ecology* 22(7):959–972.

Gomez-Limon, J. and J. V. de Lucio. 1999. Changes in use and landscape preferences on the agricultural-livestock landscapes of the central Iberian Peninsula (Madrid, Spain). *Landscape and Urban Planning* 44(4):165–175.

Gómez Manzaneque, F., M. Génova, and P. Regato. 2005. Los pinares de *Pinus nigra* del sistema central. In: *Los pinares de Pinus nigra Arn. en España: Ecología, uso y gestión*, eds. M. A. Grande and A. García-Abril. Madrid, Spain: Fundación Conde del valle de Salazar.

González-Alonso, S. et al. 1989. *Guías Metodológicas para la elaboración de estudios de impacto ambiental. Repoblaciones forestales*. Madrid, Spain: MOPU.

González Bernáldez, F. 1981. *Ecología y paisaje*. Barcelona, Spain: Blume.

Graefe, A. R. and J. J. Vaske. 1987. A framework for managing quality in the tourist experience. *Annals of Tourism Research* 14(3):390–404.

Gruehn, D. and M. Roth. 2010. Landscape preference study of agricultural landscapes in Germany. *Journal of Landscape Ecology* 9(Special Issue):67–78.

Gulinck, H., M. Mugica, J. V. de Lucio, and J. A. Atauri. 2001. A framework for comparative landscape analysis and evaluation based on land cover data, with an application in the Madrid region (Spain). *Landscape and Urban Planning* 55(4):257–270.

Gulinck, H. and T. Wagendorp. 2002. References for fragmentation analysis of the rural matrix in cultural landscapes. *Landscape and Urban Planning* 58(2–4):137–146.

Gundersen, V. and L. H. Frivold. 2011. Naturally dead and downed wood in Norwegian boreal forests: public preferences and the effect of information. *Scandinavian Journal of Forest Research* 26(2):110–119.

Gustafsson, L., S. C. Baker, J. Bauhus, W. J. Beese, A. Brodie, J. Kouki et al. 2012. Retention forestry to maintain multifunctional forests: A world perspective. *Bioscience* 62(7):633–645.

Gustafsson, L., J. Kouki, and A. Sverdrup-Thygeson. 2010. Tree retention as a conservation measure in clear-cut forests of northern Europe: A review of ecological consequences. *Scandinavian Journal of Forest Research* 25(4):295–308.

Haider, W. and L. Hunt. 2002. Visual aesthetic quality of northern Ontario's forested shorelines. *Environmental Management* 29(3):324–334.

Halpern, C. B. and T. A. Spies. 1995. Plant species diversity in natural and managed forests of the Pacific northwest. *Ecological Applications* 5(4):913–934.

Hands, D. E. and R. D. Brown. 2002. Enhancing visual preference of ecological rehabilitation sites. *Landscape and Urban Planning* 58(1):57–70.

Harshaw, H. W., R. A. Kozak, and S. R. J. Sheppard. 2006. How well are outdoor recreationists represented in forest land-use planning? Perceptions of recreationists in the Sea-to-Sky Corridor of British Columbia. *Landscape and Urban Planning* 78(1–2):33–49.

Harvey, C. A., D. Sánchez-Merlo, Arnulfo Medina, S. J. Vílchez-Mendoza, B. Hernández, A. Pérez et al. 2003. Contribución de las cercas vivas a la productividad e integridad ecológica de los paisajes agrícolas en América Central. *Agroforistería en las Américas* 10(39/40):30–39.

Harvey, C. A., D. Sánchez-Merlo, A. Medina, S. J. Vílchez-Mendoza, B. Hernández, A. Pérez et al. 2005. Contribution of live fences to the ecological integrity of agricultural landscapes. *Agroforistería en las Américas* 111(1/4):200–230.

He, H. S. 2008. Forest landscape models: Definitions, characterization, and classification. *Forest Ecology and Management* 254(3):484–498.

Hilty, J. A., W. Z. Lidicker, and A. M. Merenlender. 2006. *Corridor Ecology: The Science and Practice of Linking Landscapes for Biodiversity Conservation.* Washington, DC: Island Press.

Hull, R. B. and W. P. Stewart. 1992. Validity of photo-based scenic beauty judgments. *Journal of Environmental Psychology* 12(2):101–114.

Humphrey, J. W., S. Davey, A. J. Peace, R. Ferris, and K. Harding. 2002. Lichens and bryophyte communities of planted and semi-natural forests in Britain: The influence of site type, stand structure and deadwood. *Biological Conservation* 107:165–180.

Jacobsen, M. K. 2001. History and principles of close to nature forest management: A Central European perspective (Denmark). In: *Tools for Preserving Woodland Biodiversity*, eds. H. F. Read, A. S., R. Marciau, H. Paltto, L. Andersson, and B. Tardy. London, U.K.: Europe's Woodland Heritage, pp. 56–60.

Jessel, B. 2006. Elements, characteristics and character—Information functions of landscapes in terms of indicators. *Ecological Indicators* 6(1):153–167.

Johnson, K. N., P. Bettinger, J. D. Kline, T. A. Spies, M. Lennette, G. Lettman et al. 2007. Simulating forest structure, timber production, and socioeconomic effects in a multi-owner province. *Ecological Applications* 17(1):34–47.

Kaltenborn, B. P. and T. Bjerke. 2002. Associations between landscape preferences and place attachment: A study in Roros, Southern Norway. *Landscape Research* 27(4):381–396.

Kaplan, R. and S. Kaplan. 1989. *The Experience of Nature; a Psychological Perspective.* Cambridge, U.K.: Cambridge University Press.

Karjalainen, E. and M. Komulainen. 1999. The visual effect of felling on small- and medium-scale landscapes in north-eastern Finland. *Journal of Environmental Management* 55(3):167–181.

Keneeshaw, D., Y. Bergeron, and T. Kuuluvainen. 2011. Forest ecosystem structure and disturbance dynamics across the circumboreal forest. In: *The SAGE Handbook of Biogeography*, eds. A. Millington, M. Blumler, and U. Schickhoff. London, U.K.: SAGE Publications, pp. 263–280.

Keeton, W.S. 2006. Managing for late-successional/old-growth forest characteristics in northern hardwood-conifer forests. *Forest Ecology and Management* 235(1–3):129–142.

Kerr, G., G. Morgan, J. Blyth, and V. Stokes. 2010. Transformation from even-aged plantations to an irregular forest: The world's longest running trial area at Glentress, Scotland. *Forestry* 83(3):329–344.

Kimmins, J. P. 2001. Visible and non-visible indicators of forest sustainability: Beauty, beholders and belief systems. In: *Forest and Landscapes: Linking Ecology, Sustainability and Aesthetics*, eds. S. R. J. Sheppard and H. W. Harshaw. Cambridge, U.K.: CABI Publishing, pp. 43–56.

Kimmins, J. P. 2002. Future shock in forestry—Where have we come from; where are we going; is there a "right way" to manage forests? Lessons from Thoreau, Leopold, Toffler, Botkin and Nature. *Forestry Chronicle* 78(2):263–271.

Kohm, K. A. and J. F. Franklin. 1997. *Creating a Forestry for the 21st Century*. Washington, DC: Island Press.

Koop, H. 1987. Vegetative reproduction of trees in some European natural forests. *Plant Ecology* 72:103–110.

Koper, N., D. J. Walker, and J. Champagne. 2009. Nonlinear effects of distance to habitat edge on Sprague's pipits in southern Alberta, Canada. *Landscape Ecology* 24(10):1287–1297.

Korpel, S. 1982. Degree of equilibrium and dynamical changes of forest on example of natural forest of Slovakia. *Acta facultatis forestalis zvolen* XXIV: 9–31.

Korpel, S. 1987. Dynamics of the structure and development of natural beech forest in Slovakia. *Acta facultis forestalis zvolen* 29:59–85.

Korpel, S. 1995. *Die Urwälder der Westkarpaten*. Stuttgart, Germany: Gustav Fischer Verlag.

Kuiper, J. 2000. A checklist approach to evaluate the contribution of organic farms to landscape quality. *Agriculture Ecosystems & Environment* 77(1–2):143–156.

Kumar, B. M. and P. K. R. Nair, eds. 2006. *Tropical Homegardens: A Time-Tested Example of Sustainable Agroforestry*. Dordrecht, the Netherlands: Springer.

Kuuluvainen, T. 2002. Natural variability of forests as a reference for restoring and managing biological diversity in boreal Fennoscandia. *Silva Fennica* 36(1): 97–125.

Kuuluvainen, T., K. Aapala, P. Ahlroth, M. Kuusinen, T. Lindholm, T. Sallantaus et al. 2002. Principles of ecological forest restoration of boreal forested ecosystems: Finland as an example. *Silva Fennica* 36(1):409–422.

Kuuluvainen, T. 2009. Forest management and biodiversity conservation based on natural ecosystem dynamics in Northern Europe: The complexity challenge. *Ambio* 38(6):309–315.

Kuuluvainen, T. and T. Aakala. 2011. Natural forest dynamics in boreal Fennoscandia: A review and classification. *Silva Fennica* 45(5):823–841.

Laiho, O., E. Lahde, and T. Pukkala. 2011. Uneven- vs even-aged management in Finnish boreal forests. *Forestry* 84(5):547–556.

Lassauce, A., Y. Paillet, H. Jactel, and C. Bouget. 2011. Deadwood as a surrogate for forest biodiversity: Meta-analysis of correlations between deadwood volume and species richness of saproxylic organisms. *Ecological Indicators* 11(5):1027–1039.

Lim, S. S. 2012. *Development of Forest Aesthetic Indicators in Forest Sustainable Standards*. Vancouver, British Columbia, Canada: University of British Columbia.

Lindenmayer, D. and J. F. Franklin. 2002. *Conserving Forest Biodiversity: A Comprehensive Multiscaled Approach*. Washington, DC: Island Press.

Litton, R. B. 1972. Aesthetic dimensions of the landscape. In: *Natural Environments: Studies in Theoretical and Applied Analysis*, eds. J. V. Krutilla. Baltimore, MD: Johns Hopkins University Press, pp. 262–291.

Lucas, O. W. R. 1991. *The Design of Forest Landscapes*. New York: Oxford University Press.

Luymes, D. 2001. The rhetoric of visual simulation in forest design: some research directions. In: *Forest and Landscapes: Linking Ecology, Sustainability and Aesthetics*, eds. S. R. J. Sheppard and H. W. Harshaw. Cambridge, U.K.: CABI Publishing, pp. 191–204.

Llausas, A. and J. Nogue. 2012. Indicators of landscape fragmentation: The case for combining ecological indices and the perceptive approach. *Ecological Indicators* 15(1):85–91.

Malezieux, E. 2012. Designing cropping systems from nature. *Agronomy for Sustainable Development* 32(1):15–29.

Malezieux, E., Y. Crozat, C. Dupraz, M. Laurans, D. Makowski, H. Ozier-Lafontaine, et al. 2009. Mixing plant species in cropping systems: Concepts, tools and models. A review. *Agronomy for Sustainable Development* 29(1):43–62.

Margaleff, R. 1993. *Teoría de los sistemas ecológicos.* Barcelona, Spain: Universidad de Barcelona.

Margaleff, R. 2002. Diversidad y biodiversidad. In: *La diversidad biológica en España,* eds. F. Díaz Pineda. Madrid, Spain: Prentice Hall, pp. 3–5.

Martínez-Falero, E. and S. González-Alonso. 1995. *Quantitative Techniques in Landscape Planning.* New York: Lewis.

Martínez-Falero, E., S. Martin-Fernandez, and A. García-Abril. 2010. *SILVANET, participación pública para la Gestión Forestal Sostenible.* Madrid, Spain: Fundación Conde del Valle de Salazar.

Martins, H. and J. G. Borges. 2007. Addressing collaborative planning methods and tools in forest management. *Forest Ecology and Management* 248(1–2):107–118.

Mason, B. and G. Kerr. 2004. *Transforming Even-Aged Conifer Stands to Continuous Cover Management. Forestry Commission Information Note 40.* Edinburgh, Scotland: Forestry Commission.

Mason, B., G. Kerr, and J. Simpson. 1999. *What is Continuous Cover Forestry? Forestry Commission Information Note 29.* Edinburgh, Scotland: Forestry Commission.

Matthews, J. D. 1991. *Silvicultural Systems.* Oxford, U.K.: Oxford University Press.

Matthews, P. 2002. Scientific knowledge and the aesthetic appreciation of nature. *The Journal of Aesthetics and Art Criticism* 60(1):37–48.

McComb, W. C., M. T. McGrath, T. A. Spies, and D. Vesely. 2002. Models for mapping potential habitat at landscape scales: An example using northern spotted owls. *Forest Science* 48(2):203–216.

McDonald, G. T. and M. B. Lane. 2004. Converging global indicators for sustainable forest management. *Forest Policy and Economics* 6(1):63–70.

MCPFE. 2003. *Implementation of MCPFE Commitments. National and Pan-European Activities 1998–2003.* Vienna, Austria: Ministerial Conference on the Protection of Forests in Europe Liaison Unit in Vienna.

Mediene, S., M. Valantin-Morison, J. P Sarthou, S. de Tourdonnet, M. Gosme, M. Bertrand et al. 2012. Agroecosystem management and biotic interactions: A review. *Agronomy for Sustainable Development* 31(3):491–514.

Meitner, M. J., H. W. Harshaw, S. R. J. Sheppard, and P. Picard. 2006. Criterion 9: Quality-of-life indicators, arrow innovative forest practices agreement series: Extension note 8 of 8. *BC Journal of Ecosystems and Management* 8:99–105.

Messier, C. and K. J. Puettmann. 2011. Forests as complex adaptive systems: Implications for forest management and modelling. *L'Italia Forestale e Montana* 66(3):249–258.

Millennium Ecosystem Assessment. 2005. *Ecosystems and Human Well-Being: Biodiversity Synthesis.* Washington, DC: World Resources Institute.

Miller, J., J. Ahern, A. Boquetila-Leitao, and K. McGarigal. 2006. *Measuring Landscapes: A Planner's Handbook.* Washington, DC: Island Press.

Mladenoff, D. J. 2004. LANDIS and forest landscape models. *Ecological Modelling* 180(1):7–19.

Montero, G., A. San Miguel, and I. Cañellas. 2000. *Systems of Mediterranean Silviculture. "La Dehesa".* Madrid, Spain: Grafistaff. S.L.

Müller, J. and R. Bütler. 2010. A review of habitat thresholds for dead wood: A baseline for management recommendations in European forests. *European Journal Forest Resource* 129:981–992.

Nair, P. 1993. *An Introduction to Agroforestry.* Dordrecht, the Netherlands: Kluwer Academic Publishers.

Nair, P. and K. Ramachandran. 2007. The coming of age of agroforestry. *Journal of the Science of Food and Agriculture* 87(9):1613–1619.

Nasar, J. L. and M. Li. 2004. Landscape mirror: The attractiveness of reflecting water. *Landscape and Urban Planning* 66(4):233–238.

Nassauer, J. I. 1995. Messy ecosystems, orderly frames. *Landscape Journal* 14:161–170.

Nilsson, S. G., M. Niklasson, J. Hedin, G. Aronsson, J. M. Gutowski, P. Linder et al. 2002. Densities of large living and dead trees in old-growth temperate and boreal forests. *Forest Ecology and Management* 161(1–3):189–204.

Nonaka, E. and T. A. Spies. 2005. Historical range of variability in landscape structure: A simulation study in Oregon, USA. *Ecological Applications* 15(5):1727–1746.

Núñez, M. V., M. D. Velarde, and A. García-Abril. 2010. Landscape integration of Mediterranean reforestation: Identification of best practices in Madrid Region. Paper read at *IUFRO Landscape Ecology International Conference*, Bragança, Portugal.

Ode, A. and G. Fry. 2006. A model for quantifying and predicting urban pressure on woodland. *Landscape and Urban Planning* 77(1–2):17–27.

Ode, A., G. Fry, M. S. Tveit, P. Messager, and D. Miller. 2009. Indicators of perceived naturalness as drivers of landscape preference. *Journal of Environmental Management* 90(1):375–383.

Ode, A., M. S. Tveit, and G. Fry. 2008. Capturing landscape visual character using indicators: Touching base with landscape aesthetic theory. *Landscape Research* 33(1):89–117.

Oku, H. and K. Fukamachi. 2006. The differences in scenic perception of forest visitors through their attributes and recreational activity. *Landscape and Urban Planning* 75(1–2):34–42.

Oldeman, R. A. A. 1983. Tropical rain forest, architecture, silvigenesis and diversity. In: *Tropical Rain Forest Ecology and Management*, eds. S. L. Sutton, T. C. Whitmore, and A. C. Chadwick. London, U.K.: British Ecological Society, pp. 139–150.

Oldeman, R. A. A. 1990. *Forests: Elements of Silvology*. Berlin, Germany: Springer Verlag.

Oliver, C. D. and B. C. Larson. 1990. *Forest Stand Dynamics*. New York: McGraw-Hill.

Opdam, P., R. Foppen, and C. Vos. 2001. Bridging the gap between ecology and spatial planning in landscape ecology. *Landscape Ecology* 16(8):767–779.

Otto, H. J. 1998. *Écologie Forestière*. París, France: Institut pour le Développement Forestier.

Oxbrough, A., V. French, S. Irwin, T. C. Kelly, P. Smiddy, and J. O'Halloran. 2012. Can mixed species stands enhance arthropod diversity in plantation forests? *Forest Ecology and Management* 270:11–18.

Palmer, J. F. 2004. Using spatial metrics to predict scenic perception in a changing landscape: Dennis, Massachusetts. *Landscape and Urban Planning* 69(2–3):201–218.

Parsons, R. 1995. Conflict between ecological sustainability and environmental aesthetics: Conundrum, canard or curiosity. *Landscape and Urban Planning* 32(3):227–244.

Pearson, D. M. 2002. The application of local measures of spatial autocorrelation for describing pattern in north Australian landscapes. *Journal of Environmental Management* 64(1):85–95.

Peron, E., A. T. Purcell, H. Staats, S. Falchero, and R. J. Lamb. 1998. Models of preference for outdoor scenes—Some experimental evidence. *Environment and Behavior* 30(3):282–305.

Perry, D. A. and M. P. Amaranthus. 1997. Disturbance, recovery and stability. In: *Creating a Forestry for the 21st Century: The Science of Ecosystem Management*, eds. K. A. Kohm and J. F. Franklin. Washington, DC: Island Press, pp. 31–56.

Pitkänen, A., P. Huttunen, K. Jugnes, and K. Tolonen. 2002. A 10000 year local fire history in a dry heath forest site in eastern Finland, reconstructed from charcoal layer records of a small mire. *Canadian Journal of Forest Research* 32(10):1875–1880.

Pollo, C. J., L. Robles, F. Ballesteros, and J. R. Obeso. 2005. El hábitat del urogallo en la cordillera cantábrica. In: *Manual de conservación y manejo del hábitat del urogallo cantábrico*, eds. B. F., L. Robles. Madrid, Spain: Organismo Autónomo de Parques Nacionales, pp. 25–34.

Pommerening, A. and S. T. Murphy. 2004. A review of the history, definitions and methods of continuous cover forestry with special attention to afforestation and restocking. *Forestry* 77(1):27–44.

Puettmann, K. J., K. D. Coates, and C. C. Messier. 2011. *A Critique of Silviculture: Managing for Complexity*. Washington, DC: Island Press.

Pukkala, T. and K. von Gadow (eds.) 2012. *Continuous Cover Forestry*. Dordrecht, the Netherlands: Springer.

Purcell, A. T., R. J. Lamb, E. M. Peron, and S. Falchero. 1994. Preference or Preferences for Landscape. *Journal of Environmental Psychology* 14(3):195–209.

Ramos, A. 1979. *Planificación Física y Ecológica. Modelos y Métodos*. Madrid, Spain: E.M.E.S.A.

Ramos, A. 1993. *Por qué la Conservación de la Naturaleza?* Discurso leído en el acto de su recepción por el Excmo. Sr. D. Ángel Ramos Fernández y contestación del Excmo. Sr. D. Salvador Rivas Martínez, el día 28 de abril de 1993. Madrid, Spain: Fundación Conde del Valle de Salazar.

Ramos, A. et al. 1986. *Curso monográfico sobre restauración del paisaje*. Madrid, Spain: Fundación Conde del Valle de Salazar. Escuela Técnica Superior de Ingenieros de Montes.

Ratnadass, A., P. Fernandes, J. Avelino, and R. Habib. 2012. Plant species diversity for sustainable management of crop pests and diseases in agroecosystems: A review. *Agronomy for Sustainable Development* 32(1):273–303.

Remm, J. and A. Lohmus. 2011. Tree cavities in forests—The broad distribution pattern of a keystone structure for biodiversity. *Forest Ecology and Management* 262(4):579–585.

Ribe, R. G. 1989. The aesthetics of forestry: What has empirical preference research taught us? *Environmental Management* 13:55–74.

Ribe, R. G. 2005. Aesthetic perceptions of green-tree retention harvests in vista views: The interaction of cut level, retention pattern and harvest shape. *Landscape and Urban Planning* 73(4):277–293.

Ribe, R. G. 2006. Perceptions of forestry alternatives in the US Pacific Northwest: Information effects and acceptability distribution analysis. *Journal of Environmental Psychology* 26(2):100–115.

Rondeux, J., R. Bertini, A. Bastrup-Birk, P. Corona, N. Latte, R. E. McRoberts et al. 2012. Assessing deadwood using harmonized National Forest Inventory data. *Forest Science* 58(3):269–283.

Roth, M. and D. Gruehn. 2010. Methods and data to describe agricultural landscapes and their cultural values on national level in Germany: Confusing coexistence or multilayered complexity? *Journal of Landscape Ecology* 9(Special Issue):53–63.

Ruiz Sánchez, M. A. and I. Cañas Guerrero. 2001. Método de valoración del impacto paisajístico. In: *Gestión sostenible de paisajes rurales: técnicas e ingeniería*, ed. F. Ayuga Téllez. Madrid, Spain: Mundi Prensa y Fundación Alonso Martín Escudero, pp. 53–80.

San Miguel, A. 1994. *La dehesa española. Origen, tipología, características y gestión*. Madrid, Spain: Fundación Conde del Valle de Salazar.

Santos, T., J. L. Telleria, and R. Carbonell. 2002. Bird conservation in fragmented Mediterranean forests of Spain: Effects of geographical location, habitat and landscape degradation. *Biological Conservation* 105(1):113–125.

Sayadi, S., M. C. Gonzalez Roa, and J. Calatrava Requena. 2005. Ranking versus scale rating in conjoint analysis: Evaluating landscapes in mountainous regions in southeastern Spain. *Ecological Economics* 55(4):539–550.

Scheller, R. M. and D. J. Mladenoff. 2007. An ecological classification of forest landscape simulation models: Tools and strategies for understanding broad-scale forested ecosystems. *Landscape Ecology* 22(4):491–505.

Schüpbach, B. 2002. Methods for indicators to assess landscape aesthetic. In: *Agricultural Impacts on Landscapes: Developing Indicators for Policy Analysis, NIJOS/OECD Expert Meeting, Agricultural Indicators*, eds. W. Dramstad and C. Sogge. Oslo, Norway, pp. 270–281.

Schutz, J. 1991. *Sylviculture 1*. Lausanne, Switzerland: Presses Polytechniques et Universitaires Romandes.

Schutz, J. 1997. *Sylviculture 2. La gestion des forêts irrégulieres et melangées*. Lausanne, Switzerland: Presses Polytechniques et Universitaires Romandes.

Schütz, J. P., T. Pukkala, P. J. Donoso, and K. von Gadow. 2012. Historical emergence and current application of CCF. In: *Continuous Cover Forestry*, eds. T. Pukkala and K. von Gadow. Dordrecht, the Netherlands: Springer, pp. 1–28.

Sheppard, S. R. J. 2001. Beyond visual resource management: Emerging theories of an ecological aesthetic and visible stewardship. In: *Forests and Landscapes—Linking Ecology, Sustainability and Aesthetics*, eds. S. R. J. Sheppard and H. W. Harshaw. Wallingford, U.K.: CABI Publishing, pp. 149–172.

Sheppard, S. R. J. and H. W. Harshaw. 2001. Landscape aesthetics and sustainability: An introduction. In: *Forests and Landscapes: Linking Ecology, Sustainability, and Aesthetics*, eds. S. R. J. Sheppard and H. W. Harshaw. Wallingford, Oxon; New York: IUFRO & CABI, pp. 3–12.

Sheppard, S. R. J., H. W. Harshaw, and J. R. McBride. 2001. Priorities for reconciling sustainability and aesthetics in forest landscape management. In: *Forest and Landscapes: Linking Ecology, Sustainability and Aesthetics*, eds. S. R. J. Sheppard and H. W. Harshaw. Cambridge, U.K.: CABI Publishing, pp. 263–288.

Shorohova, E., T. Kuuluvainen, A. Kangur, and K. Jogiste. 2009. Natural stand structures, disturbance regimes and successional dynamics in the Eurasian boreal forests: A review with special reference to Russian studies. *Annals of Forest Science* 66(2):201.

Silvennoinen, H., J. Alho, O. Kolehmainen, and T. Pukkala. 2001. Prediction models of landscape preferences at the forest stand level. *Landscape and Urban Planning* 56(1–2):11–20.

Silvennoinen, H., L. Tahvanainen, and T. Pukkala. 2002. Effect of cuttings on the scenic beauty of a tree stand. *Scandinavian Journal of Forest Research* 17(3):263–273.

Smith, D. M., B. C. Larson, M. J. Kelty, and P. M. S. Ashton. 1997. *The Practice of Silviculture. Applied Forest Ecology*. New York: John Wiley & Sons.

Smith, R. L. and T. M. Smith. 2001. *Ecology and Field Biology*, ed. A. Wesley. New York: Collins College Publishers.

Spies, T. 1997. Forest stand structure, composition and function. In: *Creating a Forestry for the 21st Century: The Science of Ecosystem Management*, eds. K. A. Kohm and J. F. Franklin. Washington, DC: Island Press, pp. 11–30.

Spies, T. A. and S. L. Duncan. 2009. *Old Growth in a New World: A Pacific Northwest Icon Reexamined*. Washington, DC: Island Press.

Spies, T. A. and J. F. Franklin. 1996. The diversity and maintenance of old-growth forests. In: *Biodiversity in Managed Landscapes: Theory and Practice*, eds. R. C. Szaro and D. W. Johnson. Oxford, New York, pp. 296–314.

Spies, T. A., K. N. Johnson, K. M. Burnett, J. L. Ohmann, B. C. McComb, G. H. Reeves et al. 2007a. Cumulative ecological and socioeconomic effects of forest policies in Coastal Oregon. *Ecological Applications* 17(1):5–17.

Spies, T. A., B. C. McComb, R. S. H. Kennedy, M. T. McGrath, K. Olsen, and R. J. Pabst. 2007b. Potential effects of forest policies on terrestrial biodiversity in a multi-ownership province. *Ecological Applications* 17(1):48–65.

Stankey, G. H., R. N. Clark, and B. T. Bormann. 2005. *Adaptive Management of Natural Resources: Theory, Concepts, and Management Institutions, General Technical Report. PNW-GTR-654.* Portland (OR, U.S.): Department of Agriculture, Forest Service, Pacific Northwest Research Station.

Stölb, W. 2005. *Forest Aesthetics—Over Forestry, Nature Protection and the People Soul.* Remagen, Germany: Publishing house boiler.

Suchant, R. and R. Baritz. 2000. A species-habitat-model for the improvement and monitoring of biodiversity in modern ecological—Capercaillie (*Tetrao urogallus*) in the black forest. In: *Criteria and Indicators for Sustainable Forest management at the Forest Management Unit Level.EFI Proceedings No. 38,* eds. A. Franc, O. Laroussinie, and T. Karjalainen. Joensuu, Finland: European Forest Institute, pp. 108–122.

Swanwick, C. 2002. *Landscape Character Assessment: Guidance for England and Scotland.* London, U.K.: The Countryside Agency and Scottish Natural Heritage.

Tahvanainen, L., L. Tyrvainen, M. Ihalainen, N. Vuorela, and O. Kolehmainen. 2001. Forest management and public perceptions—Visual versus verbal information. *Landscape and Urban Planning* 53(1–4):53–70.

Tapias, R. and l. Gil-Sánchez. 2005. Estrategias regenerativas del *Pinus nigra.* Comparación con los otros pinos españoles. In: *Los pinares de Pinus nigra Arn. en España: Ecología, uso y gestión,* eds. M. A. Grande and A. García-Abril. Madrid, Spain: Fundación Conde del valle de Salazar, pp. 127–150.

Tappeiner, J. C. et al. 1997. Silvicultural systems and regeneration methods: Current practices and new alternatives. In: *Creating a Forestry for the 21st Century: The Science of Ecosystem Management,* eds. K. A. Kohm and J. F. Franklin. Washington, DC: Island press, pp. 151–165.

Tarrant, M. A. and H. K. Cordell. 2002. Amenity values of public and private forests: Examining the value-attitude relationship. *Environmental Management* 30(5):692–703.

Taylor, P. D. 2002. Fragmentation and cultural landscapes: Tightening the relationship between human beings and the environment. *Landscape and Urban Planning* 58(2–4):93–99.

Tellería, J. l. and T. Santos. 2001. Fragmentación de hábitat forestales y sus consecuencias. In: *Ecosistemas mediterráneos. Análisis funcional.,* eds. R. Zamora and F. I. Pugnaire. Madrid, Spain: C.S.I.C., A.E.E.T.

Thomas, N. E., C. Huang, S. N. Goward, S. Powell, K. Schleeweis, and A. Hinds. 2012. Validation of North American Forest Disturbance dynamics derived from Landsat time series stacks. *Remote Sensing of Environment* 115(1):19–32.

Thompson, C. W., P. Aspinall, S. Bell, and C. Findlay. 2005. It gets you away from everyday life: Local woodlands and community use—What makes a difference? *Landscape Research* 30(1):109–146.

Thompson, J. R., K. N. Johnson, M. Lennette, T. A. Spies, and P. Bettinger. 2006. Historical disturbance regimes as a reference for forest policy in a multiowner province: A simulation experiment. *Canadian Journal of Forest Research-Revue Canadienne De Recherche Forestiere* 36(2):401–417.

Tindall, D. B. 2001. Why do you think that hillside is ugly? A sociological perspective on aesthetic values and public attitudes on forests. In: *Forest and Landscapes: Linking Ecology, Sustainability and Aesthetics,* eds. S. R. J. Sheppard and H. W. Harshaw. Cambridge, U.K.: CABI Publishing, pp. 57–70.

Tíscar, P. A. 2005a. Composición, estructura y función de un bosque natural viejo de *Pinus nigra ssp. Salzmannii.* Enseñanzas para la gestión forestal sostenible. In: *Los pinares de Pinus nigra Arn. en España: Ecología, uso y gestión,* eds. M. A. Grande and A. García-Abril. Madrid, Spain: Fundación Conde del valle de Salazar, pp. 613–632.

Tíscar, P. A. 2005b. Propuestas para la aplicación de una nueva selvicultura en el Parque Natural de las Sierras de cazorla, Segura y las Villas. In: *Los pinares de Pinus nigra Arn. en España: Ecología, uso y gestión*, eds. M. A. Grande and A. García-Abril. Madrid, Spain: Fundación Conde del valle de Salazar, pp. 585–612.

Tress, B., G. Tress, H. Decamps, and A. M. d'Hauteserre. 2001. Bridging human and natural sciences in landscape research. *Landscape and Urban Planning* 57(3–4):137–141.

Tress, B., G. Tress and G. Fry. 2005. Researchers' experiences, positive and negative, in integrative landscape projects. *Environmental Management* 36(6):792–807.

Trotsiuk, V., M. L. Hobi, and B. Commarmot. 2012. Age structure and disturbance dynamics of the relic virgin beech forest Uholka (Ukrainian Carpathians). *Forest Ecology and Management* 265:181–190.

Turner, M. G, R. H. Gardner, and R. V. O'Neill. 2003. *Landscape Ecology in Theory and Practice: Pattern and Process*. New York: Springer.

Tveit, M. S. 2009. Indicators of visual scale as predictors of landscape preference; a comparison between groups. *Journal of Environmental Management* 90(9):2882–2888.

Tveit, M., A. Ode, and G. Fry. 2006. Key concepts in a framework for analysing visual landscape character. *Landscape Research* 31(3):229–255.

Tyrvainen, L., H. Silvennoinen, and O. Kolehmainen. 2003. Ecological and aesthetic values in urban forest management. *Urban Forestry and Urban Greening* 1(3):135–149.

Ulrich, R. S. 1986. Human responses to vegetation and landscapes. *Landscape and Urban Planning* 13:29–44.

USDA Forest Service. 1974. *National Forest Landscape Management*. Vol. 2, Agriculture Handbook number 462, Washington, DC: USDA Forest Service.

USDA Forest Service. 1995. *Landscape Aesthetics. A Handbook for Scenery Management*. Vol. 2, Agriculture Handbook number 701, Washington, DC: USDA Forest Service.

Vallauri, D., J. André, and J. Blondel. 2002. *Le bois mort, un attribut vital de la biodiversité de la forêt naturelle, une lacune des forêts gérées, Bois morts et à cavités - Une clé pour des forêts vivantes*. Chambéry, France: WWF France.

Van Mansvelt, J. D. and J. Kuiper. 1999. Criteria for the humanity realm: Psychology and physiognomy and cultural heritage. In: *Checklist for Sustainable Landscape Management*, eds. D. van Mansvelt and M. J. van der Lubbe. Amsterdam, the Netherlands: Elsevier Science, pp. 116–134.

Velarde, M. D., G. Fry, and E. Framstad. 2005. Collecting evidence on the effects of landscape change on biodiversity. Paper read at Planning, people and practice: the landscape ecology of sustainable landscapes. *Proceedingns of the 13th Annual IALE (UK) Conference*, University of Northampton, Northampton, England.

Velarde, M. D., M. V. Núñez Martí, A. García-Abril, and M. A. Ruíz Sánchez. 2013. *Integración paisajística de las repoblaciones forestales., Serie Técnica de Medio Ambiente*. Madrid, Spain: Dirección General de Medio Ambiente. Consejería de Medio Ambiente, Vivienda y ordenación del Territorio de la Comunidad de Madrid.

Velarde, M. D. and M. A. Ruíz. 2007. El Paisaje en la Gestión de los Espacios Forestales: Donde Ecología y Estética Coinciden. *Revista Montes* 89:26–31.

Vila, J., D. Varga Linde, A. Llausàs Pascual, and A. Ribas Palom. 2006. Conceptos y métodos fundamentales en ecología del paisaje (landscape ecology). Una interpretación desde la geografía. *Documents d'anàlisi geogràfica* 48:151–156.

von Oheimb, G., C. Westphal, H. Tempel, and W. Hardtle. 2005. Structural pattern of a near-natural beech forest (Fagus sylvatica) (Serrahn, North-east Germany). *Forest Ecology and Management* 212(1–3):253–263.

von Salisch, H. 1885. *Forstästhetik*. Berlin, Germany: Springer.

Wang, X. L., B. Song, J. Q. Chen, D. L. Zheng, and T. R. Crow. 2006. Visualizing forest landscapes using public data sources. *Landscape and Urban Planning* 75(1–2):111–124.

Weinstoerffer, J. and P. Girardin. 2000. Assessment of the contribution of land use pattern and intensity to landscape quality: Use of a landscape indicator. *Ecological Modelling* 130(1–3):95–109.

Weitkamp, G., A. E. Van den Berg, A. K. Bregt, and R. J. A. Van Lammeren. 2012. Evaluation by policy makers of a procedure to describe perceived landscape openness. *Journal of Environmental Management* 95(1):17–28.

Williams, H. J. H. and J. Cary. 2002. Landscape preferences, ecological quality, and biodiversity protection. *Environment and Behavior* 34:257–274.

Wirth, C., G. Gleixner, and M. Heimann. 2009. *Old-Growth Forests: Function, Fate and Value*. Vol. 207, Ecological Studies. Berlin, Germany: Springer.

Wissen, U., O. Schroth, E. Lange, and W. A. Schmid. 2008. Approaches to integrating indicators into 3D landscape visualisations and their benefits for participative planning situations. *Journal of Environmental Management* 89(3):184–196.

Woodall, C. W. and J. V. Monleón. 2007. *Sampling Protocol, Estimation and Analysis Procedures for the Down Woody Materials Indicator of the FIA Program. United States*. Newton Square, PA: Department of Agriculture Forest Service. Northern Research Station. General Technical Report NRS-22.

Woodbury, P. B., R. M. Beloin, D. P. Swaney, B. E. Gollands, and D. A. Weinstein. 2002. Using the ECLPSS software environment to build a spatially explicit component-based model of ozone effects on forest ecosystems. *Ecological Modelling* 150:211–238.

World Commission on Environment and Development. 1987. *Our Common Future*. New York: Oxford University Press.

Wu, J. G. 2006. Landscape ecology, cross-disciplinarity, and sustainability science. *Landscape Ecology* 21(1):1–4.

Wu, J. 2008. Landscape ecology. In: *Encyclopedia of Ecology*, ed. S. E. Jorgensen. Oxford, U.K.: Elsevier.

Yorke, M. 1998. *Continuous Cover Silviculture: An Alternative to Clear Felling*. Tywyn, North Wales: Tyddyn Bach.

Zhou, G. and A. M. Liebhold. 1995. Forecasting the spatial dynamics of gypsy moth outbreaks using cellular transition models. *Landscape Ecology* 10(3):177–189.

7 Assessment of Sustainability Based on Individual Preferences

Eugenio Martínez-Falero,
Susana Martín-Fernández, and Antonio Orol

CONTENTS

7.1 INTRODUCTION

The assessment is a result of comparison: identical values are allocated to similar alternatives and greater values to those that are more favorably considered in the comparison process.

However, comparison is only evident when systems are entirely described through a single variable. In this case, there is a globally accepted logic of comparison that is defined by the relation "greater than or equal to" over the single measure of the variable representing the whole value of the system.

Difficulties arise when working with complex systems. These systems are composed of subsystems interacting with each other, thereby giving additional value to the mere integration of the parties. Furthermore, complex systems are generally described by measuring some of their multiple consequences, and these consequences are not usually susceptible to an overall measure.* There is, therefore, no single universal logic for comparison, and each observer must define his or her own patterns for comparison, which derive from his or her individual preferences (i.e., from his or her opinions).

7.1.1 CHAPTER CONTENT

As mentioned previously, the comparison of complex alternatives is not as intuitive as it might appear. In general, we can identify two means of comparison (Vincke, 2001). On the one hand, the evaluator directly compares pairs of alternatives to establish (in his or her view) a preference relation between them; this is obtained only from the global knowledge the evaluator has about alternatives (*pair-wise comparison*). In the other means of comparison, the preference is induced from the knowledge of both the measures of several consequences of the alternatives being compared and the meaning that the evaluator attributes to the interrelationship between these consequences (*aggregation of criteria*). In either case, a numerical representation of the evaluator's preferences is inferred from the relations of preference between alternatives. Sections 7.2 and 7.3 describe the procedures for evaluating alternatives that are based on pair-wise comparison and aggregation of criteria, respectively. In Section 7.4, we propose an alternative valuation method that uses the combined information from both systems of comparison.

Outcomes from Sections 7.2 through 7.4 transform opinions into an assessment, but they do not provide the evaluator with further information other than what he or she has prior to the evaluation. However, the opinion is regulated by laws that change depending on the degree of knowledge of the system. As a result, actually giving the evaluator information is a major step forward in the assessment. In the process of *providing information*, two issues in particular must be considered: the type of rationality or coherence in the opinions of each individual and the depth of knowledge that the individual has about the system to be evaluated. Section 7.5 deals with a characterization of these topics.

---

* A clear example occurs when analyzing sustainability: it is possible to measure partial aspects (indicators), but there is no overall measure of sustainability.

Finally, in Section 7.6, the methodology described in Sections 7.2 through 7.5 is applied to the assessment of forest sustainability.

7.1.2 State of the Art of the Methodologies to Describe Opinions

The basis for describing opinions is the homogeneous representation of individual preferences. The representation must be homogeneous in order to allow the preferences of each individual to be compared with those of other evaluators.

A first set of procedures for describing opinions derives from a simplified representation of preferences through modeling. *Preference modeling* (see Öztürk et al., 2005, for a detailed description of the state of the art of this methodology) includes the identification of the type of preference structures reflecting the behavior of a decision maker, obtaining a numerical representation of preferences, and recognizing the logic of preference.

Other procedures to describe opinions include *computing of utility* (first formalized by Arrow and Debreu, 1954) and *analysis of past decisions* (developed from the works of Arrow, 1959). Utility continues to be the foundation of the classic theory of demand: under certain conditions, each individual has a value function related to his or her preference that governs his or her rational choices. However, people do not always behave as "*maximizers of utility*," as decision making usually includes other factors such as the capacity to notice the difference between options, the relationship between the alternatives being compared, and even the influence of the general context of the decision making. When this happens, the analysis of past decisions (through the characterization of choice functions) provides a guideline to describe the opinion of any individual.

It is worth noting that the outcomes arising from the implementation of the three preceding procedures are interchangeable (see Aleskerov et al., 2007 for a detailed analysis of the current state of the art on this subject). This fact makes it possible to exchange the concepts of preference, utility, and choice (see, among other authors, Sen (1987) and Suzumura (1983) for the integration of the available procedures). It is also possible to apply the computing capabilities of any of these methodologies to get outcomes from the others.

7.2 ASSESSMENT FROM PAIR-WISE COMPARISON OF ALTERNATIVES

This section describes a model to represent the preferences of an individual from direct comparison between pairs of alternatives (see Figure 7.1 for a description of the main steps in a standard pair-comparison process).

As with any modeling, the representation of preferences is affected by uncertainty. This arises because the individual who compares alternatives is unable to make a clear formulation of his or her preferences. So when an evaluator is asked whether he or she prefers one alternative to another, he or she not only answers "yes," "no," or "don't know" but also yes, "no," "I'm not sure," and numerous other options. Therefore, the modeling of preferences must incorporate the capacity to consider as many scenarios as possible in order to achieve an adequate representation of preferences.

Pair-wise comparison:

(a) Selection of a set of meaningful alternatives:

$\Omega =$

| Alternative 1 | Alternative 2 | Alternative 3 | Alternative 4 | ... |

(b) Figure out all feasible pairs of alternatives in Ω.

(c) For every pair in (b), ask for the decision maker preference between the alternatives within the pair: which of the following alternatives do you consider to be more preferred?

(d) Register the answers of the decision-maker.

1: Row alternative preferred 2: Column alternative preferred 3: Both alternatives are indifferent 4: No preference stated by the decision maker	...	Alternative $i{-}1$	Alternative i	Alternative $i{+}1$	...
...	...	...	...	...	...
Alternative $i{-}1$	...	3	3	1	...
Alternative i	...	3	3	4	...
Alternative $i{+}1$	...	2	4	3	...
...	...	...	...	...	...

| Alternative 4 | Alternative 1 |

Alternative 4

Alternative 1

They are equally preferred

I do not know

FIGURE 7.1 **(See color insert.)** Steps in a pair-wise comparison process.

7.2.1 MODELING OF PREFERENCES

7.2.1.1 By Applying Classical Logic

The outcome of a pair-wise comparison process (such as described in Figure 7.1) is the introduction of a binary relation (P) in the set of alternatives (Ω) with $P \subseteq \Omega \times \Omega$. Henceforth, we shall use $(x, y) \in P$, xPy, or $P(x, y)$ interchangeably to represent any ordered pair of alternatives $[(x, y) \in \Omega]$ satisfying the P relation, where xPy means that alternative x is strictly preferred to alternative y, for the individual who has made the comparison.*

The characterization of binary relations as sets allows operations to be defined on binary relations through operations of sets. Some operations on binary relations are presented in Table 7.1.

Preference relations (P) are usually represented by a square matrix (M^P), with a number of rows and columns equal to the number of alternatives in Ω. The M^P_{xy} element of the matrix (intersection of the row associated with alternative x and the column associated with alternative y) is 1 if xPy, and 0 otherwise. Hereafter, we shall refer to this matrix by the letter that identifies the relationship (P), without

---

* Although the formalization of binary relations dates from the late nineteenth century with the works of De Morgan and Peirce, the first studies of preference relations appear well into the twentieth century, with Dushnik and Miller (1941) and Scott and Suppes (1958).

TABLE 7.1

Some Basic Operations for Two Binary Relations (*P* and *Q*) Defined on the Same Set of Alternatives (Ω)

Operations	Notation	Description
The union	$Q \cup P$	$\{(x, y)/xQy \text{ or } xPy, \forall x, y \in \Omega\}$
The intersection	$Q \cap P$	$\{(x, y)/xQy \text{ and } xPy, \forall x, y \in \Omega\}$
The relative product	$Q \cdot P$	$\{(x, y)/\exists z \in \Omega: xQz \text{ and } zPy, \forall x, y \in \Omega\}$

having to write M^P, so the notation of relations of preference is unified (whether they are represented by a matrix or by any other system). The matrix representation also simplifies the calculation of operations between preference relations. In fact, the matrices describing the preference relations resulting from the operations in Table 7.1 can also be obtained by the matrix operations of union, intersection, and product applied to the matrices describing the original preference relations.

The compliance of preference relations (P) with specific mathematical properties (such as those defined in Table 7.2) can be used to determine particular types of systems of preferences. Table 7.3 (based on the classification proposed by Aleskerov et al., 2007) shows several systems of preference.

Relations of preference between alternatives are not only ruled by the P relation. The indifference relationship (I) can also be used to represent individual preferences. Here, xIy means that alternatives x and y are indifferent for the individual whose preferences are being analyzed.

As happens with any other type of binary relations, I relation can be represented by a square matrix (M^I) with a number of rows and columns equal to the number of alternatives in Ω. In this case, the M^I_{xy} element (intersection of the row associated with the alternative x and the column associated with the alternative y) is 1 if xIy, and 0 otherwise. As in the case of preference relations, the square matrix M^I will, henceforth, be referred to as I.

TABLE 7.2

Some Properties of Preference Relations (Only Those Necessary to Define the Systems of Preferences in Table 7.3)

Properties	Description
Irreflexive	$no(xPx), \forall x \in \Omega$
Negatively transitive	$xPy \Rightarrow xPz \text{ or } zPy, \forall x, y, z \in \Omega$
Connected	$xPy \text{ or } yPx, \forall x, y \in \Omega \ (x \neq y)$
Semi-transitive	1: xPy and $zPt \Rightarrow xPt$ or $tPy, \forall x, y, z, t \in \Omega$ and also 2: $xPyPz \Rightarrow xPt$ or $tPy, \forall x, y, z, t \in \Omega$
Strong intervality	xPy and $zPt \Rightarrow xPt$ or $tPy, \forall x, y, z, t \in \Omega$
Transitive	xPy and $yPz \Rightarrow xPz, \forall x, y, z \in \Omega$

TABLE 7.3

Some Types of Systems of Preferences

Systems of Preferences	Properties on *P*
Linear order	Irreflexive, connected, and transitive
Weak order	Irreflexive, negatively transitive, and transitive
Semiorder	Irreflexive, semi-transitive, and strong intervality
Interval order	Irreflexive and strong intervality
Biorder	Strong intervality
Partial order	Irreflexive and transitive

Source: After Aleskerov, F. et al., *Utility Maximization, Choice and Preference,* Springer, Berlín, Germany, 2007.

P and *I* are associated, so *I* is equal to the symmetric part of the complement of *P* ($I = s[c(P)]$):

$$s(Q)(\text{symmetric part of } Q) = \left\{ \frac{(x,y)}{xQy \text{ and } yQx} \right\} \tag{7.1}$$

$$c(Q)(\text{complement of } Q) = \left\{ \frac{(x,y)}{\text{no}(xQy)} \right\} \tag{7.2}$$

This means that if *P* is a linear order, then *I* is reflexive, symmetric, and transitive and is known as a "relation of equivalence" (henceforth denoted by *E*). In the case that *P* is a linear order (or, equivalently, that *I* is a relation of equivalence), then the uncertainty incorporated is the smallest that can be added through the process of modeling preferences. Therefore, it would appear desirable for the system of preferences to be a linear order. Fortunately, it is possible to associate an equivalence relation (*E*) to any preference relation (*P*): this is done by incorporating any alternative ($y \in \Omega$) to the indifference class containing another alternative (x, $\forall x \in \Omega$) when

$$xEy \equiv \forall z \in \Omega, \text{ it happens that } xPz \Leftrightarrow yPz \text{ and also } zPx \Leftrightarrow zPy \tag{7.3}$$

However, building relations by applying the preceding procedure may lead to the construction of relations of equivalence with such a reduced number of indifference classes that they do not provide information about the preferences of the evaluator. It is, therefore, preferable to apply transformations that tend to transform *P* in a linear order relation, although this does not always ensure that this system of preferences is attained (it is only assured when *P* is a semiorder). We recommend the set of transformations proposed by Fishburn (1985): the transitive closure of *I* (I^k) and the "sequel" [$S_i(P)$] changes in *P*, where

$$S_0(P) = [I \cdot P \cup P \cdot I] \cap cd[I \cdot P \cup P \cdot I]; \ldots; S_i(P) = S_0\left[S_{i-1}(P)\right], \quad \text{for} \quad i = 1,2,\ldots$$

$$(7.4)$$

where $Q \cdot P$ is as defined in Table 7.1, $cd(Q) = c[d(Q)]$, $c(Q)$ is as defined in expression (7.2), and $d(Q) = \{(x, y)/yQx\}$.

The aforementioned transformations do not incorporate any additional knowledge to that expressed by the individual in the pair comparison. Hence, the matrix resulting from the aforementioned transformations in P continues to reflect the preferences of the evaluator.

Within the field of classical logic, a final useful relationship is the characteristic relation (R) or relation "*at least as preferable as.*" It is defined as

$$R = P \cup I \qquad (7.5)$$

and relations of P and I with R are as follows:

$$xPy \Leftrightarrow xRy \text{ and } y\left[c(R)\right]x\left[c(R)\right] \text{ defined as in expression } (7.2) \qquad (7.6)$$

$$xIy \Leftrightarrow xRy \text{ and } yRx \qquad (7.7)$$

7.2.1.2 Use of Fuzzy Logic

The aforementioned binary relations have been defined on sets (and applied to set operations) in which the membership of a set is completely specified. However, the membership relation is either vaguely perceived or imprecisely known. Poor perception usually occurs in semantic logic relations (closely connected to opinions), and imprecise knowledge is caused by a deficiency in the consistency of personal opinions. Fuzzy sets (Zadeh, 1965, 1975) have emerged to incorporate and to manage this uncertainty:

Definition of Fuzzy Subset of a Set A. A fuzzy subset (F) of a set A is the result of the application $\mu_F: A \rightarrow [0,1]$, where $\forall x \in A$, $\mu_F(x)$ is the membership degree of x to F.

The intensity of preference for one alternative x over another y, the number of individuals who prefer x to y, and other magnitudes related to modeling preferences can be described through a membership function. Hereafter, and for the sake of simplicity, we shall use $R(x, y)$ instead of $\mu\{R(x, y)\}$ to denote the membership function when there is no doubt that it is used in a fuzzy environment. In this context, the "at least as preferable as" relation [$R(x, y)$] takes not only the values 1 (if x is at least as preferred as y) and 0 (otherwise) but also a range of values (between 0 and 1) representing the degree of belief that the alternative x is at least as preferred as the alternative y.

However, before applying the outcomes of fuzzy logic in modeling preferences, it is necessary to clarify some concepts related to operations in fuzzy sets (Dubois and Prade, 1980a,b; Fodor et al., 1998). First, there is no single definition for the union and intersection of fuzzy sets. Instead there are two basic classes of operators: triangular

TABLE 7.4

Examples of t-Norms and t-Conorms ($\forall a,\ b \in [0,1]$)

Name	*t-Norm* Description	*t-Conorm* Description
Zadeh	$\min(a,\ b)$	$\max(a,\ b)$
Lukasiewicz	$\max(a+b-1,\ 0)$	$\min(a+b,\ 1)$
Yager ($p>0$)	$1-\min\{1,\ [(1-a)^p+(1-b)^p]^{1/p}\}$	$\min\{1,\ [a^p+b^p]^{1/p}\}$
Hamacher ($\gamma \geq 0$)	$ab/[\gamma+(1-\gamma)(a+b-ab)]$	$[a+b-ab-(1-\gamma)\,ab]/[1-(1-\gamma)\,ab]$

norms or *t-norms* for the intersection* and triangular conorms or *t-conorms* for the union.† Because of their axiomatic definitions, there are many operations that satisfy these definitions (some of them are presented in Table 7.4).

In the second place, in order to carry out suitable operations, any pair of a t-norm and a t-conorm, together with a strict negation function (which acts as the complement in the fuzzy sets), must satisfy the De Morgan law:

Definition of De Morgan Triplets. Suppose that T is a *t-norm*, S is a *t-conorm*, and n is a strict negation. $\langle T, S, n \rangle$ is a De Morgan triplet if and only if

$$n\Big[S(a,b)\Big] = T\Big[n(a),n(b)\Big], \forall a,\ b \in [0,1]$$

where $n:[0,1] \rightarrow [0,1]$ is a non-decreasing function with $n(0)=1$ and $n(1)=0$.

De Morgan triplets are used to characterize fuzzy preference systems (Fodor and Roubens, 1994; Öztürk et al., 2005; Perny and Roubens, 1998). Thus, Table 7.5 defines [$\forall x, y, z, v \in \Omega$] some properties for fuzzy characteristic relations that are required to define fuzzy systems of preferences (Table 7.6). As can be seen, depending on the t-norm and t-conorm adopted (De Morgan triplet), the same relation can either satisfy a property or not satisfy it.

As in classical logic, it is also possible to work with fuzzy relations of strict preference and indifference. The following expressions show the relationships between preference relations:

---

* T-norm. Let us suppose $\mu_{F \cap G}(x) = T[\mu_F(x), \mu_G(x)]$, then $T:[0,1] \times [0,1] \rightarrow [0,1]$ is a t-norm), when it satisfies the following conditions, $\forall a,b,c \in [0,1]$:
- t1. Concordance in the neat case: $T(0,1)=T(0,0)=T(1,0)=0;\ T(1,1)=1$
- t2. Commutative: $T(a,b)=T(b,a)$
- t3. Associative: $T[a,T(b,c)]=T[T(a,b),c]$
- t4. Identity: $T(a,1)=a$
- t5. Monotony: if $a \leq a'\ b \leq b'$, then $T(a,b) \leq T(a',\ b')$

† T-conorm. Let us suppose $\mu_{F \cap G}(x) = S[\mu_F(x), \mu_G(x)]$, then $S:[0,1] \times [0,1] \rightarrow [0,1]$ is a t-conorm), when it satisfies the following conditions, $\forall a,b,c \in [0,1]$:
- s1. Concordance in the neat case: $S(0,1)=S(0,0)=S(1,0)=1;\ S(0,0)=0$
- s2. Commutative: $S(a,b)=S(b,a)$
- s3. Associative: $S[a,S(b,c)]=S[S(a,b),c]$
- s4. Identity: $S(a,0)=a$
- s5. Monotony: if $a \leq a'\ b \leq b'$, then $S(a,b) \leq S(a',\ b')$

TABLE 7.5

Some Properties for Characteristic Relations in Valued Models (Only those that are Necessary to Define the Preference Systems in Table 7.6)

Properties	Description
Reflexive	$R(x, x) = 1$
T-antisymmetric	$x \neq y \Rightarrow T[R(x, y), R(y, x)] = 0$
S-strongly complete	$S[R(x, y), R(y, x)] = 1$
T-transitive	$T[R(x, z), R(z, y)] \leq R(x, y)$
T-S Ferrers relation	$T[R(x, y), R(v, z)] \leq S[R(x, z), R(v, y)]$

TABLE 7.6

Fuzzy Systems of Preference

Systems of Preference	Properties on R
Fuzzy total order	Antisymmetric, strongly complete, and transitive
Fuzzy weak order	Strongly complete and transitive
Fuzzy semiorder	Strongly complete, Ferrers relation, and transitive
Fuzzy interval order	Strongly complete and Ferrers relation
Fuzzy partial order	Antisymmetric, reflexive, and transitive
Fuzzy partial preorder	Reflexive and transitive

Source: Oztürk, M. et al., Preference modelling, In *Multiple Criteria Decision Analysis: State of the Art Surveys*, eds. J. Figueira, S. Greco, and M. Ehrgott, Springer Verlag, Boston, MA, pp. 27–72.

$$P(x, y) = T\left[R(x, y), n\{R(y, x)\}\right] \tag{7.8}$$

$$I(x, y) = T\left[R(x, y), R(y, x)\right] \tag{7.9}$$

$$R(x, y) = S\left[P(x, y), I(x, y)\right] \tag{7.10}$$

However, for the same De Morgan triplet, it is impossible for the three relations (strict preference, indifference, and characteristic) to satisfy expressions (7.8) through (7.10) simultaneously (Fodor and Roubens, 1995a,b).

7.2.1.3 Preference Structures

Definition. A preference structure is a collection of binary relations defined on Ω such that $\forall \, x, y \in \Omega$, one relation of the collection, and only one, is satisfied.

The mathematical properties of all the binary relations in the structure provide a complete description of the preferences. Thus, we can distinguish the following:

7.2.1.3.1 Traditional Models

These are entirely explained through $\langle P, I \rangle$ relations. In these models, the characteristic relation (R) and the preference (P) and indifference (I) relations are uniquely related through expressions (7.5) through (7.7) and include linear orders, weak orders, semiorders, and interval orders (see Table 7.3).

7.2.1.3.2 Extended Models

These models incorporate preferences that do not use fuzzy relations and are not entirely explained through P, I. There are two main ways of extending $\langle P, I \rangle$ structures to more general cases of preferences; these are

1. Inclusion of preference relations representing (one or more) intermediate situations (usually of doubt) between strict preference and indifference
   a. Technically, the new preference relations are represented by one or more additional binary relations (Q). In these structures, it is not possible to obtain a single definition of the R relation from P, Q, and I. Such structures give rise to $\langle P, Q, I \rangle$ interval orders and semiorders (see Tosoukiàs and Vincke, 2003) and to double threshold orders (see Tosoukiàs and Vincke, 1997) or to $\langle P, Q_1, \ldots, Q_n, I \rangle$ multiple threshold orders (Cozzens and Roberts, 1982).
2. Inclusion of one or more situations of incomparability
   b. There are situations where comparison of alternatives is impossible. In these cases, we use another relation (symmetric and irreflexive), which is usually represented by J [$xJy \equiv$ no(xPy), no(yPx), no(xIy), no(xQy), and no(yQx)]. To work with structures of the type $\langle P, Q, J, I \rangle$, it is necessary to define new systems of preference (Roubens and Vincke, 1985). These systems are named by adding the word "partial" to the systems of preferences they come from. We, thus, obtain partial orders, partial preorders, partial interval orders, and pseudo-partial orders.

7.2.1.3.3 Valued Models

These are preference structures that make use of fuzzy relations in their binary relations of preference.

7.2.2 Numeric Representation of Preferences

A value function (v) is an application of the set of alternatives (Ω) into $\mathbb{R}^+$ (so that $\forall x \in \Omega$, then $v(x) \in \mathbb{R}^+$). It is specific to each individual (evaluator) and observes the preferences that the evaluator has formulated in the pair-wise comparison process, because

$$v(x) \geq v(y) \Leftrightarrow xPy \text{ or } xIy, \forall x, y \in \Omega \tag{7.11}$$

In general, utility maximization remains the foundation of the classical theory of demand. Under certain conditions, each individual has a value function (v) related to his or her preferences and governing his or her rational choices. So, an individual chooses an alternative x before y if and only if he or she prefers x to y (xPy), which supposes $v(x) > v(y)$. However, contrary to expectation, it has been experimentally proved that people do not always choose the alternative with the highest utility or value (Kahneman and Tversky, 2000).

To model the behavior of individual choices, it is necessary to incorporate a new concept, which we refer to as the threshold (ε). This concept relates choices and utility: thus, a person will choose an alternative (x) over another (y) when $v(x) \geq v(y) + \varepsilon$, that is, when the value assigned to the preference of one alternative over another exceeds the "*noticeable difference*" (Rand, 1912) between alternatives.

Threshold varies from one individual to another because it is related to the individual's ability to discriminate between two alternatives. So, if $\varepsilon \to 0$, one can assume the existence of a high capacity of discrimination (in this case, the evaluator is a "*maximizer of utility*"). But this is not the norm. It is even unusual for the threshold to be a constant ($\varepsilon > 0$), since the capacity of discrimination usually depends on the alternative that is being valued [$\varepsilon = \varepsilon(x)$], the alternative with which it is compared [$\varepsilon = \varepsilon(x, y)$], and even on the context in which the comparison is made: $\varepsilon = \varepsilon(x, y, \Omega)$.

In addition to its relationship with utility, modeling the behavior of choices is also related to modeling of preferences (Aleskerov et al., 2007). All procedures to describe opinions (modeling of preferences, optimization of utility, and analysis of past choices) are, therefore, integrated through the threshold. Specifically, this makes it possible to exchange the outputs from the aforementioned procedures to transform opinions into values. Thus, the numerical representation of individual preferences can be characterized by the threshold, so when dealing with $\langle P, I \rangle$ structures, Table 7.7 shows the conditions of the *numerical representation of preference structures* in *traditional models*.

There are also specific representation theorems and construction methods for the *numerical representation of preference structures* in *extended models*. Both (theorems and construction methods) are described in Ngo The and Tsoukiàs (2005) for the case of interval orders and in Vincke (1998) for double threshold orders. In the case of incomparability, $\langle P, Q, J, I \rangle$ preference structures have similar representation theorems to those of the "non-partial" preference systems, and the functional representation has the same expressions. However, the equivalents are reduced to implications. Thus, in the case of partial orders, it means that

$$xPy \Rightarrow v(x) > v(y) \left[\text{it does not occur that } xPy \Leftarrow v(x) > v(y) \right] \qquad (7.12)$$

To conclude this section, we describe the *numerical representation of preference structures* in *valued models*. In this case, fuzzy sets are used to define the value of an alternative related to a criterion.* In the ordered pair, $\{a, \mu_j^x\}$, a ($a \in \mathbb{R}$) represents the

---

* A criterion is every aspect of reality that determines the advantages or drawbacks of considering any alternative as the solution to a complex problem.

TABLE 7.7

Numerical Representation of Traditional Preference Structures

Structures and Threshold	Representation Theorems	Algorithms for Numerical Representation of Preferences
PI—linear or weak order $[\varepsilon = 0]$	$\exists v: \Omega \rightarrow \mathbb{R}^+, \forall x, y \in \Omega$: i. $xPy \Leftrightarrow v(x) > v(y)$ ii. $xIy \Leftrightarrow v(x) = v(y)$ iii. $xRy \Leftrightarrow v(x) \geq v(y)$	Allocate to each alternative the following quantity: the number of alternatives that are at least as preferred as the one analyzed in the pair-wise comparison.
PI—semiorder $[\varepsilon(cte.) > 0]$	$\exists v: \Omega \rightarrow \mathbb{R}^+$ and $\varepsilon > 0$ that, $\forall x, y \in \Omega$: i. $xPy \Leftrightarrow v(x) > v(y) + \varepsilon$ ii. $xIy \Leftrightarrow |v(x) - v(y)| \leq \varepsilon$ iii. $xRy \Leftrightarrow v(x) \geq v(y) - \varepsilon$	Operating procedures: the values are derived from computing the number of arcs in all the circuits in (Ω, R) (see Pirlot, 1990).
PI—interval order $[\varepsilon = \varepsilon(x) \geq 0]$	$\exists v, \varepsilon: \Omega \rightarrow \mathbb{R}^+$ that, $\forall x, y \in \Omega$: i. $xPy \Leftrightarrow v(x) > v(y) + \varepsilon(y)$ ii. $xIy \Leftrightarrow v(x) \leq v(y) + \varepsilon(y)$ and $v(y) \leq v(x) + \varepsilon(x)$	In the case of Ω finite, it is enough to obtain $v: \Omega \rightarrow A$ and $u: A \rightarrow A$, with $A = \{[0, 2 \times \text{Card}(\Omega) - 1] \cap \mathbb{N};\}$, such that $xRy \Leftrightarrow v(x) + u[v(x)] \geq v(y)$ (Fishburn, 1985).

feasible values that the x alternative can take in j and $\mu_j^x(a)$ the degree of membership of a for alternative x in j criterion.

The credibility of the preference of x over y is obtained from comparison of the fuzzy membership functions with some conditions: the fuzzy set is assumed to be normal $(\sup_a[\mu_j^x(a)] = 1)$ and convex $(\forall a, b, c \in \mathbb{R}, b \in [a, c], \mu_j^x(b) \leq \min[\mu_j^x(a), \mu_j^x(c)])$. In this case, Fodor and Roubens (1994) propose using the following expression for the degree of credibility that x is "at least as preferred as" y on a certain criterion (j):

$$\Pi_j(x, y) = \sup_{a \geq b} \left[\min\left\{ \mu_j^x(a), \mu_j^y(b) \right\} \right] \tag{7.13}$$

which also determines the degree of credibility for the strict preference between alternatives within a specific criterion (j). So

$$P_j(x, y) = 1 - \Pi_j(y, x) \tag{7.14}$$

The way to compute the degree of credibility for the overall preference (when all criteria are considered as significant $(j \in \mathcal{J})$) depends on the type of dependence between criteria (see Section 7.3.2). So, if the type of relation among criteria admits an additive utility function, then

$$\pi(x,y)=\sum_{j\in J} P_j(x,y)\,\lambda_j \tag{7.15}$$

where

$\pi(x, y)$ is the overall degree of credibility for the preference of x over y

λ_j is the relative importance of the j criterion

From the degree of credibility for the overall preference and by comparing each alternative with the remaining alternatives in Ω, it is possible to construct a preference structure on which to define a numerical representation of preferences. Thus, by comparing one alternative with the other, we obtain

$$\varphi^+(x)=\left[\frac{1}{(n-1)}\right]\sum_{a\in\Omega\backslash x}\pi(x,a)\ \text{and also}:\varphi^-(x)=\left[\frac{1}{(n-1)}\right]\sum_{a\in\Omega\backslash x}\pi(a,x) \tag{7.16}$$

Next, similar to the procedure in preference ranking organization method for enrichment of evaluations (*PROMETHEE*)-*I* (Brans, 1982; Brans and Mareschal, 2005), expression (7.16) can be transformed into a preference structure, of the type $\langle P, I, J\rangle$,* as shown in the following:

$$xPy\,(Preference)\quad\Leftrightarrow\quad\begin{cases}\varphi^+(x)>\varphi^+(y)\ \text{and}\ \varphi^-(x)<\varphi^-(y),\ \text{or}\\[4pt]\varphi^+(x)=\varphi^+(y)\ \text{and}\ \varphi^-(x)<\varphi^-(y),\ \text{or}\\[4pt]\varphi^+(x)>\varphi^+(y)\ \text{and}\ \varphi^-(x)=\varphi^-(y)\end{cases}$$

$$xIy\,(Indifference)\quad\Leftrightarrow\quad\varphi^+(x)=\varphi^+(y)\ \text{and}\ \varphi^-(x)=\varphi^-(y) \tag{7.17}$$

$$xJy\,(Incomparability)\quad\Leftrightarrow\quad\begin{cases}\varphi^+(x)>\varphi^+(y)\ \text{and}\ \varphi^-(x)>\varphi^-(y),\ \text{or}\\[4pt]\varphi^+(x)<\varphi^+(y)\ \text{and}\ \varphi^-(x)<\varphi^-(y)\end{cases}$$

Now, the procedures for the numerical representation of preferences on extended models (as described in Ngo The and Tsoukiàs, 2005; Vincke, 1998) are applicable.

7.3   ASSESSMENT FROM AGGREGATION OF MULTIPLE CRITERIA

So far, all the information for modeling preferences has been obtained from the pairwise comparison of alternatives. A limiting factor in this process is the number of questions that can be answered by an individual before his or her attention wanes.

---

* xJy occurs when x is nice in criteria where y is bad and, contrary, y is nice in criteria where x is bad. The greater the number of incomparable alternatives, the greater the risk accepted when choosing alternatives from the numerical representation of their preferences (see Section 3.4).

When the number of selected alternatives is six, the number of judgments is 15. Beyond this number of comparisons, attention declines considerably (Miller, 1956). However, the existing variability in the whole set of alternatives is barely reflected with the value of just six alternatives.

An additional way of modeling preferences is by characterizing all alternatives through the value of their measurable attributes on a set of significant criteria (see Note 4) and then obtaining an overall value by aggregating the information provided by the criteria. Thus, each alternative is characterized by a vector describing the "performances" of the alternative for each criterion,* $x = (x_1, x_2, \ldots)$. If there were n significant criteria, and given that performances are measured in real numbers, then the set of alternatives (Ω) could be represented by $\mathbb{R}^n$.

This section describes the methodology to aggregate the information from significant criteria in order to obtain an overall assessment of each alternative.

But first we must consider that aggregation of criteria implies a new type of uncertainty—exogenous uncertainty—which stems from our limited knowledge of the world. This uncertainty arises from incomplete information (such as failure to consider criteria that are meaningful to the overall assessment) or ambiguous information (such as a poor knowledge of the value of the performance of alternatives in the criteria). Preference modeling must incorporate this inaccurate information and restrict the exogenous uncertainty in the application of the model. To minimize exogenous uncertainty, the information must be systematized, which requires

- Structuring of the objectives (Belton and Stewart, 2001; Keeney, 1988; Martínez-Falero and González-Alonso, 1994; ORWorld, 2002; Pöyhönen et al., 2001; Reynolds et al., 2007). This process seeks to build a hierarchy of the elements in the problem. It starts with the global definition of the problem to assess (overall objective), then extends to cover more understandable subproblems (subobjectives) and concludes with the identification of measurable performances or attributes.†

- Description and identification of the criteria (Bouyssou, 2001; Keeney, 1981; Keeney and Raiffa, 1976; Martínez-Falero and González-Alonso, 1994; Roy, 1989, 1996). This includes the identification of the attributes, which are appropriate for measuring the degree of achievement of each objective, the elimination of repetitive attributes, and the modeling and measurement of attributes.

- Communicating, in plain language, the meaning of significant criteria and attributes to the evaluators (Schiller et al., 2001).

---

* We use the term "performance" to differentiate it clearly from value. For example, the maximum speed of a vehicle is a universal measurement of an attribute (perhaps with some degree of precision) and is "performance." The value attributed to this performance is different for each person: an individual whose preferences are oriented toward an ATV will give a different value to that attribute from the value given by another person preferring a sports car.

† The quantitative techniques available for structuring objectives include several methods such as soft systems methodology [SSM] (Checkland, 1981), strategic choice approach [SCA] (Friend and Hickling, 1987), and strategic options development and analysis [SODA] (Eden, 1989). Cognitive mapping [CM] is an essential part of SODA and supports the structuring of the problem by providing visualization in the form of loops, linkages, and trade-offs between the concepts (Axelrod, 1976).

Throughout the aforementioned steps, any evaluator will have a clear idea of the information available to make decisions regarding the assessment of complex alternatives. In this chapter, we shall not pursue this issue any further, given that the first two parts of this work are devoted to describing the entire process of systematization of information for the assessment of forest sustainability.

7.3.1 PERFORMANCE AND VALUE

The assessment of an alternative requires the transformation of a performance (x_i) into a value $[v_i(x_i)]$, where x_i is a measurable attribute and $v_i(x_i)$ the meaning attached to that attribute by the evaluator. In general, this is done by applying general tools to build the membership functions of a fuzzy set (see Figure 7.2). Thus, Edwards (1977) and Fishburn (1967) have proposed nonspecific procedures to assess general reference conditions.

Furthermore, Kirkwood and Sarin (1980) have established specific types of value for certain families of attributes. So, if delta property is satisfied, then

$$v_i\left(x_i\right) = \alpha_i + \beta_i x_i \text{ or } v_i\left(x_i\right) = \alpha_i + \beta_i \exp\{k_i x_i\} \tag{7.18}$$

and if the delta-proportional property is satisfied, then the value is of type

$$v_i\left(x_i\right) = \alpha_i + \beta_i \ln(x_i - x^*) \text{ or } v_i\left(x_i\right) = \alpha_i + \beta_i(x_i - x^*), \forall x^* \in \mathbb{R}^+ \tag{7.19}$$

7.3.2 CONSTRUCTION OF THE VALUE FUNCTION

This section maintains the definition of value function given in Section 7.2 [expression (7.11)], with the difference that the set of possible alternatives is represented by $\mathbb{R}^n$. A different notation is also adopted to refer to preference relations. Thus, when binary relations are defined on the basis of performance in different criteria, we shall use

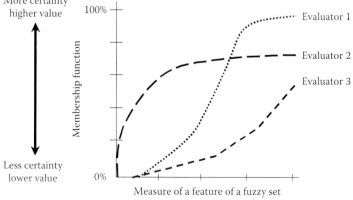

FIGURE 7.2   Construction of 1D value function as membership function of a fuzzy set.

$$x \succeq y \ (x \text{ is as least as preferable as } y)$$

$$x \succ y \left(x \text{ is strictly preferred to } y\right) \tag{7.20}$$

$$x \sim y \ (x \text{ and } y \text{ are indifferent})$$

and *P*, *I*, and *R* will represent preferences defined in direct pair comparisons of alternatives. Under these conditions, a value function (*v*) is an application $v: \mathbb{R}^n \to \mathbb{R}^+$ such that

$$x \succeq y \Leftrightarrow v\left(x\right) \geq v\left(y\right), \forall x, y \in \mathbb{R}^n \tag{7.21}$$

As discussed in Section 7.2.3, not all binary relations can be expressed by a value function (see, e.g., Briges and Metha, 1995, for a summary of the requirements). In general, it demands that the relationship $\succeq$ be complete and transitive. In particular, it requires further verification of the existence of "enough" real numbers to distinguish all possible alternatives. Fishburn (1974) has formulated the necessary and sufficient conditions for expression (7.21), Arrow and Debreu (1954) the sufficient conditions.

Once the existence of the value function has been accepted, the next step is its construction. The analytical expression of *v* depends on the type of dependency existing among the criteria to be aggregated. Thus, additive value function is related to preferential independence (*PI*) among criteria:

Theorem (Chankong and Haimes, 1983): Θ_0 is preferentially independent (*PI*) of its complement in Θ [where $\Theta = \{\theta_1, ..., \theta_n\}$ is the set of all criteria] $\forall \Theta_0 \subset \Theta$ (the set of objectives is mutually and preferentially independent) $\Leftrightarrow v(x) = k_1 v_1(x_1) + ... + k_n v_n(x_n)$.

Definition of PI: Θ_a is *PI* of Θ_b ($\Theta_a, \Theta_b \subseteq \Theta$ and $\Theta_a \cap \Theta_b = \Phi$) $\equiv \forall x_{\Theta a}^1, x_{\Theta a}^2 \in \Theta_a$ and $\forall x_{\Theta b}^+ \in \Theta_b$ such that $(x_{\Theta a}^1, x_{\Theta b}^+) \succeq (x_{\Theta a}^2, x_{\Theta b}^+) \Rightarrow (x_{\Theta a}^1, x_{\Theta b}^*) \succeq (x_{\Theta a}^2, x_{\Theta b}^*), \forall x_{\Theta b}^* \in \Theta_b$.

Preferential independence is related to the shape of the conditional value functions: it is necessary for the cross section of all conditional functions [$v(x_{\Theta b}/x_{\Theta a}), \forall x \Theta_a \in \Theta_a$] to be simultaneously increasing or decreasing in all the measurements taken at Θ_b [$\forall x \Theta_b \in \Theta_b$] (Martínez-Falero and González-Alonso, 1994). Figure 7.3 shows an example and a counterexample to this behavior of the conditional marginal value functions: it is a simplified representation of the conditional functions (geology and climate criteria) to assess the productivity of the vegetation. From the simultaneous growth and decline of all the conditional value functions, we can deduce that climate is *PI* of geology in order to assess productivity. However, the presence of certain geological functions (for a fixed climate) that increase while other functions (for another conditioning climate) decrease—in the same geological value—determines that geology is not *PI* of climate. The reason is clear: under a moderate climate (c_1), a limestone substrate (g_1) provides greater productivity than a granite substrate (g_2). Therefore, $(g_1, c_1) \succeq (g_2, c_1)$. In contrast, soils on a limestone substrate lose a considerable amount of calcium ions under an extreme climate (c_2), which causes productivity in these areas to be lower than with a granite substrate. That means that $(g_1, c_2) \prec (g_2, c_2)$ and, in consequence, $(g_1, c_1) \succeq (g_2, c_1) \not\Rightarrow (g_1, c_2) \succeq (g_2, c_2)$. Therefore, preferential independence conditions are not satisfied.

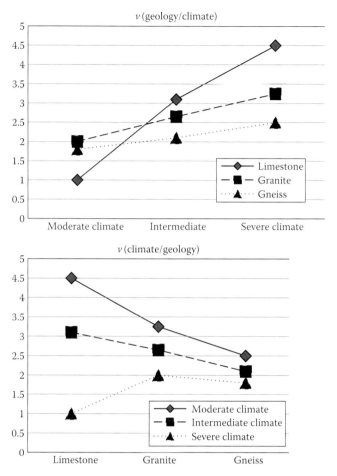

FIGURE 7.3 Graphical testing of non-PI between two attributes: At the top, the behavior of all 1D functions is monotonic. At the bottom, in the case of severe weather, the value function first rises and then falls, while in the other cases, the 1D value functions are always decreasing for the assortment of types of geology shown in the figure.

Other types of value functions correspond to other concepts of dependence and independence among criteria such as weak-difference independence (*WDI*) and Thomsen condition. So, for the most common value functions

$$v(x) = k \, v_1(x_1) \times \ldots \times v_n(x_n) \quad \text{(multiplicative)} \tag{7.22}$$

$$v(x) = k \left[v_1(x_1) \right]^{\alpha 1} \times \ldots \times \left[v_n(x_n) \right]^{\alpha n} \quad \text{(polynomial)} \tag{7.23}$$

$$v(x) = [k_1 v_1(x_1) + \ldots + k_{n-2} v_{n-2}(x_{n-2})] \times v_{n-1}(x_{n-1}) \times v_n(x_n) \quad \text{(partially additive)} \tag{7.24}$$

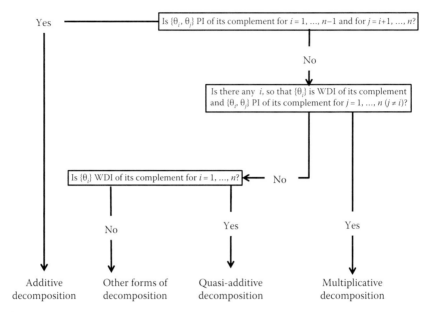

FIGURE 7.4 Key to identify the type of breakdown of the value function according to the evaluator's preferences. PI: preferentially independent, WDI: weak difference independent, n: number of criteria.

$$v(x) = k_0 + \sum_{i=1} k_i v_i (x_i) + \sum_{i=1} \sum_{j>i} k_{ij} v_i (x_i) \times v_j (x_j)$$

$$+ \sum_{i=1} \sum_{j>i} \sum_{k>j} k_{ijk} v_i (x_i) \times v_j (x_j) \times v_k (x_k) + \ldots$$

$$+ k_{12\ldots n} v_1 (x_1) \times v_2 (x_2) \times \ldots \times v_n (x_n) \quad (\text{quasi-additive}) \qquad (7.25)$$

there are specific decomposition theorems. Some of them are summarized in Figure 7.4 (deduced by the systematization carried out by Chankong and Haimes (1983)).

Once the analytical form of the value function is known and the marginal functions $[v_i(x_i)]$ have been computed, it remains to estimate the weights (k_i) in order to arrive at the complete identification of the value function. This estimation can be done by two main methods (Otero, 1979).

The first method involves defining a system of as many equations as number of weights to be determined. Each equation in the system is deduced from a pair of indifferent alternatives. Thus, for an additive value function,

$$a \sim b \Rightarrow k_1 v_1 (a_1) + \ldots + k_n v_n (a_n) = k_1 v_1 (b_1) + \ldots + k_n v_n (b_n) \text{ Equation No. 1}$$

$$c \sim d \Rightarrow k_1 v_1 (c_1) + \ldots + k_n v_n (c_n) = k_1 v_1 (d_1) + \ldots + k_n v_n (d_n) \text{ Equation No. 2}$$

$$\ldots$$

$$(7.26)$$

Solving the preceding equations is intended to determine a set of weights that are consistent with the preferences of the individual who formulated the pairs of indifferent alternatives.

A second procedure is based on a double ordering of weights (by importance and by the difference in their importance). If the value function is additive, then the double ordering causes the values of the weights to be arranged into a convex. Only the values within this set are consistent with the preferences of the individual who ordered the weights (see Martínez-Falero and González-Alonso, 1994):

$$\left.\begin{array}{l} k_1 \leq \ldots \leq k_n \\ k_1 - k_2 \leq \ldots \leq k_{n-1} - k_n \end{array}\right\} \rightarrow \left(k_1, \ldots, k_n\right) \subset \text{CONVEX SET} \qquad (7.27)$$

This procedure does not pursue unique values for the weights, but when additional restrictions are added (e.g., that any weight can be three times bigger than another), the set of values that are compatible with the preferences of the evaluator is greatly reduced.

7.3.3 SPECIFIC METHODS FOR REPRESENTING ADDITIVE PREFERENCES

Several authors consider that additivity has to be accepted in most cases and advise against overemphasizing the counterexamples, as they consider examples are due to poor structuring of criteria (Bouyssou and Pirlot, 2005; Keeney, 1992; Roy, 1985; von Winterfeldt and Edwards, 1986). However, when it comes to aggregation of information in environmental assessments, we do not agree with this assertion* and we recommend verifying the existence of mutual and preferential independence before applying procedures based on additive integration of information.

There are several procedures to systematize the computing of weights and marginal values for each criterion in the case of additive decomposition of the value function. The most representative of these procedures is probably analytic hierarchy process (*AHP*) and its extensions: analytic network process (*ANP*) and neural network process (*NNP*), all developed by Saaty (1994, 1996, 2000a,b, 2009, 2010). These methods are the most commonly used procedures in multicriteria assessments (Olson, 2008) and require the evaluator to make the following judgments:

1. To define a hierarchical structure for the decision problem, identifying (1) global objective and subobjectives, (2) criteria ($i = 1, \ldots, n$), and (3) alternatives ($j = 1, \ldots, m$)
2. To complete, for each one of the criteria, the matrix of pair-wise comparison between alternatives. When the first pair alternative shows preference

---

* As seen, geology is not *PI* from climate to assess the productivity of the vegetation; the reader can also check the non-independence of slope and orientation to assess the incident solar radiation and to find that the most commonly used criteria for assessing the fragility of the landscape are not preferentially independent (MOPT, 1993), etc. In general, environmental factors interact to produce different consequences to the mere integration of elements, which means that in many cases they are not preferentially independent of each other.

or indifference to the second, the evaluator is asked for his or her intensity of preferences between alternatives, in order to determine whether the preference is

a. Indifference
b. Moderately preferred
c. Strongly preferred
d. Very strongly preferred
e. Extremely preferred

3. To complete, for each alternative, the matrix of comparison between criteria

The application of *AHP* requires the number of criteria to be small as a direct comparison must be made of all possible pairs of criteria. The need for a small number of alternatives could similarly be assumed; however, in this case, the fulfillment of the pair-wise comparison matrix can be substituted by a value of the performances of each alternative on each criterion (Section 7.3.1) and setting the scale of step b as the differences in that value.

7.3.4 OUTRANKING METHODS

An outranking relation (S) is a relationship "at least as preferable as" defined in terms of risk, so that, for a certain level (α),

$$aSb \equiv P[a \succeq b] \geq \alpha \qquad (7.28)$$

The non-transitivity of S can clearly be verified: let us suppose a certain level of risk (e.g., $\alpha = 0.95$), that $P[a \geq b] = 0.95$ and $P[b \geq c] = 0.95$ and also that "$a \geq b$" and "$b \geq c$" are independent events. Thus, without additional information,

$$\left. \begin{array}{l} P[a \succeq b] = 0.95 \Rightarrow aSb \\ P[b \succeq c] = 0.95 \Rightarrow bSc \end{array} \right\} \Rightarrow P[a \succeq c] = 0.95 \times 0.95 < 0.95 \nRightarrow aSc \qquad (7.29)$$

The non-transitivity implies that many alternatives remain incomparable to each other, which increases the difficulty of assigning a value to each alternative. If the risk of the decision is increased (by reducing α), then it is possible to increase the number of relations among alternatives and simplify the allocation of a value to the alternatives. However, acting this way increases the likelihood of error when assuming the preference between two alternatives that does not actually occur. In any case, the construction of a value from an outranking relation is not immediate but is based on building a *<P, I, J>* binary relationship such that

$$aPb\,(a \text{ is strictly preferred to } b) \quad \Leftrightarrow \quad aSb \text{ and } \text{no}\,(bSa)$$

$$aIb\,(a \text{ and } b \text{ are indifferent}) \quad \Leftrightarrow \quad aSb \text{ and } bSa \qquad (7.30)$$

$$aJb\,(a \text{ and } b \text{ are incomparable}) \quad \Leftrightarrow \quad \text{no}\,(aSb) \text{ and } \text{no}\,(bSa)$$

and then on the application of procedures for building the numerical representation of preferences on extended models (Section 7.2.3).

The first difficulty in working with outranking relations is to choose the probability that truly reflects the likelihood that an alternative is at least as preferred as another by an evaluator*. For example, if $\mathcal{J}$ is the set of criteria, $v_i(a_i)$ the marginal value of the a alternative on the i criterion, and w_i the weight assigned to the i criterion ($\Sigma_{i \in \mathcal{J}} w_i = 1$ and $w_i \geq 0$, $\forall i \in \mathcal{J}$), then it is possible to verify that the following expression (Roy, 1968) is a probability for the occurrence of the event "$a \succcurlyeq b$," $\forall a, b \in \Omega$:

$$c_{(a,b)} = \sum_{i \in \mathcal{J} : v_i(a_i) \geq v_i(b_i)} w_i \cdot \tag{7.31}$$

This expression sets one alternative to be preferred over another when there is a clear preference relation $[v_i(a_i) \geq v_i(b_i)]$ for a representative majority of criteria. Specifically, this condition is known as concordance. So, if the evaluator considers that his or her choice is only related to concordance, then $aSb \Leftrightarrow c_{(a,b)} > \alpha$.

However, there may be other evaluators who are more interested in the nonexistence of inconsistencies. This fact requires none of the criteria to be largely opposed to accepting $a \succcurlyeq b$. Thus, with the information provided by the following index:

$$d_{(a,b)} = \max_{\{i \in \mathcal{J} : v_i(a_i) \leq v_i(b_i)\}} \{v_i(b_i) - v_i(a_i)\}, \text{ with } v_i(x_i) \in [0,1] \tag{7.32}$$

these evaluators may consider that the way to assign probabilities that are consistent with their preferences is the $1 - d_{(a,b)}$, which shows the likelihood of the occurrence of events of the type "$a \succcurlyeq b$," $\forall a, b \in \Omega$.

The selection of the best outranking method is as difficult as determining the probability distribution that best describes a random experiment. Moreover, the difficulty of selecting a method increases because the available techniques have not been developed to reflect specific patterns of preferences but have emerged as an isolated solution to specific problems. Moreover, the adoption of a rule to select the best method is even more complicated, as the formulation of available outranking procedures incorporates multiple outranking relations (which can be based on explicit relationships, whether traditional, expanded, or diffuse, or on embedded relations). In fact, expressions (7.31) and (7.32) configured together define the method elimination and choice expressing reality (*ELECTRE*)-*I* (Roy, 1968), which states

$$aSb_{(after\ ELECTRE\ I)} \equiv c_{(a,b)} \geq s \text{ and } d_{(a,b)} \leq v \tag{7.33}$$

where s and v are, respectively, known as levels of concordance and discordance.

---

* It should be remembered that probability is an axiomatic measure on the set of events, meaning that the probability of any event must be greater than or equal to zero, the probability of the union of disjointed events is the sum of their probabilities, and the probability of a sure event is one. So, any measure satisfying these three conditions is a probability.

Although it is difficult directly to choose the best procedure, it is possible to make progress in recognizing the outranking relation that best describes the preferences of an evaluator by analyzing the fit of the constructive features of each method with the preferences. However, this approach is not very practical, since a simple analysis of the available methods is an arduous task.

Greatly simplified, the review of available methods includes, first, the *ELECTRE* method (see, e.g., Figueira et al., 2005a,b, for a detailed description of all extensions). This is possibly the most commonly used outranking method. All its versions are an extension of *ELECTRE-I*, by incorporating fuzzy relations and other relations embedded in the preferences.

The main competitor of *ELECTRE*, regarding the extension of its applications, is the *PROMETHEE* (see Brans and Mareschal (2005) for a systematic description of its different versions). It has two notable features. On the one hand, it uses fuzzy relations (from its first formulations) in order to define the degree of credibility of preference for one alternative over another, and this is computed as the difference in value within each criterion. So

$$P_j\left(a,b\right) = F_j\left[v_j\left(a_j\right) - v_j\left(b_j\right)\right] \tag{7.34}$$

where $F_j(x)$ is a fuzzy membership function for j criterion. The degree of credibility for the overall preference is defined by additive* aggregation

$$\pi\left(a,b\right) = \sum_{j\in J} P_j\left(a,b\right)w_j \tag{7.35}$$

The second feature involves assigning a value to each alternative from the number of alternatives outranked by the one analyzed, when the alternative to be assessed is compared with the other (see the explanation of expressions (7.16) and (7.17) in Section 7.2.3).

Although the aforementioned methods are more often cited than others in the scientific literature, they are not the only ones. In Martel and Matarazzo (2005) alone, there are 12 other methods in addition to *ELECTRE* and *PROMETHEE*.

Given the difficulty of progressing toward the definition of a rule for the selection of an outranking method based on the patterns of preferences, indirect processes have been developed to assign the best outranking method to each evaluator. These procedures explore other features that provide acceptable dependence between types of outranking methods and the specific behaviors of evaluators. Among these we can highlight the results of Ramos (1982), which identify the relationship between the differentiation of alternatives recognized by the application of different outranking methods and the evaluators' ability to discriminate alternatives directly. These authors set the selection rule described in Figure 7.5 for a population of impact assessment evaluators in Spain in the early 1980s.

---

* As with the AHP, PROMETHEE applications do not incorporate systems to verify additivity. It is, therefore, advisable to verify this condition prior to any application.

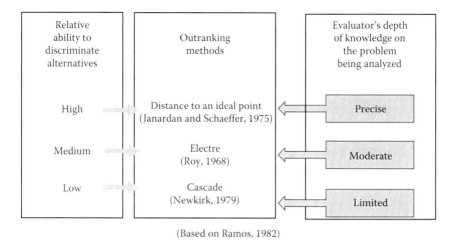

(Based on Ramos, 1982)

FIGURE 7.5   Key to identify the outranking method closer to the evaluator's preferences.

7.4   ASSESSMENT FROM INFORMATION DEDUCED FROM BOTH PAIR-WISE COMPARISON AND AGGREGATION OF CRITERIA

As stated previously, assessment of preferences from pair-wise comparison is not practical for assessing more than six alternatives. On the other hand, assessment through aggregation of criteria incorporates the uncertainty resulting from limited human perception (exogenous uncertainty) or the uncertainty deriving from making judgments (endogenous uncertainty). To reduce the influence of the overall uncertainty in evaluation, we propose a methodology that uses the best aspects of both evaluation systems.

The idea is to make a first assessment based on the aggregation of criteria for all the available alternatives and a second assessment from a pair-wise comparison on a reduced number of representative alternatives. The value obtained by pair-wise comparison is the framework where the assessment of the alternatives that are not subject to direct comparison is fitted. Thus, the thoughtful judgments required in the pair-wise comparison set the general pattern of value, and the valuation obtained from the aggregation of criteria incorporates the variability present in all alternatives to the evaluation process.

7.4.1   OBTAINING A VALUE FOR ALL ALTERNATIVES FROM PAIR-WISE COMPARISON OF A FEW

As explained in Section 7.3, the alternatives can be represented through their performance in n criteria [$x = (x_1, ..., x_n)$], and it is also possible to obtain an assessment of the type

$$g : \mathbb{R}^n \rightarrow \mathbb{R}^+, \text{ with } g(x) \in \mathbb{R}^+; \forall x \in \mathbb{R}^n \tag{7.36}$$

which expresses the value the evaluator allocates to each alternative (for the Ω set of all alternatives). For our purpose, all we need from the aforementioned expression is a measure of the separation between two alternatives from their performances in the n significant criteria:

$$d:\Omega\times\Omega \to \mathbb{R}^+,$$

where $d(x,y)$ is a measure of the separation between x and y;

$$\forall x, y \in \mathbb{R}^n \tag{7.37}$$

This measure can be obtained from expression (7.36), as $d(x, y) = |g(x) - g(y)|$, although, as we shall see in Section 7.6, the separation between two alternatives can be obtained by other means.

On the other hand, we know it is possible to obtain a numerical representation of the preferences of any evaluator through pair-wise comparison of alternatives through the implementation of the procedures described in Section 7.2. So, in $\forall x \in \Omega'$ (the set of the representative alternatives submitted to pair comparison), it is possible to obtain

$$u:\Omega' \to \mathbb{R}^+, \text{ with } u(x) \in \mathbb{R}^+; \forall x \in \Omega' \tag{7.38}$$

which expresses the value the evaluator allocates to each alternative within the set of alternatives selected for direct comparison.

In order to extend the value obtained from the comparison of a few pairs of meaningful alternatives (Ω') to the set of all possible alternatives (Ω)—where there has been no comparison—the distance between each alternative in Ω and each one in Ω' is calculated (using a distance of the type presented in expression (7.37), based on the performances of each alternative in all criteria). Then, for each new alternative to be assessed ($y \in \Omega$), y_1 is the closest alternative within Ω', $u(y_1)$ is its value (calculated as defined in expression (7.38)), and $d(y, y_1)$ is the distance from y to y_1 (as defined in (7.37)) and y_2 is the second-closest alternative within the set of those compared, $u(y_2)$ is its value, and $d(y, y_2)$ is the distance from y to y_2. Thus, $v(y)$ can be calculated by linear interpolation between $u(y_1)$ and $u(y_2)$ as

$$v(y) = \min\left\{u(y_1), u(y_2)\right\} + \frac{d}{d(y,y_1) + d(y,y_2)}\left|u(y_1) - u(y_2)\right| \tag{7.39}$$

$$\text{where } : d = \begin{cases} d(y,y_1) & \text{if } u(y_1) < u(y_2) \\ d(y,y_2) & \text{otherwise} \end{cases}$$

7.4.2   ANALYTICAL EXPRESSION OF THE VALUE

The importance of assigning a value to each occupation goes far beyond merely comparing alternatives, since the way in which the variables of the value function are combined provides information on each individual's rational decision-making process. Let us consider, for example, an individual (A) whose global sustainability value depends on the value taken by three sustainability indicators (I_1, I_2, I_3), so that $v_A = I_1 I_2 + I_3$. It can be deduced that the sustainability assessed by A is such that a value of virtually zero for I_1 is sufficient to invalidate any possible sustainability produced by the value taken by I_2. If I_1 were "forest structure" and I_2 were "timber yield," A would consider as sustainable those points with high values in both forest structure and timber yield. The relationship of the first two sustainability indicators with I_3 is different. In this case, there is a clear substitutability between them. If, for example, I_3 were "biomass," A could replace a reduced value in forest structure and timber yield with a high value in biomass and vice versa. Billot (2003) explains these qualitative relationships in terms of individual tastes, which enables the knowledge of the value derived from subjective judgments about preferences to be transformed into the knowledge of why an individual prefers one alternative over another.

The analytical expression for the value function can be obtained from the marginal value in each of the criteria and the global value allocated to each alternative by a multiple linear regression model. The independent variables in this model are all the possible combinations of performances in criteria, and the dependent variable is the global value (obtained by applying expression (7.39)). In order to systematize the process, it is necessary to take the following steps before computing the regression model:

a. The value of both dependent and independent variables is typified.
b. The value taken by a combination of criteria at each alternative is computed by multiplying the marginal value of each criterion that outlines the combination.
c. The *Smirnov–Kolmogorov* two-sample test is applied to eliminate colinear independent variables.

The regression model has been computed without considering the meaning that the individual gives to the relationships among the sustainability indicators. However, the analytical expression of the value function depends on the dependence–independence relationships among the criteria to be integrated. Indeed, the existence of a quasi-additive decomposition ((7.25) expression)—as in the applied regression model—depends on any criterion being weak-difference independent *WDI* of the remaining criteria for the evaluator (see Section 7.3.2, Figure 7.4, and Dyer and Sarin, 1979). A criterion (I) is WDI of the remaining criteria ($\bar{I}$) if for any set of four values in I (w_I, x_I, y_I, z_I), such that

$$\left(w_I, t_{\bar{I}}^0\right) \circ \left(x_I, t_{\bar{I}}^0\right) \succeq^* \left(y_I, t_{\bar{I}}^0\right) \circ \left(z_I, t_{\bar{I}}^0\right) \text{ for some } t_{\bar{I}}^0 \in \bar{I} \tag{7.40}$$

then

$$\left(w_I,t_I\right)\circ\left(x_I,t_I\right)\succeq^{*}\left(y_I,t_I\right)\circ\left(z_I,t_I\right)\text{ for all } \dot{\upsilon}_I \in \overline{I} \tag{7.41}$$

where $x\circ y$ is the difference in value between alternative x and alternative y and symbol $\succeq^{*}$ indicates whether the evaluator assigned a difference of value to the pair on the left, which is equal to or greater than the pair on the right.

WDI is a strong condition. It is, therefore, very difficult to satisfy conditions (7.40) and (7.41) for every combination of performances of criteria in a huge set of alternatives. For this reason, the percentage of combinations satisfying *WDI* conditions is a measurement of the suitability of the regression model for the application. To a certain extent, this parameter is also a descriptor to characterize the preferences of the evaluator.

7.5 FURTHER CONSEQUENCES OF MODELING PREFERENCES

This section describes other the characteristics of the evaluators that follow from modeling his or her preferences. These aspects are very important in designing the processes for communicating information to the evaluator (on the issue that is subject to opinion) and also in helping the final decision maker to determine the quality of the opinion of each evaluator in order to consider it in the ultimate decision.

7.5.1 RATIONALITY OF INDIVIDUAL PREFERENCES

As mentioned in Section 7.2.3, the procedures available to describe (and predict) the opinion of any individual have various concepts and results in common. These procedures are the representation of individual preferences, the applications are derived from the theory of value and the analysis of individual past decisions. The reader should be familiar with the first two procedures. To introduce the concepts related to the analysis of past decisions, we shall proceed to the idea of choice function. C is a choice function that applies to any subset of alternatives $X \subseteq \Omega$ if

$$C\left(X\right)=\left\{y\in\frac{X}{\forall x}\in X:u\left(y\right)\geq u\left(x\right)\right\} \tag{7.42}$$

where
    Ω is the set of alternatives
    u is a value function ($u\colon \Omega \to \mathbb{R}^{+}$)

The importance of choice functions is that different types correspond to particular individual behaviors in decision making. In general, the behavior of an individual is characterized by the pattern of changes in his or her decisions when the range of options is modified by increasing or decreasing it. An example of a type of rational behavior is described by Arrow's postulate. This postulate requires the alternatives chosen in the expanded set to be included in the reduced group; these alternatives, and only these, will be the ones selected in the contracted set.

The following are the definitions for the main types of choice functions.* Thus, a choice function C can satisfy the following axioms:

Arrow

$\forall X, X' \subseteq \Omega \ (X \neq \varnothing, X' \neq \varnothing)$, such that $X' \subset X$, it happens that

$$C(X) = \varnothing \Rightarrow C(X') = \varnothing, \text{ and also that} \tag{7.43}$$

$$C(X) \cap X' = \varnothing \Rightarrow C(X') = C(X) \cap X' \tag{7.44}$$

Jamison–Lau–Fishburn

$$C(X') \cap X'' \neq \varnothing \Rightarrow \left[X \setminus C(X)\right] \cap C(X'')$$

$$= \varnothing, \left\{\forall X, X', X'' \subseteq \Omega, X \subseteq X' \setminus C(X')\right\} \tag{7.45}$$

Heredity

$$X \subseteq X' \Rightarrow \left[C(X) \cap X'\right] \subseteq C(X') \text{ for all } X, X' \subseteq \Omega \ (X \neq \varnothing, X' \neq \varnothing) \tag{7.46}$$

Concordance

$$C(X') \cap C(X) \subseteq C(X' \cup X), \left\{\forall X, X' \subseteq \Omega \ (X \neq \varnothing, X' \neq \varnothing)\right\} \tag{7.47}$$

We already know that existing methodologies to describe and predict opinions (preference modeling, value and decision analysis) are integrated through the concept of threshold. As mentioned, the threshold characteristics are unique to each individual and are related to his or her discriminating power to recognize the difference between two alternatives. Therefore, a proper definition of the threshold provides a broad understanding of each person's process of rational choices and enables the type of rationality of any individual to be characterized.

Table 7.8 (after Aleskerov et al., 2007) proposes a characterization of the rationality of any individual in four types. This characterization is obtained from the threshold that simultaneously defines the system of preferences, the type of utility, and the behavior in past choices. This table is completed by the relationship between the threshold and the procedures of numerical representation of preferences of any person (see, e.g., Table 7.7 and other developments in Section 7.2.3).

---

* It is clear that an individual does not always act the same way in his or her choices. Over time, or when making decisions on various issues, the decision maker adopts different types of behavior, which, in turn, are described through different types of choice functions.

TABLE 7.8

Characterization of Rationality of Individual Preferences (after Aleskerov, et al., 2007)

Type of Rationality	Threshold Characterization	Choice Characterization	Type of Preference Order Relation (P) on the Area of Application	
Type I	$\varepsilon = 0$	Arrow's choice axioms	Linear or weak order	
Type II	$\varepsilon = \text{constant} > 0$	Jamison–Lau–Fishburn condition	Semiorder	
Type III	Depending on the alternative on which the global value is calculated (the analyzed value or this value and the value with which it is compared)	Heredity and concordance conditions	Interval order Biorder Partial order	$\varepsilon = \varepsilon(x) \geq 0$ $\varepsilon = \varepsilon(x)$ $\varepsilon = \varepsilon(x, y) \geq 0$ and $\varepsilon(x, z) \leq \varepsilon(x, y) + \varepsilon(y, z)$
Type IV	Depending on the context $\varepsilon = \varepsilon(x, y, \Omega)$		Acyclic weak biorders (or none)	

Source: After Aleskerov, F. et al., *Utility Maximization, Choice and Preference*, Springer, Berlín, Germany, 2007.

Accordingly, either by identifying the preferences of a person or by knowing the procedure for the allocation of value, or even after becoming aware of the way a person conducts his or her decision making, it is possible to characterize the rationality of the actions and the procedure for arranging the opinions of any person.

7.5.2 DEGREE OF KNOWLEDGE OF THE SYSTEM FOR EXPRESSING OPINION

Whatever the type of rationality with which an individual acts, the evaluator can also be characterized by the depth of his or her knowledge on the system being evaluated. In fact, there are individuals who are able to distinguish only the very good from the very bad alternatives, while there are others who are able to capture many other nuances when analyzing a system. This ability to discern is directly related to the number of indifference classes an individual is able to distinguish in the equivalence relation (E) associated with his or her preferences (see expression (7.3)).

If it is accepted that the greater the number of differences that an individual is able to perceive, the greater his or her depth of knowledge of the system evaluated (and vice versa), then by counting the number of equivalence classes in the equivalence relation associated to his or her system of preferences, we can obtain an operational procedure to determine the evaluator's depth of knowledge.

7.6   ASSESSING SUSTAINABILITY IN FOREST MANAGEMENT

In this section, we apply the concepts described earlier to assess sustainability in forestry operations so that the allocated value is consistent with the preferences of the sustainability evaluator. This section also describes a methodology for building communities of individuals with similar systems of preferences.

7.6.1   ALTERNATIVES, CRITERIA, AND EVALUATORS

The feasible *alternatives* are all the possible environmental conditions existing in the area where the assessment is carried out. Thus, alternatives are identified with each one of the locations (points) of the forest exploitation to be evaluated, and the assessment assigns a value of sustainability to each point in the forest.

For the verification of the proposed methodology, the assessment has been applied to a real forest. This is the area highlighted in Figure 7.6 (located in the *Fuenfría Valley, Madrid, Spain*, at coordinates: 40° 45′N, 4° 5′W) with elevations ranging from 1310 to 1790 m. The average annual temperature of the area is 9.4°C, the average annual rainfall is 1180 mm, and the predominant tree species

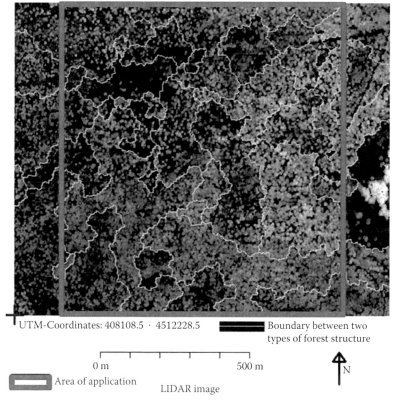

FIGURE 7.6   (See color insert.) LiDAR image of the DCHM of the area selected for the case of application.

is Scots pine (*Pinus sylvestris*). The area can be grouped into five different types of forest structures (Pascual et al., 2008):

- Uneven-aged forest (multilayered canopy) with very high crown cover
- Multi-diameter forest with high crown cover
- Multi-diameter forest with medium crown cover
- Even-aged forest (single story) with low crown cover
- Zones with scarce tree coverage

Each point in the forest is characterized from the spatial distribution of the height of the trees included within a circular plot with a diameter of 60 m around each one of the vertices of a square grid (1 m side), superimposed on the forestry operation.* In order to calculate this distribution, we have used the information encoded in a light detection and ranging (*LiDAR*) image: In August 2002, TopoSys GmbH surveyed the study area with a *LiDAR* TopoSys II sensor and a digital canopy height model (*DCHM*) was obtained after image processing.† The position and height of trees over 3 m tall was estimated from the information provided by this image (Figure 7.7 shows the position of trees in a part of the analyzed area, estimated through the application of an algorithm to calculate tree location as described in Martínez-Falero et al., 2010).

The *criteria* used for assessing forest sustainability included some of the pan-European indicators for sustainable management. Currently, sustainability indicators (Cabot et al., 2009) are a powerful tool for sustainable forest management (Wijewardana, 2008), and there are regional processes worldwide to design indicators suited to the characteristics of each region (Tamubula and Sinden, 2000; Barbati et al., 2007; Freer-Smith and Carnus, 2008; Hickey and Innes, 2008; Makropoulos et al., 2008).

In this chapter, we have used three indicators that were calculated from the available information: structural diversity, I_1 (by comparing tree height distribution at each point with distribution in the ideal case); timber yield, I_2 (computed following Martín-Fernández and García-Abril (2005)); and amount of biomass, I_3 (computed from the crown volume and from the total height of each tree). The three indicators corresponded to indicators 1.3, 3.1, and 1.4, respectively, of the pan-European sustainable forest management indicators.‡ The scale of measurement for the three indicators

---

* Although this is a one-species forest, it is clear that a more comprehensive characterization would be required for a complete assessment of real sustainability (among other characteristics and qualities, it would be necessary to consider soil type, slope, floristic cortege, and amount of dead material at each point of the forest).

† The TopoSys II *LiDAR* system recorded first and last returns with a footprint diameter of 0.95m; the average point density was 5 points/m²; the raw data (x, y, z coordinates) were processed into two digital elevation models by TopoSys using as the interpolation algorithm a special local adaptive median filter developed by the data provider. The digital surface model (DSM) was processed using the first pulse reflections, and the digital terrain model (DTM) was constructed using the last returns. Horizontal positional accuracy was 0.5 m and vertical accuracy was 0.15 m for both the DSM and DTM. To obtain a DCHM, the DTM was subtracted from the DSM. The vertical accuracy for the DCHM under forest canopy was 1.3 m.

‡ The authors would like to highlight that these indicators were selected only for the purpose of verifying the proposed methodology. The objective was not to assess their importance in the evaluation of sustainability for forest management.

Distribution of the height of the trees around each point (points 1 m away in a square grid)										
Analysis of the trees included within a circular plot of 30 m diameter around each point										
Point		Number of feet/hectare (in each class of height)								
UTM-x	UTM-y	0 – 3	3 – 6	6 – 9	9 – 12	12 – 15	15 – 18	18 – 21	21 – 24	24 – 27
408108,5	4513184,5	0	5,63	28,38	20,81	28,38	32,17	17,03	1,89	0
408108,5	4513185,5	0	5,68	28,36	20,98	28,62	32,43	15,26	1,91	0
408108,5	4513186,5	0	5,72	28,86	21,16	28,86	32,7	15,39	1,92	0
408108,5	4513187,5	0	5,77	29,37	21,35	29,11	31,05	15,52	1,94	0
408108,5	4513188,5	0	5,82	29,37	21,54	29,37	31,33	13,71	1,96	0
408108,5	4513189,5	0	5,87	29,64	21,74	29,64	31,62	13,83	1,98	0
408108,5	4513190,5	0	5,93	29,93	21,95	29,93	31,92	11,97	1,98	0
408108,5	4513191,5	0	5,99	30,22	22,16	30,22	32,24	12,09	2,01	0
408108,5	4513192,5	0	6,04	30,53	22,39	30,53	32,53	12,21	2,04	0
408108,5	4513193,5	0	6,11	30,85	22,62	30,85	30,85	12,34	2,06	0
408108,5	4513194,5	0	6,23	30,47	22,78	31,17	30,85	12,47	2,08	0
408108,5	4513195,5	0	6,3	29,41	21,01	31,51	29,41	12,61	2,1	0
408108,5	4513196,5	0	6,37	29,74	21,24	31,86	29,74	12,74	2,12	0
408108,5	4513197,5	0	6,44	27,93	21,48	32,22	27,93	12,89	2,15	0
408108,5	4513198,5	0	6,52	28,25	21,73	32,59	28,25	13,04	2,17	0
408108,5	4513199,5	0	6,6	28,58	21,99	30,78	28,58	13,19	2,2	0
408108,5	4513200,5	0	6,68	28,93	22,26	31,16	26,71	13,35	2,23	0
408108,5	4513201,5	0	6,76	29,29	22,53	29,29	27,38	13,52	2,25	0
408108,5	4513202,5	0	6,85	29,67	20,54	29,67	27,38	13,69	2,28	0
408108,5	4513203,5	0	6,93	30,05	20,8	30,05	27,74	13,87	2,31	0
408108,5	4513204,5	0	7,03	30,44	21,08	30,44	28,1	14,05	2,34	0
408108,5	4513205,5	0	7,13	30,86	21,36	30,86	28,48	14,24	2,37	0
408108,5	4513206,5	0	7,22	31,27	19,24	31,27	28,87	14,43	2,41	0
408108,5	4513207,5	0	7,32	31,71	19,51	31,71	29,27	14,63	2,44	0
408108,5	4513208,5	0	7,42	32,15	19,79	32,15	29,68	14,84	2,47	0
408108,5	4513209,5	0	7,53	32,61	20,07	32,61	29,68	15,05	2,51	0
408108,5	4513210,5	0	7,64	33,1	20,37	28	22,91	15,27	2,55	0
408108,5	4513211,5	0	7,75	33,59	20,67	28,42	23,26	15,5	2,58	0
408108,5	4513212,5	0	7,87	31,48	20,99	28,86	23,61	15,74	2,62	0
408108,5	4513213,5	0	7,99	31,97	21,32	29,31	23,98	15,99	2,66	0
408108,5	4513214,5	0	8,12	32,01	21,66	29,78	24,36	15,54	2,71	0
408108,5	4513215,5	0	8,25	33,51	21,51	30,26	22,01	13,76	2,75	0
408108,5	4513216,5	0	8,39	33,56	13,98	30,76	22,31	13,98	2,8	0
408108,5	4513217,5	0	8,53	34,12	14,22	31,28	22,75	11,37	2,84	0
408108,5	4513218,5	0	8,68	34,71	14,46	31,82	23,14	11,57	2,89	0
408108,5	4513219,5	0	8,83	32,38	14,72	32,38	20,61	11,78	2,94	0
408108,5	4513220,5	0	8,99	29,97	14,98	32,96	20,98	11,99	3	0
408108,5	4513221,5	0	9,15	30,52	12,21	33,57	21,36	12,21	3,05	0
408108,5	4513222,5	0	9,33	31,08	12,43	34,19	18,65	12,43	3,11	0
408108,5	4513223,5	0	9,5	31,68	12,67	31,68	19,01	12,67	3,17	0
408108,5	4513224,5	0	9,69	32,29	12,92	32,29	19,37	12,92	3,23	0
408108,5	4513225,5	0	9,88	32,93	13,17	32,93	19,76	13,17	3,29	0
408108,5	4513226,5	0	10,08	30,23	13,44	30,23	20,16	13,47	3,36	0
408109,5	4512230,5	0	5,44	8,16	27,21	40,82	54,42	21,77	0	0
408109,5	4512231,5	0	5,34	8,01	29,36	40,03	56,04	21,35	0	0
408109,5	4512232,5	0	5,24	7,86	28,8	41,9	54,99	20,95	0	0
408109,5	4512233,5	0	5,14	7,71	28,27	41,12	56,54	20,56	0	0
408109,5	4512234,5	0	5,04	7,57	30,28	40,37	55,51	20,19	0	0
408109,5	4512235,5	0	4,96	7,43	32,22	39,65	54,52	19,83	0	0

DCHM

Identified trees

FIGURE 7.7 Tree identification in one subarea of the area of application and information reported on each pixel of 1 m².

was the percentage of the maximum value that could be taken for the ecological characteristics of each point. For this purpose, yield tables for *P. sylvestris L.* in the *Sistema Central* mountain range (García Abejón and Gómez Loranca, 1984) were used to compute the value of the three sustainability indicators under the best sustainability conditions (ideal point (*IP*)). The *DCHM* obtained from the *LiDAR* image had a pixel of 1 m × 1 m. However, for the purpose of simplifying the presentation of results, we adopted a pixel of 20 m × 20 m to show the results instead of using the one-meter pixel employed in making the calculations (Figure 7.8). Finally, field data were used to validate the indicators. Ten plots were obtained by systematic sampling, two per each type of forest structure. The analysis of the data (Martínez-Falero et al., 2010) shows there are no significant differences for the respective values of the indicators measured both directly from the field data and computed solely from the *LiDAR* image.

Contrary to what occurs with indicators, there is no single measure for the sustainability to be universally accepted. Furthermore, as it is a complex concept, experts cannot provide this overall measure, which has to be developed by the interested parties for it to be applicable (Vainikainen et al., 2008). This means that personal opinion must be reflected by measuring sustainability. Moreover, in any case, for the assessment of sustainability to be useful for incorporation, it must consider the views of as many stakeholders as possible. However, this fact complicates the evaluation, as when there are multiple *evaluators*, it is impossible to ensure the existence of a transitive preference system for all of them (theorem of Arrow (Arrow et al., 2002; Dietrich and List, 2007)).

The need to work with multiple evaluators has encouraged the development of different lines of work that basically involve

- The development of procedures for designing a minimum common system of preferences that can be acceptable to most of the evaluators in a group, independently of whether they lead to non-transitive systems (with the consequent loss of information and increase in the risk of acceptance of the assessment).
- The application of methods for the convergence of preferences based on the expectation that the components of a group will reach a consensus (bands of indifference, DELPHI, or other methods as described in Morton et al., 1999 and 2001).
- Developments based on artificial intelligence, specifically through the design of intelligent agents, which has led to the problem being addressed by defining communities of users with similar preference systems, and by the development of participatory computing, which promotes social decisions to enable collaboration and interaction between groups (see, e.g., Wang et al., 2007, for a description of the development of social computing). The available tools make it possible to model artificial societies, and computational experiments also allow groups of people with similar views to be identified and characterized (Wang, 2008).

It is important to promote the interaction between evaluators in order to reduce the variability of views, so that each evaluator is aware of the needs and opinions of the

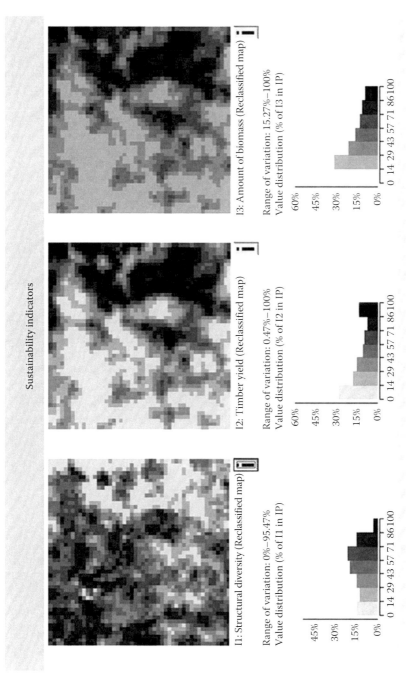

FIGURE 7.8 Spatial distribution of the indices of sustainability in the area of application (Figure 7.6) on pixels of 20×20 m^2.

others. In order to facilitate interaction, the personal data of each evaluator must be considered. The minimum data must be included (see Figure 7.9): profession, educational level, type of residence, age, sex, and stakeholder category (local resident, forest entrepreneur, landowner, academic expert, ecologist, and others). Using this information, evaluators can access the historical answers and verify how other evaluators compare sustainability scenarios, as well as the quality of the other evaluators' opinions (through the consistency and depth of knowledge). All this information enables each evaluator to rethink his or her judgments and probably the convergence of preferences.

Although the assessment of multiple evaluators is developed in Chapter 9, Section 7.6.3 identifies and characterizes communities of evaluators with similar systems of preferences based on individual preferences. It also describes the incorporation of any new evaluator to this community of evaluators with preferences that are closest to those of the new evaluator.

7.6.2   PERSONAL ASSESSMENT OF SUSTAINABILITY

The value of sustainability is deduced from the modeling of individual preferences so that the assessment fits the expressed preferences with regard to sustainability. As is known, modeling of preferences follows a process whereby the evaluator compares pairs of locations in the territory. Thus, the importance of selecting appropriate locations for the comparison of sustainability is evident, especially in view of the fact that comparisons can only be made on a limited number of points.

The selected locations have been chosen to ensure that they are representative of the variability of the forest analyzed. To do so, we measured the difference between the value of the sustainability indicators at any spatial point in the area of application and at an *IP* (see the resulting map in Figure 7.10). At the *IP*, all the sustainability indicators would have the greatest possible value. We adopted a statistical distance, based on the Mahalanobis (1936) distance, to measure the difference between any *x* point and the *IP*:

$$S_{x \to IP} = 100 \times \frac{\sqrt{(x-IP)\sum^{-1}(x-IP)^T}}{D_{max}} \qquad (7.48)$$

where
> D_{max} is the maximum value of the numerator in the preceding expression for all points in the study area
>
> $x = (x_1, x_2, \ldots, x_n)$ is the vector of the value of the *n* sustainability indicators at the *x* point. Their values are ranked from 0 to 100 and refer to the minimum and maximum value of each sustainability indicator in the area of application
>
> $IP = (100, 100, \ldots, 100)$ is the vector of the value of the *n* indicators at the ideal point
>
> $\sum$ is the correlation matrix between each pair of sustainability indicators for all points in the area of application

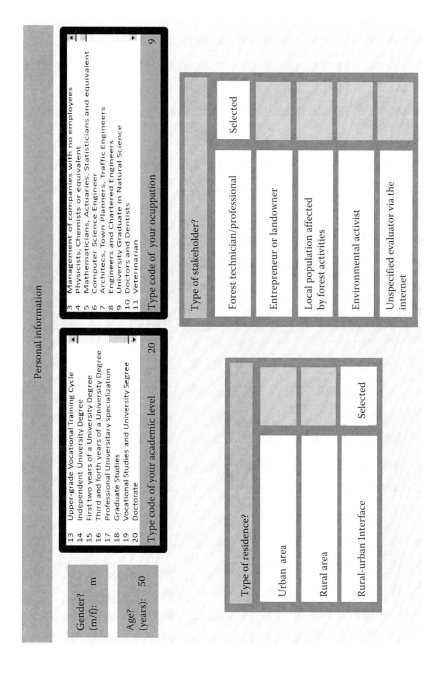

FIGURE 7.9 Example of data entry screen for the introduction of personal information of each evaluator.

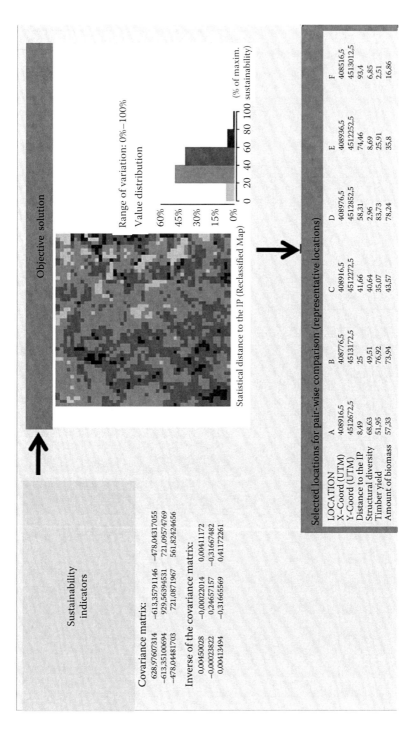

FIGURE 7.10 Spatial distribution of objective value of sustainability in the area of application and selection of the most significant territorial points for pair-wise comparison.

Locations were considered representative in the case that both the relative value of the previous statistical distance between consecutive points remained constant and when the locations covered most of the variety in the area of application. As we could only choose up to six sites, those selected had a distance to the *IP* as close as possible to 8.333%, 25%, 41.667%, 58.333%, 75%, and 91.667%, respectively. These closest points and distances were Point A, 1.16%; Point B, 23.77%; Point C, 43.46%; Point D, 60.91%; Point E, 62.87%; and Point F, 89.11%.

Once the spatial locations to be compared had been selected, the individual could be directly asked which spatial location he or she considered to be more sustainable from each pair formed from the selected locations (pair-wise comparison). Thus, each evaluator was asked a question (of the type shown in Figure 7.11a) for each pair of locations to be compared. To assist in the answer, the evaluator was provided with information on the spatial locations to be compared, such as descriptive information in plain language (Schiller et al., 2001) and an explanation of the performance of the sustainability indicators at each point (Figure 7.11b). Historical information on what the other evaluators had decided when asked about the same pair of locations is also shown (Figure 7.11c).

The next step is to build the assessment. If the preferences follow a complete preorder and the value function supports an additive decomposition, then there are operational methodologies, which, when applied to the preferences, give a value that is consistent with them. Here, we should highlight the methodologies presented in Section 7.3.3, some of those mentioned in Section 7.3.4, and other proven processes, such as Macbeth (see, e.g., Bana et al., 2005), that introduce qualitative judgments about the overall value differences between the pairs of alternatives compared.

However, the existence and the additivity of the value function are not assured. As we have to work with information that comes from the pairs of compared locations,

(a)

FIGURE 7.11 (**See color insert.**) Example of data entry screen for pair-wise comparison of sustainability of each evaluator and information provided for the comparison: (a) real image of the compared points.

(*continued*)

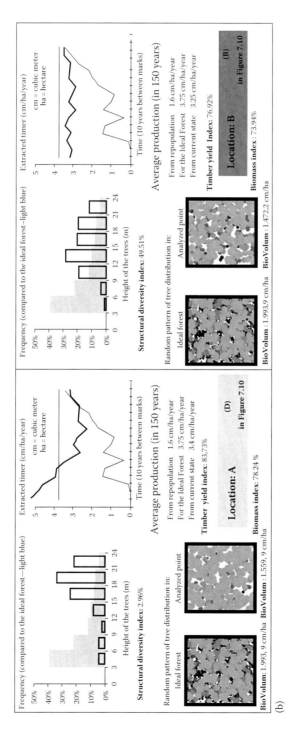

(b)

FIGURE 7.11 (continued) **(See color insert.)** Example of data entry screen for pair-wise comparison of sustainability of each evaluator and information provided for the comparison: (b) information about the indices of sustainability on each point to compare.

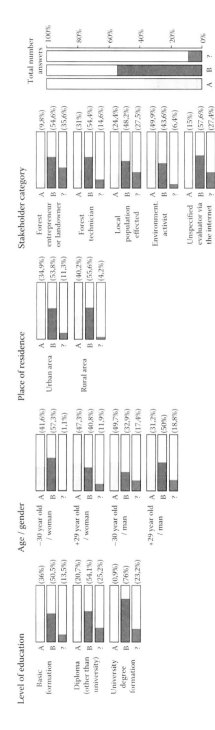

Fecha da la última actualization del histórico de respuestas: 20/10/2009

(c)

FIGURE 7.11 (continued) **(See color insert.)** Example of data entry screen for pair-wise comparison of sustainability of each evaluator and information provided for the comparison: (c) statistics of previous answers provided by other evaluators.

we took $u(x)$ (expression number (7.38)) as the number of pairs in which x both appeared and was considered by the evaluator to be more sustainable than the other element of the pair. The mapping of u to the set of meaningful spatial locations would belong to the set of natural numbers (from 0 to the number of selected meaningful locations minus one). In the case of P being a linear order, then $u(x) \geq u(y) \Leftrightarrow x \gtrsim y$, $\forall x, y \in \Omega$, and $u(x) \geq u(y) \Leftrightarrow x \succeq y, \forall x, y \in \Omega$; therefore, $u(x)$ (number of pairs of points where x is preferred) is a value function. When P is not a linear order, linearization procedures may be applied to force P to converge to a linear order: following Fishburn (1985), we applied the transitive closure of I (I^k) and sequel-type sequential transformations (see expression number (7.4)).

To resolve ties, we add the preceding computed sustainability [$u(x)$] and the statistical distance of the analyzed point (x) to the point of maximum sustainability (computed as 100 minus expression (48), but ranked from 0 to 1). This operation does not change the order of the locations that do not have ties, since the result of applying $u(x)$ belongs to the set of natural numbers and the added amount has a range from 0 to 1.

The spread of the assessment at all points in the area of application is made by applying expression (7.39). The statistical distance used is based on the Mahalanobis distance (expression (7.48)), although for this calculation, the vector representing the performances at the *IP* has been replaced by the values of the sustainability indicators at each point of the pair-wise comparison.

The outcomes of a process of sustainability assessment are shown in Figure 7.12. They correspond to the preferences of an evaluator whose response to the pair-wise comparison is also recorded in the same figure.

The next step is to obtain the analytical expression for the value, which involves applying the methodology described in Section 7.4.2. Table 7.9 shows the outcomes of the regression model in the area of application, and for the evaluator whose preferences are modeled in Figure 7.12. As can be seen, it is not possible to accept that the contribution of independent variables is zero (*F-ratio*), so regression analysis can be applied to determine the analytical expression of value. In Table 7.9, the most influential variables are indicated by an arrow, and thus, the sustainability value is essentially a linear combination of the timber yield (with negative influence) and the amount of biomass. Less influence can be seen in other factors:

$$Sustainability_{\text{(for preferences in Figure 7.12)}} \simeq -5.4\,I_2 + 7.4\,I_3 + I_1\left(-1.1 I_2 + 1.2 I_3\right) \quad (7.49)$$

Figure 7.13 compares the maps obtained by assessing sustainability through the proposed methodology (which has given expression (7.39)) and by applying the regression model described in Table 7.9 (in this case, for all parameters, not only the most meaningful). As can be seen, both maps are almost identical. This fact is consistent with the proportion of combination of values that satisfies the *WDI* condition (expressions (7.40) and (7.41)), which is over 90% of the feasible combinations of indicator performances on the forest operation analyzed.

We also want to highlight the characterization of rationality by analyzing the properties of the original array of preferences (Figure 7.12). In this case, the properties that meet the preferences contained in the pair-wise comparison process define a weak order,

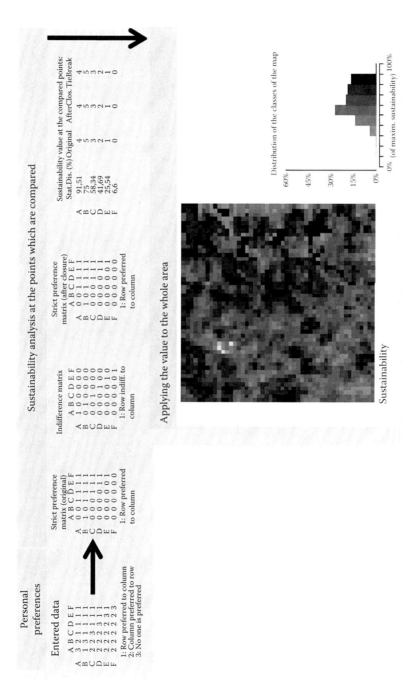

FIGURE 7.12 Steps in the computing of individual value of sustainability on each point of the area of application for the pair-wise comparison of sustainability made by a specific evaluator.

TABLE 7.9
Analysis of the Multiple Regression Model

Parameter	Estimation	Standard Error	t-Statistic	P-Value
Constant	+0.000023	0.000139	0.1657	0.8683
I1	+0.958533	0.002043	469.0316	0
I2	−5.387192	0.002647	−2034.6470	0
I3	+7.383588	0.003018	2445.7779	0
I1 × I2	−1.080177	0.003524	−306.4836	0
I1 × I3	+1.188482	0.004217	278.1796	0
I2 × I3	−0.682679	0.001490	−457.9039	0
I1 × I2 × I3	+0.056520	0.000730	77.3858	0

Analysis of the variance: F-ratio$(8, 2329) = 5873018.84$
R-square (adjusted to the degrees of freedom) $= 0.9999489215$
$$\text{Sustainability} = 0.000023 + 0.958533 \times v_{I1} - 5.387192 \times v_{I2}$$
$$+ 7.383588 \times v_{I3} - 1.080177 \times v_{I1} \times_{I2} + 1.188482 \times v_{I1} \times_{I3}$$
$$- 0.682679 \times v_{I2 \times I3} + 0.056520 \times v_{I1 \times I2 \times I3}$$

so it is a rationality of type I. Moreover, the matrix of indifference (see also Figure 7.12) shows that there are six indifference classes (as compared to the number of alternatives), and therefore, the extent of the assessor's knowledge of the system is high.

7.6.3   GROUPING PEOPLE WITH SIMILAR SYSTEMS OF PREFERENCES

The representation of preferences also allowed us to outline a measurement of the distance between individual preferences, which enabled people with homogenous preferences to be grouped together. This is a typical classification problem that could be addressed by applying the techniques of data mining or knowledge discovery in databases (Dunham, 2002; Mligo and Lyaruu, 2008). However, the existence of quantitative, qualitative, and nominal variables makes it advisable to group individual preferences by using divisive and polythetic clustering (Martínez-Falero and González-Alonso, 1994) applied to a number of descriptors in the classification process. In our case, we considered the following descriptors (see Table 7.10):

- The proximity between the sustainability value assigned by any individual (Equation 7.39) and the objective sustainability value computed from expression (7.48). The descriptors were the presence or absence of a specific level of similarity between these values, and the levels of proximity were very far from the sustainability assessed by the objective procedure (0%–20%), low–medium distance from the objective sustainability value (20%–40%), close (40%–60%), and very close to the objective value (>60%).
- The "*taste*" of each individual for sustainability. Here, descriptors indicated the presence or absence of the independent variables in the analytical expression of the value function.

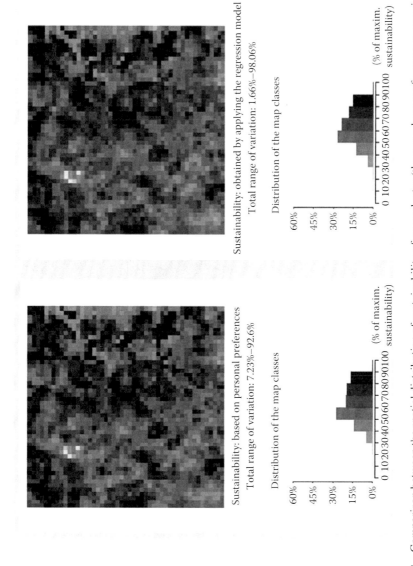

FIGURE 7.13 Comparison between the spatial distribution of sustainability of an evaluator (the one whose preferences appear in Figure 7.12) in two cases: computed directly from his or her pair-wise comparison of sustainability and computed from the analytical form of his or her sustainability value function.

TABLE 7.10

Descriptors for Clustering Evaluators' Preference

Proximity to Objective Sustainability (%)	Type of Order Relation	Most Significant Regression Parameters	Percentage of Linearization	Depth of Knowledge
1. (80, 100)	6. Linear order or	12. I1	19. 100%	23. High
2. (60, 80)	weak order	13. I2	20. 66.66%–100%	24. Medium
3. (40, 60)	7. Semiorder	14. I3		25. Low
4. (20, 40)	8. Interval order	15. I1 × I2	21. 33.33%–66.66%	
5. (10, 20)	9. Biorder	16. I1 × I3		
	10. Partial order	17. I2 × I3	22. 0%–33.33%	
	11. Acyclic weak biorder (or none)	18. I1 × I2 × I3		

- The level of linearization of individual value function. Descriptors expressed the level of linearization of the value function (percentage of combinations satisfying *WDI* conditions—expressions (7.40) and (7.41)—ranked to very high, high, low, and very low).
- The type of order of the preference relation (see Table 7.2).
- The depth of knowledge of the evaluator: high, medium, and low (see Section 7.5.2).

The first step in the classification process was the division of the initial set of individuals into two groups that were described by the presence in each one of a small number of significant descriptors. The process was repeated for each group obtained; all the groups contained at least two individuals. However, having a large number of groups of individuals makes it difficult to reach final decisions. For this reason, the classification process was halted as soon as it reached the maximum number of groups of individuals previously indicated. The chosen clusters could belong to previous levels, provided that the variation of descriptors between groups at the same level, as compared to the variation within groups, was the maximum.

The aforementioned classification process ensured that the individuals belonging to the same group had similar systems of preferences. However, the level of proximity may not have been sufficient to allow the preferences of all individuals in the group to be reflected using the same map of sustainability. Therefore, the map representing the whole group could be obtained as an average of the sustainability values assigned to each point in the territory by all the individuals in the group, and the classification process would stop when a homogeneity threshold within groups was reached. Otherwise, it would be necessary to continue the process of division until the variability of the maps in the group enabled all the preferences to be represented with a single map. Additionally, in order to facilitate the decision making (and also the interaction among potential evaluators), statistical information about the personal characteristics of the members of the group was also provided.

However, clustering processes are computationally slow and require a representative sample of individual preferences, as otherwise the preferences will form different groups each time the clustering is done. Thus, a previous step in this process was to generate a number of communities that represent all the types of preferences. For these reasons, the homogeneous communities were precalculated from a simulated random sample of 5000 individuals with different types of preferences (5000 different 6×6 matrices with values from 1 to 3, as shown in Figure 7.12). As a result, 53 groups of evaluators were obtained and characterized by applying the methodology described.

Each new evaluator is assigned to the most likely group after applying discriminatory analysis. The goodness of allocating an individual to his or her group can be seen in Figure 7.13. This figure shows the sustainability maps for two preference schemes: one corresponding to the current evaluator (with preferences expressed in Figure 7.12) and the other for "*the most characteristic individual*" in the group to which the first individual belongs. The most characteristic individual in a group is the one whose descriptors, integrated according to the most meaningful factor in the classification process, are the closest to the average value in his or her group.

7.6.4   DISCUSSION AND CONCLUSIONS OF THE PROPOSED METHODOLOGY

The most important contributions of the proposed methodology are

1. It provides a numerical representation of preferences at representative points of the territory, by modeling an individual's preferences based on pair-wise comparison of sustainability. The sustainability value is then allocated to all the points in the forest site by applying a statistical distance between every point and the most representative points.
2. It estimates the global sustainability value function in order to determine the importance assigned by any evaluator to each sustainability indicator. We have verified both the existence and form of the analytical expression of the function for describing sustainability at each point of the *forest operation*.
3. It uses parameters that describe the modeling of preferences of each individual to define groups of evaluators with analogous systems of preferences for the assessment of sustainability.

In conclusion, this chapter integrates the works on sustainability assessment from the values of particular criteria or attributes, with works based on a small number of pair-wise comparisons of global sustainability (Figure 7.14).

Previous works on the assessment of forest and natural resource management by different stakeholders have focused particularly on assessment criteria and attributes. The most commonly applied assessment method is multicriteria decision making (*MCDM*) and particularly (Korhonen and Wallenius, 2001) *AHP*, a method that generally requires a large number of pair-wise comparisons. As an example, Mendoza and Prabhu (2000) applied *AHP* at the indicator level to a process in which a panel of experts assessed the sustainability of Indonesian forests. These experts felt uncomfortable with pair-wise comparisons due to the amount of one-on-one judgments they had to make. Moreover, Lahdelma et al. (2000) assert that pair-wise comparison

(Descriptors)

	1	2	3	4	5	6	7	8	9	10	11	12	13	14	15	16	17	18	19	20	21	22	23	24	25
	Prox. to ob. sustainability					Type of order relation						Regression parameters							Linearization				D.Know.		
(Individual)	1	0	1	0	0	0	1	0	0	0	0	0	1	1	0	0	0	0	0	0	0	1	1	0	0
(Group N°33)	0	-1	0	0	0	-1	-1	-1	0	-1	-1	0	1	-1	1	0	-1	-1	0	0	0	-1	-1	1	-1

1. Presence of the descriptor in the individual

1. Significant presence of this description in the group (P > 0.45)
-1. Significant absence of this descriptor in the group (P > 0.90)

Properties of the group to which this individual belongs:

Internal variability of the group: 4,137
(Average of the standard deviation of the value of sustainability at the six most significant points of the area of application for all the individuals in the group)

Proportion of individuals in this group: 0,28% of the total number which has participated
identification of the group in the classification: 33 (out of a total number of 53 groups)

Most common characteristics of the preferences of the individuals which belongs to this group:

Sustainability value far from the objective value
Rationality of the decisions with coherence
Average proportion of linearization in the analytical expression of the value function
Medium depth of knowledge

FIGURE 7.14 Automatic description of the main characteristics and qualities of an individual as evaluator of sustainability (the one whose preferences appear in Figure 7.12) and of the group of evaluators he or she belongs to.

methods tend to lose their efficiency as the number of criteria increases. Regarding voting methods, Mendoza and Prabhu (2000) found these *MCDM* methods were not sufficiently refined to reflect the degrees of importance or relative significance of each criterion and indicator.

The application of *MCDM* methods makes it possible to identify directly how different stakeholders value different objectives. This provides valuable information when resolving conflicts between stakeholders. In our case, the importance the individual awards to each indicator is shown in the regression coefficient of the value function, as it is estimated based on the value of the indicators at each point in the territory and from the sustainability value obtained. However, it is only possible to determine the stakeholder's evaluation of each indicator when the linearization coefficient to assess the validity of the value function is high.

In most cases, *MCDM* methods such as *AHP* formulate an additive value function without analyzing whether or not the value function exists. The analysis of the preferences matrix allows us to ensure the existence of a value function when the preferences are a linear or a weak-order relationship. Otherwise, we apply a transformation that tends to establish this system of preference. Furthermore, the knowledge of the properties of the preference matrix provides far more information about the internal consistency of the decision maker than any quantitative measure, such as those usually applied in *AHP*, whose consistency index only measures the transitivity of the preference. With regard to the form of decomposition of the value function, most multicriteria methods implicitly assume the existence of some kind of value function, but none of them recommends testing the conditions for the existence of the necessary analytical form. In the proposed methodology, the value function is derived from the relation between the global value of sustainability (dependent variable) and all the feasible combinations of the sustainability indicators. This replaces the need to verify the existence of a specific form of decomposition for the value function. An index of the suitability of the quasi-additive decomposition is also applied to all cases. This confers an important advantage over other methods.

An approach based on a reduced number of comparisons of global sustainability is proposed by Reichert et al. (2007) to identify river rehabilitation actions. In this work, preferences were elicited for scenarios generated according to likely distributions of attributes, which was considered to be more satisfying to evaluators than directly assessing alternative actions. However, unlike our proposed method, Reichert et al. (2007) do not provide a value at territory level. In addition, our approach emphasizes the comparison of scenarios at a local level as well as the need for iterative learning among participants. The consequences of the first point are clear: sustainability has to be valued at the local level; we, therefore, agree with Sheppard and Meitner (2005), who argue that the practice of public involvement requires the development of techniques that can be used by local managers for operational decision making, rather than through the establishment of regional strategies. With regard to the second point, the scientific literature suggests that any participation process should emphasize iterative learning among the participants (Chase et al., 2004; Johnson et al., 2004; Lynam et al., 2007). Our method complies with this requirement, since every evaluator has access to the historical answers of the others (Figure 7.15).

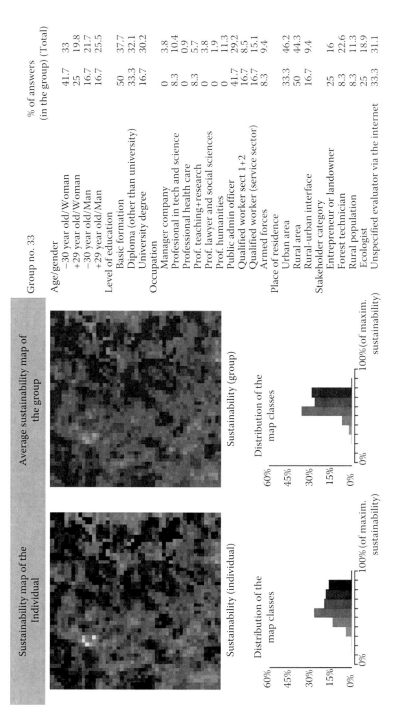

Group no. 33	% of answers (in the group)	(Total)
Age/gender		
−30 year old/Woman	41.7	33
+29 year old/Woman	25	19.8
−30 year old/Man	16.7	21.7
+29 year old/Man	16.7	25.5
Level of education		
Basic formation	50	37.7
Diploma (other than university)	33.3	32.1
University degree	16.7	30.2
Occupation		
Manager company	0	3.8
Profesional in tech and science	8.3	10.4
Professional health care	0	0.9
Prof. teaching+research	8.3	5.7
Prof. lawyer and social sciences	0	3.8
Prof. humanities	0	1.9
Public admin officer	0	11.3
Qualified worker sect 1+2	41.7	29.2
Qualified worker (service sector)	16.7	8.5
Armed forces	16.7	15.1
	8.3	9.4
Place of residence		
Urban area	33.3	46.2
Rural area	50	44.3
Rural–urban interface	16.7	9.4
Stakeholder category		
Entrepreneur or landowner	25	16
Forest technician	8.3	22.6
Rural population	8.3	11.3
Ecologist	25	18.9
Unspecified evaluator via the internet	33.3	31.1

FIGURE 7.15 Comparison between the spatial distribution of sustainability of an evaluator (the one whose preferences appear in Figure 7.12) and the average distribution of sustainability of the group to which he or she belongs to.

Finally, one of the current problems in public participation is the need to empower participants through participation. According to Reed (2008), one of the ways of solving this problem is by engaging the evaluator in the decision-making process. When the decisions are technically complex as in forest management, Reed (2008) highlights the need to educate participants and to develop their knowledge in order for them to engage meaningfully in the process. The proposed methodology has been developed using a software application that can be implemented on the web. This application contains scientific information on indicators, sustainable forest management, and preference analysis, expressed in everyday language. In addition, the final decision maker has information on the degree of consistency and depth of knowledge of the groups of evaluators, which facilitates the decision as to the choice of management plan.

REFERENCES

Aleskerov, F., D. Bouyssou, and B. Monjardet. 2007. *Utility Maximization, Choice and Preference*. Berlín, Germany: Springer.

Arrow, K.J. and G. Debreu. 1954. Existence of an equilibrium for a competitive economy. *Econometrica* 22: 265–290.

Arrow, K.J. 1959. Rational choice functions and orderings. *Economica* 26: 121–127.

Arrow, K., A.K. Sen, and K. Suzumura. 2002. *Handbook of Social Choice and Welfare*. Amsterdam, the Netherlands: North Holland:Elsevier.

Axelrod, R. 1976. *Structure of Decision*. Princeton, NJ: University of Princeton Press.

Bana, E., C.A. Costa, J.M. de Corte, and J.C. Vansnick. 2005. On the mathematical foundation of MACBETH. In *Multiple Criteria Decision Analysis: The State of the Art Surveys. International Series in Operations Research & Management Science*, eds. J. Figueira, S. Greco, and M. Ehrgott. New York: Springer. pp. 409–442.

Barbati, A., P. Corona, and M. Marchetti. 2007. A forest typology for monitoring sustainable forest management: The case of European forest types. *Plant Biosystems* 141: 93–103.

Belton, V. and T.J. Stewart. 2001. *Outranking Methods, Multiple Criteria Decision Analysis: An Integrated Approach*. Boston, MA: Kluwer Academic Publishers.

Billot, A. 2003. The deep side of preference theory. *Theory and Decisions* 53: 243–270.

Bouyssou, D. 2001. Outranking methods. In *Encyclopedia of Optimization,* eds. C.A. Floudas and P.M. Pardalos. Boston, MA: Kluwer Academic Publishers.

Bouyssou, D. and M. Pirlot. 2005. A characterization of concordance relations. *European Journal of Operational Research* 167:427–443.

Brans, J.P. 1982. L'ingénièrie de la décision; Elaboration d'instruments d'aide à la décision. La méthode PROMETHEE. In *L'aide à la décision: Nature, Instruments et Perspectives d'Avenir*, eds. R. Nadeau and M. Landry, pp. 183–213. Québec, Canada: Presses de l'Université Laval.

Brans, J.P. and B. Mareschal. 2005. PROMETHEE methods. In *Multiple Criteria Decision Analysis: State of the Art Surveys,* eds. J. Figueira, S. Greco, and M. Ehrgott, pp. 163–196. Boston, MA: Springer Verlag.

Briges, D.S. and G.B. Mehta. 1995. *Representations of Preference Orderings*. Berlin, Germany: Springer-Verlag.

Cabot, J., S. Easterbrook, J. Horkoff, J.N. Mazón, L. Lessard, and S. Liaskos. 2009. *Integrating Sustainability in Decision-Making Processes: A Modelling Strategy*. Toronto, Ontario, Canada: ICSE'09, International Conference on Software Engineering.

Chankong, V. and Y.Y. Haimes. 1983. *Multiobjective Decision Making Theory and Methodology*. New York: Elsevier Science.

Chase, L.C., D.J. Decker, and T.B. Lauber. 2004. Public participation in wildlife management: What do stakeholders want? *Society & Natural Resources* 17: 629–639.

Checkland, P. 1981. *Systems Thinking, Systems Practice.* New York: John Wiley & Sons.

Cozzens, M.B. and F.S. Roberts. 1982. Double semiorders and double indifference graphs. *SIAM Journal on Algebraic and Discrete Methods* 3: 566–583.

Debreu, G. 1954. Representation of a preference ordering by a numerical function. In *Decision Processes*, eds. C.H. Thrall and R. Davies, pp. 159–175. New York: John Wiley & Sons.

Dietrich, F. and C. List. 2007. Arrow's theorem in judgment aggregation. *Social Choice and Welfare* 29: 19–33.

Dubois, D. and H. Prade. 1980a. *Fuzzy Sets & Systems: Theory and Applications.* New York: Academic Press (APNet).

Dubois, D. and H. Prade. 1980b. The advantages of fuzzy set approach in OR/MS demonstrated on two examples of resource allocation problem. In *Progress in Cybernetics and Systems Research*, eds. R. Trappl, G.J. Klir, and F.R. Pichler. Viena, Austria: Hemisphere Publishing Corporation.

Dunham, M.H. 2002. *Data Mining, Introductory and Advanced Topics.* New York: Prentice Hall.

Dushnik, B. and E.W. Miller. 1941. Partially ordered sets. *American Journal of Mathematics* 63: 600–610.

Dyer, J.S. and K.S. Sarin. 1979. Measurable multiattribute value functions. *Operations Research* 27: 810–822.

Eden, C. 1989. Strategic options development and analysis—SODA. In *Rational Analysis in a Problematic World*, ed. J. Rosenhead, pp. 21–42. London, U.K.: John Wiley & Sons.

Edwards, W. 1977. How to use multiattribute utility measurement for social decision making. *IEEE Transactions on Systems, Man, and Cybernetics* 7: 326–340.

Figueira, J., S. Greco, and M. Ehrgott. 2005a. *Multiple Criteria Decision Analysis: State of the Art Surveys.* Boston, MA: Springer Verlag.

Figueira, J., V. Mousseau, and B. Roy. 2005b. ELECTRE methods. In *Multiple Criteria Decision Analysis: State of the Art Surveys*, eds. J. Figueira, S. Greco, and M. Ehrgott, pp. 133–162. Boston, MA: Springer Verlag.

Fishburn, P.C. 1967. *Additive Utilities with Incomplete Product Set: Applications to Priorities and Assignments.* Baltimore, MD: Operations Research Society of America (ORSA).

Fishburn, P.C. 1974. Lexicographic orders, utilities and decision rules: A survey. *Management Science* 20:1442–1471.

Fishburn, P.C. 1985. *Interval Orders and Interval Graphs.* New York: John Wiley and Sons.

Fodor, J.C., S. Orlovski, P. Perny, and M. Roubens. 1998. The use of fuzzy preference models in multiple criteria choice, ranking and sorting. In *Handbook of Fuzzy Sets and Possibility Theory, Operations Research and Statistics*, eds. R. Slowinski, D. Dubois, and H. Prade, pp. 69–101. Boston, MA: Kluwer Academic Publishers.

Fodor, J.C. and M. Roubens. 1994. *Fuzzy Preference Modelling and Multicriteria Decision Support.* Dordrecht, the Netherlands: Kluwer Academic Publishers.

Fodor, J.C. and M. Roubens. 1995a. Possibilistic mixtures and their applications to qualitative utility theory I: Aggregation of possibility measures. In *Foundations and Applications of Possibility Theory*, eds. G. de Cooman, D. Ruan, and E.E. Kerre. Singapore: World Scientific.

Fodor, J.C. and M. Roubens. 1995b. Structure of transitive valued binary relations. *Mathematical Social Sciences* 30: 71–94.

Freer-Smith, P. and J.M. Carnus. 2008. The sustainable management and protection of forests: Analysis of the current position globally. *Journal of the Human Environment* 37: 254–262.

Friend, J.K. and A. Hickling. 1987. *Planning Under Pressure: The Strategic Choice Approach.* New York: Pergamon Press.

García Abejón, J.L. and J.A. Gómez Loranca. 1984. *Tablas de producción de densidad variable para Pinus sylvestris L. en el sistema central.* Madrid: Instituto Nacional de Investigaciones Agrarias, Ministerio de Agricultura, Pesca y Alimentación, Serie Recursos Naturales. No. 29.

Hickey, G.M. and J.L. Innes. 2008. Indicators for demonstrating sustainable forest management in British Columbia, Canada: An international review. *Ecological Indicators* 8: 131–140.

Janardan, K. and D.J. Schaeffer. 1975. *Illinois Water Quality Inventory Report.* Chicago, IL: Environmental Protection Agency.

Johnson, N., N. Lilja, J.A. Ashby, and J.A. Garcia. 2004. Practice of participatory research and gender analysis in natural resource management. *Natural Resources Forum* 28: 189–200.

Kahneman, D. and A. Tversky. 2000. *Choices, Values, and Frames.* Cambridge, U.K.: Cambridge University Press.

Keeney, R.L. 1981. Measurement scales for quantifying attributes. *Behavioral Science* 26: 29–36.

Keeney, R.L. 1988. Structuring objectives for problems of public interest. *Operations Research* 36:396–405.

Keeney, R.L. 1992. *Value-Focused Thinking. A Path to Creative Decision Making.* Cambridge, MA: Harvard University Press.

Keeney, R.L. and H. Raiffa. 1976. *Decisions with Multiple Objectives: Preferences and Value Tradeoffs.* New York: John Wiley & Sons.

Kirkwood, C.W. and R.K. Sarin. 1980. Preference conditions for multiattribute value functions. *Operations Research* 28: 225–232.

Korhonen, P. and J. Wallenius. 2001. On using the AHP in multiple objective linear programming. In *The Analytic Hierarchy Process in Natural Resources and Environmental Decision Making,* eds. D. Schmoldt, J. Kangas, G. Mendoza, and M. Pesonen. New York: Kluwer Academic Publishers.

Lahdelma, R., J. Hokkanen, and P.S. Alminen. 2000. Using multicriteria methods in environmental planning and management. *Environmental Management* 26:595–605.

Lynam, T., W. De Jong, D. Sheil, T. Kusumanto, and K. Evans. 2007. A review of tools for incorporating community knowledge, preferences, and values into decision making in natural resources management. *Ecology & Society* 12: 5 [online]. http://www.ecologyandsociety.org/vol12/iss1/art5/.

Mahalanobis, P.C. 1936. On the generalized distance in statistics. *Proceedings of the National Institute of Sciences of India* 2: 49–55.

Makropoulos, C.K., K. Natsis, S. Liu, K. Mittas, and D. Butler. 2008. Decision support for sustainable option selection in integrated urban water management. *Environmental Modelling & Software* 23: 1448–1460.

Martel, J.M. and B. Matarazzo. 2005. Other outranking approaches. In *Multiple Criteria Decision Analysis: State of the Art Surveys,* eds. J. Figueira, S. Greco, and M. Ehrgott. pp. 197–264. Boston, MA: Springer Verlag.

Martínez-Falero, E., E. Ayuga, and C. González García. 1992. Estudio comparativo de distintas funciones núcleo para la obtención del mejor ajuste según el tipo de datos. *Qüestiió: Quaderns d'Éstadística i Investigació Operativa* 16: 3–26.

Martínez-Falero, E. and S. González-Alonso. 1994. *Quantitative Techniques in Landscape Planning.* Boca Raton, FL: CRC Press.

Martín-Fernández, S. and A.García-Abril. 2005. Optimization of spatial allocation of forestry activities within a forest stand. *Computers and Electronics in Agriculture* 49: 159–174.

Martínez-Falero, E., S. Martín-Fernandez, and A. García-Abril. 2010. *Participación Pública para la Gestión Forestal Sostenible.* Madrid, Spain: Fundación del Conde del Valle de Salazar.

Mendoza, G.A. and R. Prabhu. 2000. Development of methodology for selecting criteria and indicators for sustainable forest management: A case study on participatory assessment. *Environmental Management* 26: 659–673.

Miller, G.A. 1956. The magical number seven, plus or minus two: Some limits on our capacity for processing information. *Psychological Review* 63: 81–97.

Mligo, C. and H.V.M. Lyaruu. 2008. The impact of browsing and grazing pressure on vegetation community, composition and distribution pattern in Ikona wildlife. *Botany Research Journal* 1: 1–8.

MOPT. 1993. *Guía metodológica para la elaboración de estudios del medio físico.* España, Spain: Centro de publicaciones Secretaría Técnica MOPT.

Morton, A., F. Ackermann, and V. Belton. 1999. *Delphic SODA: A New Approach to Distributed Group Decision Support.* Glasgow, Scotland: Department of Management Science, University of Strathclyde.

Morton, A., F. Ackermann, and V. Belton. 2001. *Distributed Group Decision Support. A Study of Some Key Concepts.* Glasgow, Scotland: Department of Management Science, University of Strathclyde.

Newkirk, R.T. 1979. *Environmental Planning for Utility Corridor.* Ann Arbor, MI: Ann Arbor Science.

Ngo The, A. and A. Tsoukiàs. 2005. Numerical representation of PQI interval orders. *Discrete Applied Mathematics* 147: 125–146.

Olson, D.L. 2008. Multicriteria decision support. In *Handbook on Decision Support Systems*, eds. F. Burstein and C. Holsapple, pp. 299–314. Heidelberg, Germany: Springer.

ORWorld, 2002. Learning and teaching operations research and management science with a web-based hypermedia learning environment. EU-Research Project No: IST-1999-11124, January 4, 2000–January 10, 2002. ftp://ftp.cordis.europa.eu/pub/ist/docs/ka3/eat/OR-WORLD.pdf

Otero, I. 1979. *El Análisis Cualitativo de los Elementos el Medio Natural en Orden a la Planificación Física.* Madrid.

Oztürk, M., A. Tsoukiàs, and P. Vincke. 2005. Preference modelling. In *Multiple Criteria Decision Analysis: State of the Art Surveys*, eds. J. Figueira, S. Greco, and M. Ehrgott, pp. 27–72. Boston, MA: Springer Verlag.

Pascual, C., A. García-Abril, L.G. Garcia-Montero, S. Martin Fernandez, and W.B. Cohen. 2008. Object-based semi-automatic approach for forest structure characterization using LIDAR data in heterogeneous Pinus sylvestris stands. *Forest Ecology and Management* 255: 3677–3685.

Perny, P. and M. Roubens. 1998. Preference modelling. In *Handbook of Fuzzy Sets and Possibility Theory, Operations Research and Statistics*, eds. R. Slowinski, D. Dubois, and H. Prade. pp. 3–30. Boston, MA: Kluwer Academic Publishers

Pirlot, M. 1990. Minimal representation of a semiorder. *Theory and Decision* 28(2): 109–141.

Pöyhönen, M., H. Vrolijk, and R. Hämäläinen. 2001. Behavioral and procedural consequences of structural variation in value trees. *European Journal of Operational Research* 134: 216–227.

Ramos, A. 1982. *Evaluación Integrada de Espacios Naturales. Aplicación a los Espacios Arbolados de Madrid.* Madrid, Spain: Instituto Tecnológico Geominero de España.

Rand, B. 1912. *The Classical Psychologists: Selections Illustrating Psychology from Anaxagoras to Wundt.* Boston, MA: Houghton Mifflin.

Reed, M.S. 2008. Stakeholder participation for environmental management a literature review. *Biological Conservation* 141: 2417–2431.

Reichert, P., M. Borsuk, M. Hostmann, S. Schweizer, C. Spörri, K. Tockner, and B. Truffer. 2007. Concepts of decision support for river rehabilitation. *Environmental Modelling and Software* 22: 188–201.

Reynolds, K.M., A. J. Thomson, M. Köhl, M. A. Shannon, D. Ray, and K. Rennolls. 2007. *Sustainable Forestry from Monitoring and Modelling to Knowledge Management & Policy Science*. Wallingford, U.K.: CAB International.

Roubens, M. and Ph. Vincke. 1985. *Preference Modeling*. LNEMS 250. Berlin, Germany: Springer Verlag.

Roy, B. 1968. Classement et choix en présence de points de vue multiples: La méthode ELECTRE. *Revue Francaise d'Informatique et de Recherche Opérationnelle* 8: 57–75.

Roy, B. 1989. Main sources of inaccurate determination, uncertainty and imprecision. *Mathematical and Computer Modelling* 12: 1245–1254.

Roy, B. 1985. *Méthodologie multicritère d'aide à la décision*. Paris, France: Economica.

Roy, B. 1996. *Multicriteria Methodology for Decision Aiding*. Dordrecht, the Netherlands: Kluwer Academic.

Saaty, T.L. 1994. *Fundamentals of Decision Making with the Analytic Hierarchy Process*. Pittsburgh, PA: RWS Publications.

Saaty, T.L. 1996. *The Analytic Network Process: Decision Making with Dependence and Feedback*. Pittsburgh, PA: RWS Publications.

Saaty, T.L. 2000a. *Models, Methods, Concepts and Applications of the Analytic Hierarchy Process*. Boston, MA: Kluwer Academic Publisher.

Saaty, T.L. 2000b. *The Brain, Unraveling the Mystery of How it Works: The Neural Network Process*. Pittsburgh, PA: RWS Publications.

Saaty, T.L. 2009. Extending the measurement of tangibles to intangibles, (with Mujgan Sagir). *International Journal of Information Technology and Decision Making* 8: 7–27.

Saaty, T.L. 2010. *Principia Mathematica Decernendi: Mathematical Principles of Decision Making*. Pittsburgh, PA: RWS Publications.

Schiller, A., C.T. Hunsaker, M.A. Kane, A.K. Wolfe, V.H. Dale, G.W. Suter, C.S. Russell, G. Pion, M.H. Jensen, and V.C. Konar. 2001. Communicating ecological indicators to decision makers and the public. *Conservation Ecology* 5(1): 19. http://www.consecol.org/vol5/iss1/art19

Scott, D. and P. Suppes. 1958. Foundational aspects of theories of measurement. *The Journal of Symbolic Logic* 23: 113–128.

Sen, A. 1987. Rational behaviour. In *The New Palgrave Dictionary of Economics, IV,* ed. J.L. Eatwell, M. Milgatey, and P. Newman, pp. 68–76. London, U.K.: Macmillan.

Sheppard, S.R.J. and M. Meitner. 2005. Using multi-criteria analysis and visualisation for sustainable forest management planning with stakeholder groups. *Forest Ecology and Management* 207: 171–187.

Suzumura, K. 1983. *Rational Choice, Collective Decisions and Social Welfare*. Cambridge, New York: Cambridge University Press.

Tamubula, I. and J.A. Sinden. 2000. Sustainability and economic efficiency of agroforestry systems in Embu District, Kenya: An application of environmental modeling. *Environmental Modelling & Software* 15: 13–21.

Tsoukiàs, A. and Ph. Vincke. 1997. Extended preference structures in MCDA. In *Multi-criteria Analysis*, ed. J. Clímaco. pp. 37–50. Berlin, Germany: Springer Verlag.

Tsoukiàs, A. and Ph. Vincke. 2003. A characterization of PQI interval orders. *Discrete Applied Mathematics* 127: 387–397.

Vainikainen, N., A. Kangas, and J. Kangas. 2008. Empirical study on voting power in participatory forest planning. *Journal of Environmental Management* 88: 173–180.

Vincke, P. 1998. *Robust Solutions and Methods in Decision-Aid*. Brussels, Belgium: Université Libre de Bruxelles.

Vincke, P. 2001. Preferences and numbers. In *A-MCD—Aide Multi Critère à la Décision—Multiple Criteria Decision Aiding*, eds. A. Colorni, M. Paruccini, and B. Roy, pp. 343–354. Brussels, Belgium: Joint Research Center, The European Commission.

Von Winterfeldt, D. and W. Edwards. 1986. *Decision Analysis and Behavorial Research.* Cambridge, U.K.: Cambridge University Press.

Wang F.Y. 2008. Social computing: Fundamentals and applications. *IEEE International Conference on Intelligence and Security Informatics.* Tucson, AZ.

Wang, F.Y., K. M. Carley, and W. Mao. 2007. Social computing: From social informatics to social intelligence. *IEEE Intelligent Systems* 22: 79–83.

Wijewardana, D. 2008. Criteria and indicators for sustainable forest management: The road travelled and the way ahead. Ecological indicators: Integrating, monitoring, assessment and management. *Evaluating Sustainable Forest Management* 8: 115–122.

Zadeh, L. 1965. Fuzzy sets. *Information and Control* 8: 338–353.

Zadeh, L. 1975. Fuzzy logic and approximate reasoning. *Synthese* 30: 407–428.

8 Optimization Methods to Identify the Best Management Plan

Susana Martín-Fernández, Eugenio Martínez-Falero, and Miguel Valentín-Gamazo

CONTENTS

8.1   INTRODUCTION

A management plan involves the optimization of the use of scarce natural resources to ensure the sustainability of the current multifunctional role of the forest. This chapter describes the methodologies most commonly applied to optimize the sustainable use of forest resources, including an explanatory application of each one to certain stages of forest management. The chapter ends with a case of application that incorporates personal preferences to identify the best forest plan.

The earliest optimization methodology applied to forest planning was linear programming (LP) in the late 1960s, which was used to resolve small or moderate planning problems (Weintraub et al., 2000). The main planning application was in timber harvesting, by means of systems such as RAM (Navon, 1971) and FORPLAN (Jones et al., 1986), which explicitly incorporated environmental issues such as water sedimentation, wildlife, and erosion, and MELA (Siitonen, 1993) and WOODSTOCK (Walters, 1993), developed with the same focus. From the management point of view, timber harvesting involved two approaches to scheduling: the area-restriction model (ARM) and the unit-restriction model (URM), which prohibits two adjacent units from being harvested simultaneously. Both approaches can be formulated as either an integer-linear or mixed-integer programming problem (Murray, 1999; Öhman, 2002).

However, the application of LP methods revealed various weaknesses, such as the size of the problem in the case of URM, which is difficult to resolve; the nonlinear form of the constraints in the case of ARM models, for which finding a solution with exact methods is very limited (see Baskent and Keles (2005) for more details); or the fact that considerations of the adjacency of forest units could not be included. An illustrative example at the stand level is during the clear-cutting activity of one stand or harvest unit, which may expose a neighboring stand or stands to wind damage, bark injuries, soil problems, and site class deterioration (Snyder and ReVelle, 1996; Tarp and Helles, 1997; Malchow-Moller et al., 2004). Another example is at the tree level, in close-to-nature management, where cutting a tree may affect the process of regeneration, reduce competitiveness among trees, and modify the future quality of the wood in the forest (Otto, 1997; De Turckheim, 1999). However, these considerations do not rule out the use of LP and other procedures derived from it. As we shall see, LP is now used in conjunction with more complex methodologies and also in the early stages of optimization (especially for the characterization of the best solution).

Currently, forest planning is perceived to be a complex process that must necessarily include both natural complexity and the complexity arising from the participation of stakeholders—and society in general—in forest management and planning decisions. In consequence, and in order to design a forest plan, management is broken down into more simple subtasks with their own self-organization programs. The scientific literature highlights the use of heuristic—and also non-heuristic—methods to simulate self-organization.

8.1.1 Heuristic Methods to Identify the Forest Management Plan

The term "heuristic" refers to experience-based techniques for problem solving, learning, and discovery. When an exhaustive search is impractical, heuristic methods are used to speed up the process of finding a satisfactory solution. By extension, "metaheuristic" designates a computational method that solves a problem by iteratively improving a candidate with regard to a given measure of quality.

Early optimization studies such as Lockwood and Moore (1993), Baskent et al. (2000), Falçao and Borges (2002), Palahí (2002), Pukkala (2002), and Li et al. (2010) confirmed that both combinatorial optimization methods and artificial neural networks (ANNs) are an alternative to LP. According to Li et al. (2010), the methods most commonly applied to forest planning are the heuristic processes: Monte Carlo integer programming (MCIP), minimum spanning tree, simulated annealing (SA), tabu search (TS), genetic algorithms (GAs), and ANNs in its optimization approach.

The main advantages of these methods are as follows:

- They have been demonstrated to be efficient and quick at finding the desired solution in a reasonable computational time.
- The problem formulation is generally simple and flexible.
- They are able to solve large-scale multi-period forest planning problems.

On the other hand, these heuristic methods have various drawbacks:

- They do not ensure the attainment of the global optimum.
- The techniques are highly parameterized, so the quality of the solution depends on the setting of the parameters.
- The parameterization is problem specific. Therefore, there is an initial stage in the whole process in which the problem has to be analyzed in depth in order to be structured and parameterized according to the requirements of the method applied.

The literature contains the following examples of applications of the heuristic method in forest planning: Monte Carlo simulation (O'Hara et al., 1989; Clements et al., 1990; Nelson and Brodie 1990), SA (Lockwood and Moore, 1993; Ohman and Ericsson, 2002; Crowe and Nelson, 2005; Martin-Fernández and García-Abril, 2005), TS (Bettinger et al., 1997, 2007; Brumelle et al., 1998; Richards and Gunn, 2000, 2003; Díaz et al., 2007), and GAs (Lu and Eriksson, 2000; Falcão and Borges, 2002; Ducheyne et al., 2004; Thompson et al., 2009).

Furthermore, various other studies have compared the performance of these methods in forest management optimization. Table 8.1 summarizes the main conclusions of these studies.

TABLE 8.1
Performance of Heuristic Methods in Forest Management Optimization

Authors	Methods	Problem	Conclusion
Nelson and Brodie (1990)	MCIP versus mixed-integer programming	Harvest scheduling and transportation planning plus adjacency constraints	MCIP identified solutions within 10% of the true optimum.
Dahlin and Salinas (1993)	MCIP versus SA versus prebiased random search method	Harvest scheduling plus adjacency constraints	SA found the best solutions.
Murray and Church (1995a)	SA versus TS	Operational forest planning	SA quicker, TS better solutions
Mullen and Butler (1997)	MCIP versus GA	Spatially constrained harvest-scheduling model	GA solutions 3.5% better than MCIP
Boston and Bettinger (1999)	MCIP versus SA versus TS	Spatial harvesting problems	SA best solutions (96.5% of the optimum), TS (93%) MCIP, the lowest values. SA and MCIP the quickest
Boston and Bettinger (2001)	Two-stage method (1st LP, 2nd TS-GA algorithm) versus one-stage method TS-GA algorithm	Spatial forest planning	Two-stage method better solutions
Öhman and Eriksson (2002)	LP and SA versus SA	Long-term forest planning with spatial, even-flow and inventory constraints	SA and LP better solutions than SA alone
Bettinger et al. (2002)	Random search, SA, great deluge, threshold accepting (TA), TS with 1-opt moves, TS with 1-opt and 2-opt moves, GA, hybrid TS/GA	To maximize the amount of land in certain types of wildlife habitat	Very good, SA, TA, great deluge, TS with 1-opt and 2-opt moves, and TS/GA; adequate, TS with 1-opt moves, GA; and less than adequate, RS
Heinonen and Pukkala (2004)	Random ascent, HERO, SA, TS	One- and two-compartment neighborhoods in harvest-scheduling problems including a spatial objective variable	SA and TS were the best methods.
Li et al. (2010)	Comparison of search behavior of SA, TA, tabu, and the raindrop method, combination of these heuristics into 12 2-algorithm metaheuristics and 24 3-algorithm metaheuristics	Maximization of the NPV of planned management activities	Metaheuristics that combine the beneficial aspects of standard heuristics will generally produce consistently better solutions than standard heuristics alone.

8.1.2 Some Non-Heuristic Methods to Identify the Best Management Plan

The main application in forestry of non-heuristic methods such as neural networks has been as a means of choice in cases with a great diversity of data and where the relationships between variables are only vaguely understood. Examples of their application to natural resource topics include modeling complex biophysical interactions for resource planning applications (Gimblett and Ball, 1995), generating terrain textures from a digital elevation model and remotely sensed data (Alvarez, 1995), modeling individual tree survival probabilities (Guan and Gertner, 1995), using geographic information systems (GISs) to develop computer-aided visualization of proposed road networks (Harvey and Dean, 1996), predicting future grassland community composition from current knowledge of composition and climatic factors (Tan and Semeins, 1996), and using networks for developing a vegetation management plan (Deadman and Gimblett, 1997).

Recent comparisons in which ANNs performed favorably against conventional statistical approaches include Reibnegger et al. (1991), Patuwo et al. (1993), Yoon et al. (1993), Marzban and Stumpf (1996), Paruelo and Tomasel (1996), Pattie and Haas (1996), and Marzban et al. (1997; Neural networks versus Gaussian discriminant analysis). Some of the advantages of the application of ANN in natural resource modeling can be summarized as follows (Schultz et al., 2000):

- It includes both quantitative and qualitative data and merges information that is difficult to handle with conventional simulation models.
- It can be simply used to model complicated phenomena in natural systems; a priori analytical knowledge is not necessarily required for its implementation.
- It combines linear and nonlinear responses.
- It has a continuous and compensating behavior that corresponds to general ecological principles at higher organizational levels.

On the other hand, there are some network properties and demands that hinder the application and reduce its impact. These are the following:

- It is difficult to include the knowledge of an ecological process in a direct manner; this means, to a certain extent, that a priori knowledge has to be voluntarily ruled out.
- As neural networks are mainly data driven, they need a large volume of representative data to be trained in a general manner.
- When there is a lack of suitable data, neural networks can rapidly become oversized or overtrained.
- Due to their compensating behavior, it is difficult to account for qualitative behavior leaps as they may sometimes appear with elementary dependences between ecological factors.
- It is more difficult to extract new knowledge from trained networks, compared to other modeling approaches.

One current tendency when applying neural technology is not only to maximize the (apparent) advantages but also to avoid the (hidden) obstacles, for example, to use networks, and to combine them with other artificial intelligence (AI) techniques or with classical data analysis procedures. Examples are the combination of neural networks with fuzzy techniques, GAs, expert systems, combining expert systems, and neural networks for learning site-specific conditions (Broner and Comstock, 1997) or regression models (Mann and Benwell, 1996). Moreover, for certain spatial application areas, networks are combined with GISs (Gimblett and Ball, 1995; Mann and Benwell, 1996).

8.2   LP APPLIED TO FOREST MANAGEMENT

Forest management involves a variety of different objectives, including production of wood and non-wood products, protection of biodiversity, and conservation of soils and watersheds, among others. Social aspects such as provision of recreational areas are also considered. This list of objectives entails a high number of activities that make forest management complex. However, a number of strategic, tactical, and operational models have been developed to support decision making. LP has been applied to a wide range of problems since the first forest applications, such as finding the harvesting schedule that minimizes yield consequences (Nautiyal and Pearse, 1967), to more recent applications, such as finding the land acquisition pattern to maximize the area protected (Constantino et al., 2008). However, as highlighted in the Introduction section, LP has some limitations for its application to optimization problems when neighboring alternatives have to be considered or when the objective function or constraints are not linear, causing other optimization methods to be applied. But even in these cases, LP represents a useful support since it can provide guidance in finding the value of the global optimum.

8.2.1   Examining Three Types of Forest Models

8.2.1.1   Strategic Forest Management Models

Strategic forest management models focus on long-term interactions between forest management decisions, such as harvest and silviculture scheduling, and issues, such as sustainability and economic returns from the forest (Gunn, 2007). The system most widely used by the USDA Forest Service is FORPLAN; it was developed in 1991 by Kent et al., and its evolution was continued with SPECTRUM (Greer and Meneghim, 2000). Gunn (2007) classifies these models into (1) models of the ecosystem that usually simulate detailed ecosystem processes and (2) models of the economic system; some of them focused on stand-level economics. These economic models seek to obtain the rotation age that maximizes the net present value (NPV) per hectare. It is clear that different strategies will produce different rotation ages, and sustainability requirements such as water quality and quantity, forest cover, and habitat necessities for certain species are not considered in the stand-level analysis. Other models of the economic system are focused on forest product markets. These models aim to balance forest management strategies and economic development strategies; (3) the last type of model is the forestland and ecosystem management group. It is in this group that LP plays an important role. According to Dantzig (1982), LP is especially useful in this regard because of its unambiguous calculation of feasibility or unfeasibility and due to its ability to show how much this type of constraint costs at the margin.

LP strategic models follow three different approaches: those that model the process of forest growth and management, those that model the sustainability of forest products, and a third approach that models the requirements to provide certain types of forest cover. There are three modeling approaches to forest growth and management: the well-known Model I and Model II (Davis et al., 2001) and the less common Model III (García, 1990).

The Three Models

- Model I: It can be applied to an aggregated or individual stand. If aggregated, then all stands of a given age class are aggregated. It is easy to model a variety of forestry management regimes in this model: regeneration, precommercial thinning, commercial thinning and regeneration harvest, and subsequent treatments. All the decision variables are defined at the outset. They must consider all the management treatments. Once defined, they will not be changed during each period.
- Model II and Model III: The management units can change during the period. In each period, the land in an age class is either harvested, reverting to the regeneration age class, or not harvested, thus becoming one age class older. The main difference between these models is that Model II does not include all the prescriptions that Model III does. The process of growth and harvesting can be represented as the flow through a network where the regeneration stage is a node. In Model II, there can be several alternate paths from one regeneration node to the next. In Model III, separate networks are created for each silviculture treatment. Models II and III usually have fewer decision variables but many more constraints, such as flow constraints to ensure regularity of the harvest, habitat constraints, and forest cover constraints.

8.2.1.2 Application of LP to Short-Term Operational Models

The second group of forest models includes the operational models that deal with forest operations over a week, one season or two, up to perhaps a decade (Church, 2007). These models aim to solve the problem of which cutting units should be harvested in each time period, the machinery to be used, roads needing to be built, transportation schedule, and the environmental constraints that must be considered (Epstein et al., 2007).

8.2.1.3 Application of LP to Tactical Models

Tactical models have two roles. The first is to translate the decisions made at the strategic level into feasible targets at the operational level (Nelson et al., 1991). The second is to identify the impacts on forestland for maintaining specific levels of biodiversity protection (Nalle et al., 2002; Fischer and Church, 2003). There are several tactical systems such as bridging analysis model A (BAM-A) and the more flexible version BAM-B whose objective is to allocate strategic-level prescriptions at subunit level and maintain all threshold conditions among small spatial units (see Church et al. (2000) and Weintraub et al. (1986) for more details).

8.2.2 Operating with LP Models

When LP is applied, there are two types of functions to formulate in LP. These are the objective function to be maximized or minimized—for example, minimum cost, maximum level of protection, and minimum environmental impact—and the constraints, such as maximum budget, minimum protection level, and maximum acceptable impact. For example, in Model I, the formulation of a general problem after defining the prescriptions is

$$\text{Max} \sum_{i=1}^{S} \sum_{j=1}^{P_i} C_{ij} x_{ij} \tag{8.1}$$

subject to

$$\sum_{j=1}^{P_i} x_{ij} = A_i$$

where
  x_{ij} is the area dedicated to prescription j in the ith area of analysis
  C_{ij} is the NPV of all future returns for this prescription j and ith area of analysis
  A_i is the total surface of the ith analysis area

In general, each of these constraints in an LP problem defines a half-plane (or half-space when there are more than two variables) of possible solutions. For the problems that take into account two or three main variables, the optimum can be found graphically. In other cases, it is necessary to use an algorithm such as the simplex method to find the optimum (Dantzig, 1982).

The fundamental assumptions of LP are

1. There is only one objective. When there are several objectives in an LP problem, only one of them can be present in the objective function. The others have to be formulated as constraints. For instance, the revenues from timber sales of a forest, and water protection, are to be maximized. In this case, both goals are conflictive, so the most important goal is the objective function, and the other is formulated as a constraint. For problems with multiple objectives that cannot be formulated in the form of constraints, it is advisable to use other methods, such as goal programming or multi-criteria analysis.
2. Proportionality: the value of the objective function and the response of each resource are proportional to the value of the variables.
3. Additivity: the contribution of all variables to the objective function is the sum of the contributions of each variable. In other words, there is no interaction between the effects of different activities.

4. Divisibility: the variables can take fractional values, not only integer values. For problems that do not fulfill this property, it is advisable to use other techniques, such as integer programming.
5. If the optimum is unique, it is located in a vertex of the polyhedron; otherwise, it is a segment between two optimal vertexes. This is the fundamental theorem of LP, whose proof is beyond the scope of this book.
6. If it is a minimization problem, the optimum is close to the origin of coordinates and a straight line can be drawn from one to the other without crossing the feasible region. When it is not possible to draw such a line, that vertex cannot be considered as a possible optimal solution. In this case, the feasible region is not necessarily a closed polygon. It can be an open domain.

8.3   OPTIMIZATION METHODS BASED ON NATURAL PROCESSES

8.3.1   SIMULATED ANNEALING METHOD

Kirkpatrick et al. (1983) presented the concept of SA, based on the work of Ising (1925) in thermodynamics. There is a deep and useful connection between statistical mechanics (the behavior of systems with many degrees of freedom in thermal equilibrium at a finite temperature) and multivariate or combinatorial optimization (finding the minimum of a given function depending on many parameters). A detailed analogy with annealing in solids provides a framework for optimization of the properties of very large and complex systems.

Origin of the simulated annealing method

To make the method of SA more understandable, it is necessary to introduce first the concepts of Markov random fields and the Gibbs distribution:

1. Markov random fields

Markov random fields appeared in the work of Ernest Ising (1925) when experiments related to the behavior of ferromagnetic materials were explained. Consider n spins on a line, 1, 2…n. At any given time, the spin can be "up" or "down" (see Figure 8.1).

We can define a sample space Ω as all the feasible configurations:

$$w = \{w_1, w_2, w_3, \ \ldots, w_n\}$$

where $w_j = +$ if the spin is "up" and $w_j = -$ if it is "down."

We can also define the discrete random variable $\sigma : \Omega \to R^n / \forall w \in \Omega$; $\sigma(w) = \{\sigma_1(w_1), \sigma_2(w_2), \ldots, \sigma_n(w_n)\}$

where $\sigma_j(w) = 1$ if $w_j = +$ and $\sigma_j(w) = 0$ if $w_j = -$.

FIGURE 8.1   Example of a configuration of spins in n points.

To each configuration w, an energy function, $E(w)$, is defined:

$$E(w) = -J\sum_{i,j}\sigma_i(w)\sigma_j(w) - mH\sum_i\sigma_i(w)$$ (8.2)

where
   the first term expresses interaction between neighbors
   J depends on the material
   the second term represents the effect of an external field

Ising et al. (1925) assumed that only actions between neighboring spins need to be taken into account.

A probability measure can be introduced in this sample space, Ω. This probability measure $P(w)$ on Ω can be defined as the Boltzmann distribution:

$$P(w) = \frac{e^{-E(w)/kT}}{Z}$$ (8.3)

where
   T is the temperature
   k is a universal constant

$$Z = \sum_w e^{-E(w)/kT}$$ (8.4)

A probability measure of the form of (8.3), defined by an energy function, is a Gibbs measure. The importance of this type of measure is the following:

1. It is related to entropy. Entropy, in statistical mechanics, is a function of the distribution of the system on its microstates. The fundamental postulate in statistical mechanics states that a system in equilibrium does not have any preference for any of its available microstates. Given Ω microstates at a particular energy, the probability of finding the system in a particular microstate is $p = 1/\Omega$. The entropy may be interpreted as the amount of uncertainty in the outcome. For any probability measure $p(w)$ on a finite space Ω, the entropy $S(p)$ is defined by

$$S(p) = -\sum_w p(w)\log(w)$$ (8.5)

2. If we want to assign a probability to the sample space Ω representing all the alternatives that cannot be observed—but we know the expected value of the energy function, $U(w) = e$—then the Gibbs measure (8.5) is the one that maximizes entropy among all measures that make the expected value of the energy agree with the estimated value e.

3. The Gibbs measure* has the following Markov property in relation to the interaction between neighbor alternatives:
Let N_j be the neighbors of spin j; then

$$P(\sigma_j = a/\sigma_k, \quad k \neq j) = P(\sigma_j = a/\sigma_k, \quad k \in N_j) \tag{8.6}$$

That is, the probability that spin j has position a, considering the positions of the rest of the spins, is the probability that this spin has such a position, considering only the positions of its N_j neighbors.

A measure with this property is a Markov random field. From two dimensions and above, these measures can be considered as a generalization of Markov processes to spatial situations.

Metropolis algorithm

In the earliest days of scientific computing, Metropolis et al. (1953) developed the algorithm that can be used to provide an efficient simulation of a collection of atoms in equilibrium at a given temperature, based on the early Monte Carlo techniques. It generates a sequence of states of the solid in the following way: Given a state w_k, with energy $E(w_k)$, an atom is given a small random displacement to state w_{k+1} and energy $E(w_{k+1})$, and the resulting change, ΔE, in the energy of the system is computed. The acceptance criterion of state $k+1$ is the following:

a. If $\Delta E = E(w_{k+1}) - E(w_k) \leq 0$, the displacement is accepted, and the configuration with the displaced atom is used as the starting point of the next step.
b. If $\Delta E = E(w_{k+1}) - E(w_k) > 0$, this case is treated probabilistically: the probability that the configuration is accepted is $P(E(w_{k+1}) - E(w_k)) = \exp(-\Delta E/k_B T)$, where T denotes the temperature of the heat bath and k_B is the Boltzmann constant. A random number q, in the interval (0, 1), is selected and compared with $P(\Delta E)$.
   • If $q < P(\Delta E)$, the new configuration is accepted.
   • If $q \geq P(\Delta E)$, the original configuration is used to start the next step.

By repeating the basic step many times, one simulates the thermal motion of atoms in thermal contact with a heat bath at temperature T. This choice of $P(\Delta E)$ has the consequence that the system evolves into a Boltzmann distribution.

8.3.1.1   Simulated Annealing Algorithm

Kirkpatrick et al. (1983) introduced a generalization of the Metropolis algorithm. This algorithm can be applied to generate a sequence of solutions to a combinatory optimization problem. For this purpose, we assume an analogy between a physical many-particle system and a combinatorial problem (see Table 8.2).

---

* The *Gibbs measure* is the measure associated with the Boltzmann distribution and generalizes the notion of the canonical ensemble. Importantly, when the energy function can be written as a sum of parts, the Gibbs measure has the Markov property. In addition, the Gibbs measure is the only measure that maximizes the entropy for a given expected energy (Kindermann and Snell, 1980; Georgii, 1988).

TABLE 8.2

Relationship between Metropolis Criterion Features and SA Optimization Method

Metropolis Criterion	Combinatory Optimization	Parameter
State of the solid	Feasible alternative or solution	w_i
Set of feasible states or configurations/(size)	Set of feasible solutions/(size)	$\Omega/(S)$
Energy	Value function	E
Feasible state changes in w_i	w_i's neighboring feasible solutions	$N(i)$ or $N(w_i)$
Temperature	Control parameter	T
Frozen state	Optimum solution	w^*
Atom j	Individual or decision unit j	I_j
Collection of atoms	Set of decision units	D
State of atom i_j when state of solid is w_j	State or value of decision unit i_j when the alternative is w_j	$X_{ij}(j)$
Set of feasible states of an atom	Set of feasible values of a decision unit	V

The following example belongs to an optimization process in which the actions to cut or not to cut are assigned to the trees in a stand. The objective is to optimize the values of the remaining trees in the forest. In this case, the decision units I_j are trees and the set of feasible values of a decision unit V is $V = \{$to cut, not to cut$\}$. Figure 8.2a shows the set D of decision units.

Figure 8.2b represents a feasible solution w_i in which the states of the decision units are the following: $w_i = \{X(1) = $ not cut; $X(2) = $ not cut; ..., $X(9) = $ cut,..., $X(26) = $ not cut$\}$. If we define the set of w_i neighbor solutions, $N(i) = \{$The feasible solutions in which all the trees have the same state as w_i except one tree which has changed its state$\}$. Figure 8.2c and d shows different neighboring solutions of alternative w_i.

The idea of introducing temperature and simulating annealing is due to Cerny (1985) and Kirkpatrick et al. (1983), both of whom used it for combinatorial optimization.

Let Ω be the set of feasible alternatives with an a priori probability $P(X)$.

Let $E: \Omega \to R$ be the energy function defined on the solution space. The goal is to find a global minimum, w^* (i.e., $w^* \in \Omega$ such that $E(w) \geq E(w^*) \ \forall w \in \Omega$). The iterative method generates a sequence of alternatives that monotonically decreases the energy, or the posterior distribution. On the other hand, a stochastic relaxation process allows changes that increase the posterior distribution as well. These are made on a random basis, seeking to avoid convergence to local minimum.

This stochastic relaxation algorithm can be described as follows:

8.3.1.1.1 Sampling Process

Define $N(w)$ as the neighborhood function $\forall w \in \Omega$, $N: \Omega \to P(\Omega)$. A local change is made in the current solution in the immediate neighborhood. This change is random and is generated by sampling from a local conditional probability distribution. This sampling method is called Gibbs sampler (Geman and Geman, 1984):

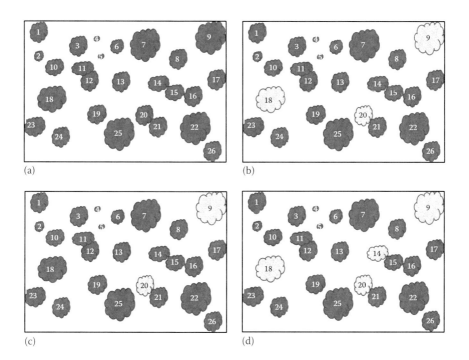

(a) (b)

(c) (d)

FIGURE 8.2 (a) Set D of decision units, (b) alternative w_i (in light gray, trees to be cut, and in dark gray, trees not to be cut), (c) w_j neighbor solution (tree 18 has changed its state), and (d) w_j neighbor solution (tree 14 has changed its state).

- Given an initial arbitrary configuration or alternative $w_0 = X(0) = \{X_{I1}(0), X_{I2}(0), \ldots, X_{IN}(0)\}$ where $X_{Ij}(0)$ represents the state of the unit or individual I_j in the iteration 0, $\forall I_i \in D$ (where D is the set of individuals or decision units). This alternative can be chosen randomly as the optimum solution does not depend on it.
- At each iteration k, a sample is obtained from the local characteristics of the Gibbs distribution. Only one individual undergoes a change, so $X(k-1)$ and $X(k)$ can differ in at most one individual or unit; $I = n_k$ and $X_{Ii}(k) = X_{Ii}(k-1)$ if $I_i \neq n_k$. This individual will have another value v of the set of values V. Therefore, in the new configuration, the states of the individuals are $X_{nk}(k) = v$ and $X_{Ii}(k) = X_{Ii}(k-1) \ \forall I_i \neq n_k$.

Let $n_1, n_2, \ldots, n_z$ be the sequence of individuals that are analyzed to change their value in the z total iterations; thus, $n_k \neq D$ and $X_{Ii}(k) = X_{Ii}(k-1)$ if $I_i \neq n_k$.
So the election of the new solution w_k will depend on the previous solution w_{k-1}, and the transition probability is

$$P\left(w_k\right) = P\left(w_k = X\left(k\right)/w_{k-1} = X\left(k-1\right), \quad \forall w_{k-1} \in N_{w_k}\right) = \frac{1}{Z_s} e^{-E\left(w_k = X(k)\right)/T} \quad (8.7)$$

$$Z_s = \sum_{k=1}^{s} e^{-E(w_k)/T} \qquad (8.8)$$

Therefore, SA can be formulated as a Markov chain. Furthermore, in the case of SA, it can be demonstrated that the set of outcomes (or solutions) in each iteration is finite (see Aarts and Lenstra (2003) for more details).

The acceptance probability is defined by

$$P(\Delta E) = \begin{cases} \exp\left(-\dfrac{E(w_k) - E(w_{k-1})}{T}\right) & \text{if} \quad E(w_k) - E(w_{k-1}) > 0 \\ 1, \dagger \text{ otherwise} \end{cases} \qquad (8.9)$$

8.3.1.1.2 Cooling Schedule

One of the most important processes in the design of SA is the cooling schedule. Romeo and Sangiovanni-Vincentelli (1991) note that an effective cooling schedule is essential to reduce the amount of time required by the algorithm to find an optimal solution. At low temperatures, the local conditional distributions concentrate on states that increase the objective function, whereas at high temperatures, the distribution is essentially uniform. The limiting cases, $T=0$ and $T=\in$, correspond, respectively, to greedy algorithms (such as gradient ascent) and undirected (i.e., "purely random") changes.

On the other hand, in order to avoid local maxima, it is highly recommended to begin at high temperatures where many of the stochastic changes may decrease the objective function (Geman and Geman, 1984). As the relaxation proceeds, temperature is gradually lowered and the process behaves increasingly like iterative improvement. The algorithm generates a Markov chain that converges in distribution to the uniform measure over the minimal energy configurations (Aarts and Lenstra, 2003). Contrary to other methods, there is a general convergence of results for SA, which states that under certain mild conditions, an optimal solution is found with probability 1 (Aarts and Ten Eikelder, 2002).

Aarts et al. (2005) define "cooling schedule" as the specification of a finite sequence of values of the control parameter and a finite number of transitions at each value of the control parameter. Specifically, a cooling schedule involves

- An initial value of the temperature, T_0
- A function to decrease the value of the control parameter
- A final value of the control parameter specified by a stop criterion
- A finite length of each homogeneous Markov chain, that is, the number of iterations with the same temperature

Cooling schedules are grouped into two classes: *static* schedules, which must be completely specified before the algorithm begins, and *adaptive* schedules, which adjust the rate of temperature decrease from information obtained during the execution of the algorithm.

An example of static cooling schedules is the geometric schedule (Kirkpatrick et al., 1983):

1. Initial value of the control parameter T: To ensure a sufficiently large value of T_0, we can choose T_0 as

$$T_0 = \max_{i \in S} \max_{j \in N(i)} \left\{ \left| E(\dot{w}_i) - E(w_j) \right| \right\} \qquad (8.10)$$

where
    S is the set of feasible solutions
    $N(i)$ is the set of i neighbors

If the calculus of T_0 is very time consuming, an estimation of its value may be sufficient.
2. Decrement function for the temperature value:

$$T_{k+1} = \alpha T_k \qquad (8.11)$$

where α is a constant close to 1, in general between 0.8 and 0.99.
3. The final value can be related to the smallest possible difference of the value function between two neighboring alternatives.
4. The number of iterations for a specific temperature may be related with the number of neighbors in the problem at hand.

An example of a static cooling schedule is the one applied to the optimal assignment of public investments to the networking of rural roads in San Luis de Potosí in Mexico (Morales, 2011). In this case, the initial value of the parameter T was $35 million, which was the difference in energy functions when all the roads were improved and when none was improved.

Regarding the decrement function, five values of parameter α were tested: 0.8, 0.85, 0.9, 0.95, and 0.99. The final value or stop criterion was when a solution equivalent to $350 was found.

Figure 8.3 shows the relationship between temperature parameter and number of iterations for different cooling speed parameters.

The lower the cooling speed parameter, the lower the number of iterations needed to reach the same temperature decrease.

In contrast, when the cooling speed is low ($\alpha = 0.95$), the calculus time needed to obtain a solution is lower than when the cooling speed is higher ($\alpha = 0.9$, 0.8). However, the total time to obtain the optimum is higher (see Figure 8.4).

Another example of how to obtain the initial value of T is the method proposed by Martínez-Falero et al. (2010). This consists of applying the Kolmogorov–Smirnov fit test between a sampling distribution and theoretical distributions. 4×10^6 random alternatives were generated, the difference in the energy function between every pair was calculated, and the sampling distribution was obtained as a result. A number of theoretical distributions were generated for different T. The lowest Kolmogorov–Smirnov statistic between the sampling distribution and each theoretical distribution

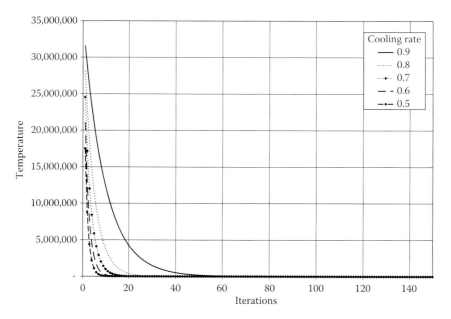

FIGURE 8.3 Relationship between temperature parameter and number of iterations for different cooling speed parameters.

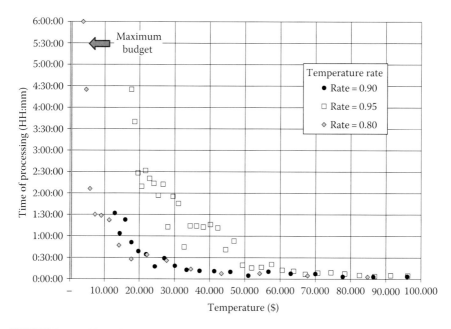

FIGURE 8.4 Effect of the temperature decrease on processing time.

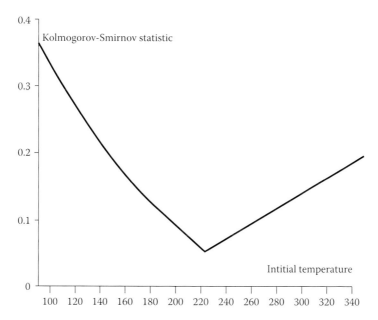

FIGURE 8.5   Initial temperature according to the Kolmogorov–Smirnov statistic.

indicated the value of T_0. Figure 8.5 shows the relationship between this statistic and the initial value of the temperature parameter.

In this case, the decrement function for the temperature value was, again, $T_{k+1} = \alpha T_k$

Finally, in the work of Geman and Geman (1984) on the Bayesian restoration of images, the $T(k)$ employed in executing the kth iteration satisfied the bound

$$T(k) = \frac{c}{\log(1+k)} \tag{8.12}$$

where c is a constant independent of k.

In summary, SA can be stated as follows:

1. Select initial solution $w_0 \in \Omega$.
2. Select the temperature change counter $s=0$.
3. Select a temperature cooling schedule, T_s.
4. Select an initial temperature $T_0 \geq 0$.
5. Select a repetition schedule, M_s, that defines the number of iterations executed at each temperature T_s.
6. Repeat.
   a. Set repetition counter $k=0$.
   b. Repeat.
      i. Generate a solution $w_{k+1} \in N(w_k)$.
      ii. Calculate ΔE.

iii. If $\Delta E \leq 0$, then $w_k \rightarrow w_{k+1}$.

iv. If $\Delta E > 0$, then $w_k \rightarrow w_{k+1}$ if $P(\Delta E) = \exp\left(-\dfrac{E(w_k) - E(w_{k-1})}{T}\right) > q$,

where q is a random number and $q \in [0, 1]$; otherwise, go to b.

v. $k = k + 1$.

7. Until $k = M_s$.

8. $s = s + 1$.

9. Until stopping criterion is met.

8.3.1.2 Convergence of Simulated Annealing

Convergence results for SA have typically taken one of the two directions; the algorithm has been modeled either as a sequence of homogeneous Markov chains or as a single inhomogeneous Markov chain.

The homogeneous Markov chain approach (it does not depend on the iteration) (see, e.g., Aarts and van Laarhoven (1985), Lundy and Mees (1986), Mitra et al. (1986), Rossier et al. (1986), Faigle and Kern (1991), Granville et al. (1994), Johnson and Jacobson (2002)) assumes that each temperature T_k is held constant for a sufficient number of iterations m such that the stochastic or transition matrix P_k (the matrix of probabilities $P(w_k)$, Equation 8.7) can reach its stationary (steady state) distribution, $\pi w'$.

SA and the homogeneous convergence theory are based on the work of Metropolis et al. (1953), which addresses problems in equilibrium statistical mechanics (Hammersley and Handscomb, 1964). To see this relationship, consider a system in thermal equilibrium with its surroundings, in solution (state) w with energy $E(w)$. The probability density in phase space of the point representing w (see Equation 8.14) is proportional to

$$\exp\left(\frac{-E(w)}{k_B T}\right) \tag{8.13}$$

Therefore, the proportion of time that the system spends in solution w_{k+1} is proportional to (8.13) (Hammersley and Handscomb, 1964); hence, the equilibrium probability density for all $w \in \Omega$ is the stationary (steady state) function:

$$\pi_w = \frac{\exp\left(-E(w)/k_B T\right)}{\int \exp\left(-E(w)/k_B T\right) dw} \tag{8.14}$$

The expectation, E, is

$$E[E(w)] = \frac{\int E(w) \exp\left(-E(w)/K_B T\right) dw}{\int \exp\left(-E(w)/K_B T\right) dw} \tag{8.15}$$

Unfortunately, for many solution functions, (8.15) cannot be evaluated analytically. However, Metropolis et al. (1953) solve this problem by first discretizing the solution space, such that the integrals in (8.14) and (8.15) are replaced by summations over the set of discrete solutions, and then by constructing an irreducible, aperiodic Markov chain with transition probabilities $P(w_{k+1})$ such that

$$\pi\left(w_{k+1}\right)=\sum_{w\in\Omega}\pi\left(w_k\right)P\left(w_{k+1}\right) \quad \forall w_k \in \Omega \tag{8.16}$$

where

$$\pi_{w_k} = \frac{\exp\left(-E(w_k)/k_BT\right)}{\sum_{w\in\Omega}\exp\left(-E(w)/k_BT\right)} \tag{8.17}$$

Hammersley and Handscomb (1964) show that Metropolis et al. (1953) accomplish this by defining $P(w_{k+1})$ as the product of the probability of generating a candidate solution w_{k+1} from the neighbors of solution w_k:

$$g(w_k,w_{k+1})=P\left\{generate\ w_{k+1}\mid X(k)=w_k\right\} \tag{8.18}$$

where

$$\sum_{w_{k+1}\in N(w_k)} g\left(w_k,w_{k+1}\right)=1, \quad \text{for all} \quad w_k \in \Omega$$

and the acceptance ratio $\pi_{w_{k+1}}/\pi_{w_k}$ that is $P\{\text{accept } w_{k+1}\mid X(k)=w_k\}$, so

$$\frac{\pi_{w_{k+1}}}{\pi_{w_k}}=\exp\left(-\frac{E\left(w_{k+1}\right)-E(w_k)}{k_BT}\right) \tag{8.19}$$

Therefore, $P(w_{k+1})$ can be formulated as

$$P\left(w_{k+1}\right)=\begin{cases}\dfrac{g\left(w_k,w_{k+1}\right)\pi\left(w_{k+1}\right)}{\pi\left(w_k\right)} & \text{if } \dfrac{\pi\left(w_{k+1}\right)}{\pi\left(w_k\right)}<1, w_{k+1}\neq w_k \\[3mm] g\left(w_k,w_{k+1}\right) & \text{if } \dfrac{\pi\left(w_{k+1}\right)}{\pi\left(w_k\right)}\geq 1, w_{k+1}\neq w_k \\[3mm] g\left(w_k,w_{k+1}\right) & \\ +\displaystyle\sum_{\substack{w_{k+2}\in\Omega\\ \pi(w_{k+2})<\pi(w_k)}} g\left(w_k,w_{k+2}\right)\left(1-\dfrac{\pi\left(w_{k+1}\right)}{\pi\left(w_k\right)}\right) & \text{if } w_{k+1}=w_k\end{cases}$$

$$\tag{8.20}$$

Asymptotic convergence of SA can be proved with a model in which the algorithm is viewed as a sequence of homogeneous Markov chains of infinite length. In such a homogeneous Markov chain, the value of the temperature T and the transition probability between two iterations k and $k+1$ are independent of k, that is, $T_k = t$ and $P(k) = P$ for all k. This leads to the following result.

Theorem 8.1

Let (w, U) be an instance of a combinatorial optimization problem, N a neighborhood function, and P the transition matrix of the homogeneous Markov chain associated with the SA algorithm defined by (8.15) and (8.16), with $T_k = t$, for all k. If the neighborhood graph is connected, the associated homogeneous Markov chain has a stationary distribution π_k, whose components are given by

$$\pi_{w_k} = \frac{\exp\left(-E(w_k)/k_B T\right)}{\displaystyle\sum_{w \in \Omega} \exp\left(-E(w)/k_B T\right)} \tag{8.21}$$

As a consequence of Theorem 8.1, we have

$$\pi_{w_k}^* \overset{\text{def}}{=} \lim_{T \to 0} \pi_{w_k} = \begin{cases} \dfrac{1}{\Omega^*} & \text{if} \quad w_k \in \Omega^* \\ 0 & \text{otherwise} \end{cases} \tag{8.22}$$

where Ω^* denotes the set of optimal solutions. This implies that

$$\lim_{T \to 0} \lim_{k \to \infty} P\left\{X(k) \in \Omega^*\right\} = 1$$

8.3.1.3 Case of Application in Forest Management

The aim of this study was to assign tree-level forestry actions in an uneven-aged forest of *Pinus sylvestris* in the Sierra de Guadarrama, Madrid. This assignment had to generate an economic maximum for the remaining trees in the stand, in terms of the amount of low- and high-quality timber to be harvested using the constraints: forest cover, biodiversity, and regeneration (see Martín-Fernandez et al. (2005) for more details).

The actions to be taken at tree level may include cutting, pruning, and fertilizing. The number of actions is close to 10, and if there are 15,000 trees in the stand, the number of alternatives is $10^{15,000}$. Figure 8.6 shows a schema of the objectives of the study.

We selected a 0.25 ha area within 300 ha of Scots pine showing an uneven-aged structure for the application.

The management strategy applied to the stand was close-to-nature silviculture (De Turckheim, 1992). To give a brief summary of the characteristics of this type of forest management, each tree has an individual treatment according to the functions

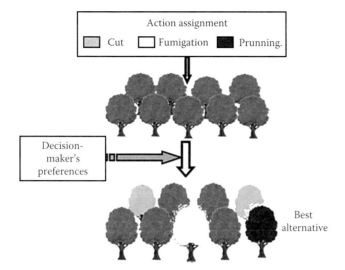

FIGURE 8.6   Schema of the objectives of the optimization process.

assigned at the ecosystem level; natural regeneration is promoted under large trees across great areas; rotation time is short, 6–12 years; and cutting is essentially restricted to trees of sufficient diameter determined by maximum economic timber value for the species and site. A tree can remain in the stand as long as it gains market value or fulfills a protection, landscape, or diversity function. Diseased trees can be removed, but dead trees are of great ecological value and some should be preserved.

Close-to-nature management ensures that in well-structured forests, regeneration occurs at a sufficient rate, adjacent large trees contribute to the self-thinning process, and natural pruning and intervention are not required to carry out activities other than felling the best-quality trees (De Turckheim, 1992). Hence, the only action that we considered was whether or not to cut a tree. In applying this silviculture strategy to our case study, we classified each tree in the stand as "main" or "complementary." "Main" trees in the forest were classified into one of three categories: stabilizer tree, a dominant and codominant tree, whose purpose is the production of high-quality timber, forest stabilization, and triggering regeneration; sprinter tree, a tree that is exceptional due to its height, an intermediate state between regeneration and stabilizer; and "complementary" trees, the remaining stems that are needed to support and improve the "main" trees, provide soil cover, and protect other trees.

All trees were classified according to dendrometric variables (e.g., diameter, height, crown diameter, and crown height), along with descriptive variables, such as health condition and harvesting impact and environmental variables including altitude, slope, and orientation (De Turckheim, 1992).

8.3.1.3.1   *Optimization Process*

The optimization method was the iterated conditional mode algorithm, a relaxation of SA optimization. This iterated conditional mode algorithm defines an iterative operational process in which, instead of simply changing the decision unit with the greatest probability of change as in SA, it alters all the points simultaneously with their most probable action. Therefore, a new action is computed for each tree using the old value of the neighboring points, and thus, all forestry actions are updated at the same time. This process is repeated until no further changes occur (Besag, 1986).

Using the principles of close-to-nature silviculture, the value expression of the value function was

$$U(w) = \sum_{i=1}^{n_1} \sum_{k=1}^{4} V_{ik} P_k - \sum_{d=1}^{CD} |TN_d - SN_d| - \left(\sum_{j=1}^{n_2} \sum_{k=1}^{4} V_{jk} \left(1+cP(c)\right)^n P_k \frac{(1+p)^n}{(1+r)^n} - V_{jk} P_k \right)$$

(8.23)

where
V_{ik} represents the quality-k timber volume of tree i that is not cut
P_k is the price per m³ of quality-k timber
n_1 is the number of stems that remain in the forest
TN_d is the percentage of diameter class-d trees, according to the balanced diameter distribution of *P. sylvestris* for the same site quality
SN_d is the percentage of diameter class-d trees that remain in the stand after cutting
V_{jk} represents the quality-k timber volume of tree j that is cut
n_2 is the number of cut stems
c is the current annual growth of the tree according to its diameter class
$P(c)$ is the probability that the growth ratio will be maintained for n years
n is the number of years between fellings (10 years in this case)
p is the increase in the timber-price ratio
r is inflation

The energy function also has to fit a number of constraints defined by the decision maker. In our case, these were

1. The maximum volume extracted per hectare should not exceed 60 m³ to avoid severe impact.
2. The volume of low-quality timber to be felled must be 24–36 m³/ha (improvement felling).
3. The volume of high-quality timber to be felled must be 16–24 m³/ha.
4. The largest gap size produced when consecutive trees are cut must be less than 200 m² (to preserve cover and shade).
5. Trees with protected birds' nests should not be felled (to preserve biodiversity).
6. Regeneration and sprinters should not be cut (to promote regeneration and improve the diameter class balance).

Before starting the optimization process, two algorithms were calculated previously:

- Algorithm to find and order each tree's neighboring trees. Each tree's neighbors need to be identified to apply the iterative conditional mode in which, given a tree i, the next alternative is sought among its neighbors. Tree i's neighboring trees are those whose distance from tree i is less than their total height, since this distance affects the tree. The neighborhood is considered a symmetric property. The order of neighbors is taken as the angle between the line joining tree i with its neighbor and the x-axis.

- Assigning a probability $P(c)$ to the feasible growth of each tree. This value was assigned by an expert according to the relationship of dominance between each tree and its neighbors. It depends on the distance D_{ij} between the trees and the relative position of the crowns.

Optimization iterative process:

- T_0 was the minimum decrement of energy obtained from a sample of 100 alternatives.
- $T_s = 0.85\, T_{s-1}$.
- First iteration: The optimization process commenced with an initial random solution from the set of trees.

Eight stabilizer and 28 complementary trees were selected. The value of the timber left in the forest was 1983.67 euros; 6.2 and 9.5 m³ of high- and low-quality timber, respectively, were felled. The differences between the number of stems per diameter class left in the forest and the balanced diameter distribution for the same area were 24 stems in class 2, 29 in class 3, 16 in class 4, and 3 in class 5 (DC1, < 10 cm; DC2, 10–20 cm; DC3, 20–30 cm; DC4, 30–40 cm; DC5, > 40 cm).

- Iteration i: The m intermediate solutions derived from the neighbors of the trees in the starting solution of the iteration were compared in every iteration.
- Final process: The process finishes when the energy function becomes stable.

In the optimum solution, seven stabilizer, 16 complementary, and two regenerated trees infected with *Peridermium pini* were selected. The economic value of the timber left in the forest was 2064 euros, which represents 77.4% of its value before felling, that is, 2666 euros. The volumes of high- and low-quality timber felled were 6.1 and 7.9 m³, respectively. The differences between the number of stems per diameter class left in the forest and the balanced diameter distribution for the same area were 21 stems in class 2, 24 in class 3, 15 in class 4, and 3 in class 5.

The optimal solution represents an improved productive capacity of the stand, since more high-quality trees remained in the forest. Thus, the model fulfills the objective of close-to-nature management (De Turckheim, 1992), and harvesting costs are reduced since fewer trees need to be extracted.

8.4 GAs

8.4.1 INTRODUCTION

GAs are a family of computational models inspired by evolution. These algorithms encode a potential solution to a specific problem on a specific chain data structure and apply recombination operators to these structures so as to preserve critical

FIGURE 8.7   Mapping of two variables, x_1 and x_2, onto a chromosome structure.

information. It was in the 1960s when evolutionary algorithms based on natural processes were first applied to optimization (Mühlenbein, 2003). These evolutionary algorithms have been successfully applied to combinatorial optimization. One of their advantages is that they are easily implemented, although they are difficult to justify mathematically. According to Mühlenbein (2003), the main difficulty lies in the fact that the algorithms combine two different search strategies: a random search by mutation and a biased search by recombination of strings or alternatives in the population.

At each generation, a new set of approximations is created by the process of selecting individuals according to their level of fitness in the problem domain and breeding them together using operators borrowed from natural genetics. This process leads to the evolution of populations of individuals that are better suited to their environment than the individuals they were created from, just as in natural adaptation. Individuals, or current approximations, are encoded as strings, *chromosomes*, composed over various codes, so that the *genotypes* (chromosome values) are uniquely mapped onto the decision variable (*phenotypic*) domain. The most commonly used representation in GAs is the binary code {0, 1}, although other representations can be used, for example, ternary, integer, and real valued. For example, a problem with two variables, x_1 and x_2, may be mapped onto the chromosome structure in the following way (Figure 8.7):

where x_1 is encoded with 10 bits and x_2 with 15 bits. At the end of this section, there are examples of integer and real-valued representation. Examining the chromosome string in isolation yields no information about the problem we are attempting to solve. It is only with the decoding of the chromosome into its phenotypic values that any meaning can be applied to the representation. However, the search process will operate on this encoding of the decision variables.

Having decoded the chromosome representation into the decision variable domain, it is possible to assess the performance, or fitness, of individual members of a population. This is done through an objective function that characterizes an individual's performance in the problem domain. This value is used to select the most highly fit individuals as parents of the next generation. In the natural world, this would be an individual's ability to survive in its present environment. Thus, the individuals with a higher fitness value have more probability of being chosen.

Genetic operators are used to produce the next generation and to exchange genetic information between pairs or larger groups of individuals.

Consider the two parent strings:

$$P_1 = 100011110$$
$$P_2 = 110010000$$

If we select randomly a position i, between chromosome 1 and $l - 1$, where l is the length of the chain, for example, $i = 5$, the two offspring could be

$$O_1 = 100010000$$
$$O_2 = 110011110$$

The pairs of parents are chosen with probability P; therefore, not all the individuals are chosen.

A further genetic operator, mutation, can be applied to the new generation. Again a probability P_m is applied. Mutation causes changes in the individual's string according to a probabilistic rule. The objective of mutation is to avoid local optimum.

If we wanted to mutate individual O_2, and the position randomly chosen is the 7th, then the new individual would be

$$O_{2m} = 110011010$$

When a new population is obtained, the process begins again, and the process continues to subsequent generations until some criteria are satisfied. Every new generation is supposed to have individuals with a higher performance than the previous one, as good individuals are preserved and the less fit individuals die out. Compared to other heuristic methods, the four most significant differences are

1. GAs search a population of points in parallel, not a single point.
2. GAs do not require derivative information or other auxiliary knowledge; only the objective function and corresponding fitness levels influence the directions of search.
3. GAs use probabilistic transition rules, not deterministic ones.
4. GAs work on an encoding of the parameter set rather than the parameter set itself (except where real-valued individuals are used).

It is important to note that the GA provides a number of potential solutions to a given problem and the choice of final solution is left to the user from among the individuals in the last generation. In cases where a particular problem does not have one individual solution, for example, a family of Pareto-optimal solutions, as is the case in multi-objective optimization and scheduling problems, then the GA is potentially useful for identifying these alternative solutions simultaneously.

8.4.2 Stages in GAs

The stages in GA can be summarized as follows:

1. Define a genetic representation of the problem.
2. Obtain the initial population: $w_{01}, w_{02}, \ldots, w_{0N}$.
3. Formulate the objective function and the fitness function.
4. Select the subset of parents, $P_1, P_2, \ldots, P_k$.
5. Obtain their fitness value.
6. Apply crossover with certain probability to each pair and other genetic operators such as mutation and reinsertion, obtaining the new population.
7. If the stopping criterion is not fulfilled, go to 4.

Another interesting approach of GA is the parallel GA. Its main advantage is the decrease in the computing time. It is a highly synchronized algorithm where a distributed selection scheme is used. Each individual or alternative makes the selection by itself. It looks for a partner in its neighborhood only. As a result, the set of neighborhoods defines a spatial population structure and each individual is active and not acted on. The schema of PGA is (Mühlenbein, 2003)

1. Define a genetic representation of the problem.
2. Obtain the initial population w_{01}, w_{02},..., w_{0N} and its structure.
3. Each individual does local hill climbing.
4. Each individual selects a partner for mating in its neighborhood.
5. An offspring is created using genetic operators such as mutation and reinsertion, obtaining the new population.
6. The offspring does local hill climbing. It replaces the parent if it is better than some criteria (acceptance).
7. If the stopping criterion is not fulfilled, go to 4.

8.4.2.1   Population Representation and Initialization

Populations are made by the potential solutions or alternatives. Typically, populations are composed of between 30 and 100 individuals. Every individual is a string obtained by encoding each decision variable in the parameter set and then linking them.

Although binary-coded GAs are the most commonly used, there are other alternatives such as integer and real-valued representation. There are some advantages in the use of real-valued representation that were pointed out by Wright (1991):

* No need to convert binary information into real information.
* No loss in precision by discretization to binary or other values.
* There is greater freedom in the process.

Having decided on the representation, the first step is to create the population. This is usually achieved by generating the required number of individuals using a random number generator that uniformly distributes numbers in the desired range. Therefore, if the number of individuals is N and the number of bits is L, a total number of $N * L$ random numbers would be produced.

Other option to generate the populations is "the extended random initialization procedure" (Bramlette, 1991). In this case, every individual is generated a number of times, and the one selected is the one with the best performance. In some cases, when the problem is well-known beforehand, individuals close to the global optimum can be chosen.

8.4.2.2   Formulation of the Objective and the Fitness Function

The objective function, as in previous methods, is used to provide the performance of every alternative. In GA, it is also very useful in the calculus of the relative performance of each solution. This transformation of the objective function is the fitness function.

Let $f(x_1, x_2, \ldots x_n)$ be the objective function, where $x_1, x_2, \ldots x_n$ are the decision variables (i.e., the phenotypic value). We can define the fitness function as

$F(x_1, x_2, \ldots x_n) = g(f(x_1, x_2, \ldots x_n))$, where g transforms the value of f into a non-negative number. As an example, the fitness function for individual i could be

$$F\left(x_{i1}x_{i2}, \ldots, x_{in}\right) = \frac{f\left(x_{i1}x_{i2}, \ldots, x_{in}\right)}{\displaystyle\sum_{i=1}^{N} f\left(x_{i1}x_{i2}, \ldots, x_{in}\right)} \tag{8.24}$$

This measure gives the probability of reproducing according to the relative fitness of individual i. In the case the objective function has negative values, a linear transformation is often used (Goldberg, 1989):

$$F\left(x_1, x_2, \ldots x_n\right) = af\left(x_1, x_2, \ldots x_n\right) + b \tag{8.25}$$

where a is a positive scaling factor if the optimization is maximizing and negative if we are minimizing. The offset b is used to ensure that the resulting fitness values are nonnegative.

This linear scaling transformation can trigger a rapid convergence to the optimum. As there is no constraint on an individual's performance in a given generation, highly fit individuals in early generations can dominate the reproduction, causing rapid convergence to possibly local optimal solutions. Baker (1987) suggests limiting the number of offspring, in which case the individuals are assigned a fitness according to their performance rank.

The linear model would be

$$F\left(k_i\right) = F_{\min} + F_{\text{Max}} \frac{k_i - 1}{N - 1} \tag{8.26}$$

where k_i is the rank of individual i. The user will define the values of $F_{\min}$ and F_{Max} as the minimum and maximum value of the fitness function.

8.4.2.3 Parent Selection

This selection determines the number and characteristics of the offspring an individual will produce.

First, it is necessary to transform the fitness values into an individual's probability of reproducing. Second, the individuals for reproduction will be probabilistically selected. This selection will take into account the fitness of individuals relative to one another. According to Baker (1987), bias, spread, and efficiency parameters measure the efficiency of each method.

Bias is the absolute difference between an individual's actual and expected selection probability that indicates the accuracy, spread is a range of an individual's possible trials, and efficiency is related to the time consumed in the process.

Some of the most commonly applied methods appear in Table 8.3.

TABLE 8.3
Parents Selection Methods

Selection Method	Description	Parameters
Roulette wheel (find modifications of the method in Chipperfield et al. (1992))	1. Select Sum as parameter a or b. 2. Calculate interval $[0, Sum]$. 3. Map individuals into contiguous intervals of size equal to its fitness value in the range $[0, Sum]$. 4. Obtain a random value in the interval $[0, Sum]$. 5. The chosen individual is the one whose interval contains the random value. 6. Repeat 4–5 until the desired number of individuals is selected.	a. $Sum = \sum_{i=1}^{N} ePi$ $ePi \equiv$ expected selection probability b. $Sum = \sum_{i=1}^{N} f(xi)$ $f(x_i) \equiv$ raw fitness value
Stochastic universal sampling	1. Shuffle the population randomly. 2. Calculate $I \equiv [0, Sum/s]$. 3. Generate a random number, rn, in I. 4. Generate the s pointers $[rn, rn+1, \ldots, rn+s-1]$. 5. Choose the s-pointed individuals whose fitness spans the positions of the pointers.	Sum as previous method, s number of selections required
Binary tournament selection	1. Shuffle the population randomly. 2. Select 2^n individuals from the population. 3. Form couples. From each couple, select the best individual, according to fitness function and value function. 4. With the individuals obtained in step 3, repeat step 3. 5. Breed individuals obtained in step 4, applying the user's chosen method.	N number of individuals in the offspring

8.4.2.4 Recombination

Crossover produces individuals that have some parts of both parents' genetic material.

- Multiple-Point Crossover
  Select m crossover positions with no duplicates and sorted into ascending order, and exchange the information between two sequential positions. For example, if two parent strings are considered,

$$P1 = 101011110$$
$$P2 = 110010001$$

  If $m = 3$, three random numbers between 1 and 8 (number of bits-1) are selected. In this case, {2, 5, 7} is as follows (Figure 8.8):
  This type of crossover encourages the exploration of all the space and makes it more robust (Spears and De Jong, 1991).
  Other types of crossover such as uniform crossover can be seen in Syswerda (1989) and Caruana et al. (1989).

- Intermediate Recombination
  Given a real-valued encoding of the chromosome structure, intermediate recombination is a method of producing new phenotypes around and between the values of the parents phenotypes (Mühlenbein and Schlierkamp-Voosen, 1993). Offspring are produced according to the rule:

$$O1 = P1 * \alpha(P2 - P1) \tag{8.27}$$

  where
  α is a scaling factor chosen uniformly at random over some interval, typically $[-0.25, 1.25]$
  $P1$ and $P2$ are the parent chromosomes (see, e.g., Mühlenbein and Schlierkamp-Voosen (1993))

In the general version, each variable in the offspring is the result of combining the variables in the parents according to the aforementioned expression. A new α is chosen for each pair of parent genes. There is a particular version in which α is constant. As a result of the method, in geometric terms, intermediate recombination is capable of producing new values of the variables other than those defined by the parents but constrained by the range of α, as shown in Figure 8.9, when there are only two possible variables.

P1 = 101011110		O1 = 110011001
P2 = 110010000	Offspring →	O2 = 101010110

FIGURE 8.8 Example of a multiple-point crossover method of obtaining new individuals with GAs.

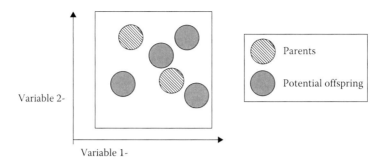

FIGURE 8.9 Graphic example of offspring and parents when a scaling factor is applied in the offspring determination.

8.4.2.5 Mutation

The definition of mutation is a random process where one chromosome is replaced by another to produce a new genetic structure. The role of mutation is often seen as providing a guarantee that the probability of searching any given string will never be zero and as acting as a safety net to recover good genetic material that may be lost through the action of selection and crossover (Goldberg, 1989).

Many variations on the mutation operator have been proposed, for example, biasing the mutation toward individuals with lower fitness values to increase the exploration in the search without losing information from the fitter individuals (Davis, 1989) or parameterizing the mutation such that the mutation rate decreases with the population convergence (Fogarty, 1989a); *trade* mutation (Lucasius and Kateman, 1992), whereby the contribution of individual genes in a chromosome is used to direct mutation toward weaker terms; and *reorder mutation* (Lucasius and Kateman, 1992) that swaps the positions of bits or genes to increase diversity in the decision variable space. Comparing real coding to binary coding, real-coded GAs may take advantage of higher mutation rates than binary-coded GAs, increasing the possible exploration of the search space without adversely affecting the convergence characteristics (Janikow and Michalewicz, 1991 and Wright, 1991).

8.4.2.6 Reinsertion

Once a new population has been produced, the fitness of the individuals in the new population may be determined. In the case where the number of new individuals produced in each generation is one or two, the GA is said to be *steady state* (Whitley, 1989), otherwise *incremental* (Huang and Fogarty, 1991). If one or more of the fittest individuals are deterministically allowed to propagate through successive generations, then the GA is said to use an *elitist strategy*.

To maintain the size of the original population, the new individuals have to be reinserted into the old population. Likewise, if not all the new individuals are to be used at each generation or if more offspring are generated than the size of the old population, then a reinsertion scheme must be used to determine which individuals are to exist in the new population.

When selecting which members of the old population should be replaced, the most apparent strategy is to replace the least fit members deterministically.

However, in studies, Fogarty (1989b) has shown that no significant difference in convergence characteristics was found when the individuals selected for replacement were chosen with inverse proportional selection or deterministically as the least fit. He further asserts that replacing the least fit members effectively implements an elitist strategy, as the most fit will probabilistically survive through successive generations. Indeed, the most successful replacement scheme was one that selected the oldest members of a population for replacement. This is reported as being more in keeping with generational reproduction, as every member of the population will, at some time, be replaced. Thus, for an individual to survive successive generations, it must be sufficiently fit to ensure propagation into future generations.

8.4.2.7 Termination of the GA

Like other heuristic methods, it is difficult to formally specify convergence criteria. As the fitness of a population may remain static for a number of generations before a superior individual is found, the application of conventional termination criteria becomes problematic. A common practice is to terminate the GA after a prespecified number of generations and then test the quality of the best members of the population against the problem definition. If no acceptable solutions are found, the GA may be restarted or a fresh search initiated.

8.4.2.8 Response to Selection

Applying GA to general combinatorial problems leads to the genetic representation of the problem. Unique and problem-specific mutation and recombination operators have to be applied. In this regard, Mühlenbein (2003) proposes a general principle to maximize the response. Specifically, he proposes maximizing the product of the realized heritability and the standard deviation of the offspring:

Let R be the response to selection. R is defined as the difference in the population mean fitness of generation $t+1$ and generation t. So it represents the expected progress of the population:

$$R(t) = \bar{F}(t+1) - \bar{F}(t) \tag{8.28}$$

Let S be the selection differential. This is the difference between the mean fitness of the selected parents and the mean fitness of the population in the same generation:

$$S(t) = \bar{F}_s(t) - \bar{F}(t) \tag{8.29}$$

Response to selection can be predicted from the selection differential as

$$R(t) = b(t)S(t) \tag{8.30}$$

where $b(t)$ is the realized heritability. $b(t)$ can be obtained in previous generations or be estimated by different methods (Crow, 1986). In general, it is assumed to be constant:

$$R(t) = bS(t) \tag{8.31}$$

But when GA is applied, it is more interesting to predict the cumulative response for k generations

$$R_k = \sum_{t=0}^{k} R(t) \tag{8.32}$$

To obtain this parameter, it has been found that the selection intensity I provides a more convenient measure of the strength of the selection:

$$I = \frac{S(t)}{\sigma(t)} \tag{8.33}$$

where
   $\sigma(t)$ is the standard deviation of the fitness of the individuals
   I can be computed analytically if the fitness values are normally distributed

In any event, simulations have demonstrated that the aforementioned expression is valid for many practical applications. So R can be obtained as (Falconer, 1981)

$$R(t) = bI\sigma(t) \tag{8.34}$$

The designer of the GA problem has to find a recombination operator to maximize this expression. An application of this principle can be found in Voigt et al. (1995), where a fuzzy recombination operator is shown to be superior to others.

The response to selection can also be used for analyzing selection methods. If all the selection methods have the same I, the best one will be that which selects parents with the highest standard deviation. For example, Blickle and Thiele (1995) demonstrated that tournament selection is better than truncation selection in this regard.

If we have a binary string of size n, that is, (101011110), if the population is large enough to converge to the optimum, and $I > 0$, then

$$R(t) = \frac{1}{\sqrt{n}} \sqrt{p(t)(1 - p(t))} \tag{8.35}$$

where $p(t)$ is the probability that there is a 1 at a position t.

The number of generations needed until convergence is proportional to $\sqrt{n}$ and inversely proportional to the selection intensity. Another question is to determine the minimum size of the population that allows the process to converge to the optimum. This depends on the size of the problem n, the selection intensity I, and $p(t=0)$ (see Mühlenbein and Schlierkamp-Voosen (1994) for more details).

8.4.3 CASE OF APPLICATION IN FOREST MANAGEMENT

The following example is a simplification of a study on land-use optimization in northwest Spain.

TABLE 8.4

Constraints for Forest Species Growth in a GA Optimization Case

Species	Maximum Area (ha)	Minimum Area	Income/ha Shady Site	Income/ha Sunny Site
Species 1	250	100	5	10
Species 2	350	—	20	50
Species 3	300	—	20	60

We assume an area of 650 ha, of which 500 ha is in a sunny site and 150 ha is in a shady site. Three forest species will be grown with the constraints in Table 8.4:

What would be the best distribution of the 650 ha among these three species in order to maximize the benefit?

The approach to the problem is the following:

Let S_1 be the area of species 1 in the sunny site, S_2 the area of species 1 in the shady site, S_3 the area of species 2 in the sunny site, S_4 the area of species 2 in the shady site, S_5 the area of species 3 in the sunny site, and S_6 the area of species 3 in the shady site. The objective function is

$$\text{Max} Z = 10 * S_1 + 5 * S_2 + 50 * S_3 + 20 * S_4 + 60 * S_5 + 20 * S_6, \text{ such that } S_1 \epsilon [0, 250],$$

$$S_2 \epsilon [0,150], S_3 \epsilon [0,350], S_4 \epsilon [0,150], S_5 \epsilon [0,300], S_6 \epsilon [0,150] \tag{8.36}$$

With the following constraints:

$$S_1 + S_3 + S_5 \leq 500$$
$$S_2 + S_4 + S_6 \leq 150$$
$$S_1 + S_2 \geq 100$$
$$S_1 + S_2 \leq 250$$
$$S_3 + S_4 \leq 350$$
$$S_5 + S_6 \leq 300$$

The constraint function is

$$U = S_1 + S_3 + S_5 - 500 + S_2 + S_4 + S_6 - 150 - (S_1 + S_2 - 100)$$

$$+ S_1 + S_2 - 250 + S_3 + S_4 - 350 + S_5 + S_6 - 300 \tag{8.37}$$

The higher the value of U, the worse.

In this example, a population of 36 individuals is generated randomly. The values of the objective and fitness functions are calculated for each individual (see Table 8.5).

The parents are obtained by the binary tournament selection with replacement method. First, 80 individuals are obtained randomly from the population and grouped into couples; the best individual is chosen from every couple, first according to its

TABLE 8.5

Value of Independent Variables, Objective, and Constraint Functions for Every Individual of the Population

	Population							
Individual	S_1	S_2	S_3	S_4	S_5	S_6	Objective Function	Constraint Function
1	82	68	126	97	35	43	12360	−698
2	159	130	231	102	31	117	20030	−199
3	222	47	2	125	224	110	20695	−259
4	77	125	34	53	176	61	15935	−600
5	190	44	45	45	193	88	18610	−474
6	145	67	332	92	277	69	38225	302
7	105	129	219	114	198	139	29585	124
8	240	8	283	57	278	34	35090	102
9	229	139	249	86	21	81	20035	−208
10	51	50	328	10	70	12	21800	−509
11	61	71	4	86	259	145	21325	−330
12	196	72	236	97	9	42	17440	−414
13	214	11	90	72	211	24	21275	−431
14	212	137	213	129	23	142	20255	−87
15	66	77	323	122	200	22	32075	27
16	108	83	47	1	216	137	19565	−457
17	75	6	163	19	296	143	29930	−127
18	10	74	154	43	255	51	25350	−360
19	103	146	125	124	226	102	26090	−47
20	10	48	269	133	173	95	28730	−52
21	21	72	238	17	38	76	16610	−619
22	186	45	21	102	171	21	15855	−589
23	231	20	94	100	125	118	18970	−325
24	100	2	183	137	29	73	16100	−504
25	158	123	146	7	164	120	21875	−295
26	52	117	26	67	32	44	6545	−943
27	167	81	216	50	63	131	20275	−282
28	152	16	341	25	132	6	27190	−274
29	122	14	226	104	79	73	20870	−350
30	14	145	310	59	38	112	22065	−253
31	150	67	127	127	27	77	13885	−517
32	141	1	83	86	81	86	13865	−636
33	59	142	142	118	96	149	19500	−239
34	7	137	155	112	197	99	24545	−180
35	128	100	200	68	193	55	25820	−190
36	198	17	196	50	164	62	23945	−291

fitness function and then to its objective function. This process is repeated again, until the last 20 individuals will be the parents of the offspring. Table 8.6 shows this process.

The offspring were obtained by applying the uniform crossover method. The following Tables 8.7 and 8.8 show the results of the process.

The next stage was mutation of the offspring. In this case, the fourth chromosome was chosen randomly and an increase of 5% applied. The final offspring appear in Table 8.8.

Finally, the initial population was replaced by the offspring. The individuals 11, 12, 16, 19, 21, 28, 33, 35, and 36 were randomly chosen.

The second example we present in this section is part of a study in which GA and SA were combined to evaluate the performance of this integration compared to applying only SA.

This case shows the process of land-use assignment in the municipality of Huejotzingo, on the sides of the Popocatepetl volcano, State of Puebla, Mexico (Perez-Ramírez, 2007).

Seventy-two percent of the population lives in the city of Heroica Puebla de Zaragoza in the eastern part of the municipality. This population works in the service and industrial sectors, while the rest of the population works in the agriculture sector. The main problem for the farmers is their low crop yields. The objective of this case study is to determine the best land uses according to social, environmental, and economic requirements. A survey was conducted in this municipality, and the proposed land uses are shown in Table 8.9.

The first stage of the process involved obtaining homogeneous land units by overlaying maps of slope, climate, soil quality, water availability, and urban areas (see Figure 8.10).

Table 8.10 describes the main characteristics of the homogeneous units in the study area.

The most frequent units were 1332, 1331, 2331, 4332, 1312, and 1112, which accounted for up to 90% of the area.

GAs were applied in these units to see how the best solution with GA could improve the performance of SA. The surface constraints were not considered at this stage.

The objective function was

$$U(S) = \left[\left[I_{ik} - \left(PC_{ik} + TC_{ik} \right) \right] + \Delta OM_{ik} \right] \left[\frac{SP_{ik}}{EI_{ik}} \right] \quad (8.38)$$

where

S is alternative
I is income
PC is production costs
TC is transformation costs
ΔOM is increase in employment
SP is stakeholder perception index
EI is ecological index
i is píxel
k is land use

TABLE 8.6
Binary Tournament Selection in GA Optimization

Binary Tournament Selection with Replacement

Trial	First Round	Second Round	Parents	Trial	First Round	Second Round	Parents
1	17	17		41	16	16	16
2	24			42	25		
3	11	11	11	43	4	4	
4	12			44	14		
5	31			45	12		
6	16	16	16	46	18	18	18
7	14			47	16	16	
8	20	20		48	33		
9	3			49	2		
10	32	32		50	35	35	
11	33			51	2		
12	16	16	16	52	22	22	22
13	22	22	22	53	7		
14	35			54	36	36	36
15	9	9		55	7		
16	35			56	22	22	
17	2	2		57	19		
18	33			58	16	16	16
19	24	24	24	59	5	5	
20	11			60	12		
21	34			61	21	21	
22	27	27	27	62	28		
23	15			63	12	12	12
24	2	2		64	31		
25	10	10	10	65	35	35	
26	24			66	35		
27	22			67	18	18	18
28	16	16		68	19		
29	5	5		69	20		
30	19			70	1	1	10
31	17			71	10	10	10
32	30	30	30	72	14		
33	15			73	32		
34	11	11	11	74	31	31	
35	23	23		75	28	28	28
36	26			76	21		
37	28			77	33		
38	18	18	18	78	36	36	36
39	19			79	9		
40	17	17		80	22	22	

TABLE 8.7
Obtaining New Offspring through the Uniform Crossover Method

Parents	Chromosomes						Offspring					
11	61	71	4	86	259	145						
							61	83	4	1	259	137
16	108	83	47	1	216	137						
16	108	83	47	1	216	137						
							108	45	47	102	216	21
22	186	45	21	102	171	21						
24	100	2	183	137	29	73						
							100	81	183	50	29	131
27	167	81	216	50	63	131						
10	51	50	328	10	70	12						
							51	145	328	59	70	112
30	14	145	310	59	38	112						
11	61	71	4	86	259	145						
							61	74	4	43	259	51
18	10	74	154	43	255	51						
16	108	83	47	1	216	137						
							108	74	47	43	216	51
18	10	74	154	43	255	51						
22	186	45	21	102	171	21						
							186	17	21	50	171	62
36	198	17	196	50	164	62						
16	108	83	47	1	216	137						
							108	72	47	97	216	42
12	196	72	236	97	9	42						
18	10	74	154	43	255	51						
							10	50	154	10	255	12
10	51	50	328	10	70	12						
28	152	16	341	25	132	6						
							152	17	341	50	132	62
36	198	17	196	50	164	62						

The constraints were related with the compatibility among homogeneous units and land uses, shown in Table 8.11. In this case, all the neighbor possible uses were compatible. In this example, other constraints such as the geometry of the use, surface of the uses, and size of a single use were not considered.

In order to simplify the problem, a population of 15 individuals was generated. Eight individuals were randomly chosen to find the parents among them. Table 8.12 shows the population and the feasible parents in gray.

TABLE 8.8
Result of the Random Mutation Process in the GA Uniform Crossover Method

Individual	Chromosome						Obj. Func.	Constraint Func.
	S_1	S_2	S_3	S_4	S_5	S_6		
1	61	83	4	1.05	259	137	19526	−503.9
2	108	45	47	107.1	216	21	19177	−514.8
3	100	81	183	52.5	29	131	15965	−478
4	51	145	328	61.95	70	112	25314	−110.1
5	61	74	4	45.15	259	51	18643	−596.7
6	108	74	47	45.15	216	51	18683	−549.7
7	186	17	21	52.5	171	62	15545	−634
8	108	72	47	101.85	216	42	19627	−456.3
9	10	50	154	10.5	255	12	23800	−527
10	152	17	341	52.5	132	62	28865	−106

TABLE 8.9
Land Uses Proposed in the Municipality of Huejotzingo, State of Puebla, Mexico

N°	Land Use	N°	Land Use
1	Prunus persica	15	Malus domestica–Zea mays
2	Prunus armeniaca	16	Crataegus mexicana–Zea mays
3	Malus domestica	17	Prunus virginiana–Zea mays
4	Crataegus mexicana	18	Juglans regia–Zea mays
5	Prunus virginiana	19	Grain, Zea mays
6	Juglans regia	20	Forage, Zea mays
7	Prunus persica—Phaseolus vulgaris	21	Phaseolus vulgaris
8	Prunus armeniaca—Phaseolus vulgaris	22	Grain, Vicia faba
9	Malus domestica—Phaseolus vulgaris	23	Vicia faba
10	Crataegus mexicana—Phaseolus vulgaris	24	Vegetable
11	Prunus virginiana—Phaseolus vulgaris	25	Medicago sativa
12	Juglans regia—Phaseolus vulgaris	26	Capsicum sp.
13	Prunus persica—Zea mays	27	Gladiolus sp.
14	Prunus armeniaca—Zea mays	28	Forest

To obtain the offspring, two methods were applied:

1. Binary tournament: Table 8.13 shows the process of parent selection. In order to simplify, only one couple was obtained.

   In this example in Table 8.14, the offspring were obtained by randomly choosing chromosomes 5 and 6 and exchanging its value between parents.

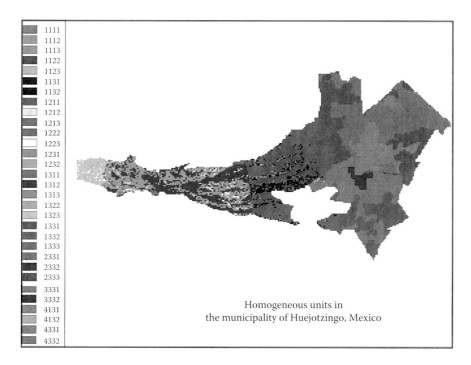

FIGURE 8.10 (See color insert.) Homogeneous land units in the municipality of Huejotzingo, Mexico.

2. Parent random selection: The couples were selected randomly. The land uses of every child are selected as in the previous method; for every child, the value function was also calculated. Table 8.15 shows these results. Compared to the previous method, good solutions were found in both cases, although the second method is easier to implement.

8.4 CA

Cellular automata (CA) were first proposed by von Neumann and Ulman in the 1950s (von Neumann, 1966). This is a mathematical interactive method, based on decentralized self-organization systems that can describe complex systems with simple rules.

Many complex natural systems such as insect colonies or immunological systems are the result of the association of homogeneous, simple elements that work synchronously (Crutchfield et al., 1998). CA are based on these systems. They consist of cells whose state can change with every time interval, according to a set of local rules that depends on the neighborhood of the cell.

The most common neighborhoods used with 2D grids are the von Neumann and Moore neighborhoods (Wolfram, 1984, 2002). Figure 8.11 shows both approaches.

In general, CA can be of any dimension. The parameters that characterize them are the dimension, D; the radius of the neighborhood, r; the number of feasible states, k; the shape of the cells, s; and a set of rules that determine the state of the cells in time.

TABLE 8.10

Characteristics of the Homogeneous Units in the Municipality of Huejotzingo, Mexico

		Description				Area	
N°	Code	Humidity	% Slope	Climate	Soil Quality	ha	%
1	1111	Dry land	15–45	Semicold	High	15	0.09
2	1112	Dry land	15–45	Semicold	Medium	1.643	9.47
3	1113	Dry land	15–45	Semicold	Low	15	0.09
4	1122	Dry land	15–45	Cold	Medium	5	0.03
5	1123	Dry land	15–45	Cold	Low	392	2.26
6	1131	Dry land	15–45	Mild	High	359	2.07
7	1132	Dry land	15–45	Semicold	Medium	222	1.28
8	1211	Dry land	45–89	Mild	Medium	2	0.01
9	1212	Dry land	45–89	Semicold	Medium	140	0.81
10	1213	Dry land	45–89	Semicold	Low	6	0.03
11	1222	Dry land	45–89	Cold	Medium	1	0.01
12	1223	Dry land	45–89	Cold	Low	48	0.28
13	1231	Dry land	45–89	Mild	High	55	0.32
14	1232	Dry land	0–15	Mild	Medium	34	0.20
15	1311	Dry land	0–15	Semicold	High	15	0.09
16	1312	Dry land	0–15	Semicold	Medium	1.763	10.16
17	1313	Dry land	0–15	Semicold	Low	1	0.01
18	1322	Dry land	0–15	Semicold	Medium	2	0.01
19	1323	Dry land	0–15	Cold	Low	58	0.33
20	1331	Dry land	0–15	Mild	High	2.943	16.95
21	1332	Dry land	0–15	Mild	Medium	3.198	18.42
22	1333	Dry land	0–15	Mild	Low	39	0.22
23	2331	Irrigated land	0–15	Mild	High	2.841	16.37
24	2332	Irrigated land	0–15	Mild	Medium	5	0.03
25	2333	Irrigated land	0–15	Mild	Low	67	0.39
26	3331	Urban	0–15	Mild	High	325	1.87
27	3332	Urban	0–15	Mild	Medium	184	1.06
28	4131	Marshy	15–45	Mild	High	3	0.02
29	4132	Marshy	15–45	Mild	Medium	15	0.09
30	4331	HR	0–15	Mild	High	1.064	6.13
31	4332	HR	0–15	Mild	Medium	1.898	10.93
Total						17.358	100.0

There may be only one table of rules (uniform CA) or a set of different tables of rules for different cells (nonuniform CA).

For example, if $D=1$ and $r=1$, the neighborhood of any cell will be the cells located closest to its right and left. If $r=2$, the neighboring cells will be the two closest cells located on its right and left sides.

TABLE 8.11

Compatibility between Homogeneous Units and Land Uses and Value of the Objective Function per ha

HU	LU	VF	HU	LU	VF	HU	LU	VF
1112	3	20476.74	1331	14	9451.23	2331	8	25462.81
1112	4	4328.1	1331	15	25839.66	2331	9	70021.95
1112	5	20110.05	1331	16	9.668.829	2331	10	12485.36
1112	6	11050.64	1331	17	28936.5	2331	11	39583.82
1112	15	12919.83	1331	18	26805.82	2331	12	39974.82
1112	16	4834.415	1331	19	22650.2	2331	13	85871.34
1112	17	14468.25	1331	21	18471.33	2331	14	30206.83
1112	18	13402.91	1331	22	36449.54	2331	15	68581.27
1112	28	56257.94	1331	23	52408.48	2331	16	17524.77
1312	3	40953.49	1331	28	28128.97	2331	17	38503.31
1312	4	8.656.199	1332	3	40953.49	2331	18	38894.31
1312	5	40220.09	1332	4	8656.199	2331	20	40257.15
1312	6	22101.29	1332	5	40220.09	2331	21	27034.76
1312	15	25839.66	1332	6	22101.29	2331	22	39251.66
1312	16	9668.829	1332	9	26594.94	2331	23	82176.23
1312	17	28936.5	1332	10	10424.11	2331	24	127957
1312	18	26805.82	1332	11	29691.78	2331	25	95609.09
1312	19	22650.2	1332	12	27561.1	2331	26	113758.4
1312	28	28128.97	1332	15	25839.66	2331	27	540224.3
1331	1	66274.95	1332	16	9668.829	2331	28	28128.97
1331	2	16316.72	1332	17	28936.5	4332	3	55969.49
1331	3	40953.49	1332	18	26805.82	4332	4	4.398.854
1331	4	8656.199	1332	19	22650.2	4332	5	32182
1331	5	40220.09	1332	21	18471.33	4332	6	19953.34
1331	6	22101.29	1332	28	28128.97	4332	15	43420.55
1331	7	42582.62	2331	1	151155.7	4332	17	18472.58
1331	8	9.806.512	2331	2	27934.65	4332	18	18121.76
1331	9	26594.94	2331	3	101883.9	4332	19	24757.46
1331	10	10424.11	2331	4	16253.5	4332	20	17729.36
1331	11	29691.78	2331	5	48841.5	4332	25	82602.18
1331	12	27561.1	2331	6	36107.55	4332	28	28128.97
1331	13	35914.22	2331	7	80112.02			

An example of a table of rules when $D=1$ and $r=1$ and $k=2$ is shown in Table 8.16. In this example, if both neighbors have the same state, the central cell changes its state; it does not change in other cases.

If we apply this table of rules to a set of seven cells, the result is the one shown in Figure 8.12.

The total possible number of rules when $k=2$ and $r=1$ is 256 for one dimension. The best-known 2D CA is Life (Conway, 1960). In this case, every cell can have alive or dead states, $r=1$, and the rule is as follows: one live cell dies if there are fewer

TABLE 8.12

Population with Parents in Gray and the Land Uses Assigned to the Homogeneous Units

Individual	Land Use Assigned to the Most Frequent Homogeneous Units						Parents' Obj. Function
	1332	1331	2331	4332	1312	1112	
1	17	17	12	11	10	3	
2	15	19	12	12	18	4	115548.51
4	5	6	3	3	27	25	746944.8
5	18	6	8	16	17	19	9869871.75
3	6	4	7	17	2	6	
6	15	18	18	5	5	6	8744621.64
7	16	16	11	12	12	20	143216.66
13	3	3	14	5	4	3	
14	18	6	16	4	24	4	
8	18	5	26	10	1	15	
9	28	3	15	28	26	4	8775360.38
10	16	15	22	12	22	28	149111.41
11	4	28	10	19	11	19	
12	15	18	15	25	25	25	
15	5	5	7	9	14	17	202965.14

TABLE 8.13

Parents Selection by the Binary Tournament Method

Trial	First Round	Second Round	Parents
1	9		
2	5	5	5
3	4	4	
4	2		
5	10		
6	15	15	
7	6	6	6
8	7		

TABLE 8.14
Offspring Obtained from the Random Selection of the Parent's Chromosomes

5	18	6	8	16	17	19	Offspring	18	6	8	16	17	6
6	15	18	18	5	5	6		15	18	18	5	5	19

TABLE 8.15
Offspring Obtained from a Parent Random Selection

Parents	Offspring						VF
9	28	3	15	28	17	19	214440.8
5	18	6	8	16	26	4	24023458
4	5	6	3	3	18	4	4561867
2	15	19	12	12	27	25	713518.7
10	16	15	22	12	14	17	4972945
15	5	5	7	9	22	28	196888.3
6	15	18	18	5	12	20	164455.7
7	16	16	11	12	5	6	14629292

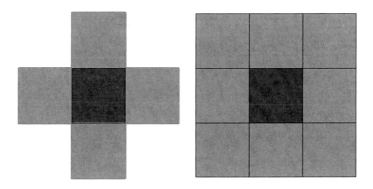

FIGURE 8.11 On the left, Neumann's neighborhood; on the right, Moore's neighborhood.

TABLE 8.16
Example of a Table of Rules of a CA

Time Interval	New State of the Central Cell							
t	000	100	010	001	110	101	011	111
$t+1$	1	0	0	0	1	1	1	0

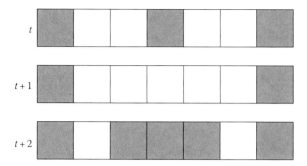

FIGURE 8.12 Application of the table of rules (Table 8.16) to a set of seven cells.

than two neighboring live cells or more than three; a dead cell revives if exactly three neighboring cells are alive.

The analysis of the behavior of CA shows that after a number of time intervals, some reach a fixed configuration, others show a periodical behavior, some converge to a complex configuration, and the rest do not show a pattern. Wolfram (1984) proposed a classification of the behavior of CA:

Fixed behavior: All the initial configurations converge to the same final configuration.

Periodical behavior: After some iterations, all the active cells are in the same location.

Complex behavior: CA converge to a complex structure that is maintained over time.

Chaotic behavior: The configuration of convergence cannot be predicted.

Wolfram (1984) demonstrated that as the number of k or r increases, fewer CA converge. For example, if $k=2$ and $r=1$, 50% of CA have a fixed behavior, none have a complex behavior, and 25% have a chaotic behavior. However, if $k=2$ and $r=3$, 9% have a fixed behavior and 73% a chaotic one. In this regard, Langton (1991) defined the parameter λ as the percentage of states differing from a specific state in the initial table of rules. For example, if $k=2$ and 0 is the reference state, λ will be the percentage of 1's. The table of rules of CA is more heterogeneous, since λ evolves from 0 to $[1 - 1/k]$ and the behavior of CA changes in the same sense: fixed->periodical->complex->chaotic.

In conclusion, if a suitable table of rules is designed, CA are able to carry out complex tasks. However, it is only with simulation that we can check whether the table of rules allows us to reach this objective. GA can be applied to obtain a population of adapted CA that can develop the preestablished function. In this regard, the table of rules would be the chromosome chain. As a result, more adapted individuals or tables of rules will be obtained from generation to generation.

8.4.1 APPLICATION OF CA TO FORESTRY

A particular characteristic of spatial optimization problems is the relation between local interactions and global system behavior and the simulation of the spatial dynamics of certain phenomena such as wildfires or wind effects. These considerations lead to the application of CA to the spatial analysis of ecosystems. Hogeweg (1988) used them to simulate changes in landscape. Green et al. (1985), Karafyllidis and Thanailakis (1997), Karafyllidis (2004), Supratid and Sadananda (2004), and Hernández Encinas et al. (2007) employed CA to simulate the spread of forest fires, while Balzter et al. (1998) and Jennerette and Wu (2001) studied land-use dynamics from this standpoint. Sole and Manrubia (1995) simulated the dynamics of forest openings by means of CA. Strange et al. (2002), Heinonen and Pukkala (2007), and Mathey et al. (2008) evaluated the effectiveness of CA in solving this type of problems in forest planning. Zeng et al. (2010) also applied CA to minimize the risk of wind damage in forest planning. Moreover, CA have been used for simulation of succession and spatial analysis of vegetation growth (Colasanti and Grime, 1993). Other applications related to forest and land-use management include management of groundwater aquifers and water allocation (Sidiropoulos and Tolikas, 2008), reservoir management (Afshar and Shahidi, 2009), landscape dynamics in an Amazonian colonization area (Silveira et al., 2002), and land-use assignment in afforestation areas (Strange et al., 2002).

Most of the studies apply a regular grid structure; for example, Sidiropoulos and Tolikas (2008) in their study of water management use a 2D grid of square cells to represent the terrain, the neighborhood is defined in the sense of von Neumann (Figure 8.11), and the states of the cells are whether or not to establish a well. Heinonen and Pukkala (2007) used hexagonal cells with a hexagonal neighborhood to avoid single points of contact between neighboring cells in their study of forest planning. In this case, the states were the treatment alternative. However, Flache and Hegselmann (2001) pointed out that the regular grid structure limits the application of CA in practice, since some natural variables are not well explained in this type of cells. They successfully applied Voronoi polygons to migration simulation, and the neighbors were searched using Euclidean distances between central points. Zeng et al. (2010) used the forest stands and their shape directly as CA cells. The neighbor polygons were searched based on the topology between stands and edges. The states of the cells were clear-cutting schedules, and the number of neighboring cells changes from one cell to another; this approach avoided upward bias in the evaluation of ecological processes (McGarigal and Marks, 1995).

Regarding CA rules, Sidiropoulos and Tolikas (2008) sought a nonconstant rule in water management optimization. Instead, GAs were embedded into the CA in order to guide its evolution. More specifically, two types of GA were implemented: the operative GA, which defined a renewed rule each time for synchronous changes to each cell on the basis of the neighboring states, and the natural GA endowed with a neighborhood rule. This rule will operate on a neighborhood level and on the basis of the local values of the objective function for the purpose of enhancing the performance of the natural GA. The natural GA works on the whole configuration, and its genetic operators are not based on local interactions among neighboring cells.

According to the rules, the first state is assigned randomly. The mutation and innovation of the cells are evaluated in view of their probabilities, which were established experimentally (Heinonen and Pukkala 2007).

There are two different methods for updating cells: parallel (cells are updated simultaneously) or sequential (cells are evaluated and updated one after another). In asynchronous updating, the order in which cells are evaluated must be determined, that is, randomly or according to coordinates. Most CA use parallel updating; however, studies like Zeng et al. (2010) obtained equally good results with sequential updating and with parallel updating.

A basic issue in spatial optimization concerns the formulation of local objectives in relation to the overall global objectives. Local objectives permit the design of local transition rules in CA. In the case of water resources management, Sideropoulos and Talikas (2008) treated the optimization problem by means of operators applied to the individual cells in the local sense, without decomposing the objective function into local contributions. Another approach would be to define local components of the objectives and then attempt to reduce the overall problem to the solution of the partial corresponding problems at the neighborhood level of each cell. Other authors such as Strange et al. (2001) considered the individual cell contributions of the objective function. Optimization is performed on the basis of optimizing each one of these cell components. In Heinonen and Pukkala (2007), the objectives are distinguished into local and global, and a composite objective function is used. This function consists of the weighted sum of a local and a global term. The weighting coefficient of the global term is gradually increased in the course of a successive local and global solution of the optimization problem. The local part is treated by means of updating via mutations in the cells of an underlying automaton. Zeng et al. (2010) designed the dynamics of CA by means of a global function that stems from a local function. The risk of wind damage was minimized for each stand by selecting the schedule that induced the shortest weighted length of the edges at risk. In a grid search approach, Seppelt and Voinov (2002) present a clear distinction between a local and a global method. Their objective function consists of a sum of cell-dependent terms, and the solution consists of two stages, a local and a global one. The latter is performed through a GA. Although a grid forms the basis of the problem formulation, no complete CA characteristics, such as local transition, appear in the whole treatment.

Zeng et al. (2010) compared CA with other heuristic methods, concluding that CA usually provided at least as good results as SA, AG, or TS. In the study on minimizing wind risk in forest planning, CA had an output of a shorter length of edges at risk than the GA but longer vulnerable edges than SA. In general, CA better fulfilled the objective of minimizing the risk of wind damage, although it did not fulfill the even-flow timber harvest objective as well as the other heuristic methods. The optimization was mainly obtained in the first iteration. Heinonen and Pukkala (2007) compared CA solutions with LP and indicated that CA had a good performance. CA reached the solution with less iterations than SA, and both SA and CA showed good performance regarding the value of the NPV.

8.5 ARTIFICIAL NEURAL NETWORK METHOD IN OPTIMIZATION PROBLEMS

8.5.1 Introduction

The origin of ANNs goes back to McCulloch and Pitts's studies in the 1940s. They established the similarity between the response of any neuron and a proposition that proposed its adequate stimulus (McCulloch and Pitts, 1943). It is well known that the nervous system is a network of neurons, each having one soma, one axon, and dendrites. The synapses or junctions (links in the network) are always between the axon of the transmitter and one dendrite of the receiver. The effect of several simultaneous signals arriving at the dendrites is usually almost linearly additive, whereas the resulting output is a strongly nonlinear, all-or-none process. The high computational power and speed of the nervous system are due to its capacity of parallel processing. McCulloch and Pitts (1943) proposed that there may be a correspondence between this parallel organization and relations among propositions. The idea that this large degree of local connectivity between the simple processing units (neurons) is an important contribution to the computational power of the nervous system motivated the study of the general properties of neural networks (Hopfield and Tank, 1985). This parallelism between the nervous system and computing led to the application of the ability in computation to adjust simultaneously and self-consistently to many interacting variables. ANNs have been widely applied to pattern recognition and prediction problems. In these cases, ANNs constitute a nonlinear extension of conventional linear interpolation/extrapolation methods. The approach varies when they are applied to combinatorial optimization problems. According to Peterson and Söderberg (2003), while heuristic methods do not fully or partially explore the different possible configurations, ANNs "feel" their way in a fuzzy manner toward the optimum. There are two main steps in the process:

Formulation of the problem as the minimization of a feedback ANN function $E(s_1, s_2, \ldots, s_N)$ where the neurons s_i (or decision units) encode possible solutions

Finding of an approximate solution by iteratively solving the corresponding mean field (MF) equations

ANN Parameters

The neurons v_i can normally take real values within the interval [0, 1] or [−1, 1]; $i = 1,2,\ldots, N$. Sometimes, it can be simpler with discrete neurons s_i, with $s_i \in \{0, 1\}$ or $\{-1, 1\}$.

The local updating rule of the value of the neuron is usually

$$v_i = g\left(\sum_{j=1}^{N} w_{ij}v_j - \theta_i \right) \tag{8.39}$$

where
  $w_{ij} \in R$ are the weights (synapses)
  they are nonzero only for the neurons v_j connected with the dendrites of neuron v_i
  these weights can have both positive values to excite the neuron v_i or negative to inhibit it
  θ_i is a threshold corresponding to the membrane potential in a biological neuron. If the integrated input signal is larger than θ_i, the neuron changes its state

The nonlinear transfer or activation function $g:R \rightarrow [0, 1]$ is typically a sigmoid-shaped function such as

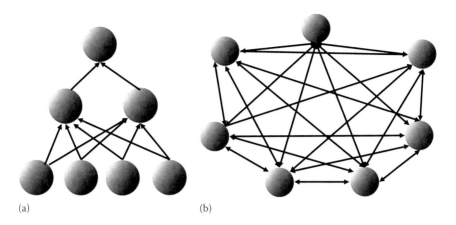

(a)   (b)

FIGURE 8.13   (a) An example of a feedforward architecture; (b) an example of a feedback architecture.

$$g(x) = \frac{1}{2}\left(1 + \tanh\left(\frac{x}{c}\right)\right) \qquad (8.40)$$

where
> the parameter $c>0$ sets the inverse gain
> a lower value generates a steeper transfer function, in the limit, when $c\rightarrow 0$ the
> function corresponds to discrete neurons

There are two types of architecture in neural network modeling: feedforward and feedback (see Figure 8.13).

Feedforward networks process signals from the bottom layer of neurons to the top in one direction, using the local updating rule. They have two major applications: feature recognition and function approximation that are beyond the scope of this book.

On the other hand, neurons in feedback networks continue processing signals until a steady state is reached. Feedback networks are used in optimization problems (Hopfield and Tank, 1985; Peterson and Söderberg, 1989), in feature recognition using the Boltzmann machine (Ackley et al., 1985), and in MF approximation (Peterson and Hartman, 1989).

Simple models for magnetic systems have much in common with feedback networks (Peterson and Söderberg, 2003). The Ising models described in the SA section offer an easy example for an understanding of the relation between magnetic systems and feedback networks.

In Ising model, a set of spins s_i, (with $s_i \equiv \sigma_i$) $i=1,\ldots, N$, can have two states, $\{1, -1\}$; the expression of the energy function that governs the state is

$$E(s) = -\frac{j}{2}\sum_{i,j} s_i s_j \qquad (8.41)$$

where i and j are neighbor spins. The lowest energy is reached by iterative updating of the spins according to

$$s_i = sgn\left(J \sum_{j \in N(s_i)} s_j \right) \tag{8.42}$$

where

s_j are all the neighbors of s_i
sgn is the algebraic sign function
J is the constant attractive coupling of strength

A state is reached when all the spins have the same sign and are therefore all aligned. The probability of any configuration follows the Boltzmann distribution (8.3), which depends on the temperature T. The degree of order of the spins also depends on it. At very high temperatures, there is no alignment, all the spins are completely random, and there is a transition point T', critical temperature, from which the spins are aligned. The transition into an order phase plays an important role in feedback networks. A generalization of the Ising model is the spin glass system, where there are nonlocal interactions (i.e., s_i cannot interact with itself), so

$$E(s) = -\frac{1}{2} \sum_{i \neq j} w_{ij} s_i s_j \tag{8.43}$$

This is the basis of the Hopfield model.

Hopfield Model

Hopfield (1982) built a network model with a number of simplifications that made it possible to obtain information analytically on the characteristics of the system. Hopfield rediscovered the self-associative networks that have different behavior from feedforward networks such as Adaline/Madaline or Perceptron. The basis of his model was on the spin glass system.

Initially, Hopfield (1982) developed a discrete version of his model with binary neurons {1,−1}, where the energy function was (8.43). He later developed the continuous version of the model where the neurons can have any value in the interval [−1, 1] or in [0,1] (Hopfield, 1988; Hilera and Martinez, 1995).

When the appropriate weight parameters w_{ij} are chosen, the main goal of the model is to let the system function like an associative memory. A dynamic that locally minimizes (8.43) is given by

$$s_i = sgn\left(\sum_{j \neq i} w_{ij} s_j \right) \tag{8.44}$$

In the case of the discrete approach, the transfer function $g(x)$ that determines the new value of every neuron in a new iteration follows a step function such as

$$g(x) = \begin{cases} +1 & x > \theta_i \\ -1 & x < \theta_i \end{cases} \tag{8.45}$$

where
   x is the value of $E(s)$
   $g(x)$ is the new value of the neuron

If $x = \theta_i$, then the neuron does not changes its value, and θ_i is the threshold.
   The value of the threshold θ_i in the discrete case is usually

$$\theta_i = k \sum_{j=1}^{N} w_{ji} \tag{8.46}$$

If the binary values of the neurons are −1 or +1, usually, $\theta_i = 0$, When the binary values are 0 and 1, then, $\theta_i = 1/2$.
   In the case of the continuous model, the transfer function if the interval of possible values of the neurons is [−1, 1] is

$$g(x - \theta_i) = htg(\alpha(x - \theta_i)) = \frac{e^{\alpha(x-\theta_i)} - e^{-\alpha(x-\theta_i)}}{e^{\alpha(x-\theta_i)} + e^{-\alpha(x-\theta_i)}} \tag{8.47}$$

where α is the slope of the function.

How Does This Model Work?

One of the characteristics of the Hopfield model is that it is a self-associative model. This means that during the training stage, different patterns or information can be stored in the network as if it worked like a memory.
   These patterns can be expressed as $R^{(p)} = \left(r_1^{(p)}, r_2^{(p)}, \ldots, r_N^{(p)} \right)$, where p is the number of the pattern, with $p = 1, \ldots, N_p$, and t_i^p is the value of the pattern p for neuron i and $r_i^{(p)} \in \{-1, 1\}$.
   The process is as follows:

1. For $t = 0$, every neuron s_i will have some input information, $s_i = t_i$.
2. For $t = 1$, every neuron will receive as input the sum of the output of the other neurons multiplied by the weights, $\sum_{j=1}^{N} w_{ij} s_j$, and the transfer function $g(x)$ is applied to this value.
3. For any iteration $t = k + 1$,

$$s_i(k+1) = g\left(\sum_{j=1}^{N} w_{ij} s_j(k) - \theta_i \right) \quad \forall i = 1, \ldots, N \tag{8.48}$$

4. The process continues until $s_i(k+1) = s_i(k)$. The values of the neurons in the last iteration will be the output generated by the network and will correspond to some pattern of the training stage.

As stated before, the expression of the transfer function $g(x)$ is

$$s_i(t+1) = g(x) = \begin{cases} +1 & x = \sum_{j=1}^{N} w_{ij}s_j(t) > \theta_i \\ s_i(t) & x = \sum_{j=1}^{N} w_{ij}s_j(t) = \theta_i \\ -1 & x = \sum_{j=1}^{N} w_{ij}s_j(t) < \theta_i \end{cases} \tag{8.49}$$

As mentioned earlier, the Hopfield model has a learning stage. During this learning part, the weights of the process are established according to the value of the patterns. Specifically, Hopfield adopted the Hebb rule (Hebb, 1949) to obtain these parameters:

$$w_{ij} = \sum_{p=1}^{N_p} r_i^{(p)} r_j^{(p)} \tag{8.50}$$

If $i = j$, then $w_{ij} = 0$.

Therefore, the matrix of the weights, W, is

$$W = \sum_{p=1}^{N_p} \left(R_p^T R_p - I \right) \tag{8.51}$$

where R_p has the values of the pth pattern.

An example of this method could be a network that has to learn two patterns (Hilera and Martinez, 1995). These patterns appear in Figure 8.14.

The trees (neurons) in dark gray indicate trees that are going to be cut and are assigned a value +1. Trees in light gray are not cut, and the value is −1. So the value of the parameters is $N_p = 2$ and $N = 4$ and the input vectors

$$R_1 = \{-1, -1, 1, 1\}; \quad R_2 = \{1, 1, -1, -1\}$$

The first step in the learning stage is to obtain W:

$$W = \begin{bmatrix} w_{11} & w_{12} & w_{13} & w_{14} \\ w_{21} & w_{22} & w_{23} & w_{24} \\ w_{31} & w_{32} & w_{33} & w_{34} \\ w_{41} & w_{42} & w_{43} & w_{44} \end{bmatrix} = R_1^T R_1 - I + R_2^T R_2 - I$$

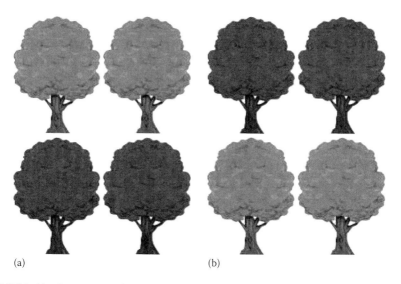

FIGURE 8.14   Patterns to train a network in the Hopfield model: (a) pattern 1 of the network, (b) pattern 2 of the network.

$$R_1^T R_1 - I = \begin{bmatrix} -1 \\ -1 \\ 1 \\ 1 \end{bmatrix} \begin{bmatrix} -1 & -1 & 1 & 1 & 1 \end{bmatrix} - \begin{bmatrix} 1 & 0 & 0 & 0 \\ 0 & 1 & 0 & 0 \\ 0 & 0 & 1 & 0 \\ 0 & 0 & 0 & 1 \end{bmatrix}$$

$$= \begin{bmatrix} 0 & 1 & -1 & -1 \\ 1 & 0 & -1 & -1 \\ -1 & -1 & 0 & 1 \\ -1 & -1 & 1 & 0 \end{bmatrix}$$

$$R_2^T R_2 - I = \begin{bmatrix} 1 \\ 1 \\ -1 \\ -1 \end{bmatrix} \begin{bmatrix} 1 & 1 & -1 & -1 \end{bmatrix} - \begin{bmatrix} 1 & 0 & 0 & 0 \\ 0 & 1 & 0 & 0 \\ 0 & 0 & 1 & 0 \\ 0 & 0 & 0 & 1 \end{bmatrix}$$

$$= \begin{bmatrix} 0 & 1 & -1 & -1 \\ 1 & 0 & -1 & -1 \\ -1 & -1 & 0 & 1 \\ -1 & -1 & 1 & 0 \end{bmatrix}$$

FIGURE 8.15 Input information of the neural network.

So, $W = \begin{bmatrix} 0 & 2 & -2 & -2 \\ 2 & 0 & -2 & -2 \\ -2 & -2 & 0 & 2 \\ -2 & -2 & 2 & 0 \end{bmatrix}$ and the learning phase is finished.

Once the learning stage is over, the network could be used as an associative memory store. Given a certain input information, the iterative process would start until the network produced the information that was most similar to the pattern.

Figure 8.15 shows an example of input information for the neurons in the network, $s = \{1, -1, -1, -1\}$.

In $t = 0$, the first iteration, the output information of every neuron, is the same as the input, $s(t = 0) = \{1, -1, -1, -1\}$.

In $t = 1$, the input information of every neuron will be $s_i(t = 1) = \sum_{j=1}^{N} w_{ij} s_j$. Therefore, for all the neurons, s_1, s_2, s_3, s_4,

$$\left(\text{Input in } t = 1\right) s\left(t = 0\right) W = \begin{bmatrix} 1 & -1 & -1 & -1 \end{bmatrix} \begin{bmatrix} 0 & 2 & -2 & -2 \\ 2 & 0 & -2 & -2 \\ -2 & -2 & 0 & 2 \\ -2 & -2 & 2 & 0 \end{bmatrix}$$

$$= \begin{bmatrix} 2 & 6 & -2 & -2 \end{bmatrix}$$

If the activation function $g(x)$ is a step function with $\theta_i = 0$, then

$$s_i(1) = g(x) = \begin{cases} +1 & x = \sum_{j=1}^{4} w_{ij} s_j(0) > 0 \\ s_i(0) & x = \sum_{j=1}^{4} w_{ij} s_j(0) = 0 \\ -1 & x = \sum_{j=1}^{4} w_{ij} s_j(0) < 0 \end{cases} \qquad (8.52)$$

Since, for neuron s_1, the input information is $2 > 0$, its output information will be $s_1(t=1) = 1$.

For the four neurons, the output information of iteration 1 is $s(1) = [1\ 1\ -1\ -1]$.

The process is repeated for iteration 2:

$$\text{Input in } (t=2) \equiv s(t=1)W = \begin{bmatrix} 1 & 1 & -1 & -1 \end{bmatrix} \begin{bmatrix} 0 & 2 & -2 & -2 \\ 2 & 0 & -2 & -2 \\ -2 & -2 & 0 & 2 \\ -2 & -2 & 2 & 0 \end{bmatrix}$$

$$= \begin{bmatrix} 6 & 6 & -6 & -6 \end{bmatrix}$$

Then, $s(2) = [1\ 1\ -1\ -1]$. The output is the same as iteration 1, so the process is finished, and the pattern more similar is R_2.

The calculus of the process is similar in the continuous case. With this value of w_{ij}, it is demonstrated that under certain conditions and when initiated at some starting value $s_i^{(0)}$, the updating rule (8.45) brings the system to the closest stored pattern $x_i^{(p)}$, which is a local minimum of the energy, E (Hopfield, 1982).

Combinatorial Optimization

The ANN has some advantages for solving combinatorial optimization problems such as the quality of the solution or its facility for parallel implementation. Peterson and Söderberg (2003) group ANN optimization algorithms into two sets:

- Pure ANN approach, based on both binary neurons (Hopfield and Tank, 1985) and multistate neurons (Peterson and Söderberg, 1989). This is a very general approach that is suitable for generic multiple choice problems such as optimal assignment or scheduling.
- Hybrid approaches, such as deformable template algorithms (Durbin and Willshaw, 1987) that introduce specific problem variables into the system, apart from the neural variables. This approach is adequate to low-dimensional problems such as the traveling salesman problem (TSP), but it is beyond the scope of this book.

Binary Neuron Approach

To apply the Hopfield model to an optimization problem implies formulating the objective function $U(x)$ to minimize. This objective function will be compared to the energy function of the Hopfield model. The weights (w_{ij}) and the thresholds (θ_i) are determined in terms of the parameters of the objective function, making both expressions equivalent. Next, the minimum value of the energy function is found in the iteration process, which matches up the minimum of the objective function.

In order to avoid local minimum, the parameters of the energy function and the activation function may be recalculated during the process.

This approach can be applied to any one of the several route problems in forestry such as the order for visiting sampling plots, collecting or delivering wood, controlling pests, and controlling wildlife. They are all based on the TSP.

To clarify this method, we present a case where five research stands ($N=5$) are visited once and the visit ends at the first stand (see Figure 8.16). In which order do we have to visit them so the distance is shortest? The number of different ways in this simple case is $5!/2*5 = 12$.

The distance in kilometers between stands is shown in Table 8.17.

The problem can be solved using the Hopfield model with N^2 neurons ($N=5$). This is a continuous case with values in the interval [0, 1]. In the optimum, the neurons will have the value 0 (inactive) or 1 (active), meaning that the value of the slope in the transfer sigmoidal function must be high.

Neurons can be organized in a matrix where rows represent stands and columns represent the place (order) of the stand in the visit (Table 8.18).

If $s_{23} = 1$, it means that stand 2 will be visited in third place.

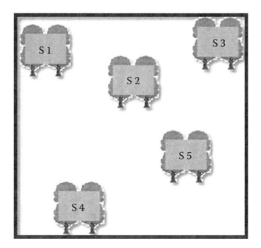

FIGURE 8.16   Location of the stands.

TABLE 8.17

Distance between Stands in Kilometers

Stand	1	2	3	4	5
1	0	2	3	10	8
2	2	0	1	9	5
3	3	1	0	15	4
4	10	9	15	0	4
5	8	5	4	4	0

TABLE 8.18

Organization of the Neurons according to Stands and Visiting Order

Stand	Stop Order				
	1	2	3	4	5
1	S_{11}	S_{12}	S_{13}	S_{14}	S_{15}
2	S_{21}	S_{22}	S_{23}	S_{24}	S_{25}
3	S_{31}	S_{32}	S_{33}	S_{34}	S_{35}
4	S_{41}	S_{42}	S_{43}	S_{44}	S_{45}
5	S_{51}	S_{52}	S_{53}	S_{54}	S_{55}

To solve the optimization problem, the objective function will be

$$U(s) = \frac{A}{2}\sum_{\substack{i=1}}^{N}\sum_{\substack{j=1}}^{N}\sum_{\substack{l=1 \\ l \neq j}}^{N} s_{ij}s_{il} + \frac{B}{2}\sum_{\substack{i=1}}^{N}\sum_{\substack{j=1}}^{N}\sum_{\substack{k=1 \\ k \neq i}}^{N} s_{ij}s_{kj} + \frac{C}{2}\left(\sum_{i=1}^{N}\sum_{j=1}^{N} s_{ij} - N\right)^2$$

$$+ \frac{D}{2}\sum_{\substack{i=1}}^{N}\sum_{\substack{j=1}}^{N}\sum_{\substack{k=1 \\ k \neq i}}^{N} d_{ik}\left(s_{ij}s_{kj+1} + s_{ij}s_{kj-1}\right) \tag{8.53}$$

where

d_{ij} is the distance between stands i and k

the constants A, B, C, D are the relative importance of the terms of the function

The first term means that every stand can only appear once on the route and the second term that one stop j is assigned only to one stand. The third term obliges all the stands N to appear on the route, and the last means that the total length of the route is the minimum.

The next step is to relate this objective function to the energy function

$$E(S) = \frac{1}{2}\sum_{i=1}^{N}\sum_{j=1}^{N}\sum_{k=1}^{N}\sum_{l=1}^{N} w_{ij,kl}s_{ij}s_{kl} + \sum_{i=1}^{N}\sum_{j=1}^{N}\theta_{ij}s_{ij} \tag{8.54}$$

To make these functions equivalent, the value of $w_{ij,kl}$ should be

$$w_{ij,kl} = -A\delta_{ik}\left(1-\delta_{jl}\right) - B\delta_{jl}\left(1-\delta_{ik}\right) - C - Dd_{ik}\left(\delta_{j,l+1} + \delta_{j,l-1}\right) \tag{8.55}$$

where δ_{xy} is the function delta of Kronecker, whose value is 1 if $x=y$ and 0 otherwise.
The threshold values of the activation functions are

$$\theta_{ij} = -CN \tag{8.56}$$

The main problem is to determine A, B, C, D, and θ_{ij} to make the iterative problem converge to the optimum. This last parameter usually has a low value in the first iterations that grows until all the neurons have the value 0 or 1.

The optimum solution is the one that appears in Figure 8.17.

The total length is 23 km. However, the Hopfield method has the disadvantage that it may end up in a local minimum close to the starting solution. To avoid this result, a stochastic algorithm can be integrated. One possibility is SA.

Peterson and Söderberg (2003) proposed an approximation to SA: MF equations. The statements of this method can be consulted in Aarts and Lenstra (2003). In this chapter, we will focus on the case that every spin can take more than two values. This is the Potts neural networks.

A k-state Potts spin is a variable that has K possible values. So spin s_i can be described as a vector $s_i = (s_{i1}, s_{i2}, \ldots, s_{ik})$ where s_{ij} can take the value 0 or 1 and for every i, only one s_{ij} is 1. For example, let S be a set of spins or territorial homogeneous

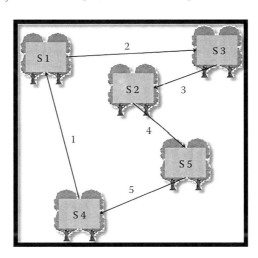

FIGURE 8.17 Optimum route among stands.

units that can have three different land uses {forest, urban, corn plantation}. If spin s_i has assigned the vector {1, 0, 0}, this means that the land use in this spin is forest.

New types of spins appear during the optimization process: v_{ij} that define a Potts neural network; these are the MF variables. The input of these neurons is the new variables u_{ij} that are acted upon by a transfer sigmoidal function. The relationship among them is the following:

$$u_{ij} = -\frac{\partial E(v)}{\partial v_{ij}} : T \tag{8.57}$$

$$v_{ij} = \frac{e^{u_{ij}}}{\sum_{j=1}^{K} e^{u_{ij}}} \tag{8.58}$$

where

$v_{ij} > 0 \; \forall \; i,j$ and $\sum_{j=1}^{K} v_{ij} = 1$

$E(v)$ is the energy function
T is the temperature of the SA method.

The different steps of a generic Potts neural network are the following:

1. Formulate the objective function.
2. Obtain the energy function from the objective function: values of parameters w_{ij}.
3. Estimate the value of T' from which no trivial results are obtained. For synchronous updating of the spins, T' is the largest eigenvalue of matrix W (positive or negative). For serial updating, T' is the largest positive eigenvalue of matrix W.
4. Initialize the neurons v_{ij} with $1/K$ random values and set $T = T'$.
5. Obtain u_{ij} Equation 8.57.
6. Obtain $\sigma = \frac{1}{N} \sum_{i,j} v_{ij}^2$.
7. Do until $\sigma \geq 0.99$.
   a. For $i = 1$ to N.
   b. For $j = 1$ to K.
      i. Obtain v_{ij} Equation 8.58.
   c. Next j.
   d. Next i.
   e. $T = 0.9\,T$.
   f. Obtain u_{ij}.
   g. Calculate σ.
   8 End do.

To make this method clearer, we propose the following example:

Land uses are to be assigned to a territory in order to maximize the benefit considering the costs of transformation from the current land use to the new.

Let $N=5$ be the number of spins that represent homogeneous territorial units in terms of slope, type of soil, and type of climate.

Let $K=3$ be the number of components of the vector of the spins that represents the number of different land uses. These land uses are forestry, urban, and flower plantation. The objective function is

$$f(s)=\sum_{i=1}^{5}\sum_{j=1}^{3}\beta_{ij}s_{ij} \tag{8.59}$$

where β_{ij} means the utility of spin or unit I when the land use is j.

The constraints are

$$\sum_{i=1}^{5}F_{i}s_{i1}\le F \quad \sum_{i=1}^{5}Ur_{i}s_{i1}\le Ur \quad \sum_{i=1}^{5}Fl_{i}s_{i1}\le Fl \tag{8.60}$$

where F_{i}, Ur_{i}, Fl_{i} are the costs of transforming the current use of spin i into forestry, urban, or flower growing and F, Ur, and Fl are the maximum amount of money dedicated to the transformation of land uses. Since there are inequality constraints in the problem, we can introduce penalty functions, such as the penalty amounts, δ, to the constraints

$$\sum_{i=1}^{5}F_{i}s_{i1}+\delta_{F}^{+}-\delta_{F}^{-}-F=0 \tag{8.61}$$

$$\sum_{i=1}^{5}Ur_{i}s_{i1}+\delta_{Ur}^{+}-\delta_{Ur}^{-}-Ur=0 \tag{8.62}$$

$$\sum_{i=1}^{5}Fl_{i}s_{i1}+\delta_{Fl}^{+}-\delta_{Fl}^{-}-Fl=0 \tag{8.63}$$

In general, the expression of these constraints could be $\displaystyle\sum_{i=1}^{5}\gamma_{ri}s_{ir}+\delta_{r}^{+}-\delta_{r}^{-}-\theta_{r}=0$ where γ_{ri} is the transforming costs and θ_{r} the maximum transformation amount for land use r, $r=1,2$ or 3.

If we can assume that E is a linear function of v_{ij} between the extreme values $[-1, 1]$ or $[0,1]$, then we can approximate $\partial E/\partial v_{ij}$ to

$$\frac{\partial E}{\partial v_{ij}}=-\beta_{ij+}+\alpha\sum_{r=1}^{3}\left(\sum_{i=1}^{5}\gamma_{ri}v_{ir}+\delta_{r}^{+}-\delta_{r}^{-}-\theta_{r}\right)_{v_{ij}=1}-\left(\sum_{i=1}^{5}\gamma_{ri}v_{ir}+\delta_{r}^{+}-\delta_{r}^{-}-\theta_{r}\right)_{v_{ij}=0} \tag{8.64}$$

where α is the weighting parameter that the decision maker assigns to the constraints.

At this point, the optimum is obtained following the steps of a generic Potts neural network, as explained earlier.

8.6   MEMORY-BASED OPTIMIZATION METHOD: TABU SEARCH

8.6.1   Introduction

TS is a general framework for a variety of iterative local search strategies for discrete optimization. Hansen (1986) sketched its basic ideas, but it was Glover (1989,1990) who first presented this method in its present form. TS uses the concept of memory by controlling the algorithm's execution via a dynamic list of forbidden moves. The method applies adaptive forms of memory, which equips it to penetrate complexities that often confound alternative approaches (Glover and Laguna, 2002). This allows the TS algorithm to intensify or diversify its search of a given problem's solution space in an effort to avoid entrapment in local optima. On the other hand, no proofs of convergence exist in the literature for the general TS algorithm. Faigle and Kern (1991) propose a particular TS algorithm, called probabilistic TS, as a metaheuristic to help guide SA.

Probabilistic TS attempts to capitalize on both the asymptotic optimality of SA and the memory feature of TS. In probabilistic TS, the probabilities of generating and accepting each candidate solution are set as functions of both a temperature parameter, T (as in SA), and information gained in previous iterations (as for TS). Specifically, they consider that at each temperature T_s, the probabilities of considering w_j, a potential successor of w_k, are given by a stochastic matrix $A(T_s) = (a_{kj}(T_s))$ having the property that there exists an ε, such that for each $T_s > 0$, $a_{kj}(T_s) > 0$ implies $a_{kj}(T_s) \geq \varepsilon$ whenever $w_k \neq w_j$. (To simplify, we write the subindices k,j instead of w_k, w_j).

Once an alternative w_j has been chosen as a potential successor of a current solution w_k, it is accepted with a probability $b_{kj}(T_s)$. The transition matrix $P_k(T_s) = (p_{kj}(T_s))$ can be defined by

$$p_{w_k w_j} = a_{w_k w_j}\left(T_s\right) b_{w_k w_j}\left(T_s\right) \quad \forall w_k, \dot{w}_j, \dot{w}_k, \neq w_j \tag{8.65}$$

Let all the transitions from w_k whenever $a_{kj}(T_s) > 0$ be considered, and let $U(w)$ be the objective function to be minimized. It is assumed that for all T_s, the transitions from w_k are the same and for any real value C, the transitions restricted to the solutions w_k for which $f(w_k) < C$ are strongly connected. If $\pi_k (T_s)$ is the stationary distribution of the transition matrix and $\lim_{T \to 0} \pi(T) = \pi^*$, under the aforementioned assumptions, $\pi_k^* > 0$ only if w_k is an optimal solution.

Since at each step of TS, the neighborhood of a solution changes, this means a change in some of the generation probabilities $a_{kj}(T_s)$. These changes may take into account information stored in the process from previous iterations. Faigle and Kern (1991) are then able to prove asymptotic convergence of their particular TS algorithm as long as probabilities are modified within the bounds $[\varepsilon, 1-\varepsilon]$.

TS contrasts with memoryless designs that heavily rely on semirandom processes that implement a form of sampling. Examples of memoryless methods are GAs, SA, or semigreedy heuristics. TS also contrasts with rigid memory designs typical of branch and bound strategies. However, some authors argue that some types of evolutionary procedures that operate by combining solutions, such as GAs, embody a form of implicit memory (Glover and Laguna, 1997).

8.6.2 Process

Let $U(w)$ be the objective function to be minimized, where $w \in \Omega$ and Ω is the set of solutions or alternatives. Therefore, $U: \Omega \rightarrow R$; the process will try to find a $w^* \in \Omega$ such that $U(w^*)$ is acceptable for some criteria. In general, $f(w^*) > f(w_i)$, $\forall w_i \in \Omega$.

If $N(w_k)$ is the set of neighbor solutions of w_k, we can define $S(w)$ as a subset of solutions in $N(w_k)$. The selection of $S(w)$ is crucial to avoid being stuck in a local minimum. Therefore, solutions like $U(w_j) > U(w_k)$ can be accepted. But there is the risk of cycling through the same solutions in the iterative process. In the case of TS, this is solved by using the information provided from the storage of the exploration process. If memory is introduced, then the set $N(w_k)$ will depend on the itinerary and the iteration in process, so it is more accurate to represent the neighborhood as $N(k, w_k)$.

The use of $N(k, w_k)$ implies that some solutions recently visited in the optimization process were removed from $N(w_k)$. These are considered tabu alternatives, which should be avoided in the next iteration. This is the first characteristic of the memory: recency, that is, whether the solution is tabu or not. This characteristic will partially prevent cycling. If at iteration k, the tabu list is TL of the last n solutions visited, then $N(k, w_k) = N(w_k) - TL$. But TL is very impractical to use, so only the actual moves performed will be tracked and stored in TL. This restriction involves a loss of information and does not guarantee that no cycle of a length of at most n will occur (Hertz et al., 2003).

Another problem of the simplification of the tabu list is that some solutions that may be unvisited are given the tabu status due to solutions being replaced by moves. The tabu status can be relaxed by introducing the aspiration criteria. A tabu move m applied to a current solution w_k may give a better solution than the best one found. The aspiration criterion is the threshold that determines which tabu alternatives can be selected if they satisfy this value.

Finally, the use of short- and long-term memory allows different search strategies to be defined. Previously, it has been seen that short-term memory prohibited some moves. Regarding long-term memory, there are two main strategies: intensification and diversification.

Intensification Strategy: This strategy searches the closest neighbors of elite solutions or also the current solution. Explicit memory is related to intensification strategy, since explicit memory searches the neighborhoods of elite solutions. This intensification can be achieved by introducing an extra term, $I(w)$, in the objective function, which penalizes solutions far from elite or current solutions. Intensification can be performed over a few iterations.

Diversification Strategy: This is a strategy to search for unvisited solutions and generate solutions that differ in various different ways from those seen before. This may be achieved by introducing a penalization term, $D(w)$, of close-to-current-solution alternatives into the objective function. So the objective function will be $U' = U + K_I I(w) + K_D D(w)$, K_I. $K_D = 0$ or 1.

The process is

1. Select an initial $w_0 \in \Omega$ and let $w^* = w_0$.
2. Count $k = 0$.
3. Begin with an empty set TL of tabu moves.
4. Set $k = k + 1$ and generate a subset V^* of solutions in $N(k, w_k)$ such that either one of the tabu conditions is violated or at least one of the aspiration conditions holds.
5. Find the best $w_j \in V^*$ with respect to U or U', $w_k = w_j$.
6. If $U(w_k) < U(w^*)$, then $w^* = w_k$.
7. Update the tabu and aspiration condition.
8. If the stopping condition is met, stop; otherwise, go to 4.

Some of the stopping conditions could be

- k has reached the maximum value.
- There is evidence the optimum has been reached.

$$N\left(k, w_k\right) = \phi$$

8.6.3 OTHER CONSIDERATIONS IN TABU SEARCH

The length of the tabu list is a parameter to be determined for every case. If this length is too short, cycling can occur, but if it is too long, it will be difficult to go outside the local optimum. The core of *TS* is its short-term memory process. The short-term memory of TS constitutes a search for the solution that corresponds to the best (highest evaluation) move possible, subject to certain constraints. These constraints can be related to the problem itself or are designed to prevent the reversal, or sometimes repetition, of certain moves to avoid a cycling behavior. These moves are added to the tabu list. An effective way of avoiding the problem of the length of *TL* is to vary its size. The alternatives will reside for a specified number of iterations bounded by given maximum and minimum values and are removed, freeing them from their tabu status. The tabu list is a circular list, adding elements in sequence in positions 1 through *t*, where *t* is the list size, and then starting over at position 1 again. The addition of each element can thus erase the element recorded in its position *t* iterations ago. Empirical results have indicated that a robust range of values exists for which such a simple tabu list performs very effectively to drive the search beyond local optima and obtain progressively improved solutions.

Elite Solution: In order to identify a set of elite solutions, a threshold is set that is connected to the objective function value of the best solution. They are usually identified during long-term memory processes.

Memory Characteristics

1. Frequency, number of times an element has been chosen to take part in any chosen solution.

2. Quality, ability to differentiate the merit of the solutions. Memory can be used to identify elements that are common to good solutions or paths that lead to these solutions and avoid poor solutions.
3. Influence. Impact of the choices made during the search not only on quality but also on structure.
4. Explicitness. *TS* memory is explicit if it records complete solutions and the value of the attributes, that is, the index of jobs, may be used as an attribute to inhibit or encourage the method to follow certain directions. Jobs with high employment demand cannot be exchanged with jobs with low demand. There is no general rule in the search process to decide which values and attributes should be applied other than experimentally.

8.6.4 Forest Management Application Case

The objective of this case is to choose which trees are going to be cut in order to maximize the benefit. This is a Scots pine forest that belongs to the forest described in the section on SA. The forest management is close-to-nature management. As we have described before, trees in the stand are classified as stabilizer, regeneration, sprinter, or complementary trees.

To formulate the example, we can define the following variables or attributes:

X_1 is the number of regeneration trees to be cut.
X_2 is the number of sprinter trees to be cut.
X_3 is the number of complementary trees to be cut.
X_4 is the number of stabilizer trees to be cut.

A move (the election of a tree) is possible if it does not belong to the tabu list or if it fulfills the aspiration criteria. The aspiration criteria are

- A tabu move is performed if it improves the best solution found so far.
- If a move has never been chosen during a large number of iterations, it is performed whatever the solution it leads to.

The restrictions are the following:

$X_1 + X_2 < 1$, violation penalty 50.
$X_4 < X_3$, violation penalty 30.

Total wood to be cut is 8 m³.

To simplify, the length of the tabu list is two. The neighbor solutions of every alternative are determined by changing one tree for another one of its neighbors. The neighborhood of a tree is the set of trees inside the circle of a radius equal or less than the height of the tree.

The volume and social category of each tree can be seen in Table 8.19.

We have assumed that the value of 1 m³ of wood is 40 euros and the cost of cutting 1 m³ of wood is 25 euros.

TABLE 8.19

Tree Data of the Study Area: Tree Id, Social Category, and Volume

Tree Id	Social Category	Volume n (m³)
T1	Complementary tree	1.5
T2	Regeneration	0.01
T3	Stabilizer tree	2
T4	Stabilizer tree	2.5
T5	Complementary tree	1.5
T6	Regeneration	0.01
T7	Sprinter tree	0.5
T8	Complementary tree	1.5
T9	Complementary tree	1
T10	Complementary tree	1
T11	Stabilizer tree	2
T12	Complementary tree	0.75
T...	...	...

Iterations from k to $k+3$ are described as follows. Figure 8.18 indicates trees to be cut and trees in the tabu list:

Iteration k

Trees T3, T4, T7, T9, and T11 will be cut. Since T7 is a sprinter tree and the number of stabilizer trees chosen is higher than the number of complementary trees, none of the restrictions apply. There is a penalty of 80 euros.

Iteration $k+1$

In this iteration, tree T7 is dropped, and one of its neighbors is added. In this case, it is tree T10, and this is added to the tabu list. Trees T3, T4, T10, T9, and T11 will be cut. Since the number of stabilizer trees chosen is higher than the number of complementary trees, the second of the restrictions does not apply. There is a penalty of 30 euros.

Iteration $k+2$

In this iteration, tree T11 is dropped from the solution, and tree T8 is added. Tree T8 joins the tabu list. In this case, the two constraints apply. So the benefit is 120 euros.

Iteration $k+3$

Tree T10 is removed from the tabu list and dropped, T8 remains in the list, and T12 is added to the solution and to the tabu list. The chosen trees are T3, T4, T9, T8, and T12, with a benefit of 116.25 euros.

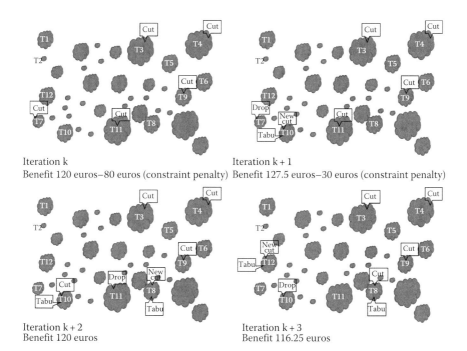

Iteration k
Benefit 120 euros−80 euros (constraint penalty)

Iteration k + 1
Benefit 127.5 euros−30 euros (constraint penalty)

Iteration k + 2
Benefit 120 euros

Iteration k + 3
Benefit 116.25 euros

FIGURE 8.18 (See color insert.) Iteration k through $k+3$ of the TS process. Crown projection of the Scots pine trees in the study area. Labeled trees to be cut or in the tabu list.

Another didactic presentation of TS with a number of applications can be consulted in Glover and Laguna (1993).

8.7 DETERMINATION OF THE BEST MANAGEMENT PLAN

A management plan involves formulating a number of spatially and temporally located actions that will take place in an area of application. The number of feasible plans is so high that it is not operative to generate all the possible options. For instance, if there were ten activities per tree and year (cutting, pruning, labor, pest treatment, etc., and a number of their combinations), the rotation period of the forest is 100 years and the number of trees 500,000; thus, the number of alternatives would be $1000^{500,000}$. In view of this, we determined the best management plan by applying algorithms of combinatory optimization (Martín-Fernández and García-Abril, 2005). Moreover, we specifically used the SA algorithm (Černý, 1985; Kirkpatrick et al., 1983; Metropolis et al., 1953), introducing the following operative simplifications in the previous algorithm before application:

- The selected decision units were the types of forest structures; in our case, five.
- If there is an initial alternative with a specific number of trees per height class and structure, its neighbor alternative would be a solution with a variation of ±5% in the number of trees in every height class. In our case, we

considered eight height classes of 3 m each; the plan is to be applied for 10 years, and the neighbor solution has to be compatible with the natural evolution of the forest.

The study area was located in the Fuenfría Valley (Madrid, Spain). It covers an area of 127.10 ha ($40°45'N$, $4°5'W$) and has altitudes ranging from 1310 to 1790 m, an average annual temperature of $9.4°C$, and average annual rainfall of 1180 mm. Its primary vegetation is Scots pine (*P. sylvestris* L.), grouped in five different types of structure (Pascual et al., 2008):

- Type 1: uneven-aged forest (multilayered canopy) with very high crown cover
- Type 2: multi-diameter forest with high crown cover
- Type 3: multi-diameter forest with medium crown cover
- Type 4: even-aged forest (single story) with low crown cover
- Type 5: zones with scarce tree coverage

The objective is to determine the action to take at tree level in order to maximize the manager's preferences in forest sustainability. We used three indicators that could be calculated from the information available: structural diversity, I_1 (by comparing tree height distribution at each point with the distribution in the ideal case); timber yield, I_2 (computed following García-Abril et al. (2005)); and amount of biomass, I_3 (computed from the crown volume and from the total height of each tree). The three indicators responded to indicators 1.3, 3.1, and 1.4, respectively, of the pan-European sustainable forest management indicators. The indices were computed in a circular plot with a 60 m diameter around each vertex of a grid (with a 1 m side) superimposed on the study area.

We would like to highlight that these indicators were selected only for the purpose of verifying the proposed methodology. Our objective was not to assess their importance in the evaluation of sustainability for forest management.

The scale of measurement for the three indicators was the percentage of the maximum value that could be taken for the ecological characteristics of each point. For this purpose, yield tables for *P. sylvestris* L. in the *Sistema Central* mountain range (García-Abejón, 1984) were used to compute the value of the three sustainability indicators under the best sustainability conditions (IP).

Additionally, a LIDAR image was used to compute the spatial value of the sustainability indicators. The digital canopy height map (DCHM) obtained from the LIDAR image had a pixel of 1 m. The description of the procedure used to acquire the image can be found in Pascual et al. (2008). Finally, field data were used to validate the indicators. Ten plots were obtained by systematic sampling, two per each type of forest structure. Table 8.20 shows the respective values of the indicators (in percentages regarding the most sustainable conditions) both measured directly from the field data and computed solely from the LIDAR image.

The application of this methodology requires two types of data from the individual:

- Personal data
- His or her preferences regarding sustainability between forest locations

TABLE 8.20

Values of the Indicators Obtained with Field Measurements and Measurements from the LIDAR Forest (in % Regarding Maximum Sustainability Conditions)

Type of Plot Structure	Indicator I1 Structural Diversity (%)		Indicator I2 Timber Yield Return (%)		Indicator I3 Biomass (%)	
	Terrain	Image	Terrain	Image	Terrain	Image
T1–P1	63.29	49.26	53.49	49.82	53.15	47.01
T1–P2	55.32	41.47	69.26	55.57	64.45	49.61
T2–P1	49.08	41.06	88.4	89.18	83.8	82.38
T2–P2	44.11	44.15	74.44	72.54	66.19	66.4
T3–P1	58.41	54.47	33.68	33.68	33.88	33.9
T3–P2	49.01	54.37	30.14	28.85	28.06	29.23
T4–P1	57.93	56.16	10.68	15.23	10.99	19.01
T4–P2	56.78	59.55	15.85	10.09	15.99	11.28
T5–P1	61.47	56.77	2.49	8.74	2.69	9.03
T5–P2	62.11	62.07	3.13	4.64	3.83	5.88

The personal data required were profession, educational level, type of residence, age, sex, and stakeholder category (local resident, forest entrepreneur, landowner, academic expert, environmentalist, others). This information allowed us to characterize homogeneous groups of evaluators.

The second type of information was required to fill out the preference matrix, which has been widely explained in Chapter 7.

The analytical expression of the value function could be obtained from the value taken by sustainability indicators at each point of the study area and the sustainability that an individual assigned to each point. It was, therefore, possible to adjust the expression that best explained the value through a linear regression model. The independent variables of this model are all the possible combinations of sustainability indicators, and the dependent variable is the global sustainability value. In order to compare outcomes, the following steps were followed before computing the regression model: (1) The value taken by a combination of sustainability indicators at each point was computed by multiplying the sustainability indicators that outline the combination, (2) the value of both dependent and independent variables was typified, and (3) since a large number of independent variables could be used to calculate the value function (e.g., 696 independent variables are included in the combinations of just 16 sustainability indicators, taken from 1 by 1 to 3 by 3), a procedure to eliminate colinearity among variables in regression was used. Colinearity was assumed between two independent variables when it was accepted that they both have the same distribution by applying the Smirnov–Kolmogorov two-sample test.

The regression model is therefore

$$v_A = k_1I_1 + k_2I_2 + k_3I_3 + k_{12}I_1I_2 + k_{13}I_1I_3 + k_{23}I_2I_3 + k_{123}I_1I_2I_3 \qquad (8.66)$$

where

k_i is the estimated value for each independent variable

I_j is the value of the jth sustainability indicator at each point to calculate sustainability

The previous step in this process was to generate a number of communities that represent all the types of preferences. Unfortunately, classification processes are computationally slow if the number of individuals and descriptors is large and, in our case, meaningful samples require a large sample size. For these reasons, the homogeneous communities were precalculated from a simulated random sample of 5000 individuals with different types of preferences (5000 different 6×6 matrices with values from 1 to 3). As a result, 53 groups of evaluators were obtained and characterized applying the polythetic algorithm.

Each new evaluator is assigned to the most likely group after applying the discriminatory analysis. The goodness of allocating an individual to his or her group can be seen in Figure 7.13. This figure shows the sustainability maps for the preferences of two individuals: for the current evaluator whose objective function is

$$v = 0.85I_1 + 4.38I_2 + 1.82I_3 - 8.62I_1I_2 + 3.77I_1I_3 - 6.94I_2I_3 + 6.14I_1I_2I_3 \qquad (8.67)$$

and for "the most characteristic individual" of the group the first individual to which belongs. The most characteristic individual of a group is the one whose descriptors, integrated according to the most meaningful factor in the classification process, are the closest to the average value in his or her group.

Despite the simplifications adopted, the number of alternatives to be compared in order to calculate the best management plan remains high. It should be recalled that a small perturbation to an initial solution was defined as a variation of between −5% and +5% in the number of trees in each height class and also that the existence of a variation of at least 1% in any of the height classes was adopted as a significant variation from one perturbation to another. Therefore, for any spatial unit of decisions, the number of feasible small perturbations is $11^8 = 214,358,881$. As the spatial unit of decisions is each one of the five classes of forest structures, the number of alternatives to be compared in each step of the optimization process is 5×11^8. Actually, this figure is lower, as any alternative to be analyzed has to be compatible with actual existing stocks in the forest (and some increments in the number of trees may be not compatible with the current stock and the 10-year plan adopted).

To obtain a solution with an acceptable waiting time, the best management plans were precalculated for the preferences of "the most characteristic individual" in each of the 53 communities. Therefore, all the individuals in the same group would have been assigned the same management plan.

The result was that the best management plan was determined from the actual existing stocks (see Table 8.21), which in this case corresponded to the group shown

TABLE 8.21

Example of Trees to Be Cut in Every Type of Forest Structure in a 10 Year Management Plan in the Study Area

					Forest Structures					
	Type 1		Type 2		Type 3		Type 4		Type 5	
Tree Heights	No. Trees/ha	10 Year Harvest (Trees/ha)	No. Trees/ha	10 Year Harvest (Trees/ha)	No. Trees/ha	10 Year Harvest (Trees/ha)	No. Trees/ha	10 Year Harvest (Trees/ha)	No. Trees/ha	10 Year Harvest (Trees/ha)
0–3	4421	0	6132	0	8558	0	7603	0	8841	0
3–6	4	0	5	0	8	0	10	0	9	0
6–9	7	7	16	0	29	0	23	0	13	0
9–12	21	0	60	18	78	20	36	15	17	17
12–15	52	5	127	97	111	57	34	17	16	7
15–18	101	14	127	112	71	24	18	0	8	8
18–21	125	20	58	16	22	0	7	2	2	0
21–24	67	36	13	3	3	0	1	0	0	0

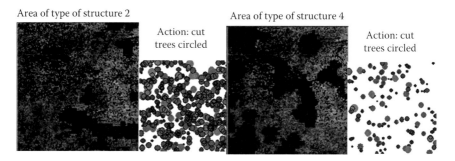

FIGURE 8.19   Evaluator's and group sustainability maps.

in Figure 8.19, 4 (group number 33). The spatial distribution of this plan is shown in Figure 8.19 for two of the existing classes of forest structure.

REFERENCES

Aarts, E. J. Korst, and W. Michiels. 2005. Simulated annealing. In *Search Methodologies Introductory Tutorials in Optimization and Decision Support Techniques*. Eds. E.K. Burk and G. Kandall. New York: Springer.

Aarts, E. and J. Lenstra. 2003. *Local Search in Combinatorial Optimization*. Princeton, NJ: Princeton University Press.

Aarts, E.H.L. and H.M.M. Ten Eikelder. 2002. Simulated annealing. In *Handbook of Applied Optimization*. Eds P.M. Pardalos and M.G.C. Resende. New York: Oxford University Press.

Aarts, E.H.L. and P.J.M. van Laarhoven. 1985. Statistical cooling: A general approach to combinatorial optimization problems. *Phillips Journal of Research* 40: 193–226.

Ackley, D.H., G.E. Hinton, and T.J. Sejnowski. 1985. A learning algorithm for Boltzmann machines. *Cognitive Science* 9: 147–169.

Afshar, M.H. and M. Shahidi. 2009. Optimal solution of large-scale reservoir-operation problems: Cellular-automata versus heuristic-search methods. *Engineering Optimization* 41(3): 275–293.

Alvarez, S. 1995. *Generation of Terrain Textures Using Neural Networks*. MSc thesis. Department of Computer Science, Colorado State University, Fort Collins, CO.

Baker, J.E. 1987. Reducing bias and inefficiency in the selection algorithm. Paper presented at the *International Conference of Genetic Algorithms*. Cambridge, MA, pp. 14–21.

Balzter, H., P. W.Braun, and W. Köhler. 1998. Cellular automata models for vegetation dynamics. *Ecological Modelling* 107: 113–125.

Baskent, E.Z., G.A. Jordana, and A.M.M. Nurullah. 2000. Designing forest landscape (ecosystems) management. *Forestry Chronicle* 76(5): 739–742.

Baskent, E.Z. and S. Keles. 2005. Spatial forest planning: A review. *Ecological Modelling* 188(2–4): 145–173.

Besag, J. 1986. On the statistical analysis of dirty pictures. *Journal of the Royal Statistical Society Series B* 48: 259–302.

Bettinger, P., K. Boston, J. Sessions, and W. Chung. 2002. Eight heuristic planning techniques applied to three increasingly difficult wildlife planning problems. *Silva Fennica* 36(2): 561–584.

Bettinger, P., K. Boston, Y.H. Kim, and J. Zhua. 2007. Landscape-level optimization using tabu search and stand density-related forest management prescriptions. *European Journal of Operational Research* 176(2): 1265–1282.

Bettinger, P., J. Sessions, and K. Boston. 1997. Using Tabu search to schedule timber harvests subject to spatial wildlife goals for big game. *Ecological Modelling* 94(2–3): 111–123.

Blickle, T. and L. Thiele. 1995. A comparison selection schemes used in genetic algorithms. TIK Report, Swiss Federal Institute of Technology, Zurich, Switzerland.

Boston, K. and P. Bettinger. 1999. An analysis of Monte Carlo integer programming, simulated annealing, and tabu search heuristics for solving spatial harvest scheduling problems. *Forest Science* 45(2): 292–301.

Boston, K. and P. Bettinger. 2001. Development of spatially feasible forest plans: A comparison of two modeling approaches. *Silva Fennica* 35(4): 425–435.

Bramlette, M.F. 1991. Initialization, mutation and selection methods in genetic algorithms for function optimization. *Procedures of the 4th International Conference on Genetic Algorithms*. San Diego, CA.

Broner, I. and C.R. Comstock. 1997. Combining expert systems and neural networks for learning site-specific conditions. *Computer Electronics and Agriculture* 19: 37–53.

Brumelle, S., D. Granot, M. Halme, and I. Vertinsky. 1998. A tabu search algorithm or finding a good forest harvest schedule satisfying green–up constraints. *European Journal of Operations Research* 106: 408–424.

Caruana, C.A., L.A. Eshelman, and J.D. Schaffer. 1989. Representation and hidden bias II: Eliminating defining length bias in genetic search via shuffle crossover. In *Eleventh International Joint Conference on Artificial Intelligence*. Ed. N.S. Sridharan. Vol. 1, pp. 750–755. Burlington, MA: Morgan Kaufmann Publishers.

Cerny, V. 1985. A thermodynamical approach to the travelling salesman problem: an efficient simulation algorithm. *Journal of Optimization Theory and Applications* 45: 41–51.

Chipperfield, A., P. Fleming, H. Pohlheim, and C. Fonseca. 1992. *Genetic Algorithm Toolbox for Use with MATLAB. Version 1.2. User's Guide*. Department of Automatic Control and Systems Engineering, University of Sheffield: Sheffield, South Yorkshire, U.K.

Church, R.L. 2007. Tactical-level forest management models. In *Handbook of Operation Research in Natural Resources*. Eds. A. Weintraib, C. Romero, T. Bjorndal, and R. Epstein. New York: Springer.

Church, R., A. Murray, and K. Barber. 2000. Forest planning at the tactical level. *Annals of Operations Research* 95: 3–18.

Clements, S.E., P.L. Dallain, and M.S. Jamnick. 1990. An operational, spatially constrained harvest scheduling model. *Canadian Journal of Forest Research* 20: 1438–1447.

Colasanti, R.I. and J.P. Grime. 1993. Resource dynamics and vegetation processes: A deterministic model using two-dimensional cellular automata. *Functional Ecology* 7: 169–176.

Constantino, M., I. Martins, and J.G. Borges. 2008. A new mixed integer programming model for harvest scheduling subject to maximum area restrictions. *Operations Research* 56: 542–551.

Conway, J.H. 1960. *Game of Life*. University of Cambridge, Cambridge, U.K.

Crow, J.F. 1986. *Basic Concepts in Population, Quatitative and Evolutionary Genetics*. New York: Freeman.

Crowe, K.A. and J.D. Nelson. 2005. An evaluation of the simulated annealing algorithm for solving a restricted harvest-scheduling model against optimal benchmarks. *Canadian Journal of Forest Research* 35(10): 2500–2509.

Crutchfield, J.P., M. Mitchell, and R. Das. 1998. *The Evolutionary Design of Collective Computation in Cellular Automata*. Working papers, Santa Fe Institute, Santa Fe, NM.

Dahlin, B. and O. Sallnas. 1993. Harvest scheduling under adjacency constraints: A case study from the Swedish sub-alpine region. *Scandinavian Journal of Forest Research* 8: 281–290.

Dantzig, G.B. 1982. Reminiscences about the origins of linear programming. *Operations Research Letters* 1(2): 43–48.

Davis, L. 1989. Adapting operator probabilities in genetic algorithms. *Procedures of the International Conference on Genetic Algorithms* 3:61–69.

Davis, L.S., K.N. Johnson, P.S. Bettinger, and T.E. Howard. 2001. *Forest Management*, 4th edn. New York: McGraw-Hill.

Deadman, P.J. and H.R. Gimblett. 1997. Applying neural networks to vegetation management plan development. *Applications of Artificial Intelligence* 11: 107–112.

De Turckheim, B. 1992. Pour une sylviculture proche de la nature, *Forêts de France* 350: 14–20.

De Turckheim, B. 1999. Bases Económicas de la Selvicultura Próxima a la Naturaleza. *Proceedings of the Technical Meeting of Pro Silva*. Valsaín, Spain.

Díaz, A., J.A. Ferland, C.C. Ribeiro, J.R. Vera, and A. Weintraub. 2007. A tabu search approach for solving a difficult forest harvesting machine location problem. *European Journal of Operational Research* 179(3): 788–805.

Ducheyne, E.I., R.R. De Wulf, and B. De Baets. 2004. Single versus multiple objective genetic algorithms for solving the even-flow forest management problem. *Forest Ecology and Management* 201(2–3): 259–273.

Durbin, R. and D.J. Willshaw. 1987. An analogue approach to the traveling salesman problem using an elastic net method. *Nature* 326: 689–691.

Epstein, R., J. Karlsson, M. Rönnqvist, and A. Wintraub 2007. Harvest operational models in forestry. In *Handbook of Operation Research in Natural Resources*. Eds. A. Weintraib, C. Romero, T. Bjorndal, and R. Epstein. New York: Springer.

Faigle, U. and W. Kern. 1991. Note on the convergence of simulated annealing algorithms. *SIAM Journal on Control and Optimization* 29: 153–159.

Falcão, A.O. and J.G. Borges. 2002. Designing an evolution program for solving integer forest management scheduling models: An application in Portugal. *Forest Science* 47(2): 158–168.

Falconer, D.S. 1981. *Introduction to Quantitative Genetics*. London, U.K.: Longman.

Fischer, D. and R.L. Church. 2003. Clustering and compactness in reserve site selection: An extension of the biodiversity management area selection model. *Forest Science* 49: 1–11.

Flache, A. and R. Hegselmann. 2001. Do irregular grids make a difference? Relaxing the spatial regularity assumption in cellular models of social dynamics. *Journal of Artificial Societies and Social Simulation* 4(4): 6. http://jasss.soc.surrey.ac.uk/4/4/6.html.

Fogarty, T.C. 1989a. Varying the probability of mutation in the genetic algorithm. *Procedures of the Third International Conference on Genetic Algorithms*. San Diego, CA, USA, pp. 104–109.

Fogarty, T.C. 1989b. An incremental genetic algorithm for real-time learning. *Procedures of the 6th International Workshop on Machine Learning*. San Diego, CA, USA, pp. 416–419.

Garcia, O. 1990. Linear programming and related approaches in forest planning. *New Zealand Journal of Forest Science* 20: 307–331.

García-Abejón, J.L. 1984. Tablas de producción de densidad variable para *Pinus sylvestris L.* en el Sistema Central. Ministerio de Agricultura, Pesca y Alimentacion. INIA.

García-Abril, A., M.A. Grande, E. Blanco, and J. Velázquez. 2005. *Tendencias de la Gestión Sostenible e Importancia de la Biodiversidad*. Ed: Junta de Comunidades de Castilla y León.

Geman, S. and D. Geman. 1984. Stochastic relaxation, Gibbs distributions, and the Bayesian restoration of images. *IEEE Transactions on Pattern Analysis and Machine Intelligence* 6(6): 721–741.

Georgii, H.-O. 1988. *Gibbs Measures and Phase Transitions*. Berlin, Germany: de Gruyter.

Gimblett, R.H. and G.L. Ball. 1995. Neural network architectures for monitoring and simulating changes in forest resource management. *Applications of Artificial Intelligence* 9: 103–123.

Glover, F. 1989. Tabu search, Part I. *ORSA Journal on Computing* 1: 190–206.

Glover, F. 1990. Tabu search, Part II. *ORSA Journal on Computing* 2: 4–31.

Glover, F. and M. Laguna. 1993. Tabu search. In: *Modern Heuristic Techniques for Combinatorial Problems*. Ed. C.R. Reeves. Oxford, U.K.: Blackwell.

Glover, F. and M. Laguna. 2002. Tabu serach. In *Handbook of Applied Optimization*. Eds. P.M. Pardalos and M.G.C. Resende. New York: Oxford University Press.

Glover, F. and M. Laguna. 1997. *Tabu Search*. Norwell, MA: Kluwer Academic Publishers.

Goldberg, D.E. 1989. *Genetic Algorithms in Search, Optimization and Machine Learning*. Boston, MA: Addison Wesley Publishing Company.

Granville, V., M. Krivanek, and J.P. Rasson. 1994. Simulated annealing—A proof of convergence. *IEEE Transactions on Pattern Analysis and Machine Intelligence* 16: 652–656.

Green, D.G., A.P. House, and S.M. House. 1985. Simulating spatial patterns in forest ecosystems. *Mathematics and Computers in Simulation* 27: 191–198.

Greer, K. and B. Meneghin. 2000. Spectrum: An analytical tool for building natural resource management models. In *Proceedings of the Seventh Symposium on Systems Analysis I Forest Resources, 1991*. Ed. J. Fried, M. Vasiewvich, and L. Leefers. St. Paul, MN: USDA, North Central Research Station.

Guan, B.T. and G.Z. Gertner. 1995. Modeling individual tree survival probability with a random optimization procedure: An artificial neural network approach. *Applications of Artificial Intelligence* 9: 39–52.

Gunn, E.A. 2007. Models for strategic forest management. In *Handbook of Operation Research in Natural Resources*. Eds. A. Weintraib, C. Romero, T. Bjorndal, and R. Epstein. New York: Springer.

Hammersley, J.M. and D.C. Handscomb. 1964. *Monte Carlo Methods*. London, U.K.: Chapman & Hall.

Hansen, P. 1986. The steepest ascent mildest descent heuristic for combinatorial programming. Talk presented at the *Congress on Numerical Methods in Combinatorial Optimization*. Capri, Italy.

Harvey, W. and D.J. Dean. 1996. Computer-aided visualization of proposed road networks using GIS and neural networks. In *First Southern Forestry GIS Conference*. Eds. G.J. Arthaud and W.C. Hubbard. Athens, GA: University of Georgia.

Hebb, D. 1949. *Optimization of Behaviour*. New York: Wiley.

Hernández Encinas, A., L. Hernández Encinas, S. Hoya White, A. Martín del Rey, and G. Rodríguez Sánchez. 2007. Simulation of forest fire fronts using cellular automata. *Advances in Engineering Software* 38(6): 372–378.

Heinonen, T. and T. Pukkala. 2004. A comparison of one- and two compartment neighborhoods in heuristic search with spatial forest management goals. *Silva Fennica* 38(3): 319–332.

Heinonen, T. and T. Pukkala. 2007. The use of cellular automaton in forest planning. *Canadian Journal of Forest Research* 37: 2188–2220.

Hertz, A., E. Taillard, and D. de Werra. 2003. Tabu search. In: *Local Search in Combinatorial Optimization*. Eds. E. Aarts and J. Lenstra. Princeton, NJ: Princeton University Press.

Hilera, J.R. and V.J. Martínez. 1995. *Redes Neuronales Artificiales: Fundamentos, Modelos y Aplicaciones*. Madrid, Spain: RA-MA Editorial.

Hogeweg, P. 1988. Cellular automata as a paradigm for ecological modeling. *Applied Mathematical and Computations* 17: 81–100.

Hopfield, J.J. 1982. Neural networks and physical systems with emergent collective computational abilities. *Proceedings of the National Academy of Sciences of the United States of America* 79: 2554–2558.

Hopfield, J.J. 1988. Neurons with graded response have collective computational properties like those of two states neurons. In *Neurocomputing*. Eds. J. Anderson and E. Rosenfeld. Cambridge, MA: MIT Press.

Hopfield, J.J. and D.W. Tank. 1985. Neural computation of decisions in optimization problems. *Biological Cybernetics* 52: 141–152.

Huang, R. and T.C. Fogarty. 1991. Adaptive classification and control-rule optimization via a learning algorithm for controlling a dynamic system. *Procedures of the 30th Conference on Decision and Control*. Brighton, England, pp. 867–868.

Ising, E. 1925. Beitrag zur theorie des ferromagnetismus. *Zeitschrift für Physik* 31(1): 253–258.

Janikow, C.Z. and Z. Michalewicz. 1991. An experimental comparison of binary and floating point representations in genetic algorithms. *Procedures of the Fourth International Conference on Genetic Algorithms.* pp. 31–36.

Jennerette, G.D. and J. Wu. 2001. Analysis and simulation of land-use change in the central Arizona-Phoenix region, USA. *Landscape Ecology* 16: 611–626.

Johnson, A.W. and S.H. Jacobson. 2002. On the convergence of generalized hill climbing algorithms. *Discrete Applied Mathematics* 119: 37–57.

Jones, J.G., J.F.C. Hyde, and M.L. Meacham. 1986. Four analytical approaches for integrating land management and transportation planning on forest lands, Research Paper. U.S. Department of Agriculture, Forest Service, Intermountain Research Station, Ogden, UT.

Karafyllidis, I. 2004. Design of a dedicated parallel processor for the prediction of forestfire spreading using cellular automata and genetic algorithms. *Engineering Applications of Artificial Intelligence* 17: 19–36.

Karafyllidis, I. and A. Thanailakis. 1997. A model for predicting forest fire spreading using cellular automata. *Ecological Modelling* 99: 87–97.

Kent, B., B.B. Bare, R. Field, and G. Bradley. 1991. Natural resource land management planning using large-scale linear programs: The USDA Forest Service experience with FORPLAN. *Operations Research* 19: 13–27.

Kindermann, R. and J.L. Snell. 1980. *Markov Random Fields and Their Applications.* Providence, RI: American Mathematical Society, ISBN 0-8218-5001-6.

Kirkpatrick, S., C.D. Gelatt, and M.P. Vecchi. 1983. Optimization by simulated annealing. *Science* 220: 671–680.

Langton, C.G. 1991. *Emergent Computation.* Cambridge, MA: MIT Press.

Li, R., P. Bettinger, and K. Boston. 2010. Informed development of meta heuristics for spatial forest planning problems. *The Open Operational Research Journal* 4: 1–11.

Lockwood, C. and T. Moore. 1993. Harvest scheduling with spatial constraints: A simulated annealing approach. *Canadian Journal of Forest Research* 23: 468–478.

Lu, F. and L.O. Eriksson. 2000. Formation of harvest units with genetic algorithms. *Ecology and Management* 130(1–3): 57–67.

Lucasius, C.B. and G. Kateman. 1992. Towards solving subset selection problems with the aid of the genetic algorithm. In *Parallel Problem Solving from Nature 2*, Eds. R. Männer and B. Manderick, pp. 239–247. Amsterdam, the Netherlands: North-Holland.

Lundy, M. and A. Mees. 1986. Convergence of an annealing algorithm. *Mathematical Programming* 34: 111–124.

Malchow-Møllera, N., N. Strangeb, and B.J. Thorsen. 2004. Real-options aspects of adjacency constraints. *Forest Policy and Economics* 6(3–4): 261–270.

Mann, S. and G.L. Benwell. 1996. The integration of ecological, neural and spatial modelling for monitoring and prediction for semi-arid landscapes. *Computer and Geosciences* 22: 1003–1912.

Martínez-Falero, E., S. Martín Fernández, and A. Orol. 2010. *Sistema para la orientación personalizada para el empleo. OPEm.* Number of reference M-00847/2009. Madrid, Spain: Comunidad de Madrid.

Martín-Fernández, S. and A. García-Abril. 2005. Optimisation of spatial allocation of forestry activities within a forest stand. *Computers and Electronics in Agriculture* 49: 159–174.

Marzban, C., H. Paik, and G. Stumpf. 1997. Neural networks versus Gaussian discriminant analysis. *Applications of Artificial Intelligence* 11: 49–58.

Marzban, H. and G. Stumpf. 1996. A neural network for tornado prediction based on Doppler radar-derived attributes. *Journal of Applied Meteorology* 35: 617–626.

Mathey, A.E., E. Krcmar, S. Dragicevic, and I. Vertinsky. 2008. An object-oriented cellular automata model for forest planning problems. *Ecological Modelling* 212: 359–371.

McCulloch, W. and W. Pitts. 1943. A logical calculus of the ideas immanent in nervous activity. *Bulletin of Mathematical Biophysics* 7: 115–133.

McGarigal, K. and B.J. Marks. 1995. Fragstats: Spatial pattern analysis program for quantifying landscape structure. Technological Report. PNW-GTR-351. Portland, OR.

Metropolis, N., A.W. Rosenbluth, M.N. Rosenbluth, A.H. Teller, and E. Teller. 1953. Equation of state calculation by fast computing machines. *Journal of Chemistry* 21: 1087–1091.

Mitra, D., F. Romeo, and A.L. Sangiovanni-Vincentelli. 1986. Convergence and finite time behavior of simulated annealing. *Advances in Applied Probability* 18: 747–771.

Morales, F. 2011. Aplicación de métodos de toma de decisiones multi-atributo en la definición de prioridades en la gestión de infraestructuras en San Luis Potosí, México. PhD thesis. Technical University of Madrid, Madrid, Spain.

Mühlenbein, H. 2003. Genetic algorithms. In *Local Search in Combinatorial Optimization*. Eds. E. Aarts and J. Lenstra. Princeton, NJ: Princeton University Press.

Mühlenbein, H. and D. Schlierkamp-Voosen. 1993. Predictive models for the breeder genetic algorithm. *Evolutionary Computation* 1(1): 25–49.

Mullen, D.S. and R.M. Butler. 1997. The design of a genetic algorithm based spatially constrained timber harvest scheduling model. *Proceedings of the 7th Symposium on Systems Analysis in Forest Resources*, Traverse City, MI. http://www.for.msu.edu/e4/e4 ssafr97.html.

Murray, A.T. 1999. Spatial restrictions in harvest scheduling. *Forest Science* 45(1): 45–52.

Murray, A.T. and R.L. Church. 1995. Heuristic solution approaches to operational forest planning problems. *OR Spektrum* 17: 193–203.

Nalle, D.J., J.L. Arthur, and J. Sessions. 2002. Designing compact and contiguous reserve networks with a hybrid heuristic algorithm. *Forest Science* 48: 59–68.

Nautiyal, J.C. and P.H. Pearse. 1967. Optimizing the conversion to sustained yield—A programming solution. *Forest Science* 13: 131–139.

Navon, D.I. 1971. Timber RAM: A long-range planning method for commercial timber lands under multiple-use management. USDA Forest Service, Research Paper PSW-70, University of Berkeley, Berkeley, CA.

Nelson, J. and J.D. Brodie. 1990. Comparison of random search algorithm and mixed integer programming for solving area-based forest plans. *Canadian Journal of Forest Research* 20: 934–942.

Nelson, J., J.D. Brodie, and J. Sessions. 1991. Integrating short-term, area-based logging plans with long-term harvest schedules. *Forest Science* 37: 101–102.

O'Hara, A.J., B.H. Faaland, and B.B. Bare. 1989. Spatially constrained timber harvest scheduling. *Canadian Journal of Forest Research* 19: 715–724.

Öhman, K. 2002. Spatial optimization in forest planning: A review of recent Swedish research. In *Multi Objective Forest Planning*. Ed. T. Pukkala. pp. 153–172. Dordrecht, the Netherlands: Kluwer Academic Publishers.

Öhman, K. and L.A. Eriksson. 2002. Allowing for spatial consideration in long-term forest planning by linking programming with simulated annealing. *Forest Ecology and Management* 161(1): 221–230.

Otto, H.J. 1997. Principes d'une sylviculture proche de la nature. *Revue Forestière Française* XLIX: 477–488.

Palahí, M. 2002. Modeling stand development and optimizing the management of even-aged Scots pine forests in Northeast Spain. Academic dissertation. Research notes 143. Faculty of Forestry, University of Joensuu, Joensuu, Finland.

Paruelo, J.M. and F. Tomasel. 1996. Prediction of functional characteristics of ecosystems: a comparison of artificial neural networks and regression models. Unpublished research paper. Department of Rangeland Ecosystem Science, Colorado State University, Fort Collins, CO.

Pascual, C., A. García-Abril, L.G. García-Montero, S. Martín-Fernández, and W.B. Cohen. 2008. Object-based semi-automatic approach for forest structure characterization using lidar data in heterogeneous *Pinus sylvestris* stands. *Forest Ecology and Management* 255: 3677–3685.

Pattie, D.C. and G. Haas. 1996. Forecasting wilderness recreation use: Neural network versus regression. *Applications of Artificial Intelligence* 10: 67–74.

Patuwo, E., M.Y. Hu, and M.S. Hung. 1993. Two-group classification using neural networks. *Decision Science* 24(4): 825–845.

Pérez-Ramirez, N. 2007. Modelo de ordenación del uso de suelo para el desarrollo sostenible: aplicación al municipio de Huejotzingo, Puebla, México. PhD thesis, Universidad Politécnica de Madrid, Madrid, Spain.

Peterson, C. and E. Hartman. 1989. Explorations of the mean field theory learning algorithm. *Neural Networks* 2: 475–494.

Peterson, C. and B. Söderberg. 1989. A new method for mapping optimization problems onto neural networks. *International Journal of Neural System* 1: 3–22.

Peterson, C. and B. Söderberg. 2003. Artificial neural networks. In *Local Search in Combinatorial Optimization*. Eds. E. Aarts and J. Lenstra. Princeton, NJ: Princeton University Press.

Pukkala, T. 2002. Introduction to multi-objective forest planning. In *Multi-Objective Forest Planning*. Ed. T. Pukkala. pp. 1–19. Dordrecht, the Netherlands: Kluwer Academic Publishers.

Reibnegger, G., G. Weiss, G. Werner-Felmayer, G. Judmaier, and H. Wachter. 1991. Neural networks as a tool for utilizing laboratory information: Comparison with linear discriminant analysis and with classification and regression trees. *Procedures of the National Academy of Sciences* 88:11426–11430.

Richards, E.W. and E.A. Gunn. 2000. A Model and tabu search method to optimize stand harvest and road construction schedules. *Forest Science* 46(2):188–203.

Richards, E.W. and E.A. Gunn. 2003. Tabu search design for difficult forest management optimization problems. *Canadian Journal of Forest Research* 33(6): 1126–1133.

Romeo, F. and A. Sangiovanni-Vincentelli. 1991. A theoretical framework for simulated annealing. *Algorithmica* 6: 302–345.

Rossier, Y., M. Troyon, and T.M. Liebling. 1986. Probabilistic exchange algorithms and euclidean traveling salesman problems. *OR Spektrum* 8: 151–164.

Schultz, A., R. Wieland, and G. Lutze. 2000. Neural networks in agroecological modelling— Stylish application or helpful tool? *Computers and Electronics in Agriculture* 29(1–2): 73–97.

Seppelt, R. and A. Voinov. 2002. Optimization methodology for land use patterns using spatially explicit landscape models. *Ecological Modelling* 151: 125–142.

Sidiropoulos, E. and P. Tolikas. 2008. Genetic algorithms and cellular automata in aquifer management. *Applied Mathematical Modelling* 32(4): 617–640.

Siitonen, M. 1993. Experiences in the use of forest management planning models. *Silva Fennica* 27(2): 167–178.

Silveira Soares-Filho, B., G. Coutinho-Cerqueira, and C. Lopes-Pennachin. 2002. Dinamica–A stochastic cellular automata model designed to simulate the landscape dynamics in an Amazonian colonization frontier. *Ecological Modelling* 154 (2002): 217–235.

Snyder, S. and C. ReVelle. 1996. Temporal and spatial harvesting of irregular systems of parcels. *Canadian Journal of Forest Research* 26: 1079–1088.

Sole, R.V. and C.S. Manrubia. 1995. Are rainforests self-organized in a critical state? *Journal of Theoretical Biology* 173: 31–40.

Spears, W.M. and K.A. De Jong. 1991. An analysis of multi-point crossover. In *Foundations of Genetic Algorithms*. Ed. J.E. Rawlins. pp. 301–315. San Mateo, CA: Kaufmann Publishers.

Strange, N., H. Meilby, and P. Bogetoft. 2001. Land use optimization using self-organizing algorithms. *Natural Resource Modeling* 14(4): 541–573.

Strange, N., H. Meilby, and J.T. Thorsen. 2002. Optimizing land use in afforestation areas using evolutionary self-organization. *Forest Science* 48(3): 543–555.

Supratid, S. and R. Sadananda. 2004. Cellular automata–critical densities on forest fire dispersion. *WSEAS Transactions on Computers* 6(3): 2043–2048.

Syswerda, K.G. 1989. Uniform crossover in genetic algorithms. *Procedures of the Third International Conference on Genetic Algorithms*. Fairfax, VA.

Tan, S.S. and F.E. Smeins. 1996. Predicting grassland community changes with artificial neural network model. *Ecological Modeling* 84: 91–97.

Tarp, P. and F. Helles. 1997. Spatial optimization by simulated annealing and linear programming. *Scandinavian Journal of Forest Research* 12: 390–402.

Thompson, M.P., J.D. Hamann, and J. Sessions. 2009. Selection and penalty strategies for genetic algorithms designed to solve spatial forest planning problems. *International Journal of Forestry Research* 2009: 1–14.

Voigt, H.M., H. Muhlenbein, and C.D. Cvetkovi. 1995. Fuzzy recombination for the breeder genetic algorithm. *Proceedings of the 6th International Conference on Genetic Algorithms*. Ed. S. Forrest. pp. 104–111. San Mateo, CA: Morgan Kaufmann.

Von Neumann, J. 1966. *Theory of Self-Reproduction Automata*. Champaing, IL: University of Illinois Press.

Walters, K.R. 1993. Design and development of a generalized forest management modeling system: Woodstock. *Proceedings of the International Symposium on Systems Analysis and Management Decisions in Forestry*. Valdivia, Chile, pp. 190–196.

Weintraub, A., R.L. Church, A.T. Murray, and M. Guignard. 2000. Forest management models and combinatorial algorithms: Analysis of state of the art. *Annals of Operations Research* 96: 271–285.

Weintraub, A., S. Guitart, and V. Kohn. 1986. Strategic planning in forest industries. *European Journal of Operational Reseach* 24: 152–162.

Whitley, D. 1989. The Genitor algorithm and selection pressure: Why rank based allocations of reproductive trials is best. *Procedures of the Third International Conference on Genetic Algorithms*. Fairfax, VA.

Wolfram, S. 1984. Universality and complexity in cellular automaton. *Physica* 10: 1–35.

Wolfram, S. 2002. *A New Kind of Science*. Champaign, IL: Wolfram Media Inc.

Wright, A.H. 1991. Genetic algorithms for real parameter optimization. In *Foundations of Genetic Algorithms*. Ed. J.E. Rawlins. pp. 205–218. San Mateo, CA: Kaufmann Publishers.

Yoon, Y., G. Swales, and T.M. Margavio. 1993. A comparison of discriminant analysis versus artificial neural networks. *The Journal of the Operational Research Society* 44(1): 51–60.

Zeng, H., T. Pukkala, H. Peltola, and S. Kellomäki. 2010. Optimization of irregular-grid cellular automata and application in risk management of wind damage in forest planning. *Canadian Journal of Forest Research* 40: 1064–1075.

9 Multiparticipant Decision-Making

Esperanza Ayuga-Téllez,
Concepción González-Garcia, and
Eugenio Martínez-Falero

CONTENTS

9.1   INTRODUCTION

Two extreme types of methodologies are recognized when making choices involving multiple decision-makers. At one extreme are the models that simulate the spread of opinions (Watts, 2003; Holme and Newman, 2006). These models focus on the process of infection or contagion, without considering the optimization of individual agents' behavior.* The other extreme is the simulation of spatial or networked games (Davidsen et al., 2002; Minnhagen et al., 2007; Baek and Bernhardsson, 2010). In this case, each agent actively tries to maximize his or her individual gain, and the success of one agent is in detriment to the others, or vice versa: there exists a conflict of interest, which requires a certain type of collaborative process to achieve a global solution.†

Collective decision-making (CDM) (Curty and Marsili, 2006; Watkins and Rodriguez, 2008) falls somewhere between the previously mentioned two extremes. As it is an intermediate methodology, some applications can be expanded to share the properties of any of the models described as extreme. However, the main characteristic of CDM is the aggregation of individuals' information to generate a global solution, taking account of individual actions and social interactions and conforming the outlook of the population on a specific issue. We have adopted the methodology of CDM to incorporate participation in forest management. Furthermore, we focus exclusively on web-based CDM systems, which are a means of incorporating the individuals who use the Internet to make decisions.

Collective web-based intelligence mitigates the effects of many of the biases that occur in individual decision-making (Myers, 2002). In the phase of solution generation, these biases include the tendency to seek information that confirms

---

* Chapter 1 introduces some of these models in order to simulate the acceptance of sustainable management in the population, as well as the effects of public participation in accomplishing the social spread of sustainable development. Most of these models are based on system dynamics tools (Holling, 1973; Harich, 2010; Smith et al., 2011).

† The need for collaboration is particularly evident in early methodologies incorporating multiple decision-makers: Group Decision Support Systems (*GDSS*) and Social Decision Support Systems (*SDSS*). GDSS seek to reach a global negotiated decision among participants through face-to-face or videoconference meetings. SDSS (see, e.g., Turoff et al., 2002) are based on visualization of the flow of the discussion through a network of statements, opinions, arguments, and comments, which helps to yield consensus prior to voting on an issue. Both suffer from lack of participation and both seek to reach a decision through consensus. Therefore, this type of approach rarely finds the best solution for the toughest problems, and this is a burden when excellence is almost a prerequisite for survival.

previous assumptions and the inertia in maintaining previous beliefs even in the face of contrary evidence. Other biases, this time in the field of evaluation of alternatives, are the habit of seeing patterns where none exists and the tendency to be influenced by the way a solution is presented. The effects of these and other biases are reduced by adopting the three types of approach, which characterize CDM (Bonabeau, 2009): *outreach* to tap into people who have not traditionally been included, *aggregation* of information from multiple sources (which reinforces the nuclear trend of the system in the context of the central limit theorem), and *self-organization* to enable interactions in which the whole is greater than the sum of its parts. These three approaches determine, respectively, the three main sections of this chapter (9.2 through 9.4).

9.1.1 OUTREACH

The mathematical formulation of outreach was first developed by Condorcet (1785)*. For n decision-makers and each decision-maker having a probability $p \in [0, 1]$ of choosing the best of two options in a decision, Condorcet's theorem states that if $p > 0.5$ and $n \to \infty$, then the probability of a majority vote outcome rendering the best decision approaches certainty at 1. In other words, if a decision-making group has a large n of reasonably informed ($p > 0.5$) and independent decision-makers, then the group increases its chances of optimal decision-making[†].

Condorcet's theorem indicates that direct democracy is the best system for yielding optimal decisions. However, the burden of constant voting and the logistical problems inherent in direct democracy make it necessary to have some type of representation. It is, therefore, critical that representatives "act in the same manner as the whole body would act if they were present" (Paine, 1776). Paine stated that representatives should maintain "fidelity to the public," something that is only possible through frequent elections.

Today, the use of the Internet affords repeated "elections." Moreover, the web allows individuals to create (or destroy) links to other people as they please. Building web-based social network systems allows citizens to choose their representatives dynamically. If a person is unable to participate in a decision-making process, then he or she may abstain from participating, in the knowledge that the underlying social network will accurately distribute their voting power to their neighbors or neighbors' neighbors. As Rodriguez and Watkins (2009) have demonstrated, a dynamic, distributed, and democratic representation allows the decision-making of the whole population to be simulated from a subset of the population. Section 9.2.1 shows the

---

* Marie Jean Antoine Nicolas de Caritat, Marquis de Condorcet (1743–1794), was a French mathematician, philosopher, and political scientist. His method of voting tally selects the candidate who would beat each of the other candidates in a runoff election. Condorcet advocated a liberal economy, public education, freedom, constitutionalism, and equal rights for women and people of all races. His ideas embody the ideals of rationalism and the age of enlightenment and remain influential today. Condorcet died a mysterious death in prison after a period as a fugitive from the French Revolutionary authorities.

[†] As an empirical verification of this theorem, Surowiecki (2005) offers a large collection of cases of application where a group of diverse, independent, and reasonably informed people outperform the best individual decisions.

algorithm for decision-making in a trust-based social network (TBSN), which used to propagate the assessment of sustainability from active decision-makers to the whole of the population.

Condorcet's theorem also holds in the reverse direction: if $p<0.5$, then as $n \rightarrow \infty$, the probability of a majority vote outcome rendering the best decision approaches 0. This means we do need not only representative participants but also reasonably informed participants ($p>0.5$). Although it is not always possible to ensure a sufficiently well-informed group of decision-makers, it is however possible to increase p. This can mainly be achieved through three actions: first, by incorporating procedures to facilitate communication to the general public of the way sustainability indicators explain sustainable development (as described in Chapters 3 through 7); second, by developing systems for aggregating information that guarantees that individual knowledge is thoughtfully applied to the decision (Sections 9.3 and 9.4); and finally, by designing dynamic social networks (Sections 9.2.1 and 9.2.2) by prioritizing the connection of inactive individuals to active voters who show a similar voting tendency but have a greater consistency in their preferences and a deeper understanding of the concept of sustainability (these characteristics—consistency and depth of knowledge—have been described in Chapter 7).

9.1.2 Aggregation of Individual Preferences

Aggregation of preferences plays a central role in operating CDM. But can a society as a whole choose between different options? The work of Arrow (1950 and 1963)—inspired by that of Borda (1784) and Condorcet (1785)—placed social choice within a structured framework. It relates social preferences (or decisions) to individual preferences through a relation known as a "social welfare function" (Sen, 1982). Unfortunately, the main result in this area is apparently pessimistic: even some very mild conditions of reasonableness could not be simultaneously satisfied by any social choice procedure.* In fact, only a dictatorship would avoid inconsistencies, which would involve insensitivity to the interests of a diverse population (Sen, 1998). In consequence, aggregation of preferences will usually involve circumventing Arrow's impossibility theorem (Arrow, 1963) by relaxing some of its applicability conditions[†].

From a historic perspective, the conditions used by Arrow confine the effect of social choice procedures to voting rules. Voting systems are, therefore, the first methods we describe (Section 9.3.1). However, there are currently other web-based aggregation mechanisms such as prediction markets.[‡]

---

* These conditions include the following: (1) Pareto efficiency (a solution is better if it increases the utility of all agents), (2) non-dictatorship, (3) independence (requiring that the social choice within any set of alternatives depends only on the preferences for these alternatives), and (4) unrestricted domain (requiring that the social preference should be a complete ordering, with full transitivity, and that it must work for every conceivable set of individual preferences).

† For example, one of the most frequently used methods in aggregating individual preferences—the Borda count—which has been applied in real situations of participatory forest management (Laukkanen et al., 2002; Vainikainena et al., 2008), satisfies the axioms of unanimity and non-dictatorship; however, it does not satisfy the independence of irrelevant alternatives.

‡ The Web 2.0 provides other systems for the aggregation of preferences including document ranking, folksonomy, recommender systems, vote systems, open software, and wiki.

Prediction markets are based on the fact that individual preferences are applied to distinguish between subjectively "better" (preferred) and "worse" situations. In economic terms, the aggregate desire becomes the market "demand," and the aggregate perception of the present situation becomes the "supply" (Heylighen, 1997). Therefore, the simulation of a market of wisdoms can be applied to the aggregation of collective preferences, although it must be noted that what is preferable for an individual is not necessarily what is preferable for a group (Heylighen and Campbell, 1995). This issue is a further verification of Arrow's results and, thus, of the need to relax some of the conditions of reasonableness for achieving a collective decision. Prediction markets are described in Section 9.3.2.

Parallel to the social choice theory that focuses on voting, the utilitarian economists (Edgeworth, 1881; Marshall, 1890)—inspired by Bentham (1789)*— used the aggregation of individual utilities to obtain evaluations of social interest. They were concerned with the total utility of the community but did not incorporate the manner in which utility is distributed or concentrated (Sen, 1998). Instead, and following the argument of "logical positivism" (Robbins, 1938), utilitarian economists made use of a single criterion of social improvement (Pareto efficiency), which states that an alternative is the best, if it increases the utility of all. This approach still failed to take account of the distribution of utility. However, the identification of individual gains and losses made it possible to progress toward the interpersonal comparison of utilities. This enables the use of many different types of welfare rules (egalitarianism, envy-freeness, etc.), which differ in the treatment accorded to fairness and efficiency. In view of this, we have also introduced a section entitled "Interpersonal Comparison of Utility" as another set of tools to aggregate individual preferences (Section 9.3.3).

The previously described aggregation mechanisms can sometimes be computed in linear (or quadratic) time from the number of participants. However, interpersonal comparisons of utility require even more complicated computational rules. Fortunately, the relatively new field of computational social choice allows the use of computationally hard aggregation rules (Chevaleyre et al., 2005). We have included a short description of these rules (Section 9.3.4), as aggregation procedures are often nondeterministic polynomial (NP) problems that require sophisticated algorithms for their solution.

9.1.3 Self-Organization

A complement to aggregation comes from the recognition of the complex and dynamic interaction that occurs in decision-making (Rodriguez and Steinbock, 2004a). This means that decision-making derives from the emergence of bottom-up processes, which include the collaborative interaction between decision-makers. The common

---

* Jeremy Bentham (1748–1832) was an English attorney. His criticism of the legal system led him to the formulation of the utilitarian doctrine. According to this doctrine, every human act must be judged by the pleasure or pain produced in people. In spite of the profound difference of his approach from the natural law defended by Rousseau, the goal of achieving "the greatest happiness for the greatest number of people" brought him into affinity with progressive and democratic political trends of the French Republic.

characteristic is that individual preferences can modify a shared aggregated decision, resulting in the construction of collective systems of preferences, which can be represented in the same way as the systems of preferences for individual decision-makers. Such self-organization is widespread and implies that the appearance of this structure is not imposed by any external agent (Heylighen, 2001).

Collective intelligence is considered to be the ability of a group to solve more problems than its individual members. It is based (Heylighen, 1999) on the idea that the obstacles created by individual cognitive limits and the difficulty of coordination can be overcome by using collective mental maps (CMMs).* In fact, bottom-up emergence systems can be found aggregating individual contributions into CMMs. As a result, the understanding of group behavior has become an issue of such importance that an increasing body of scientific literature is appearing on this subject (Klein et al., 2003; Braha and Bar-Yam, 2004, 2007; Kozlowski and Ilgen, 2006; Brodbeck et al., 2007).

In this chapter, we apply self-organization to the assessment of forest sustainability. However, before considering the construction of collective and web-based decision-making, let us remember some previous outcomes: in Chapter 7, we introduced a methodology to identify individual preferences and build individual assessments of sustainability, based on these preferences. We also defined a procedure to design the management plan that best fits the preferences of each evaluator (Chapter 8). Now, we seek to develop a methodology to describe how sharing opinions with other evaluators allows individual opinions—that is, personal preferences for sustainability assessment—to be modified. Each person can, thus incorporate the knowledge of other individuals into their own assessments. However, ensuring convergence toward acceptable solutions requires both a successful web-based application (to guarantee a suitable number of participants) and for this application to endure over time (to allow the adaptation of individual preferences). To speed up this process, we propose making a prior computer simulation of the interactions between sustainability assessments before aggregating the collective preferences. Every evaluator should have access to this information when making his or her assessment. This computer simulation is described in Sections 9.4.1 and 9.4.2. Section 9.4.1 describes the model adopted to simulate the interactions between evaluators, and Section 9.4.2 presents the application of this model to the collective assessment of forest sustainability.

To conclude this introduction section, we shall refer to the state of the art of the study of self-organizing processes in noncompetitive collective decisions. Developments in CMM arise when each individual in the group builds a model of the skills and/or preferences of the other members (Mohammed and Dumville, 2001). This is the idea behind the development of the "theory of mind" (ToM) (Watt, 1997; Ikegami and Morimoto, 2003; Pynadath and Marsella, 2005; Takano and

---

* A CMM is an external memory with shared read/write access, which represents problem states, actions, and preferences for actions. CMMs encourage understanding between members of the group with regard to reporting requirements and the need for communication and coordination (Marks et al., 2000; Mathieu et al., 2000; Baron-Cohen, 1995) and therefore play a key role in collective behavior in group decision-making.

Arita, 2006). ToM expresses the ability of one agent to perceive the model of cognition and behavior of another. In existing applications, it is often associated with evolutionary positions in competitive environments, leading to group behavior that depends on a recursive feedback between individual agents in a competitive environment. However, little work has been done on the computational modeling of CMM on collaborative group decision-making. Among the noncompetitive models, it is worth highlighting the model by Sayama et al. (2010), which we use as the basis for our developments. These authors propose a computational model to study the effects of mental model formation on the effectiveness of group discussion. In the model, each agent has his or her own unique utility function that differs from the true utility function (which is unknown to them). Each agent also has a certain amount of memory that stores the history of the group discussion and uses it dynamically to form a mental model of the others.

9.2 INVOLVING THE WHOLE POPULATION IN FOREST SUSTAINABILITY ASSESSMENT

A problem of overload occurs in CDM when a collective does not have the information-processing infrastructure to support the active participation of all its constituent members in all decision-making processes (Fischer 1999; Rodriguez 2004). To overcome this issue, societies have come to approximate full participation by using a set of decision-making representatives.

The algorithms used to obtain reliable representations are based on the dynamic delegation of proxy power across a social network. These increase the likelihood that decision outcomes will accurately reflect the opinions of the whole population.

In Section 9.2.1, we conducted a review of the state of the art of algorithms of decisions made using social networks. First, we describe a model that is currently applied to several types of networks (Grönlund et al., 2008). This model simulates CDM as a process that is both social and individual. Basically, it means that the process whereby an individual reaches a decision contains elements of both social influence and individually obtained information. We shall describe this model of decision-making without going into details of algorithms with common characteristics such as those used in social processes and algorithmic computer science, for inference problems such as belief propagation (Frey, 1998) or models of associative memory (Hertz et al., 1991). We then review models that incorporate conditional probability (Rodriguez et al., 2007) and fuzzy logic elements in graphs representing TBSNs for CDM (Rovarini et al., 2009).

Section 9.2.2 shows an application for propagating the assessment of sustainability in forest management from active decision-makers to the whole of the population. The simplest of the algorithms used to obtain this result is the dynamically distributed democracy (*DDD*) social representation algorithm (Rodriguez and Steinbock, 2004b). We have applied this in order to identify the most sustainable forest management (*SFM*) plan for a set of experts and forest users (local residents, leisure, or recreational users) in a TBSN.

9.2.1   DECISION-MAKING ALGORITHMS FOR SOCIAL NETWORKS

Social networks are used to show the relationships among agents in a population. Current social-networking software allows any user to identify those they trust most, to make that information useful online (anonymously of course), and to use it to make collective decisions. Here, we review some of the algorithms studied in recent years by different authors. In general, a social network is represented by a graph with N nodes (agents) connected by branches (the interagent relationships in a social space) that describe the underlying static social network.

9.2.1.1   Dynamic Model for Collective Decision Processes

This model is based on statistical mechanics analysis and was developed by Curty and Marsili (2006). The decision method is restricted to a binary process: the outcome of the decision is represented by +1 for the "correct" or "good" outcome and by −1 for the wrong outcome. It is assumed that agents are influenced by others in making their decision and that the agents can obtain information that may guide them toward making a correct decision. The key elements defining the model are

$S_t(i)$ is the accumulated information used by an agent i at time t of the decision-making process.

$S_0(i)$ is the initial value, picked randomly, representing direct information.

p is the probability of $S_0(i) = +1$.

$\text{sgn}(S_t(i))$ is the current choice of agent i, so that

$$\text{sgn}(x) = \begin{cases} -1 & \text{if} \quad x < 0 \\ 0 & \text{if} \quad x = 0 \\ 1 & \text{if} \quad x > 0 \end{cases}$$

where $\text{sgn}(S_t(i)) = 0$ indicates the rare situation that agent i is completely undecided at time t

At simulation, it is assumed that the social information-spreading process updates agents iteratively until all agents have reached fixed states.

Θ is a threshold, considered for simplicity to be the same for every agent, so that if the information in favor of a particular decision is strong enough, that is, $|S_t(i)| > \theta$, the agent will finalize his or her decision, and $S_t(i)$ remains fixed for the rest of the run. A higher θ implies that the system needs more time to converge to a decision.

The information exchange between two agents works as follows: an agent i (not finalized) is selected at random and activated. At the same time, also at random, a neighbor j of i is selected such that

$$S_{t+1}(i) = S_t(i) + \text{sgn}(S_t(j)) \tag{9.1}$$

The dynamic model proposed by Grönlund et al. (2008) is of the nonequilibrium type, converging to a final decision. These authors run it on random graphs and scale-free networks (Figure 9.1 shows the flow chart of the algorithm of this model).

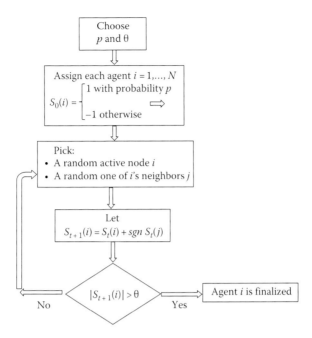

FIGURE 9.1 Flow chart of the dynamic model algorithm. (From Grönlund, A. et al., *Europhys. Lett.*, 81(2), 28003, 2008.)

Assuming the direct information is to the benefit of the agent, the reasonable range of p is [1/2, 1]. If $p = 1/2$, the initial knowledge does not guide agents toward a correct decision at all; if $p = 1$, the knowledge or preference strongly influences the final decision.

This model also considers the transfer of information, so nodes with many connections are more likely to influence others than vertices with few connections. This can represent situations where people with many social ties are more likely to function as opinion makers.

Runs of the dynamic model proposed show that systems of agents with less restricted communication patterns will improve more from sharing information. Different global network topologies ("random graphs" and "scale-free networks") do not quantitatively affect the decision ability very much. Nevertheless, local network structure makes a quantitative difference between the agents. An agent that is central in the information flow has a significantly higher chance of making the correct decision than more peripheral agents.

9.2.1.2 Trust-Based Social Networks Using Fuzzy Logic

The trust-based decision-making theory assumes the delegation of decision-making by some agents, based on trust, in others who are presumed to be more experienced or better informed to make the "right" (or "good") decision for the collective. In this case, the CDM social network must make explicit the notion of decision-making trust. For instance, if agent A connects B with a trust edge, then A is stating that he or she trusts B to make a good decision. Besides connection, the associated graph must include the weights of that trust relation ($r_{k,h}^{trust} \in \mathbb{R}$ represents the weight connecting the node a_k with the node a_h according to a predefined semantics δ, with $\delta \in \Sigma^*$, where Σ is an alphabet and Σ^* the *"Kleene star"** over Σ—the set of all strings of finite length consisting of symbols in Σ-).

Thus, in a *TBSN*, the graph representing the interagent relationships referring to a trust relation in decision-making is based on a semantic relation δ=trust. If a_k lacks expertise and believes that another agent a_h will make a better decision, then a branch is created from node a_k to node a_h with a label $r_{k,h}^{trust}$. Rodriguez et al. (2007) propose the use of conditional probabilities so as to find $r_{k,h}^{trust}$ quantitatively, considering only one domain (contextual independent):

$$r_{k,h}^{trust} = P\left[\frac{a_h \text{ is good}}{a_k \text{ trust } a_h}\right] \tag{9.2}$$

The preceding equation shows that a_k believes in the ability of a_h to make a good decision based on previous knowledge about a_h's behavior. If the information of the agents involved is multidimensional, with different beliefs and abilities over several domains, the domain in which a_k trusts a_h will be lost. It is, therefore, necessary to address the problem by assigning labels of trust according to the domains. Thus, if the domain considered is *SFM*, then the probability would be

$$r_{k,h}^{trust} = P\frac{a_h \text{ is good in } SFM}{a_k \text{ trust } a_h \text{ in } SFM} \tag{9.3}$$

Rovarini et al. (2009) posit the use of fuzzy logic as an aggregation mechanism to combine the input of individual characteristics as a way of reaching a collective decision. Their proposals take into account that the trust level of one agent in another is largely subjective, using previous knowledge extracted from a database (on historic behavior, colleague's references, management or government reports, etc.). They describe a methodology that allows crisp values to be obtained for the degrees of trust involved, from a restriction of the number of the characteristics that an agent uses when he or she wishes to make an evaluation of trust in another agent.

The degrees of trust are represented by a synthesis of the information in levels of fuzzy sets to describe linguistically particular situations (good, high, similar, very, enough, close, near, etc.).

There are several papers where these authors (Rovarini and Cerviño, 2011) show methodologies and algorithms to automate the processes of decision-making in TBSNs by using elements and concepts of fuzzy logic (fuzzy knowledge base (*FKB*), rule-based fuzzy systems (*RBFS*)).

---

* The name of this operator is in honor of the work of the mathematical logician Stephen Cole Kleene (1909–1994). It is used in mathematical logic and computer science (Kleene, 1956).

9.2.1.3   DDD Algorithm for Collective Solution Ranking

Rodriguez and Steinbock (2004b) consider the dynamic delegation of proxy power across a social network, which improves the likelihood that decision outcomes will accurately reflect the opinions of the whole population. These authors have developed the *DDD* algorithm to handle fluctuating levels of participation. *DDD* incorporates the central idea of representation as originally outlined by Paine (1776) and is an accurate way to model the collective's perspective as voter participation wanes.

The *DDD* algorithm requires a TBSN and an energy propagation algorithm. The "*social network*" component serves as the substrate for the distribution of vote power/energy during group decision-making processes. The branch A → B, between individuals A and B (nodes), is created if A believes B is "*good*" at making decisions. Social network uses are represented by link matrix $A \in [0,1]^{N \times N}$, whose elements show how good the connections are between nodes:

- For active individuals, the weight of any edge is computed as

$$A_{i,j} = 1 - |x_i - x_j| \qquad (9.4)$$

  where
  x_i is the opinion of an individual (i) on the issue to assess and is assumed to vary between 0 and 1: $x_i \in [0,1]$

- For inactive individuals who do not express their opinion, the weight of the edge is calculated from the average opinion in groups with similar external characteristics. Thus, let C_i and C_j be two groups of individuals, each of them with similar characteristics (e.g., in age, gender, jobs, place of residence, income). Then

$$A_{i,j} = 1 - |m_{Ci} - m_{Cj}| \qquad (9.5)$$

  where m_{Ck} is the average opinion of the active individuals belonging to group C_k.

To describe the "*energy propagation*" component, every individual in the group is initially supplied with an equal amount of vote power ($1/N$). This vote power is then used to vote on a particular option/solution for a particular issue/problem (e.g., one man/one vote). However, if an individual does not participate, then his/her voting power is distributed to the nearest neighbors on the social network. If any neighbor is not an active agent, then the voting power continues propagating until it finds a sink node (or active participant). Individuals with more voting power will have one major influence on the specific activity of *CDM*. The algorithm (see Rodriguez and Steinbock, 2004b; Rodriguez and Watkins, 2007, 2009; Rodriguez et al., 2009) works to compute $\pi \in \mathbf{R}_+^N$, the final vector representing the *vote power* of any individual in the group (0 for nonactive agents and a positive value for active ones, such that the total amount of vote power in the population is 1).

This representation is dynamic, distributed, and democratic, because if an individual does not participate in the decision-making process, the underlying social network will ensure that his or her vote power will be distributed to his or her neighbor or neighbor's neighbor.

9.2.2 Forest Sustainability Assessment for a Trust-Based Social Network of Multiple Users

The procedures described in the preceding section increase the likelihood that group decisions will reflect the real opinions of the whole population. In the case that a population is well defined, the problem of representation is solved: it is sufficient to build the link matrix (A), which models the TBSN for both active and nonactive participants.

In a participative process, active individuals can express their preferences on forest sustainability and obtain a value for sustainability according to these preferences. In the first step, personal preferences are transformed into assessments of sustainability in a set of representative territorial points in the forest (see Section 7.6). This makes it possible to build a vector $\mathbf{x}_i = (x_{i-1}, x_{i-2}, \ldots, x_{i-n})$ whose components show the assessment of sustainability at n representative territorial points of the forest for each active participant (i). Thus, the weight of the link connecting two active individuals—see Expression (9.4)—can be computed from their respective assessments of sustainability, and a statistical distance ($D_{ij\text{-active}}$) can be used to determine the weight of the link ($A_{ij} = 1 - D_{ij\text{-active}}$). We propose using the Mahalanobis distance (Mahalanobis, 1936):

$$D_{ij-\text{active}} = \frac{\sqrt{(\mathbf{x}_i - \mathbf{x}_j)\Sigma^{-1}(\mathbf{x}_i - \mathbf{x}_j)^T}}{D_{\max}} \qquad (9.6)$$

where
  Σ is the covariance of the assessments of sustainability at the n most representative territorial points for the data of all the active participants
  $D_{\max}$ is the maximum value of the numerator in Expression (9.6) for all pairs of active participants
  $\mathbf{x}_i$ is the vector representing the assessment of sustainability for the i participant at the n most representative territorial points

In contrast, nonactive decision-makers do not express their preferences on forest sustainability and it is, therefore, not possible to know their personal assessments of sustainability ($\mathbf{x}_i$). However, in the case of a well-defined population, we may know some personal characteristics of these individuals, and these can then be used as a basis to build a distance between each pair of individuals in the population. The procedure is clear: personal characteristics can always be reduced to dichotomy variables (presence or absence of a specific gender, age, academic level, occupation, place of residence, type of agent, or any other characteristic). It is then possible to apply multivariate techniques (such as reciprocal averaging—Hill, 1973) to these dichotomy variables in order to project each individual in the population on the gradient axis, absorbing the maximum variability of the data. The difference in the values over the gradient of two individuals ($D_{ij\text{-gradient}}$) is a measure of their respective separation. We can then compute the weight of the link between two nonactive participants or between an active and a nonactive participant as $A_{ij} = 1 - D_{ij\text{-gradient}}$.

Generally speaking, it seems reasonable to entrust democratic representation to people with analogous personal characteristics to one's own, although the identification of the similarity of a set of reported personal characteristics with equivalent assessments of sustainability must be tested. However, this is not the most critical point.

The main problem for outreach is that many participatory processes cannot be relegated to a partial and well-defined population. Environmental problems cannot be reduced either to the individuals in a particular organic community or to those paying taxes to a specific governmental body, which can provide the personal characteristics of those affected by the problem. The nature of environmental problems affects the whole of humankind, and everyone should have the right to participate in the assessments of sustainability. Unfortunately, the procedures described in the previous section are not applicable when the nonactive participants so far exceed the potentially active ones, as occurs with the whole of humankind compared with the individuals that may take part in a participatory process requiring some type of environmental and computational knowledge. Furthermore, no one has the right to force anyone to participate in a process of forest sustainability assessment: it has to be a voluntary process. These considerations point to the implementation of a free and easily accessible means of participation rather than to the development of a sophisticated representation procedure. Hence, except in the case a governmental body is legally required to enable its citizens to take part in public participation processes (as is the case in many current environmental impact assessment regulations), we have opted to develop a free participatory procedure via the Internet—with as little requirement for intellectual knowledge as possible—to facilitate participation and to consider only active participants in the final decision.

9.3 AGGREGATING ASSESSMENT FROM MULTIPLE DECISION-MAKERS

Aggregation processes are necessary in the presence of multiple decision-makers. A major feature of aggregation is the reduction of variability in the results,* which makes it easier to recognize the most representative trends in a participatory assessment and to make decisions according to these trends.

This section describes three mechanisms of aggregation: voting, prediction markets, and interpersonal comparison of utility. All of them include examples of application or a case study for the assessment of forest sustainability. And they all

---

* The best-known example of reduction of variability in an aggregation process is provided by the central limit theorem (in statistics). It ensures that the decrease in the variation of the mean for a given a set of variables—when comparing the variability of the mean with the variability of each one of the variables—is proportional to the number of variables considered. So although the variability of the potential earnings of a person playing in a casino could be large, the variability of casino losses (mean of the earnings for all the players) is small.

conclude with an analysis of the applicability of the aggregation mechanism for the assessment of sustainability.

As many of the aggregation mechanisms require hard computational rules, we have also included a description of the techniques most commonly applied to reduce the computation time.

9.3.1 Voting Systems

Voting is the natural way of expressing multiple preferences regarding decision-making in democratic systems of government. Voting procedures are characterized by how they assess votes expressing preference, by how they allocate the votes discarded in the first round, and finally, by the way they add together the voters' preferences.

Although voting systems allow large groups to be involved in decision-making, the drawback is that merely having a large group does not guarantee either the quality of the decision-making or a satisfactory result for the majority of the individuals in the group.

The following is a description of the most commonly applied voting procedures.

9.3.1.1 Plural Voting

Each decision-maker has a vote, the aggregation of preferences is done additively, and there is no reallocation of votes; thus, the most voted alternative is the chosen one.

This system is simple to conduct and easy to aggregate but has the disadvantage that, if there are more than two alternatives, the chosen alternative may not satisfy the majority of the decision-makers, as most of the individuals could have opted for the other alternatives.

Example No. 1 One hundred decision-makers have voted to select the most sustainable location within six possible locations presented to them (A, B, C, D, E, and F). The result was 27 votes for location A, 17 for B, 14 for C, 20 for D, 8 for E, and 14 for F. Thus, alternative A will be chosen (27% of the votes). However, this option has not been chosen by 73% of voters, so many of the individuals in the group will not be satisfied with the decision.

Let us now assume a simpler case:

Example No. 2 There are now just nine decision-makers (DM1, DM2,..., DM9) and three locations (A, B, and C). Each decision-maker has arranged his or her preferences as shown in Table 9.1. It can be seen (row 1) that A has four votes, B has three, and C has two. Thus, A is the most voted location, but it has less than 50% of the votes.

One way to solve this problem is to set a minimum threshold of votes for an alternative to be chosen. Suppose that this boundary is 50% of the votes, meaning that five votes are required to choose an alternative. None of the alternatives is chosen as none of them has exceeded that threshold.

Another solution is to implement a strategy to reallocate votes, consisting—for example—of a second round between the two most voted alternatives. Assuming

TABLE 9.1

Arranging of Preferences for Example No. 2

Position	DM1	DM2	DM3	DM4	DM5	DM6	DM7	DM8	DM9
1st	A	B	C	B	C	A	A	A	B
2nd	B	C	B	C	A	C	B	C	C
3rd	C	A	A	A	B	B	C	B	A

that the decision-makers who chose C first (DM3 and DM5) simply transfer their secondary preferences to the first one—that is, to B and to A, respectively—then A would obtain five votes and B four, so A is chosen.

However, it is easy to see that it would be sufficient for DM5 to have chosen B as a second option for this alternative to be chosen, as it was initially less voted than A.

9.3.1.2 Majoritarian Systems

9.3.1.2.1 Borda Count

There is evidence of the application of a variant of this voting method by the Roman Senate in about AD 105. However, its modern mathematical description has been proposed independently by at least three authors:

- Ramon Llull (1232–1315) who, after the discovery of his manuscripts "Ars notandi," "Ars Eleccionis," and "Alia ars eleccionis" (all of them in 2001), has been recognized as being the first author of the Borda count and the Condorcet criterion.
- Nicholas de Cusa (1401–1464), who in 1433 unsuccessfully advised this method for the election of the German Holy Roman Emperor.
- Jean-Charles de Borda (1733–1799), who devised this system as a fair way of selecting the members of the French Academy of Sciences in 1770. The method was first published in 1781 in the "Mémoire sur les élections au scrutin" in the Histoire de l'Académie Royale des Sciences in Paris and was used by the academy from 1784 until its abolition by Napoleon in 1800.

In the Borda method, for m alternatives, each decision-maker assigns $m - 1$ points to his or her most preferred alternative, $m - 2$ points to the second most preferred one, and—declining—0 points to the least preferred alternative. The alternative with the highest total score wins. There is no reallocation of points.

Example No. 3 A total of five decision-makers express their preferences regarding the most sustainable location by applying the Borda system to the six alternatives in Example No. 1. The results are shown in Table 9.2. It can be deduced that A is the best alternative (21 points), followed by D (18 points).

This method of voting can lead to the impossibility of choosing between alternatives, as adding points can yield equal results.

Example No. 4 With the same conditions and preferences as Example No. 2, decision-makers express their preferences on the most sustainable location from

TABLE 9.2
Scoring for Example No. 3

	Decision-Maker					Total
Alternatives	DM1	DM2	DM3	DM4	DM5	Score
A	5	5	4	4	3	21
B	2	3	3	5	2	15
C	3	2	5	2	4	16
D	4	4	2	3	5	18
E	1	0	0	1	0	2
F	0	1	1	0	1	3

TABLE 9.3
Scoring for Example No. 4

	Decision-Maker									Total
Alternatives	DM1	DM2	DM3	DM4	DM5	DM6	DM7	DM8	DM9	Score
A	2	0	0	0	1	2	2	2	0	9
B	1	2	1	2	0	0	1	0	2	9
C	0	1	2	1	2	1	0	1	1	9

three alternatives by applying the Borda system. The results are shown in Table 9.3. What is the chosen alternative in this case?

Despite its limitations, the Borda method avoids some of the failings observed in the plural voting method, which can only choose one alternative or vote, and rules out other options.

9.3.1.2.2 Condorcet's Method

Although first proposed by Ramon Llull (1299), it is named after the Marquis de Condorcet (see footnote * on page 501) who developed this method for the election of the alternative that would win by majority rule in all pairings against the other alternatives.

The alternative preferred in all pair-wise comparisons is called the *Condorcet candidate*. If the Condorcet candidate exists, then it will also be the solution of the Borda count.

The chosen alternative is the most preferred by the decision-makers in the comparisons between all the possible pairs of alternatives. Voting is done by expressing the preferences of the evaluators when faced with groups of two alternatives and additionally considers that decision-makers can also express indifference between the two alternatives compared. These processes are usually represented by voting matrices.

Example No. 5 Five decision-makers (DM1, DM2,..., DM5) choose the most sustainable location from six possibilities (A, B,..., F) by filling in a numeric range of comparison between the potential sustainability of all the pairs of alternatives, using the following code: 1 shows that the row alternative is preferred to (is more sustainable than) the column alternative, 2 shows that the column is preferred to the row, and 3 shows indifference between row and column. Outcomes are shown in Table 9.4. For example, it shows that DM1 prefers location A to the rest; D to C, B, E, and F; C to B, E, and F; B to E and F; and E to F. Similar characteristics can be deduced from the preferences of other decision-makers.

The matrices in Table 9.4 are very useful for the overall outcome of the vote; all we need is to add together the matrices of each voter to obtain the set of aggregated preferences. In the present example, the sum of matrices is shown in Table 9.5, where the cases in which the row alternative is preferred to the column alternative are marked in gray. It can be deduced from the table that alternative A is the best in the pair-wise comparison. In this case, the choice also coincides with Borda's result.

It may happen that the number of preferences over different alternatives is equal and there will therefore be no solution. In these cases, regardless of the chosen alternative, it is clear that most decision-makers prefer another option to the one finally chosen. This situation is known as the Condorcet paradox and shows that the transitivity of individual preferences does not need to lead to transitivity in collective preferences.

Although in many cases the Borda and Condorcet procedures lead to the same winner, both proposals diverge because the Borda approach is positional, while Condorcet is not. For a comparative discussion between the two methods and other alternative methods seeking convergence between these two voting systems, see the work of Martinez-Panero (2006).

9.3.1.2.3 *Proportional Representation (Single Transferable Vote)*

The single transferable vote (STV) is a voting system based on proportional representation and on preferential voting.

Although the concept of transferable voting was first proposed by Thomas Wright Hill in 1819, the British lawyer Thomas Hare is internationally recognized as the author of this voting method. The system remained untested in real elections until 1855, when a transferable vote system was applied for elections in Denmark (Tideman, 1995). Today it is used to choose public servants all over the world.

The system may require second or subsequent rounds. These are conducted using the same system as the first round, and a quota of votes is set, which must be achieved by the choices made in any round. A standard procedure to calculate the quota of votes is the one given by the following expression (Brams and Fishburn, 1991):

$$q = \left[\text{int}\left(\frac{n}{(m+1)} \right) + 1 \right] \tag{9.7}$$

where
  n is the number of voters
  m is the number of alternatives to choose
  int(x) is the integer part of x

TABLE 9.4

Pair-Wise Comparisons for Example No. 5

Decision-Maker		Voting Matrices					
DM1	Location	A	B	C	D	E	F
	A	3	1	1	1	1	1
	B	2	3	2	2	1	1
	C	2	1	3	2	1	1
	D	2	1	1	3	1	1
	E	2	2	2	2	3	1
	F	2	2	2	2	2	3
DM2	Location	A	B	C	D	E	F
	A	3	1	1	1	1	1
	B	2	3	1	2	1	1
	C	2	2	3	2	1	1
	D	2	1	1	3	1	1
	E	2	2	2	2	3	2
	F	2	2	2	2	1	3
DM3	Location	A	B	C	D	E	F
	A	3	1	2	1	1	1
	B	2	3	2	1	1	1
	C	1	1	3	1	1	1
	D	2	2	2	3	1	1
	E	2	2	2	2	3	2
	F	2	2	2	2	1	3
DM4	Location	A	B	C	D	E	F
	A	3	2	1	1	1	1
	B	1	3	1	1	1	1
	C	2	2	3	2	1	1
	D	2	2	1	3	1	1
	E	2	2	2	2	3	1
	F	2	2	2	2	2	3
DM5	Location	A	B	C	D	E	F
	A	3	1	2	2	1	1
	B	2	3	2	2	1	1
	C	1	1	3	2	1	1
	D	1	1	1	3	1	1
	E	2	2	2	2	3	2
	F	2	2	2	2	1	3

The vote of an evaluator is initially allocated to his or her favorite alternative. If it has already been chosen, then all surplus votes are transferred according to the voter's preferences to other alternatives. The system minimizes blank votes. It also provides a representation similar to the proportional system and allows individual votes to be applied to explicit alternatives instead of to a closed list. To achieve this, it is necessary to define subsets of evaluators with similar preferences and also to

TABLE 9.5

Results from the Aggregation of Preferences in Example No. 5

For All Decision-Makers	A	B	C	D	E	F
A	15	6	7	6	5	5
B	9	15	8	8	5	5
C	8	7	15	9	5	5
D	9	7	6	15	5	5
E	10	10	10	10	15	8
F	10	10	10	10	7	15

TABLE 9.6

First Round Outputs of the *STV* Voting for Example No. 6

Aggregation of Decision-Makers	Total Score	Preferences (High to Low)					
DMG1	9	A	D	B	C	E	F
DMG2	6	B	C	D	A	F	E
DMG3	2	C	B	D	A	F	E
DMG4	4	C	D	B	A	E	F
DMG5	5	D	B	C	A	F	E

allow the transfer of votes for alternatives that would otherwise be wasted in losing or winning scenarios.

Example No. 6 We are seeking the two most sustainable alternatives (territorial points) within those allocated in the points A, B, C, D, E, and F. For this, five groups of voters have been defined from the 26 voters performing *STV* polling as shown in Table 9.6. From these data, the quota will be $q = \text{int}(26/3) + 1 = 9$ votes. Accordingly, in the first round, A is the only alternative that reaches the quota and is, therefore, the winner in this case.

As the nine votes of alternative A were used, surplus votes are not transferred, and the situation now is as shown in Table 9.7. With the data in this table, no alternative obtains the necessary quota of votes, so there is no winner. In this situation, E, the least voted option, is eliminated. As this alternative has no vote winner, then neither are any votes transferred.

In the next round, we will have the same situation for the winners, and alternative F is eliminated as it is the last in the scale of preference.

In the last round, once again, no alternative exceeds the required quota. However, when removing the least preferred of the remaining alternatives (D), there are five

TABLE 9.7

Second Round Outputs of the *STV* Voting for Example No. 6

Aggregation of Decision-Makers	Total Score	Preferences (High to Low)				
DMG2	6	B	C	D	F	E
DMG3	2	C	B	D	F	E
DMG4	4	C	D	B	E	F
DMG5	5	D	B	C	F	E

TABLE 9.8

***STV* Matrix after Several Voting Rounds from Example No. 6**

Aggregation of Decision-Makers	Total Score	Preferences		
DMG2 + DMG5	6 + 5	B	C	
DMG3	2	C	B	
DMG4	4	C		B

votes transferable to the winning alternative. By transferring these votes, the feedback matrix is as shown in Table 9.8:

As can be seen in the table, the votes of group 5 (whose components prefer B to C) have been transferred to group 2 of evaluators, as they also preferred B to C. Hence, alternative B has obtained 11 votes, surpassing the quota of votes for election. Thus, A and B are the selected alternatives.

As it is a preferential method, it is easy to transform the preferential options expressed by the evaluators into a Borda count poll. In this case, alternative B is the first selected in scores ranging from highest to lowest, with 97 points, and then comes C with 87 points. The next option is D with 82 points, followed by A with 79, and finally, alternatives E and F with 13 points each. As the reader can see, *STV* winners are not Borda winners.

Although *STV* violates some of the properties of voting systems (Kelly, 1987) that are considered to be desirable in social elections (see footnote * on page 502), it has important advantages as a system of proportional representation. In particular, minorities can obtain a number of candidates that are more or less in proportion to their numbers in the electorate. Furthermore, if the voting does not ensure that a person chooses his or her first choice, he or she can still have a chance with his or her lower choices.

A comprehensive discussion on the voting system and its possible shortcomings appears in Brams and Fishburn (1991). Examples of vote transfers can be found in Tideman and Richardson (2000).

9.3.1.3   Special Systems

9.3.1.3.1   Cumulative Voting

Each voter has a fixed number of votes that can be shared among alternatives. The alternative with the most votes is chosen. Here, evaluators express the intensity of their preferences by the number of votes, instead of by ordering alternatives. Voters with minority views can be sure their preferred alternatives are included in decision-making, at least in a proportion to their size in the group, provided that they concentrate their votes in the subset of their preferred alternatives.

This system requires knowing the choices of the homogeneous group of reviewers with whom we share similar preferences.

Example No. 7 As in previous examples, we have six management alternatives and we want to choose the three most sustainable based on the vote of 300 evaluators. Each of them has three votes to distribute among their preferred alternatives.

Let us suppose there are two homogeneous groups of evaluators and one of them is twice as large (in number of evaluators) as the minority group. The group of 200 evaluators decides to allocate its three votes equally among its three most preferred alternatives: A, C, and D. Thus, each of these three alternatives obtains 200 votes each. In contrast, the minority group chooses to allocate its three votes to a single alternative (F). This alternative will receive 300 votes, so despite being a minority option, it is one of three management alternatives considered.

Although a second round could be incorporated into this method, the method of achieving this is not specified in the scientific literature. A systematic analysis of the optimal strategies in cumulative voting is described in Brams (1975), while a study on its potential to represent minority options is shown in Cooper (2007). An analysis of the potential role of this voting system can be seen in recent works such as Zhao and Brehm (2011).

9.3.1.3.2   Approval Voting

This system was introduced in 1977 by Ottewell (1977) and also by Kellett and Mott (1977), Weber (1977), and Brams and Fishburn (1978).

This is a procedure designed to prevent the election of minority preference alternatives when there are more than three alternatives. As we have seen, in plural voting systems, a clearly minority alternative can be chosen or at least have an option for a second round. The greatest drawback in systems with a second round is that even the alternative with the highest Condorcet score can be ruled out in the first round and not go through to the second round.

In this voting system, decision-makers can take on as many alternatives as they wish and each approved alternative receives one vote, with the winner being the one with the most votes. The method of conducting the second round is not specified.

Approval voting allows the decision-makers more flexibility in selecting sustainable alternatives when compared with voting systems allowing them to choose only one. It helps to select the most preferred alternative. It also makes it possible to reflect preferences for the alternatives, which are less attractive to most, provided these are important to a particular evaluator. Evaluators with minority preferences can approve majority alternatives without feeling ineffective, as they are able to vote

TABLE 9.9

Outcomes of the Approval Voting of Example No. 8

Decision-Maker	DM1	DM2	DM3	DM4	DM5	Total ADD
A		x	x	x	x	x	5
B				x	x		2
C	x			x	x		3
D	x	x		x	x		4
E				x			1
F		x		x	x		3

for minority options. The alternatives preferred by a small number of evaluators are valued by their real preference, even though they are not chosen.

Example No. 8 A total of five evaluators express their preferences on the territory they consider to be most sustainable by approval voting. The results are shown in Table 9.9: the chosen alternative is A, followed by D.

9.3.1.3.3   Mixed Systems (Additional Member Systems)

Electors vote for a group of alternatives (similar to a party in political elections) under proportional voting, and alternatives are selected provided they have a minimum threshold of votes. Furthermore, alternatives are added to those initially chosen, according to the number of votes obtained by the minority group of alternatives, so that the aggregated number of alternatives is proportional to the number of minority votes.

Voters usually have two votes, one for the group of alternatives (the party) and the second for minority alternatives (the candidate in a constituency), even if these votes are sometimes combined. It is characterized by dividing the population into subsets of equal size, and each subset may choose one or several alternatives for its preference.

This system allows voters to express their complex preferences whenever they are reasonable and proportional. Compared to other systems, however, it is complicated for both evaluators (electors) and final decision-makers, who must count the ballots for the different alternatives.

Example No. 9 There are two groups of evaluators with similar preferences. One group is significantly larger than the other, and their sizes can be quantified.

Evaluators are clustered into eight subgroups. Each group of voters must choose a single alternative. Each group contains 80% who support the majority alternatives and 20% in favor of the minority. In this case, the majority option will win in the eight subgroups. If the minority preferentially choose two different alternatives from that of the majority group, it may be of interest to increase the number of selected alternatives to 10, since the proportion of the minority opinion is 20% (and 20% of 8 alternatives is approximately 2), and these 2 alternatives are the most voted by the minority group.

9.3.1.4   Main Conclusions

It is natural to use voting in decision-making owing to the numerous features both theories have in common. However, since most voting systems address a single criterion or aggregated information summarized in a single value, if voting is to be applied to decision-making, this condition must be met. Consequently, it is more natural to apply voting systems to the integrated value of sustainability, rather than to apply them to the relative importance of several indicators of sustainability.

In the management of natural resources and in particular in forest management, voting systems have been used to select management alternatives (Laukkanen et al., 2002). However, it is very important to select the most appropriate voting system as each system is designed to solve different problems. In general,

a. Methods expressing preferences (such as Borda, Condorcet, and mixed) can be manipulated, by applying appropriate strategies, to guarantee an a priori choice of certain alternatives (see, e.g., Stensholt, 2011, for a comparison of the electoral strategies available for some voting methods).

b. On the other hand, majority voting systems may not allow selection of alternatives representing minorities. However, in the case of (minority) groups with extensive technical knowledge and who can express opinions on scientific grounds (although their preferences may not necessarily outweigh the preferences of other evaluators), it would be desirable for their preferences to be known by the other participants in the voting process.

9.3.2   Prediction Markets

Prediction markets are speculative markets, where participants trade contracts whose payoffs are tied to a future event (Wolfers and Zitzewitz, 2006). The contracts in these markets take the form: "pay \$1 if *i* contract happens"; therefore, the price— in dollars—that anyone would be willing to pay for a contract will range between 0 and 1, and the closing market price for any contract is used as an aggregate assessment of its value.

The efficiency of prediction markets for aggregating information depends on the quality of the assumed market hypothesis. Under certain conditions (such as those formulated by Grossman, 1976, or by Wolfers and Zitzewitz, 2008), prediction-market prices coincide with average beliefs among traders. Unfortunately, there is still a lack of suitable theoretical results on these topics, and much of the existing analysis simply assumes that a revealed market-based price is simply a nonrobust estimate of the real value (Manski, 2004). As occurs with voting systems, even the latest developments in prediction markets* are not able to satisfy a set of rational

---

* Hanson's logarithmic market-scoring rule (LMSR) is an automated market maker with particularly nice properties and behavior (Hanson, 2003, 2007). LMSR is used by a number of companies including Inkling Markets, Consensus Point, Yahoo, Microsoft, and the large-scale noncommercial Gates Hillman Prediction Market at Carnegie Mellon (Othman and Sandholm, 2010).

conditions simultaneously*, making it necessary to prioritize which conditions to lose (Othman et al., 2010). Nevertheless, speculative markets do a remarkable job of aggregating available relevant information into market prices (Lo, 1997). In particular, betting markets drive to acceptable price estimates (Hausch et al., 1994).

A centerpiece in any market is the market mechanism to facilitate trading (a way for traders to vie for contracts and a place to have public offers to buy or sell and the procedure to update the prices in response to trades). The simplest market mechanism can be formulated as a betting game (see, e.g., the very intuitive presentation from Rodriguez and Watkins, 2007), but prediction-market contracts have been traded in a variety of market mechanisms, including continuous double auctions, pari-mutuel pools, bookmaker-mediated betting markets, or implemented as market-scoring rules. In general, a market mechanism consists of one or a few central actors called market makers, whose work is to collect offers and buy orders and update prices. Human market makers have been found to do as well as the standard double-auction market at aggregating information (Krahnen and Weber, 1999), but automated market makers can also play this role and even improve on it. This is the case when a quick update of markets is required (mainly in markets with continuous offer and demand schedules).

Next, we apply a modification of market-scoring rules to aggregate the preference of multiple decision-makers to the assessment of forest sustainability. But prior to describing this procedure, we shall introduce the concept of scoring rules.

9.3.2.1   Scoring Rules

Let us consider an evaluator (A), whose subjective probabilities on n mutually exclusive issues ($n > 1$) are represented by the vector $\mathbf{p} = (p_1, \ldots, p_n)$. Here p_i is the private belief of A referred to the future occurrence of the question i and $\sum p_i = 1$ (for $i = 1, \ldots, n$).[†] In order to estimate $\mathbf{p}$, A is asked to declare publicly the likelihood he or she attributes to the occurrence of these same issues. The report of A is denoted by $\mathbf{r} = (r_1, \ldots, r_n)$.[‡]

The knowledge A has about the occurrence of an event is rewarded by a scoring function $R_i(\mathbf{r})$, which determines the payment made to A in the case that i issue occurs in future. Under these conditions, the average expected reward of A is given by the following expression:

$$\bar{R}\left(\frac{\mathbf{r}}{\mathbf{p}}\right) = E_\mathbf{p}\left[R_i(\mathbf{r})\right] = \sum_{i=1}^{n} p_i R_i(\mathbf{r}) \qquad (9.8)$$

---

[*] Conditions such as *path independence* (any way the market moves from one state to another state yields the same payment or cost to the traders in aggregate), *no-arbitrage* (the cost of buying a guaranteed payout of *x* always costs *x*), and *liquidity sensitive* (market makers adjust the elasticity of their pricing response based on the volume of activity in the market).

[†] The probability (**p**) that an evaluator attributes to the occurrence of different issues can be matched to the subjective value that the evaluator allocates to these issues, provided the values are nonnegative and their sum is equal to 1.

[‡] The evaluator expects to be rewarded for his or her knowledge about the issues that are going to occur in the future. Therefore, he or she can develop different strategies to maximize his or her reward, which means that **p** and **r** do not necessarily have to coincide. In addition, the evaluator may not have thought enough about his or her report and the judgments he or she has made or be unable to reproduce his or her personal knowledge regarding the issues discussed.

where

E is the mathematical expectation operator

R_i is the reward (or score) the evaluator will receive when the evaluator has reported that the likelihood for potential occurrences of issues is $\mathbf{r}$ and the question i happens

There are many scoring rules, but the most popular are the quadratic ($Q_i(\mathbf{r}) = 2r_i - \mathbf{r} \cdot \mathbf{r}$) and the logarithmic ($L_i(\mathbf{r}) = a_i + b \ln(r_i)$). In any case, since the possible number of scoring rules is unlimited, how can one know which is the best? This question is answered by determining the conditions R should have in order for the evaluator to be motivated to tell the truth. Several authors (Toda, 1963; Roby, 1965; Shuford et al., 1966; Winkler and Murphy, 1968) agree that the coincidence of $\mathbf{p}$ and $\mathbf{r}$ values identifies the maximum compensation to be perceived through R. Scoring rules satisfying this condition are called "proper scoring rules":

Definition: T is a strictly proper scoring rule $\equiv \begin{cases} \overline{T}(\mathbf{p/p}) > \overline{T}(\mathbf{r/p}) & \text{if}\dagger \quad \mathbf{r} \neq \mathbf{p} \\ \overline{T}(\mathbf{p/p}) = \overline{T}(\mathbf{r/p}) & \text{if}\dagger \quad \mathbf{r} = \mathbf{p} \end{cases}$

Proper scoring rules are used to maximize the reward in the form of average expected return. However, maximum reward only coincides with the maximum utility for the evaluator when his or her utility is a linear function of the scoring rule.* If this is not the case, then we cannot accept that $\mathbf{r} = \mathbf{p}$ implies obtaining the maximum reward. The utility function is not usually known; it is thus necessary to study the behavior of utilities in the maximum nonlinear score functions. From these studies, Bickel (2007) has found that the function, which is least affected by nonlinearity conditions, is the logarithmic score function.

9.3.2.2 Continuous Prediction Market

Since the early 1970s, it has been known that iterating with a proper scoring rule is equivalent to participating in a continuous supply and demand market (Savage, 1971). Hanson (2002, 2003) subsequently extended this result to higher dimensions by introducing the concept of market-scoring rules. Market-scoring rules are scoring rules where anyone, at any time, can change his or her report (assignment of probabilities) and be paid according to their new report, as long as everyone agrees to pay the last person reporting according to that individual's report.

So, for any new report, an evaluator should voluntarily agree to accept payment in the form

$$c_i = \Delta R_i(\mathbf{r}, \rho) = R_i(\mathbf{r}) - R_i(\rho) \tag{9.9}$$

where, for a same evaluator, $\mathbf{r}$ is the current report, ρ is the previous one, and $R_i(\cdot)$ is the scoring rule. Any evaluator can ensure his or her benefits unchanged ($\mathbf{r} = \rho \Rightarrow c_i = 0$) or maximize them (in this case, as ρ cannot be modified, then maximizing c_i means maximizing the scoring rule; therefore, $\mathbf{r} = \mathbf{p}$).

---

* If the mathematical expression of the utility function is known and admits inversion, then the scoring rule that maximizes the utility of the evaluator would be the composition of the inverse of the utility with T (Winkler, 1996).

As mentioned earlier, prediction markets are financial security markets ("pay $1 if i contract happens"). This means that traders just want to buy (sell) securities because they believe they can resell (rebuy) them later, at a better price. The closing price of the market gives information about the knowledge and beliefs of other operators, which induces the evaluators to change their beliefs about the future price of the security ($\mathbf{p}$). Therefore, the total amount of each one of the securities ($i = 1, \ldots, n$) held by all the traders in a market can be used to determine the prices. To calculate the current unit price ($\mathbf{p}$), let us consider a cost function $C(\mathbf{q})$ that records the total amount of money that traders have played in the market. This function is a mapping from $\mathbf{q} = (q_1, \ldots, q_n)$ into $\mathbb{R}$ (monetary units), where q_i is the total quantity of the i security ($i = 1, \ldots, n$) held by all the traders in the market. A trader who wants to purchase δ units of i security must pay, per unit of security, the following money:

$$\frac{C(q_1, \ldots, q_i + \delta, \ldots, q_n) - C(\mathbf{q})}{\delta} \tag{9.10}$$

Thus, the current unit price, for a tiny amount of the i security, will be $p_i = \partial C / \partial q_i$. So, in case of a logarithmic scoring rule, the corresponding cost and price functions are

$$C(\mathbf{q}) = b \ln \left(\sum_{j=1}^{n} e^{q_j / b} \right) \tag{9.11}$$

$$p_i = \frac{e^{q_i / b}}{\sum_{k=1}^{n} e^{q_k / b}} \tag{9.12}$$

where b is the parameter characterizing the logarithmic scoring rule.

The main questions a trader asks the market maker on arriving in a market can now be answered. These questions are

- I want to buy B_i units of security i and to sell S_j of security j. How much will that cost?
- What will be the unit prices of the market at the end of the earlier trades?

To answer the preceding questions, the market maker extracts the information implicit in previous trades in order to infer new rational prices. Then, under the preceding conditions—and when the amount of securities already purchased by all the traders is $\mathbf{q} = (q_1, \ldots, q_n)$—the cost of any operation can be computed as

$$C\left(q_1, \ldots, q_{i-1}, q_i + B_i, q_{i+1}, \ldots, q_{j-1}, q_j - S_j, q_{j+1}, \ldots, q_n\right) - C\left(q_1, \ldots, q_n\right) \tag{9.13}$$

The market maker will pay the preceding amount to the traders in the case that the preceding amount is positive. When it is negative, the market maker will receive money from the traders. On the other hand, if a trade changes the global amount of the securities from $\mathbf{q} = (q_1, \ldots, q_n)$ to $\mathbf{q}^* = (q_1^*, \ldots, q_n^*)$, then the final unitary prices will be

$$p_i = \frac{\exp\left\{q_i^*/b\right\}}{\sum\limits_{k=1}^{n} \exp\left\{q_k^*/b\right\}}, \quad \text{for} \quad i = 1,\dots n \tag{9.14}$$

It is very important to consider that current market prices ($\mathbf{p} = (p_1, \dots, p_n)$) only apply for trading with infinitesimal amounts of securities. In order to calculate the total cost of a finite trade, we must use the cost function $C(\mathbf{q}^*) - C(\mathbf{q})$ or calculate the integral over time (starting at t_s and ending at t_e) of the function of prices as

$$\int\limits_{t_s}^{t_e} \sum\limits_{i=1}^{n} p_i\left(\mathbf{q}(t)\right) q_i' dt \tag{9.15}$$

Market-scoring rules can be viewed as the sequential shared version of scoring rules (Pennock and Sami, 2007). Thus, the market maker begins by setting prices equal to an initial probability estimate. The first agent to arrive agrees to buy according to the scoring rule payment associated with the market maker's estimated probability (prices) and to sell securities according to the scoring rule payment associated with his or her own estimated probability (prices). Any trade modifies the market-maker prices as described through Expressions (9.13) to (9.14) or Expression (9.15). Trades continue till prices incentivize operators to reveal their true probability estimate. The final trader pays the scoring rule payment owed to the second-to-last trader and receives a scoring rule payment from the market maker.

Market-scoring rules act like a continuous automatic market maker. At any time, any person is free to change their reports, but to do so he or she has to take more risks. Thus, everyone comes to a limit where they do not want to make any further changes, at least not until they receive more information. At this point, the market can be said to be in equilibrium. At this point, the market price for any security can be used as an aggregate assessment of its value. When the goal is to select a security, then the one chosen will be the most valued.

9.3.2.3 Application of Prediction Markets in the Assessment of Forest Sustainability

As has been done with other methodological developments throughout this book, we aim to apply prediction markets to the evaluation of forest sustainability, specifically to the case study described in Section 7.7.6 in order to allocate a value of sustainability to each one of the territorial points in the forest area described in Section 7.6.1. This is an individual assessment for each evaluator, which requires the application of an assessment methodology that consists of the following steps (the reader can see the application of this methodology in Figure 7.12):

A. The first step is to select the six most meaningful territorial points for the assessment of sustainability in the forest analyzed (this point is described in Figure 7.10). Then the evaluator makes a subjective comparison of sustainability on each one of the pairs of points that can be formed with the selected points (the pair-comparison process and the information available to any evaluator for making his or her decision are described in Figure 7.11). For each of the pairs of points under comparison, the evaluator can only choose one of the two following actions: either one of the points in the pair is more sustainable than the other or he or she does not know (or does not answer) which of the points in the pair is more sustainable than the other.

B. Next, following the methodology described in Section 7.2.2, a sustainability value is allocated to each of the six selected points, according to the preferences of the evaluator.

C. Finally, as is explained in Section 7.4.1, sustainability assessment on significant points is extended to assess the sustainability of all the points in the forest area analyzed.

It is possible to move from step B to C and vice versa, but the amount of memory required to store the results of step B (it only takes six real numbers to describe the sustainability and twelve for the topographic coordinates of all the points) is much less than the memory required to store the results of step C (three times the number of points in the territory). Moreover, the expected number of evaluators is very high (and there is a different assessment for each evaluator). Therefore, the information stored at the end of the process corresponds to step B.

Now, we apply the results provided by prediction markets to the aggregate assessments of sustainability made by multiple evaluators on the same forest.

As we are simply showing an example of its application, we have generated six random numbers—with a range of variation from 0 to 1—and we have considered that each of these random numbers corresponds to the sustainability value of one of the meaningful points we have selected for the assessment of sustainability. We also assume we have only 15 evaluators, and we have therefore generated 15 sets of 6 random numbers and regard each set of numbers as the assessment of each evaluator. Figure 9.2a shows this initial information.

After analyzing the potential difficulty of participating in a prediction market, we decided not to develop a complete prediction-market software application. Instead, we opted to simulate the expected behavior of any evaluator in a prediction market from his or her decisions in comparisons of sustainability. This approach to the problem is considered advisable to avoid any agent evaluator from being deterred from participating in the process of information aggregation due to difficulties in understanding prediction markers. Thus, as the final contract is to predict the aggregate response in individual pair-wise comparison, a report of subjective probability of comparison was built. This report is consistent with all personal assessments of sustainability (see also Figure 9.2a) and determines the starting price for the evaluator.

The construction of the report is very simple. When the difference in sustainability between two points is greater than 33%, then the probability of these points being indifferent is null, and the probability of strict preference is computed from the difference in sustainability. Conversely, when the difference in sustainability between two points is less than 33%, then the probability of indifference is given by a potential function mapping on the separation between sustainability values. When both

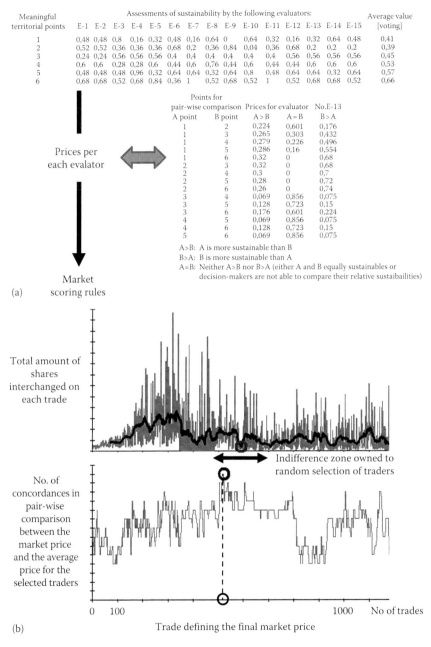

FIGURE 9.2 Aggregated assessment by applying prediction-market tools: (a) Traders' subjective assessments. (b) Market simulation and stopping rule.

(continued)

Points for pair-wise comparison		Final market prices for:			Agreggated
A point	B point	A > B	A = B	B > A	decision
1	2	0,0333	0,0064	0,006	A>B
1	3	0,0343	0,0346	0,0082	A=B
1	4	0,0087	0,0378	0,0267	A=B
1	5	0,0064	0,0198	0,0057	A=B
1	6	0,0088	0,0146	0,0057	A=B
2	3	0,0508	0,0708	0,007	A=B
2	4	0,028	0,0058	0,0091	A>B
2	5	0,0088	0,0128	0,0123	A=B
2	6	0,0299	0,0221	0,0316	B>A
3	4	0,0231	0,028	0,0369	B>A
3	5	0,0321	0,0272	0,0465	B>A
3	6	0,0057	0,0131	0,046	B>A
4	5	0,0171	0,0058	0,0469	B>A
4	6	0,0134	0,0059	0,0362	B>A
5	6	0,0158	0,0338	0,0205	A=B

A>B: A is more sustainable than B
B>A: B is more sustainable than A
A=B: Neither A>B nor B>A

Agreggated pair-wise comparison matrix:

	1	2	3	4	5	6	Agreggated value
1	#	1	#	#	#	#	0,16
2	2	#	#	1	#	2	0,2
3	#	#	#	2	2	2	0,08
4	#	2	1	#	2	2	0,28
5	#	#	1	1	#	#	0,48
6	#	1	1	1	#	#	0,68

1: Row-point sustaibale than Colum-point
2: Colum-point more sustaibale than Row-point

(c)

FIGURE 9.2 (continued) Aggregated assessment by applying prediction-market tools: (c) Final market assessment.

points have the same value of sustainability, this function cancels the probability of strict preference between them.

In market-scoring rules, traders arrive one at a time and tell the market maker how many shares they want to buy or sell of each outcome. If both the trader arriving in the market and his or her order are known (an order is a set of quantities $o = (x_{1-2}, x_{1-3}, \ldots, x_{5-6})$ to be sold $-x_{i-j} < 0$- or to be bought $-x_{i-j} > 0$-), it is then possible to run the market. So, if q is the current total amount of securities purchased by all traders in the market and we adopt a logarithmic cost function, the cost of any order will be

$$C_o = C_{q+o} - C_q \tag{9.16}$$

And, in the case of a logarithmic cost function, the new market unit prices, after processing the order o, will be

$$p^o_{i-j} = \frac{\exp\{(q_{i-j} + x_{i-j})/b\}}{\sum_{i=1}^{j} \sum_{j=i+1}^{6} \exp\{(q_{i-j} + x_{i-j})/b\}}, \quad \text{for} \quad i = 1, \ldots, 5\dagger \tag{9.17}$$

In order to determine the next trader in the market, let us calculate the relative market attractiveness to any evaluator. The attraction is proportional to the expected benefit and also to the inverse of the separation between two sets of prices: the one is deduced from the personal beliefs and the other induced by the purchase of any order. Thus, the decision to incorporate a trader–order pair into the market is made randomly, with a probability of trade proportional to the attraction of the pair (trader–order) for market entry.

As mentioned earlier, any agent, at any time, can change his or her orders in an attempt to maximize his or her payoffs. But doing so means taking risks regarding personal beliefs. Thus, everyone comes to a limit where they do not want to make any further changes. In consequence, the market continues until no more shares are exchanged (Figure 9.2b). Since this is a simulation with random incorporation of agents, it is necessary to analyze the market price in a zone of indifference around the point the transactions are being canceled.

As can be seen (Figure 9.2c), it is possible to construct a matrix of aggregated pair-wise comparison of sustainability from the final market price. In turn, this matrix also allows us to determine the aggregated value of sustainability for the significant points in the forest.

9.3.3 Interpersonal Comparison of Utility

Utility is a measure of satisfaction (or, as Bentham—see footnote on page 503— stated, the balance between the pleasure and the pain) of an individual.

It is clear that human beings can achieve satisfaction due to many factors. However, the study of utility is reductionist. It relinquishes the understanding of human behavior in order to explain the study of all sources of satisfaction under a single theory. Utility solely analyzes what a person does and whether individuals act in a consistent or inconsistent way with respect to their past behavior.

The identification of rationality with consistency of behavior has not been easily accepted. Some philosophers (especially the followers of Kant) argue that utility in some individuals appears to be more rational than in others; however, most economists today are aligned with Hume (1739) and do not analyze either the morality of actions or the behavior's rationality. For the followers of utilitarianism, there is nothing irrational in pursuing any purpose, at any time, because unlike Kant's proposal, they consider that rationality is defined by the means and not by the ends.*

---

* From the foregoing considerations, the reader could assume that the authors do not agree with this theory. However, we do concur with this approach, as utilitarian hypotheses provide an appropriate starting point for applying the scientific method to the analysis of behavior. As we shall see, the use of the capacities of the scientific method does not necessarily violate personal convictions. Even more so, utility theory needs to incorporate the concept of empathy in order to progress in joint decision-making. Through empathy, individual behavior transcends the individual and incorporates criteria from other people. Merely to step outside oneself means accepting ethical criteria to take part in decision-making. Additionally, the reader should consider that nearly all societies are educated in moral values, which will be applied in personal judgments. Putting oneself in another's shoes favors the consideration of the ends in the process of assessment of utility and takes utilitarianism beyond the mere sum of individual selfishness.

Although comparison of utilities is a human capacity, there was no procedure available for comparing sizes of utilities until the works by von Neumann and Morgenstern* (1944): It may be clear that a person prefers a tie to a bow tie, but it is not evident that the usefulness of a tie is, for example, eight times greater than the utility of a bow tie. The reason is that by applying the law of decreasing returns, when you have eight ties, the usefulness of the eighth tie is unlikely to be equal to the value of the first tie in the set (and equal to the value of the single bow tie available).

Von Neumann and Morgenstern defined the utility of a good as the measure of the risk that a person is willing to accept to obtain that good. The computing of the utility is immediate under this formulation: suppose we know the possible occurrence of an outcome (H) that, for a certain person, is a more preferred result than any other feasible outcome (T). That person can compare the occurrence of T (with probability 1) with the occurrence of H (with probability p). The value of p in which T is preferred to H determines the utility of T ($100p$ units of utility).

The use of risk in comparing outcomes makes it necessary for the utility function to be formulated on lotteries[†]. In consequence, a utility function is an application of the set of all possible lotteries (L) over the set of real numbers ($\mathbb{R}$) ($U: L \to \mathbb{R}$), which satisfies

$$\forall L_i, L_j \in L : L_i \preccurlyeq L_j \Leftrightarrow E\left[U\left(L_i\right)\right] \leq E\left[U\left(L_j\right)\right] \tag{9.18}$$

where

$$E[U(L)] = p_1 U(x_1) + \cdots + p_m U(x_m); \; L = \begin{pmatrix} p_1 p_2 \cdots & p_m \\ x_1 x_2 \cdots & x_m \end{pmatrix}$$

$U(L)$ is the utility of L lottery

$U(x_i)$ is the utility of x_i outcome

$\preccurlyeq$ is the "*at least more preferable*" relation

The existence of the utility function requires the evaluator's preferences to satisfy certain conditions[‡] (von Neumann-Morgenstern, 1944).

---

* In fact, these authors were the developers of the first relevant application for the comparison of utility under risk. Prior, Nicholas Bernoulli (1713) formulated the expected utility model, and Daniel Bernoulli (1738) later proposed the first solution to the quantitative comparison of utilities in his explanation of the St. Petersburg paradox.

† Given the set of all possible outcomes ($x_1, x_2, \ldots, x_m$), a lottery L is a specific combination of possible outcomes with specific values of their possibilities of occurrence ($p_1, p_2, \ldots, p_m$), such that $\Sigma p_i = 1$ (a lottery is a discrete probability distribution of the set of possible outcomes).

‡ These conditions are *completeness* (the evaluator has well-defined preferences for any pair of lotteries), *transitivity* (preference is consistent across any three options), *convexity/continuity* (there is a "tipping point" between *better than* and *worse than* a given middle point), and *independence for every triplet of lotteries* (the preference between two alternatives holds independently of the possibility of another outcome).

Three main consequences can be outlined from the earlier definition:

1. First, any individual that chooses according to this theory acts (to an external observer) as if he or she were always aiming to maximize the expected (average) value of something. That "something" is what is called utility.*
2. Second, if personal preferences ($\preccurlyeq$) rule decisions (a person will choose L_i instead of L_j if $L_i \preccurlyeq L_j$), then the earlier definition of utility is consistent with what a person does ($U \leftrightarrow \preccurlyeq$).
3. Finally, the von Neumann and Morgenstern theory provides a value for utility that maximizes the average utility; therefore, each unit of utility—built in this way—is "worth" the same as any other.[†] This fact allows sizes of utilities to be compared.

9.3.3.1 Utility for Multiple Decision-Makers

It has been possible to establish that each unit of utility a person receives is worth the same to him or her. But is this applicable when comparing the utilities of two or more people? In other words, is it possible to make comparisons between the sizes of utilities of two or more people? This is a relevant question that has been considered from the early developments of utility.[‡] From the beginning, utilitarianism has focused on maximizing individual utility as a moral criterion for the organization of society. The aim should be to maximize the total utility of individuals, with the goal of achieving "the greatest happiness for the greatest number of people." Further developments of utilitarianism[§] would seek to maximize the utility of the individuals with lowest utility in order to create a more equitable society.

---

* People do not always maximize the expected utility. Let us, for example, consider the choice between two scenarios: one with a guaranteed payoff and other with a random one. In the guaranteed case, the individual receives €100; in the uncertain scenario, a coin is flipped to decide whether the individual receives €200 or nothing. The expected payoff in both cases is €100. If an individual is indifferent between the bet and the sure 100 payment, then he or she is *risk-neutral* (this type of person decides according to the von Neumann–Morgenstern theory). But there are also *risk-averse* and *risk-seeking* individuals. The first will accept a sure payment lower than €100 instead of the bet; individuals in the second set will require a payment over €100 to induce them to take the guaranteed option.

† This statement would not apply if the utility is measured in money. In this case, the nonlinearity of marginal utility (satisfaction due to the last euro depending on the total amount of available money) would prevent quantitative comparisons of utility.

‡ We can see the first reasoning on social utility even in pioneering works by Bentham and Mill (considered the fathers of utilitarianism). John Stuart Mill (1806–1873) was the editor and intellectual heir of Jeremy Bentham (see footnote on page 503). Given Bentham's tendency to leave his writings unfinished, the editor in many cases had to complete the author's work. Mill is also considered an important empiricist and positivist.

§ See, for example, the works of John Rawls (1921–2002) and Harsanyi's theory that will be discussed in the next paragraph. Rawls has been revered for the widespread consensus that his theory of justice (1971) brought about the revival of political philosophy. In this work, Rawls argued heuristically for a reconciliation of the principles of freedom and equality with his famous approach to the—seemingly insurmountable—problem of distributive justice: when the parties face moderate shortages and are neither naturally altruistic nor purely selfish, then both parties choose principles of justice that are mutually acceptable.

In order to make an interpersonal comparison of utilities, the initial approach was to develop welfare functions in terms of raw commodities for consumption. However, this approach has several problems that hinder the comparison of utilities among individuals. The main difficulties concern discrepancies in the ability to access the same goods from different individuals, with differences in preferences for various goods or combinations of goods and with scarcity or abundance of goods.

An alternative procedure is to measure the welfare in terms of the benefits that individuals obtain from consumption. This could be used to know how much health, wellness, or desires are derived from a basket of consumer goods. In this sense, Rawls (1971) proposed a list of primary benefits.* However, although Rawls benefits could be measured in accurate terms, there would still be the problem of aggregating the benefits in an index. Rawls argues that in general (see footnote § on page 531) a broad consensus of opinion can be accepted on this issue. Obviously, the work of moral philosophers would be easier if there really were primary benefits about which everybody felt more or less the same, but this assumption is far from evident.

There are also other noteworthy methods of aggregating utilities based on additional paradigms to the aforementioned. These involve mainly counting perception thresholds and formulating 0–1 rules. However, the analysis of their performance also shows a lack of rationality (Binmore, 1998).[†]

Next, we introduce the procedure we have chosen for making an interpersonal comparison of utility. It is based on the method proposed by Harsanyi (1977, 1992).

9.3.3.2   Harsanyi's Theory of Interpersonal Comparison of Utility

The application of Harsanyi's developments requires additional assumptions to the von Neumann–Morgenstern theory based on the notion of empathetic preferences (Suppes, 1966; Sen, 1970; Arrow, 1978).

---

* The primary benefits Rawls proposes are "*the powers and prerogatives of the office,*" "*the social basis of self-respect,*" and "*income and wealth.*"

[†] The idea of counting perception thresholds is the same underlying the awareness of an evaluator's depth of knowledge (see Section 7.5.2 of Chapter 7). It requires observing how far a parameter needs to be changed before an individual perceives that a change has taken place. The number of perceptual jumps experienced as the parameter moves from one end of its range to the other can then be used as a measure of the intensity of the preference between the two extremes. Clearly, the number of jumps perceived by different people can be counted. But can anyone infer that a person feels less pleasure in music than another from the fact that the ear of the first person is less sensitive than the ear of the second?

Alternatively, the zero–one rule is the basis for computing cardinal utilities. Returning to the von Neumann–Morgenstern theory, let us suppose that two individuals (*A* and *B*) agree that W and H outcomes (lotteries) are respectively the worst and the best possible outcomes. The zero–one rule involves the recalibration of the utility scales so that the utility functiovn satisfies $U^A(W) = U^B(W) = 0$ and $U^A(H) = U^B(H) = 1$. The question is: Is it possible to adopt a method of utility comparison that treats the two individuals equally (one of them—e.g., *A*—seeing W only marginally worse than H and, on the other hand, another—*B*—who perceives many nuances between W and H)?

Although the composition of the preceding two procedures could lead to effective comparisons of utility, the complexity of the resulting procedure would make it very difficult to understand by most evaluators. In this case, most decision-makers could see the proposed methodology as a black box, and if they were unable to identify its contribution to added value, they would probably refrain from participating in the assessment.

A decision-maker (A) empathizes with another (B) when A sees things from the point of view of B.* This is not new in nature (mothers tend to take care of their children in many animal species), and humans also sympathize, at least, with their extended family, friends, and neighbors. There are no special difficulties in incorporating altruistic preferences into a utility function: the von Neumann–Morgenstern theory is adequate to determine interpersonal comparison, because an evaluator (A) would only need to consult his or her own sympathetic utility function to find out how many units of utility to assign to a change in B's situation when compared with the same change in his or her own situation. We shall construct this utility function in the case of application we shall discuss next.

However, empathetic identification goes even further. It is crucial for the survival of human societies (Binmore, 2005), as we do not only empathize with other: we also have empathy with our own ethical concerns. Particularly, we justify ourselves by making references to "fairness" in order to explain our behavior. When making fairness judgments, any decision-maker (A) must be able to know how much better he or she feels when identifying with another decision-maker (B) instead of identifying with himself. This means not only empathetic identification but incorporating empathy in preferences and/or decisions. Empathetic preferences provide the inputs for the selection of the assessment criteria that lead us to speak of "fairness" when explaining what we are doing.

Binmore (2005) goes a step further and considers that the "fairness" criteria can make us perceive certain types of behavior as more successful than others. As occurs in other aspects, social evolution will tend to favor the survival of whatever empathetic preferences promote the social success of those that hold them at the expense of those that do not. In this context, Binmore argues that in the medium run, equilibrium in empathetic preferences will be achieved: Everybody will have the same empathetic preferences in this equilibrium, and thus, all evaluators will share a common standard for making interpersonal comparisons of utility.

9.3.3.3 Application to Assessment of Forest Sustainability

Next, we shall explain how to obtain the empathetic utility function (*euf*) in the assessment of forest sustainability. In essence, the case we describe is similar to that explained in the previous section (application of perdition markets to obtain an aggregated value): We have 15 evaluators' preferences on forest sustainability (here randomly selected from an actual database), and we seek to obtain a joint assessment of sustainability for the group of evaluators.

We apply the same individual assessment methodology as in the case study discussed in Section 7.6 (Chapter 7). Let us recall the method whereby each evaluator allocates his or her own value of sustainability to each point of the forest: Each evaluator (A) is asked about his or her preference regarding sustainability, in particular about which point he or she considers to be most sustainable from each of the

---

* It is important to consider that to see things from others' point of view does not mean I am not able to separate my decisions from the decisions of others (we can all put ourselves in the place of another person who is ill, but if the sick individual were to die, we would not necessarily want to die too). However, as we shall see, a certain degree of empathy in preferences and decisions is necessary for a society to survive.

possible pairs of points that can be formed from the six most significant points in the forest. As explained in Sections 7.2 and 7.4 (both in Chapter 7), the result of the comparison leads to a value of sustainability in the six points compared ($u_A = (u_1A$, $u_2A, \ldots, u_6A)$), which is particular to evaluator A, and to which we shall call an assessment of sustainability. It is possible to extend the values of the coordinates of u_A to the assessment of sustainability in all the points of the forest (by applying Expression (7.39) in Chapter 7). Let $v_A(x_i)$ denote the assessment of sustainability at the spatial point x_i for evaluator A, and let v_A be the average value of sustainability for the whole production area of the forest ($v_A = (1/n)\Sigma v_A(x_i)$, for $i = 1, \ldots, n$, where n is the total number of points in the forest).* In summary, the information available from each evaluator's preferences on sustainability is both the assessment of sustainability (u_A) and the average value of sustainability for the whole production area of the forest (v_A). They both configure the opinion of an evaluator (A), which we represent by $o_A = (u_A, v_A)$.

Similarly, $o_g = (u_g, v_g)$ represents the aggregated opinion of sustainability that we shall compute late (although can be also computed as described in Sections 9.3.1 and 9.3.2), and $o_{Ob} = (u_{Ob}, v_{Ob})$ is an objective opinion of sustainability computed from experts' knowledge. This objective assessment of sustainability at the significant points is the one computed for the case study described in Chapter 7 (see Figures 7.10 and 7.12).

In the case of application that concerns us here, we have randomly chosen 15 evaluators from a database of actual evaluators (named E1, …, E15), whose assessments of sustainability at the six most significant spatial points—together with the global assessment of sustainability for the whole of the productive area of the forest—are shown in Table 9.10. For these evaluators, we have calculated the aggregate and the objective assessments of sustainability, which are also shown in Table 9.10.

In addition to personal evaluations, it is necessary to examine the conditions required for making interpersonal comparisons of utilities. Next, we analyze the Hammond conditions (Hammond, 1991). In this regard,

1. The *identity axiom* is satisfied in our case of application (any third person—
   K—always identifies the systems of preference of any two evaluators—
   A and B—in a similar way). As the assessment of any evaluator is known,
   it is then evident that

$$U^{A(K)}(x) \geq U^{B(K)}(y) \Leftrightarrow v_A(x) \geq v_B(y), \quad \forall K, A, B \in \Theta \qquad (9.19)$$

   where
   $U^{A(K)}(x)$ is the utility that the observer K assumes that evaluator A has
   attributed to the x alternative ($x \in X$)
   Θ is the set of all possible evaluators

---

* The method of computing sustainability we apply minimizes for each evaluator the number of spatial points where his or her preferences of sustainability do not correspond with the attributed values of sustainability.

TABLE 9.10

Assessments of Sustainability for the Six Most Significant Points of a Forest and Global Sustainability of the Forest

Spatial Points	Evaluators															Objective Assess.	Aggregate Assess.
	E1	E2	E3	E4	E5	E6	E7	E8	E9	E10	E11	E12	E13	E14	E15		
1	0.76	0.54	0.59	0.38	0.92	0.22	1	0.74	0.63	0.84	0.74	0.32	0.46	0.76	0.31	0.92	0.7
2	0.91	0.8	0.72	0.88	0.37	0.46	0.47	0.43	0.38	0.2	0.34	0.58	0.84	0.24	0.36	0.75	0.62
3	0.4	0.59	0.49	0.29	0.18	0.71	0.66	0.46	0.62	0.31	0.59	0.89	0.16	0.76	0.61	0.54	0.57
4	0.44	0.31	0.19	0.35	0.1	0.34	0.19	0.29	0.4	0.79	0.42	0.21	0.46	0.57	0.04	0.42	0.4
5	0.04	0.23	0.04	0.5	0.58	0.19	0.34	0.16	0.69	0.12	0.31	0.35	0.22	0.41	0.65	0.26	0.38
6	0.29	0.6	0.17	0.16	0	0.54	0.15	0.16	0.01	0.13	0.4	0.64	0.36	0.43	0.66	0.07	0.36
Global sust. in the forest	0.43	0.46	0.3	0.37	0.25	0.42	0.37	0.33	0.41	0.46	0.45	0.46	0.4	0.54	0.38	0.42	0.46

Under this condition, the utility is a monotone increasing transformation (φ) of the value: $U^A(x) = \varphi[v_A(x)]$ (in the case of application, x is a spatial point in the forest where the evaluator assesses the sustainability). Consequently, if the utility does not incorporate additional (such as ethical) criteria to those used for calculating the value, then $U^A(x)$ and $v_A(x)$ are interchangeable.

2. It is also possible to satisfy the *conditions that foster egalitarianism* in the case of application. Extreme egalitarianism uses $U^A(x) > U^B(x)$ as a justification for any action that increases person **B**'s utility, even if this may mean lowering person **A**'s utility (provided the actions do not go so far as to make $U^A(x) < U^B(x)$). In the case of application that concerns us here, if $v_A(x) > v_B(x)$, then it is expected that a transfer of knowledge from **A** to **B** allows an increase in the sustainability attributed to point x by evaluator **B**. To transfer knowledge, it is enough to identify the main differences between the pairwise comparisons made by **A** and **B** and to inform both evaluators about the differences. However, in most applications, the amount of **A**'s utility that can be sacrificed to increase **B**'s utility remains unclear.*

Therefore, we need to advance in the determination of a decision framework where interpersonal utility comparisons can be made and applied to ethical decisions. To define this framework, it is first necessary to introduce the concept of impersonality (Harsanyi, 1953, 1955), which supposes that an ethical observer must be unaware of what type of individual he or she will become as a result of the decisions he or she makes (this is so evaluators set aside selfish considerations when making moral judgments). Under this condition, Harsanyi assumes that an impersonal evaluator is a "rational Bayesian" and maximizes the expected utility of a von Neumann–Morgenstern utility function.

In order to build the Bayesian environment required for decision-making, let us denote Θ as the set of all possible evaluators and let $L = \Delta(X \times \Theta)$ denote the set of all simple probability measures (lotteries) on $X \times \Theta$, where X is the set of spatial

---

* For example, if $v_A(x)$, $v_A(y)$, and $v_B(y)$ are all greater than $v_B(x)$ and all other evaluators are indifferent between the sustainability in x and in y, then it could be socially accepted that sustainability at point y is higher than at x. This is the *equity axiom* (Hammond, 1976, 1979; Sen, 1977). In a domain with no restrictions, under the condition of independence of irrelevant alternatives and the acceptance of the Pareto indifference, the earlier situation is equivalent to the *two-person leximin rule* (Sen, 1970). In turn, both postulates have been formulated under the concept of *preference priority* (Strasnick, 1976a,b, 1977, 1979): For each pair of individuals **A**, **B** ∈ Θ and each pair of alternatives x, y ∈ X, we say that **A**'s preference for x over y takes priority over **B**'s preference for y over x if xPy when xP^Ay, yP^Bx, and all other evaluators are indifferent between x and y. The earlier arguments rely on being able to compare the utility of **A**'s loss with **B**'s gain, but this is only possible when comparisons of utility differences can be made: $U^A(x) - U^A(y) > U^B(k) - U^B(l)$, which balance the intensity of preference for x over y (for evaluator **A**) with the intensity of preference for l over k (for evaluator **B**).

Returning to the case of application, the procedure used to calculate $v_A(x)$, as well as its identification with $U^A(x)$, allows direct comparisons between utilities and between their differences. However, accepting $U^A(x) - U^A(y) > U^B(y) - U^B(x)$ and $U^A(x) > U^A(y) > U^B(y) > U^B(x)$ requires the acceptance of both $v_A(x) - v_A(y) > v_B(y) - v_B(x)$ and $v_A(x) > v_A(y) > v_B(y) > v_B(x)$ and also the social decision of whether the excess of **A**'s gain over **B**'s loss is sufficient to compensate the results produced by the change in considering point x to be more sustainable than y, instead of considering point x as more sustainable than y. Besides, tensions may appear between direct utility comparisons and comparisons of intensities of preferences (Sen, 1973), which increase the complexity of the problem.

points (x) in the forest to evaluate. Each lottery $Lin \Delta(X \times \Theta)$ is a finite collection of possible threesomes (x_i, θ_j, p_{x_i,θ_j}) consisting of a spatial point x_i ($x_i \in X$) and an evaluator θ_j ($\theta_j \in \Theta$), together with the nonnegative probability p_{x_i,θ_j} of the outcome corresponding to (x_i, θ_j) (i.e., $v_{\theta_j}(x_i)$) occurring. Under the von Neumann–Morgenstern theory, the aggregated sustainability—for a group of evaluators—of a spatial point (x_i) is expressed as

$$E\left[U^g(x_i)\right] = \sum_{x_i \in X; \theta_j \in \Theta} p_{x_i,\theta_j} \times v_{\theta_j}(x_i) \equiv v_g(x_i) \qquad (9.20)$$

where

$U^g(x_i)$ is the social utility of point x_i, whose expected value corresponds to the aggregate assessment of sustainability at point x_i

$v_{\theta_j}(x_i)$ is the assessment of sustainability in point x_i for the evaluator θ_j

To compute p_{x_i,θ_j}, we have considered that Harsanyi's ethical interpersonal comparisons of utility make it necessary to represent preferences regarding the kinds of people it is desirable to have in society: an ethical interpersonally comparable utility function measures an ethical observer's view of the *utility of a person* to society as a whole (Hammond, 1991). In this context, the likelihood of an outcome occurring is the probability of that outcome being socially accepted. Thus,

$$p_{x_i,\theta_j} = \frac{q_{x_i,\theta_j}}{\sum_{\theta_j \in \Theta} q_{x_i,\theta_j}} \qquad (9.21)$$

where

$$q_{x_i,\theta_j} = \begin{cases} 1 & \text{if} \quad v_{\theta_j}(x_i) > v_{Ob}(x_i) \\ \dfrac{v_{\theta_j}(x_i)}{v_{Ob}(x_i)} & \text{otherwise} \end{cases}$$

$v_{Ob}(x_i)$ is the fair-socially accepted outcome for x_i

For computing $v_{Ob}(x_i)$, we have used the assessment of sustainability corresponding to a pair-wise comparison according to the assessment defined in Expression (7.48). This value of sustainability derives from the statistical distance of each point in the territory to the ideal point (a point with the highest potential values of all the indices of sustainability).

In order to calculate *euf*, we force changes in the aggregated assessment of sustainability (Δu_g) and determine the consequences of these alterations in the utility of each evaluator. The comparison of the changes that occur in the utility of one evaluator (A) with those produced in the utility of any other (B)—owing to modifications in the aggregated value—provides the information needed to measure the *euf* of A regarding B.

The consideration of empathy means the utility for the sustainability of an evaluator (A) does not solely depend on his or her own opinion on sustainability (i.e., of the pair $o_A = (u_A, v_A)$), because each individual utility is built from the whole set of opinions the evaluator is able to handle: when an evaluator (A) analyzes his or her empathy regarding another evaluator (B), he or she must first consider B's opinion ($o_B = (u_B, v_B)$). But the history of the discussions for a joint assessment of sustainability must also be considered or, at least, the final aggregated assessment ($o_g = (u_g, v_g)$). Finally, in order to promote fairness of individual judgment, it is advisable to use the opinion of experts in sustainability ($o_{Ob} = (u_{Ob}, v_{Ob})$).

In consequence, each evaluator constructs its own utility function from its own understanding of the problem, which means to use all the previously mentioned opinions. If we call $U_{A \to B}(x)$ to the utility of A taking into account all the opinions conforming the empathic utility, then, for an assessment of sustainability ($x = (x_1, x_2, ..., x_6)$) of any other evaluator, the value of $U_{A \to B}(x)$ is calculated as follows:

$$U_{A \to B}(x) = \sum_{i \in \{A,B,g,Ob\}} \left(\frac{w_i(x)}{\sum_{i \in \{A,B,g,Ob\}} w_i(x)} v_i \right) \tag{9.22}$$

where $w_i(x) = \|u_i - x\|^{-2}$

The earlier expression is not yet the A *euf* relative to B. As mentioned, the role of the empathy function seeks to express the equivalence in the utilities of A and B when the aggregate assessment of sustainability (Δu_g) is modified. Now, in order to obtain *euf*, we force systematic changes in the coordinates of the vector u_g, and we then determine the changes that these modifications produce in both A and B's utilities. This is formulated as follows:

$$u_{g\langle k,d\rangle} = \left(u_{1g}, ..., u_{(k-1)g}, u_{kg} + d, u_{(k+1)g}, ..., u_{6g} \right)$$
$$e_{A \to B,kd} = U_{A \to B}(u_{g<k,d>}) - U_{A \to B}(u_g) \tag{9.23}$$

where
$k = 1,2, ..., 6$
$d \in \mathcal{D} = \{- x_{kg}, -.7x_{kg}, -.45x_{kg}, -.25x_{kg}, -.1x_{kg}, .1(1 - x_{kg}), .25(1-x_{kg}), .45(1-x_{kg}), .7(1-x_{kg}), x_{kg}\}$

Under these conditions, the *euf* of A (relative to B) is a vector of 60 coordinates (corresponding to possible range of variations of k and d, with $k = 1,...,6$ and $d \in \mathcal{D}$). Henceforth, we shall represent this vector as euf_{AB}, where $e_{A \to B}$, kd are each one of its coordinates. The whole *euf* of A will be represented by the matrix euf_A. This is a matrix whose rows represent the *euf* of A with respect to each of the M evaluators participating in the evaluation of sustainability.

The main contribution of *euf* to the aggregation of preferences is to identify the degree of convergence in the evaluators' preferences. As mentioned, the combination of "fairness" and empathy leads to a state of equilibrium. In this state, all evaluators will have the same preferences of empathy. Therefore, a measure of the similarity of the empathy for all evaluators will provide a measure of the degree of convergence in the preferences of these evaluators.

Expression (9.24), adopted to measure the similarity between empathic preferences of two evaluators (A and B), is based on the distance between two arrays. This expression gives the percentage of similarity between the utilities of two evaluators:

$$S_{AB} = 100 \times \left[1 - \frac{\|\mathbf{euf}_A - \mathbf{euf}_B\|}{\min\{M,60\}} \right]; \qquad (9.24)$$

where M is the total number of evaluators

$$\|\mathbf{euf}_A - \mathbf{euf}_B\| = \sqrt{\mathrm{tr}\left[\left(\mathbf{euf}_A - \mathbf{euf}_B\right)^T \circ \left(\mathbf{euf}_A - \mathbf{euf}_B\right) \right]}$$

∘ denotes the usual multiplication of matrices

tr() represents the trace of a matrix

The outcomes of applying Expressions (9.22) through (9.24) to the data of Table 9.10 are shown in Table 9.11.

The average of the similarities obtained by comparing all possible pairs of evaluators has been used as a measure of the similarity among all evaluators:

$$S_T = \frac{1}{M(M-1)/2} \sum_{I=1}^{M-1} \sum_{J=I+1}^{M} S_{IJ} \qquad (9.25)$$

where M and S_{IJ} are both as described in (9.24).

The applicability of these concepts is evident: When the convergence of individual preferences is small and the increase in the number of participants does not result in an increase in the similarity in empathy, then the agent ultimately responsible for decision-making must end the participation process (as it is not justifiable to root decision-making in the participatory process).

However, when increasing the number of participants involves increasing the similarity of preferences of empathy, then participatory process should be encouraged in order to achieve an acceptable similarity of preferences. In this case, we can say that the aggregated assessment is representative of the preferences of the people involved in the evaluation, and this value can be adopted to define the best joint management plan. If participants constitute a representative sample of society (see Section 9.2), then the decision will also be socially acceptable.

As can be deduced from Table 9.11, in the case of application that concerns us here, the overall similarity for the 15 evaluators analyzed is $S_T=65.8\%$. Thus, although some degree of convergence can be seen (as corresponds to actual evaluations), the global convergence among evaluators is far from the best possible similarity of preferences (100%), which means that the aggregated utility for the evaluators analyzed does not reveal an acceptable convergence of individual utilities. Moreover,

TABLE 9.11

Matrix of Similarity (%) between Evaluators

	E1	E2	E3	E4	E5	E6	E7	E8	E9	E10	E11	E12	E13	E14	E15
E1	100	85.2	45.3	78.1	42.8	96.3	77	50.2	92.6	89.4	75.5	89.9	89	57.9	82.2
E2	85.2	100	32.2	64	28.5	81.8	63.4	38.8	78.3	85.8	85.9	89.7	74.6	71.2	67.4
E3	45.3	32.2	100	66.7	86	49	68.3	86.8	52.6	36.3	30.5	35.4	56.2	3.6	60.1
E4	78.1	64	66.7	100	64.5	81.8	96.7	69.2	85.5	69.6	58.5	68.5	89.2	36.1	92.3
E5	42.8	28.5	86.0	64.5	100	46.4	64.9	73.5	50.1	35.2	24.3	33.6	53.7	0.8	59.8
E6	96.3	81.8	49	81.8	46.4	100	80.7	53.7	96.3	86.4	73.2	86.4	92.7	54.3	85.6
E7	77	63.4	68.3	96.7	64.9	80.7	100	71.8	84.3	67.8	59	67.1	87.9	35.1	89.3
E8	50.2	38.8	86.8	69.2	73.5	53.7	71.8	100	57.1	40.2	39.4	40.1	60.2	10	61.6
E9	92.6	78.3	52.6	85.5	50.1	96.3	84.3	57.1	100	83.2	70.7	82.8	96.3	50.6	88.7
E10	89.4	85.8	36.3	69.6	35.2	86.4	67.8	40.2	83.2	100	72.3	96.1	79.9	64.8	75.5
E11	75.5	85.9	30.5	58.5	24.3	73.2	59	39.4	70.7	72.3	100	76	67.8	66.5	59.5
E12	89.9	89.7	35.4	68.5	33.6	86.4	67.1	40.1	82.8	96.1	76	100	79.2	67.1	73.7
E13	89	74.6	56.2	89.2	53.7	92.7	87.9	60.2	96.3	79.9	67.8	79.2	100	46.9	91.7
E14	57.9	71.2	3.6	36.1	0.8	54.3	35.1	10	50.6	64.8	66.5	67.1	46.9	100	40.8
E15	82.2	67.4	60.1	92.3	59.8	85.6	89.3	61.6	88.7	75.5	59.5	73.7	91.7	40.8	100

if the increase in participation does not lead to a similar evolution in the measure of similarity, then it is not possible to justify that the aggregate assessment is a socially acceptable solution. Therefore, the final decision must be justified by criteria other than participation. Simply by way of example, we could mention being as close as possible to the objective sustainability (directly or with some restriction on the cost of implementation), better meeting the needs of disadvantaged groups—such as the local population (Rawls solution), or adopting other criteria considered appropriate by the landowners or public authorities.

9.3.4 COMPUTATIONALLY HARD AGGREGATION RULES

Current methods of aggregation are tasks performed by computers. The computation time of these methods is a linear or quadratic function on the number of alternatives and is usually linear in the number of voters (Chevaleyre et al., 2005). Therefore, preference aggregation rules (the opposite of political elections) are usually quite complex from the programming point of view. In the following, we refer to some of the most notable works on programming algorithms for the aggregation of multiple preferences:

- *Kemeny's aggregation*: Given a set of m total orders, called votes, over a set of n alternatives, the Kemeny optimal aggregation problem asks the voters for the total order from alternatives that minimize the sum of τ-distances from the votes, where the τ-distance between two total orders is the number of pairs of alternatives that are ordered differently in these two total orders. The computing is an NP-hard* problem. Kemeny's aggregation procedure is addressed in several studies (Bartholdi et al., 1989; Davenport and Kalagnanam, 2004; Hemaspaandra et al., 2005; Ailon et al., 2005; Conitzer et al., 2006).
- *Slater's rule*: This rule minimizes the number of inconsistencies resulting from pair-wise comparison of alternatives by defining a distance between preference matrices. Slater's rule is NP-hard. The reader can find more about the computer processing of this rule and its related computational problems in Bartholdi et al. (1989), Charon and Hudry (2000), Alon (2006), Conitzer (2006), and Hudry (2010).
- *Dodgson voting rule*: In this voting procedure—proposed in 1876 by Dodgson (Lewis Carroll)—wins the candidate for the "closest" to being a Condorcet winner: the winner needs a minimum number of elementary exchanges to become a Condorcet candidate. An elementary exchange in favor of a candidate means an improvement in the preference profile consisting of an exchange in the preferences of a voter's position with the candidate immediately above. This rule is also NP-hard (Bartholdi et al., 1989; Hemaspaandra et al., 1997).

---

* This is a class of problems, which are as hard as the hardest problems that can be verified in polynomial time by a nondeterministic Turing machine (Hochbaum, 1995; more information about Turing machines in Viso (2008)). Recognizing NP-hard winners is intractable.

- *Young's voting rule*: The winner is the candidate who needs to exclude the fewest voters to become a Condorcet candidate. Thus, Young's method (as occurs with Dodgson's method) takes into account changes in the profile of preferences for a candidate to overcome as many as possible of the others. But unlike the previous method, changes are produced by eliminating certain voters, instead of elementary exchanges in preferences. This rule is an NP-complete problem* (Rothe et al., 2003).
- *Banks collection of profiles*: A candidate is a winner if it is a top element in a maximal acyclic subgraph of the majority graph (Banks, 1985). Checking whether a candidate is a Banks winner is NP-hard (Woeginger, 2003; Hudry, 2004, 2009), so computing all Banks winners is also NP-hard. But computing only some Banks winners is easy: it is sufficient to add new alternatives—transitively and sequentially—to an initial chain so that the result of the inclusion is acyclic (the top element in the extended chain is a Banks winner).

9.3.4.1 Application of Computational Aggregation Rules to the Definition of an Operative Procedure in the Assessment of Forest Sustainability

In general, computational methods of aggregation seek to optimize the procedure that leads to the best solution, involving the fewest number of evaluators. In previous sections, especially Sections 9.3.2 and 9.3.3, we have described some of the procedures for obtaining an aggregated representation of preferences. However, if we accept the possibility of grouping individuals into categories,[†] their number is reduced, and another interpretation can be given to the basic utility function (Expression (9.20)), which will be familiar to economists:

$$E\left[U^g\left(x_i\right)\right]=\sum_{\theta\in\Theta}N(\theta)\times\overline{v}\left(x_i,\theta\right)\qquad(9.26)$$

where
  Θ is the set of θ individual personal characteristics
  $N(\theta)$ denotes the number of individuals who have personal characteristic θ
  $\overline{v}\left(x_i,\theta\right)$ is the average assessment of sustainability at point x_i for the evaluators in θ

---

* It is in the set of problems that any given solution to the decision problem can be verified in polynomial time. More information about this type of problems can be found in Garey and Johnson (1979).
† In Chapter 7, we have established two types of characteristics that can be applied to classify the potential participants in a public process for sustainability assessment. One of them concerns the personal characteristics of the evaluator (see Figure 7.9 and Section 7.6.1), namely, qualitative variables such as *gender, age class, educational level, occupation, type of stakeholder*, and *place of residence*. The other is based on the characteristics used to describe the assessments of sustainability of each evaluator, which consist of a set of descriptors arising from a range of the following variables: *proximity of the personal assessment of the objective sustainability, type of rationality, depth of knowledge on sustainability, significant indicators in assessing sustainability*, and *percentage of linearization of the individual value function* (see Figure 7.14, Table 7.10, and Section 7.6.3). The intersection of both clusters determines the set of all possible classes of evaluators (Θ).

Expression (9.26) says that the expected utility should have linear indifference curves in the space of all possible vectors, with components $N(0)$ (all $\theta \in \Theta$) and with constant marginal rates of substitution $v(x, \theta)/v(x, \theta')$ between the numbers of individuals with any pair (θ, θ') of personal characteristics. Such constant marginal rates of substitution determine, for each fixed x, an interpersonally comparable utility function $\bar{v}(x, \cdot)$ in Θ.

The descriptors we have used to characterize the assessments of sustainability of each evaluator (Figures 7.14 and 7.15 and Table 7.10) have enabled all the evaluators to be classified into 53 types (Section 7.6.3). The personal characteristics of each evaluator that we have used in our case study (Section 7.6.1 and Figure 7.9) have caused the evaluators to be clustered into another 220 possible types.* Consequently, the number of potential classes of evaluators will be $n = 53 \times 220 = 11,660$. Under these conditions, if we applied the methodology described in Section 3.3, it would then be necessary to process 11,660 matrices of the euf_A type (each with 11,660 rows and 60 columns). This information is easy to process with the computational resources available today. In consequence, the proposed aggregation leads to an operational procedure with the cluster of evaluators described in footnote † on page 542.

9.4 SELF-ORGANIZATION IN COLLECTIVE DECISION-MAKING

9.4.1 Collective Decision-Making Simulation in Noncompetitive Emergent Systems

Next, we introduce an agent-based simulation to build a joint assessment for multi-dimensional problems, which has to be acceptable to multiple decision-makers. The simulation must allow changes in each agent's assessment in view of other evaluators' opinions. Thus, the model facilitates the convergence of individual utilities toward the joint utility of the group. In the simulation, each evaluator has his or her personal utility that is different from the joint utility of the group, as well as a certain amount of memory in which the history of the group discussion is stored and used to form a dynamic mental model for all the agents.

In order to typify the problem, let us suppose a group of n individuals that is taking part in a participatory process with the following characteristics (Sayama et al., 2010):

- The members of the group (evaluators) seek to achieve a joint assessment of an *m*-dimensional problem (each dimension is called an *aspect* of the problem).
- An *individual assessment* is a set of choices (a *choice* is a value in each dimension of the problem—aspect), which is made by the agent according to

---

* This number was obtained by reclassifying the qualitative variables that describe the characteristics of each evaluator into the types shown in Figure 7.15. The existence of relationships between educational level and occupation and between place of residence and type of stakeholder has also been assumed. Thus, four types have been used for the variables gender–age class, eleven for academic level–occupation, and five for place of residence–type of stakeholder. Total, $4 \times 11 \times 5 = 220$ different types of evaluators.

his or her personal opinion. Individual assessments are specific to each agent (A) and are recognized by the m choices that A makes ($u_A = (u_{1A}, u_{2A}, ..., u_{mA})$).

- Likewise, we can speak of the *current group assessment* (u_g) and of the *final group assessment* (u_{gF}).

- A *plan* is a set of choices for all aspects of the problem, independently of both the type of agent making the choice and the moment of making it. It is a set of coordinates (i.e., a vector) in the m-dimensional problem space. Thus, personal and group assessments are plans. We also assume that the range of variation of choices is the interval [0.1].

- We assume that once a plan is defined (been u_A or u_g), then the value of its *utility* (v_A or v_g) for the agent, or the group, can also be computed. The pair of an assessment and its utility is called an *opinion* ($o_A = (u_A, v_A)$).

- The agents do not know the final aggregated assessment (u_{gF}). It is obtained by aggregating the individual assessments at the end of the simulation of the discussion process. However, the current aggregated value (u_g) is known (it is calculated by applying the aggregation tools discussed in Sections 3.1 through 3.3 of this chapter to the current individual assessment of the group). The main goal of the group is to find a joint assessment, which would have to be accepted by all the agents of the group at the end of the participatory process.

- Each individual agent has a memory in which his or her personal comprehension of the problem and the history of the group discussion are stored. The memory is defined as a list of opinions either held by the agent or expressed by other agents during the discussion. The memory of each agent (M_A) can store up to p self-opinions (his or her initial opinion plus the modifications he or she has accepted during the discussion) and up to q opinions expressed by each of the other agents ($M_A = \{o_{A,k}\}, k = 1, ..., p, p+1, ..., p+q$). The total memory of an agent is limited to a certain number of opinions; thus, $p + q \leq l$. If the number of opinions involved in the personal understanding of the problem exceeds this number, then some of the opinions in the memory must be removed.

Evaluators redefine their assessments from the set of opinions in their memory. Based on these, they modify their choices, seeking to improve the utility of their individual plans. The simulation progresses through a set of iterative steps. In each step, a randomly selected speaker analyzes which aspect of his or her individual plan has the most significant impact if incorporated into the group plan and suggests modifying this aspect. The suggestion is shared with other agents, who respond to it based on their respective individual utility. Individual responses determine the revision of the group plan and its expected utility. This cycle is repeated for a fixed number of iterations.

9.4.2   APPLICATION TO PARTICIPATIVE SUSTAINABILITY ASSESSMENT

In order to apply the earlier methodology to the assessment of forest sustainability, we shall consider—as we did in Sections 3.2 and 3.3—that each individual *plan* (for evaluator A) is the evaluator's individual assessment of sustainability in the six most significant spatial points of the forest under analysis ($\boldsymbol{u}_A = (u_{1A}, u_{2A}, \ldots, u_{6A})$) and that an *opinion* (also for evaluator A) is the pair $o_A = (\boldsymbol{u}_A, v_A)$, where v_A is the average sustainability for all the points in the forest. Thus, an *aspect* of a *plan* is the sustainability assessments for each of the most significant points in the forest under analysis. We shall also denote the objective and the aggregated opinions on sustainability $o_{Ob} = (\boldsymbol{u}_{Ob}, v_{Ob})$ and $o_g = (\boldsymbol{u}_g, v_g)$, respectively (computed as described in Section 3.3). These three opinions (o_A, o_{Ob}, and o_g) always belong to the memory of each evaluator.

As mentioned earlier, self-organization requires discussion: each individual evaluator seeks to impose his or her own individual plan on the group plan, but at the same time, the evaluator accepts changes in his or her individual plan if convinced. The discussion is an iterative process with the following phases in each iteration:

1. Each evaluator examines whether there is an easy way to improve the utility of his or her current plan.

   Each evaluator examines his or her neighbors' plan and incorporates into his or her memory the *opinion* with the highest utility for all the evaluators belonging to his or her neighborhood. An evaluator's neighbors are those evaluators with similar personal characteristics (age, gender, educational level and occupation, place of residence, and type of stakeholder). We shall denote this opinion as $o_{n(A)} = (\boldsymbol{u}_{n(A)}, v_{n(A)})$.

2. A speaker is selected at random.

   One of the 220 types of personal characteristics (see footnote * on page 542) is randomly selected. The speaker is then randomly selected from the evaluators belonging to the selected type of personal characteristics. We shall denote this speaker as S.

3. The speaker makes a suggestion for the revision of the group plan.

   The speaker (S) identifies which aspect of his or her individual plan has the most significant impact if incorporated into the group plan. This is done by returning to the concepts introduced in Section 3.3, specifically to the $\mathbf{euf}_S$ matrix, which is now computed with the speaker's utility, taking into account the opinions in the memory of all the other evaluators ($U_{S \to A}, \forall A$-evaluators):

$$U_{S \to A}(\boldsymbol{x}) = \sum_{i \in S \cup M_A} \left(\frac{w_i(\boldsymbol{x})}{\sum_{i \in S \cup M_A} w_i(\boldsymbol{x})} v_i^* \right) \qquad (9.27)$$

where
   $w_i(\boldsymbol{x})$ is as in Expression (9.22)
   S refers to the speaker's assessments
   M_A is the memory of evaluator A
   v_i^* is not merely the average sustainability for all the points in the forest (v_i)

It refers to the objective opinion as

$$v_i^* = 1 - |v_i - v_{\mathrm{Ob}}|$$ (9.28)

The coordinates of the u_g vector are then systematically modified to calculate the **euf$_s$** matrix (as in Expression (9.23)). The column in this matrix with the highest sum of its components (j) determines the suggestion of change proposed by the speaker (the column of an *euf* determines both the aspect to modify—k—and the new value of the modification—d).

The new opinion the speaker suggests to the group is denoted as $o_{g\langle j\rangle} = (u_{g\langle j\rangle}, v_{g\langle j\rangle})$.

As the number of evaluators grows, individual analyses are substituted by analyzing the average evaluator in each one of the 220 groups of the different types of personal characteristics.

4. The speaker's suggestion is evaluated by the other evaluators who respond to the suggestion at the individual level.

Each evaluator then studies the utility of the speaker's suggestion, which is done by applying Expression (9.22) to the opinions belonging to their memory:

$$U_A\left(o_{g\langle j\rangle}\right) = \sum_{p \in M_A}\left(\frac{w_p(x)}{\displaystyle\sum_{p \in M_A} w_p(x)} v_p\right)$$ (9.29)

where $w_p(x) = \|u_{g\langle j\rangle} - x\|^{-2}$ and $M_A = \{o_A\}$ is the memory of A.

If $U_A(o_{g\langle j\rangle}) > U_A(o_g)$, then evaluator A expresses support for the suggestion, and $o_{g\langle j\rangle}$ is incorporated into M_A. Otherwise, that is, $U_A(o_{g\langle j\rangle}) \leq U_A(o_g)$, then A's individual plan is not affected by the speaker's suggestion. However, A may still express support for the suggestion with some probability given by

$$P\left(o_{g\langle j\rangle}, o_g\right) = \exp\left(\frac{\left\|-U_{A\left(og\langle j\rangle\right)} - \ddot{}U_{A(og)}\right\|}{T}\right)$$ (9.30)

where T is the "temperature" of the agent's cognition (i.e., how much agents can tolerate low-utility suggestions).

For operative reasons, the total number of opinions in the memory of each evaluator (A) is limited to four: his or her own opinion (o_A), the objective opinion (o_{Ob}), the opinion with the highest utility among his or her neighbors ($o_{n(A)}$), and the last group opinion accepted by the evaluator ($o_{g(A)}$). This means that in the case of accepting the new group opinion ($o_{g\langle j\rangle}$), the previous one ($o_{g(A)}$) has to be removed from the memory.

5. Response to suggestion at group level.

The group determines whether or not the speaker's suggestion should be adopted as the new group plan by counting the number of members of the group incorporating the suggestion into their memories. If more than 50% of the evaluators include the suggestion, then the previous group opinion is replaced with the suggestion ($o_g \leftarrow o_{g(j)}$).

In order to determine the influence of self-organization, we next simulate a participatory process for the assessment of forest sustainability, and we then compute— and compare—the aggregate assessment of sustainability in two scenarios. The first one refers to a joint assessment by the simple aggregation of individual preferences obtained from a computer simulation of the preferences of 100 evaluators. The second scenario incorporates self-organization through the previously described agent-based simulation for the aggregation of the same sustainable preferences as in the first case.

9.4.2.1 Simulation of a Participatory Process that Solely Considers Personal Assessments of Sustainability

We have made different pair-wise comparisons of sustainability on the most significant territorial points of a forest, simulating the responses that could have been given by 100 evaluators*. As the number of evaluators is small, the only personal characteristics used refer to the agent typology (forestry expert, forest landowner or entrepreneur, local population affected by forestry, environmental activist, and unspecified evaluator via the Internet): the possible variability of responses considering all the analyzed personal characteristics (gender, age, occupation, etc.) cannot be represented with 100 evaluations of sustainability.

Pair-wise comparisons of territorial points were transformed into individual assessments of sustainability[†] by applying the methodology developed in Chapter 7. Figure 9.3 shows the sustainability assessments clustered by agent typology for all the evaluators.

The aggregated assessment shown in Figure 9.3 is the average of the individual assessments of sustainability for all the simulated evaluators. We have used this system of aggregation because it is the simplest. In general, a different aggregate assessment is achieved by applying other aggregation systems, but this would not affect our objective, namely, to see how self-organization modifies the aggregation of preferences (in our case, how the average assessment of sustainability is changed).

---

* Pair-wise comparisons come from class exercises conducted by students in the Master's program "Desarrollo Rural y Gestión Sostenible" at the UPM (Technical University of Madrid). Most comparisons were made by the students themselves (about 2/3 of the respondents) assuming the roles of different types of agents, while the rest come from surveys of real agents conducted by the same students.

[†] The reader should remember that an individual assessment of sustainability (a plan) is a vector whose components are the sustainability assessments of the individual at each of the significant points of the territory that are used in the pair-wise comparisons of sustainability.

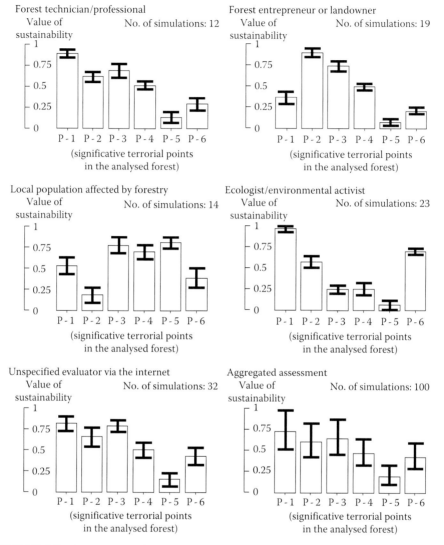

FIGURE 9.3 Assessments of sustainability in five groups for 100 simulated evaluators and aggregated assessment for all the evaluators. Top ends of bars indicate observation means for the group. Line segments represent confidence intervals $(1 - \alpha = 95\%)$ for the individual assessments.

9.4.2.2 Simulation of a Participatory Process Taking into Account the Self-Organization Process

Now, we apply the methodology described earlier in this section to trigger a self-organization process on the data represented in Figure 9.3. As the number of evaluators is small, we have used the individual assessments of sustainability throughout the different steps in the process (not the assessment of the average evaluator in each of the five groups corresponding to the different types of agents).

We have also assumed that evaluators do not usually tolerate low-utility suggestions, and we have thus adopted a low temperature for Expression (9.30). This way, convergence of opinions is not overly facilitated.

As the measure of the convergence of opinions—belonging to the self-organization—we use the distance between the last group opinion accepted by each evaluator into his or her memory ($o_{g(A)} = (\boldsymbol{u}_{g(A)}, v_{g(A)})$) and the current aggregated group opinion ($o_g = (\boldsymbol{u}_g, v_g)$). Both opinions change in each iteration of the process (i). Thus, we will rewrite them as $o_{g(A)-i}$ and o_{g-i}, respectively. So

$$CoO_i = \frac{1}{M} \sum_{A=1}^{M} \left\| \boldsymbol{u}_{g(A)-i} - \boldsymbol{u}_{g-i} \right\| \tag{9.31}$$

where
  CoO_i is the measure of the convergence of opinions in the i iteration
  M is the total number of evaluators ($M = 100$)
  The other variables are described earlier

The initial group assessment corresponds to the average assessment of sustainability shown in Figure 9.3: $\boldsymbol{u}_{g-1} = (0.733, 0.615, 0.649, 0.472, 0.205, 0.429)$. The last group opinion accepted by each evaluator in the first iteration is his or her individual opinion ($\boldsymbol{u}_{g(A)-1} = \boldsymbol{u}_A$). Figure 9.4 shows the evolution in the convergence of opinions against the number of iterations in the process. This shows that for the number of iterations performed, the maximum convergence of opinions corresponds to iteration number 198, and $\boldsymbol{u}_{g-198} = (0.92, 0.71, 0.526, 0.472, 0.205, 0)$. This is the new aggregated assessment adopted after this short self-organization process.

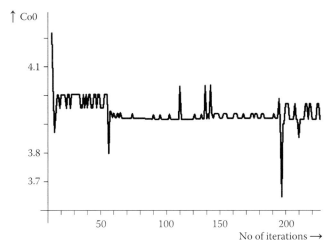

FIGURE 9.4 Convergence of aggregated opinion through a self-organization process.

REFERENCES

Ailon, N., M. Charikar, and A. Newman. 2005. Aggregating inconsistent information: Ranking and clustering. In *Proceedings of the Thirty-Seventh Annual ACM Symposium on Theory of Computing STOC-2005*. pp. 684–693. ACM, New York.

Alon, N. 2006. Ranking tournaments. *SIAM Journal of Discrete Mathematics*, 20(1): 137–142.

Arrow, K.J. 1950. A difficulty in the concept of social welfare. *Journal of Political Economy*, (August) 58(4): 328–346

Arrow, K.J. 1963. *Social Choice and Individual Values*, 2nd edn. John Wiley & Sons, New York.

Arrow, K. 1978. Extended sympathy and the problem of social choice. *Philosophia*, 7: 233–237.

Baek, S.K. and S. Bernhardsson. 2010. Equilibrium solution to the lowest unique positive integer game. *Fluctuation and Noise Letters*, 9(1): 61–68.

Banks, J. 1985. Sophisticated voting outcomes and agenda control. *Social Choice and Welfare*, 1(4): 295–306.

Baron-Cohen, S. 1995. *Mindblindness: An Essay on Autism and Theory of Mind*. MIT Press: Cambridge, MA.

Bartholdi, J., C. Tovey, and M. Trick. 1989. Voting schemes for which it can be difficult to tell who won the election. *Social Choice and Welfare*, 6(3):157–165.

Bentham, J. 1789. *An Introduction to the Principles of Moral and Legislation*. Payne, London, U.K. Republished, Clarendon Press, Oxford, 1907.

Bernoulli, N. 1713. Correspondence of Nicolas Bernoulli to Montmort, Basel, Sept. 9, 1713, *Essay d'Analysis*, 402, Richard J. Pulskamp. 1999.

Bernoulli, D. 1738. Paper presented at a gathering of mathematicians in St. Petersburg, Russia and translated by Dr. Louise Sommer. (January 1954). "Exposition of a New Theory on the Measurement of Risk". *Econometrica*, 22(1): 22–36

Bickel, J.E. 2007. Some comparisons among quadratic, spherical, and logarithmic scoring rules. *Decision Analysis*, 4(2): 49–65.

Binmore, K. 1998. *Just Playing: Game Theory and the Social Contract II*. MIT Press, Cambridge, MA.

Binmore, K. 2005. *Natural Justice*. Oxford University Press, New York.

Bonabeau, E. 2009. Decisions 2.0: The power of collective intelligence. *MIT Sloan Management Review*, 50(2): 45–52.

Borda, J.C. 1784. Mémoire sur les élections au scrutin. *Histoire de l'Académie Royale des Sciences*. Année 1781. pp. 657–665. Paris, France.

Braha, D. and Y. Bar-Yam. 2004. Topology of large-scale engineering problem-solving networks. *Physical Review E*, 69(1): 016113_1–016113_7.

Braha, D. and Y. Bar-Yam. 2007. The statistical mechanics of complex product development: Empirical and analytical results. *Management Science*, 53(7): 1127–1145.

Brams, S. 1975. *Game Theory and Politics*. Free Press, New York.

Brams, S. and P. Fishburn. 1978. Approval voting. *American Political Science Review*, 72(3): 831–847.

Brams, S. and P. Fishburn. 1991. Alternative voting systems. In *Political Parties and Elections in the United States: An Encyclopedia*, Vol. 1. Ed. L. Sandy Maisel. pp. 23–31. Garland, New York.

Brodbeck, F.C., R. Kerschreiter, A. Mojzisch, and S. Schulz-Hardt. 2007. Group decision making under conditions of distributed knowledge: The information asymmetries model. *Academy of Management Review*, 32(2): 459–479.

Charon, I. and O. Hudry. 2000. Slater orders and Hamiltonian paths of tournaments. *Electronic Notes in Discrete Mathematics*, 5: 60–63.

Chevaleyre, Y., U. Endriss, J. Lang, and N. Maudet. 2005. A short introduction to computational social choice. 2007. In *Proceedings of the 33rd, Conference on Current Trends in Theory and Practice of Computer Science, (SOFSEM-2007)*, LNCS, 4362. pp. 51–69. Springer-Verlag, Berlin, Germany.

Condorcet, M. 1785. *Essai sur l'application de l'analyse 'a la probabilité des décisius rendues á la pluralité des voix*. Paris, France

Conitzer, V. 2006. Computing slater rankings using similarities among candidates. In *Proceedings of the 21st National Conference on Artificial Intelligence AAAI-2006*, 1. pp. 613–619. Boston, MA, AAAI Press.

Conitzer, V., A. Davenport, and J. Kalagnanam. 2006. Improved bounds for computing Kemeny rankings. In *Proceedings of the 21st National Conference on Artificial Intelligence (AAAI)*. pp. 620–626. AAAI-Press, Boston, MA.

Cooper, D.A. 2007. The potential of cumulative voting to yield fair representation. *Journal of Theoretical Politics*, 19(3): 277–295.

Curty, P. and M. Marsili. 2006. Phase coexistence in a forecasting game. *Journal of Statistical Mechanics: Theory and Experiment*, 2006: P03013.

Davenport, A. and J. Kalagnanam. 2004. A computational study of the Kemeny rule for preference aggregation. In *Proceedings of the 19th National Conference on Artificial Intelligence (AAAI)*, 2004. pp. 697–702. AAAI-Press, Boston, MA.

Davidsen, J., H. Ebel, and S. Bornholdt. 2002. Emergence of a small world from local interactions: Modeling acquaintance networks. *Physical Review Letters*, 88(12): 128701-1–128701-4.

Edgeworth, F.J. 1881. *Mathematical Psychics: An Easy on the Application of Mathematics to Moral Sciences*. Kegal Paul, London, U.K.

Fischer, G. 1999. A group has no head: Conceptual frameworks and systems for supporting social interaction. *Information Processing Society of Japan (IPSJ) Magazine*, 40(6): 575–582.

Frey, B.J. 1998. *Graphical Models for Machine Learning and Digital Communication*. MIT Press, Cambridge, MA.

Garey, M.R. and D.S. Johnson. 1979. *Computers and Intractability: A Guide to the Theory of NP-Completeness*. Freeman Press, San Francisco, CA.

Grönlund, A., P. Holme, and P. Minnhagen. 2008. Dynamic scaling regimes of collective decision making. *Europhysics Letters*, 81(2): 28003.

Grossman, S.J. 1976. On the efficiency of competitive stock markets where traders have diverse information. *The Journal of Finance*, 31(2): 573–585.

Hammond, P.J. 1976. Equity, arrow's conditions, and Rawls' difference principle. *Econometrica*, 44: 793–804.

Hammond, P.J. 1979. Equity in two person situations: Some consequences. *Econometrica*, 47: 1127–1135.

Hammond, P.J. 1991. Interpersonal comparisons of utility: Why and how they are and should be made. In *Interpersonal Comparisons of Well-being*, eds. J. Elster and J. E. Roemer. Cambridge University Press, Cambridge, New York.

Hanson, R. 2002. Eliciting objective probabilities via lottery insurance games. Tech. rep. George Mason University Economics. http://hanson.gmu.edu/elicit.pdf (accessed September 19, 2012)

Hanson, R. 2003. Combinatorial information market design. *Information Systems Frontiers*, 5(1): 107–119.

Hanson, R. 2007. Logarithmic market scoring rules for modular combinatorial information aggregation. *The Journal of Prediction Markets*, 1(1): 3–15.

Harich, J. 2010. Change resistance as the crux of the environmental sustainability problem. *System Dynamics Review*, 26(1):35–72, doi: 10.1002/sdr.431.

Harsanyi, J.C. 1953. Cardinal utility in welfare economics and in the theory of risk-taking. *Journal of Political Economy*, 61: 434–435.

Harsanyi, J.C. 1955. Cardinal welfare, individualistic ethics, and interpersonal comparisons of utility. *Journal of Political Economy*, 63: 309–321.

Harsanyi, J. 1977. *Rational Behavior and Bargaining Equilibrium in Games and Social Situations*. Cambridge University Press, Cambridge, New York.

Harsanyi, J. 1992. Normative validity and meaning of Von Neumann and Morgenstern utilities. In *Studies in Logic and the Foundations of Game Theory: Proceedings of the Ninth International Congress of Logic, Methodology and the Philosophy of Science*, ed. B. Skyrms. Kluwer Academic Press, Dordrecht, the Netherlands.

Hausch, D.B., V.S. Lo, and W.T. Ziemba. 1994. *Efficiency of Racetrack Betting Markets*. Academic Press, San Diego, CA.

Hemaspaandra, E., L.A. Hemaspaandra, and J. Rothe. 1997. Exact analysis of Dodgson elections: Lewis Carroll's 1876 system is complete for parallel access to NP. *JACM*, 44(6): 806–825.

Hemaspaandra, E., H. Spakowsk, and J. Vogel. 2005. The complexity of Kemeny elections. *Theoretical Computer Science*, 349(3): 382–391.

Hertz, J., A. Krogh, and R. G. Palmer. 1991. *Introduction to the Theory of Neural Computation*. Addison-Wesley, Redwood City, CA.

Heylighen, F. 1997. Evolution and complexity: An introduction to the book. In *The Evolution of Complexity*, ed. F. Heylighen. Kluwer Academic Press, Dordrecht, the Netherlands.

Heylighen, F. 1999. Collective intelligence and its implementation on the web: Algorithms to develop a collective mental map. *Computational and Mathematical Organization Theory*, 5(3): 253–280.

Heylighen, F. and D.T. Campbell. 1995. Selection of organization at the social level: Obstacles and facilitators of metasystem transitions. *World Futures: The Journal of General Evolution*, 45: 181–212.

Heylighen, F. 2001. The science of self-organization and adaptivity. In *Knowledge Management, Organizational Intelligence and Learning, and Complexity*, ed. L. D. Kiel. In *The Encyclopedia of Life Support Systems*. EOLSS Publishers, Oxford, U.K.

Hill, M.O. 1973. Reciprocal averaging: An eigenvector method of ordination. *Journal of Ecology*, 61: 237–251.

Hochbaum, D. 1995. *Approximation Algorithms for NP-Hard Problems*. PSW Publishing Company, Boston, MA.

Holling, C.S. 1973. Resilience and stability of ecology systems. *Annual Review of Ecology and Systematics*, 4: 1–23.

Holme, P. and M.E.J. Newman. 2006. Nonequilibrium phase transition in the coevolution of networks and opinions. *Physical Review E*, 74(5):056108-1–056108-5.

Hudry, O. 2004. A note on Banks winners in tournaments are difficult to recognize by G. J. Woeginger. *Social Choice and Welfare*, 23(1): 113–114.

Hudry, O. 2009. A survey on the complexity of tournament solutions. *Mathematical Social Sciences*, 57(3): 292–303.

Hudry, O. 2010. On the complexity of Slater's problems. *European Journal of Operational Research*, 203(1): 216–221.

Hume, D. 1739. *A Treatise of Human Nature*, 2nd edn. 1978, ed. L. A. Selby-Bigge. Clarendon Press, Oxford, U.K.

Ikegami, T. and G. Morimoto. 2003. Chaotic itinerancy in coupled dynamical recognizers. *Chaos: Interdisciplinary Journal of Nonlinear Science*, 13(3): 1133–1147.

Kellett, J. and K. Mott. 1977. Presidential primaries: Measuring popular choice. *Polity*, 9(4): 528–537.

Kelly, J.S. 1987. *Social Choice Theory: An Introduction*. Springer-Verlag, New York.

Kleene, S.C. 1956. Representation of events in nerve nets and finite automata. *Automata Studies*. pp. 3–41. Princeton University Press, Princeton, NJ.

Klein, M., H. Sayama, P. Faratin, and Y. Bar-Yam. 2003. The dynamics of collaborative design: Insights from complex systems and negotiation research. *Concurrent Engineering: Research and Applications*, 11(3): 201–209.

Kozlowski, S.W.J. and D.R. Ilgen. 2006. Enhancing the effectiveness of work groups and teams. *Psychological Science in the Public Interest*, 7(3): 77–124.

Krahnen, J. P. and M. Weber. 1999. Does information aggregation depend on market structure? Market makers vs. double auction. *Zeitschrift für Wirtschafts- und Sozialwissenschaften*, 119: 1–22.

Laukkanen, S., A. Kangas, and J. Kangas. 2002. Applying voting theory in natural resource management: A case of multiple-criteria group decision support. *Journal of Environmental Management*, 64(2): 127–137.

LLull, R. 1299. *De arte electionis*. Manuscrito, Paris, France.

Lo, A.W. 1997. *Market Efficiency: Stock Market Behaviour in Theory and Practice*. Elgar, Lyme, NH.

Mahalanobis, P.Ch. 1936. On the generalised distance in statistics. *Proceedings of the National Institute of Sciences of India*, 2(1): 49–55.

Manski, C. 2004. Interpreting the predictions of prediction markets, *NBER Working Paper* 10359. National Bureau of Economic Research, Inc. http://www.nber.org/papers/w10359 (accessed September 19, 2012).

Marks, M., S.J. Zaccaro, and J.E. Mathieu. 2000. Performance implications of leader briefings and team-interaction training for team adaptation to novel environments. *Journal of Applied Physiology*, 85(6): 971–986.

Marshall, A. 1890. *Principles of Economics*, 9th edn. Macmillan. London, U.K., 1961.

Martínez-Panero, M. 2006. Métodos de votación híbridos bajo preferencias ordinarias y difusas. *Anales de Estudios Económicos y Empresariales*, 16: 187–219.

Mathieu, J.E., T.S. Heffner, G.F. Goodwin, E. Salas, and J.A. Cannon-Bowers. 2000. The influence of shared mental models on team process and performance. *Journal of Applied Physiology*, 85(2): 273–283.

Minnhagen, P., S. Bernhardsson, and B.J. Kim. 2007. Phase diagram of generalized fully frustrated XY model in two dimensions. *Physical Review B*, 76(22):224403-1–224403-5.

Mohammed, S. and B.C. Dumville. 2001. Team mental models in a team knowledge framework: Expanding theory and measurement across disciplinary boundaries. *Journal of Organizational Behavior*, 22(2): 89–106.

Myers, D.G. 2002. *Intuition: Its Powers and Perils*. New Haven, CT: Yale University Press.

Othman, A. and T. Sandholm. 2010. Automated market making in the large: The Gates Hillman prediction market. In: *Proceedings of the 11th ACM Conference on Electronic Commerce (EC 2010)*. pp. 367–376. Association for Computing Machinery, New York.

Othman, A., T. Sandholm, D.M. Pennock and D.M. Reeves. 2010. A practical liquidity-sensitive automated market maker. In *Proceedings of the 11th ACM Conference on Electronic Commerce (EC 2010)*. pp. 377–386. Association for Computing Machinery, New York.

Ottewell, G. 1977. The Arithmetic of Voting. *In Defense of Variety*, 4: 42–44.

Paine, T. 1776. Common sense. American colonies, Archiving Early America. http://www.earlyamerica.com/earlyamerica/milestones/commonsense/title.html (accessed September 19, 2012).

Pennock, D. and R. Sami. 2007. Computational aspects of prediction markets. In *Algorithmic Game Theory*, eds. N. Nisan, T. Roughgarden, E. Tardos, and V. Vazirani. pp. 651–647. Cambridge University Press, Cambridge, U.K.

Pynadath, D.V. and S.C. Marsella. 2005. PsychSim: Modeling theory of mind with decision-theoretic agents. In: *Proceedings of the 2005 International Joint Conference on Artificial Intelligence*. pp. 1181–1186. Morgan Kaufmann: San Francisco, CA.

Rawls, J. 1971. *A Theory of Justice*. Oxford University Press, Oxford, U.K.

Robbins, L. 1938. Interpersonal comparison of utility: A comment. *Economic Journal*, 47(2): 635–641.

Roby, T.B. 1965. Belief states: A preliminary empirical study. *Behavioral Science*, 10(3): 255–270.

Rodriguez, M. 2004. Advances towards a societal-scale decision-making system, UCSC Masters Thesis, University of California, Santa Cruz, CA.

Rodriguez, M.A. and Steinbock, D.J. 2004a. A social network for societal-scale decision-making systems. In *Proceedings of the North American Association for Computational Social and Organizational Science Conference*, Pittsburgh, PA.

Rodriguez, M. and D. Steinbock. 2004b. Dynamically distributed democracy: A social network for decision-making, *Human Complex Systems Conference*. California, LA.

Rodriguez, M.A., D.J. Steinbock, J.H. Watkins, C. Gershenson, J. Bollen, V. Grey, and B. deGraf. 2007. Smartocracy: Social networks for collective decision making, In *Proceedings 40th Hawaii International Conference on Systems Science (HICSS-40 2007)*, January 3–6, 2007, pp. 90–100. IEEE Computer Society, Big Island, HI.

Rodriguez, M.A. and J.H. Watkins. 2007. Distributed collective decision making: From ballot to market. In *Discovery Workshop: Applying Complexity Science to Organizational Design and Multistakeholder Systems*. National Alliance for Physician Competence, Chicago, IL.

Rodriguez, M.A. and J.H. Watkins. 2009. Revisiting the age of enlightenment from a collective decision making systems perspective. First Monday LA-UR-09–00324. *University of Illinois at Chicago Library*, 14(8): 1–7.

Rodriguez, M.A., J. Bollen, and H. Van de Sompel. 2009. Automatic metadata generation using associative networks. *ACM Transactions on Information Systems*, 27(2): 1–20.

Rothe, J., H. Spakowski, and J. Vogel. 2003. Exact complexity of the winner for Young elections. *Theory of Computing Systems*, 36(4): 375–386.

Rovarini, P. and M. Cerviño. 2011. Fuzzy knowledge base optimization in decision making systems II. *Scientia Interfluvius*, 2(1): 66–77.

Rovarini, P., M. Cerviño, and G. Juárez. 2009. Globalización: Necesidad de Toma de Decisiones Colectivas. *Primer Congreso Boliviano de Ingeniería y Tecnología* IEEEUMSE, Universidad Mayor de San Andrés, La Paz, Bolivia.

Savage, L.J. 1971. Elicitation of personal probabilities and expectations. *Journal of the American Statistical Association*, 66(336): 783–801.

Sayama, H., D. Farrell, and S.D. Dionne. 2010. The effects of mental model formation on group decision making: An agent-based simulation. *Complexity*, 16(3): 49–57.

Sen, A.K. 1970. *Collective Choice and Social Welfare*. Holden Day, San Francisco, CA.

Sen, A.K. 1973. *On Economic Inequality*. Clarendon Press, Oxford, U.K.

Sen, A.K. 1977. On weights and measures: Informational constraints in social welfare analysis. *Econometrica*, 45: 1539–1572; reprinted in Sen (1982).

Sen, A.K. 1982. *Choice, Welfare and Measurement*. Basil Blackwell, Oxford, and MIT Press, Cambridge, MA.

Sen, A.K. 1998. The possibility of social choice. *Economic Sciences*, 178–206. (Nobel lecture at Trinity College, Cambridge, CB2 1TQ, Great Britain on December 8, 1998).

Shuford, E.H., Jr., A. Albert, and H.E. Massengill. 1966. Admissible probability measurement procedures. *Psychometrika*, 31(2): 125–145.

Smith, C.L., V.L. Lopes, and F.M. Carrejo 2011 Recasting paradigm shift: "True " sustainability and complex systems. *Human Ecology Review*, 18(1): 67–74.

Stensholt, E. 2011. Voces populi and the art of listening. *Social Choice and Welfare*, 35(2): 291–317.

Strasnick, S. 1976a. Social choice and the derivation of Rawls's difference principle. *Journal of Philosophy*, 73: 85–99.

Strasnick, S. 1976b. The problem of social choice: Arrow to Rawls. *Philosophy and Public Affairs*, 5: 241–273.

Strasnick, S. 1977. Ordinality and the spirit of the justified dictator. *Social Research*, 44: 668–690.

Strasnick, S. 1979. Extended sympathy comparisons and the basis of social choice. *Theory and Decision*, 10: 311–328.

Suppes, P. 1966. Some formal models of grading principles. *Synthese*, 6:284–306.

Surowiecki, J. 2005. *The Wisdom of Crowds*. Anchor Books, New York.

Takano, M. and T. Arita. 2006. Asymmetry between even and odd levels of recursion in a theory of mind. In: *Artificial Life X: Proceedings of the Tenth International Conference on the Simulation and Synthesis of Living Systems*. pp. 405–411. MIT Press, Cambridge, MA.

Tideman, N. 1995. The single transferable vote. *The Journal of Economic Perspectives*, 9(1): 27–38.

Tideman, N. and D. Richardson. 2000. Better voting methods through technology: The refinement-manageability trade-off in the single transferable vote. *Public Choice*, 103: 13–34.

Toda, M. 1963. Measurement of subjective probability distributions. *Tech Doc Rep U S Air Force Systems Command Electronic Systems Division*, 86: 1–42.

Turoff, M., S.R. Hiltz, H.-K.Cho, Z. Li, and Y. Wang. 2002. Social decision support systems (SDSS). *Proceedings of the 35th Hawaii International Conference on System Sciences*. pp. 81–90. IEEE Computer Society, Waikoloa, HI.

Vainikainena, N., A. Kangas, and J. Kangas. 2008. Empirical study on voting power in participatory forest planning. *Journal of Environmental Management*, 88(1): 173–180.

Viso, E. 2008. *Introducción a la teoría de la computación*. UNAM. Fac. de Ciencias Autómatas y lenguajes formales.

Von Neumann, J. and O. Morgenstern. 1944. *The Theory of Games and Economic Behavior*. Princeton University Press, Princeton, NJ.

Watkins, J.H. and M.A. Rodriguez. 2008. A survey of web-based collective decision making systems. In *Studies in Computational Intelligence: Evolution of the Web in Artificial Intelligence Environments*, eds. R. Nayak, N. Ichalkaranje, and L.C. Jain. pp. 245–279. Springer-Verlag, Berlin, Germany.

Watt, S.N.K. 1997. Artificial societies and psychological agents. In: *Software Agents and Soft Computing towards Enhancing Machine Intelligence*. pp. 27–41. Springer, Berlin, Germany.

Watts, D.J. 2003. *Small Worlds: The Dynamics of Networks between Order and Randomness*. Princeton University Press, Princeton, NJ.

Weber, R.J. 1977. Comparison of voting systems. Cowles Foundation Discussion Paper No. 498. Yale University, New Haven, CT. http://cowles.econ.yale.edu/P/cd/d04b/d0498.pdf (accessed October 15, 2012).

Winkler, R.L. 1996. Scoring rules and the evaluation of probabilities. *Test*, 5(1): 1–60.

Winkler, R.L. and A.H. Murphy. 1968. "Good" probability assessors. *Journal of Applied Meteorology*, 7(5): 751–758.

Woeginger, G.J. 2003. Banks winners in tournaments are difficult to recognize. *Social Choice and Welfare*, 20(3): 523–528.

Wolfers, J. and E. Zitzewitz. 2006. Interpreting prediction market prices as probabilities. NBER Working Paper 12200. National Bureau of Economic Research, Inc. http://www.nber.org/papers/w12200.pdf (accessed September 19, 2012).

Wolfers, J. and Eric Zitzewitz. 2008. Prediction markets in theory and practice. In *The New Palgrave Dictionary of Economics*, 2nd edn. eds. L. Blume and S. Durlauf. London, U.K.: Palgrave. http://www.nber.org/papers/w12083 (accessed September 19, 2012).

Zhao, A. and A.J. Brehm. 2011. Cumulative voting and the conflicts between board and minority shareholders. *Managerial Finance*, 37(5): 465–473.

10 A Computer-Based Decision-Making Support System to Incorporate Personal Preferences in Forest Management

Eugenio Martínez-Falero,
Antonio García-Abril, Carlos García-Angulo,
and Susana Martín-Fernández

CONTENTS

10.1 INTRODUCTION

In view of the wide variety of products and services provided by forests and the complexity of the groups that benefit from them, it is necessary to develop a representative support system that provides a global solution to participatory forest management based on the integration of individual preferences on all criteria (including products and services).

To make silvicultural management work for the benefit of the population, we should clearly recognize the multiple goals of forestry. Since the relative importance people give to these goals varies according to individual preferences and over time, we need to incorporate a dynamic and participative—rather than a static—decision-making system. Thus, the developments in social networks and social communications in recent years may provide a new framework for public participation in forest management which fulfills the criteria of being both representative and changing as people change their minds. Furthermore, as stated in Chapter 9, collective web-based intelligence mitigates many of the biases that occur in making individual decisions (Myers 2002).

One of the main problems of participatory forest management is the burden of constant voting and its consequences (time lost by voters, logistical problems, costs, failure in voting, etc.); in fact, conventional participation techniques have not proved to be effective enough (Pykäläinen et al. 1999). Web-based collective decision-making systems can also reduce many of the problems associated with traditional voting techniques.

The ultimate purpose of participative decision-making systems is to increase the influence of public opinion in the management process, and at the same time to exchange information among evaluators, experts, and the public, thus increasing public knowledge of the evaluation terms and criteria and providing the experts with appropriate feedback about the methodological and technical aspects of the particular decisions involved. Such a system becomes an extendable tool which can be used to assess similar problems and can even be applied to different knowledge areas.

The theoretical issues of a decision support method were established in Chapters 7 through 9. In the current chapter, the technical limitations of its implementation must be taken into account, since computing power can rapidly become scarce given the kind of computations proposed.

10.1.1 CHAPTER CONTENT

In Section 10.2 of this chapter, we describe the design goals of a prototype intended to support the decision-making process (Martínez-Falero et al. 2010). The next paragraph is dedicated to the particular application we have developed (data and procedures, results, discussion). Finally, a road map for further development is proposed.

10.1.2 STATE OF THE ART OF THE EXISTING METHODOLOGIES

The movement toward SFM has placed a much greater emphasis on the direct inclusion of people's values in forestry decision making (Sheppard 2005). However, the methodologies used to include public participation have evolved, partly due to the availability of new platforms and partly due to the new decision methods.

Methods for integrating a range of opinions fall into the group of multi-criteria decision methods (MCDM), which also include group decision methods (GDM). We can classify multi-criteria decision methods into two groups: methods based on multi-attribute utility theory (MAUT), which include simple multi-attribute rating techniques (SMART) and analytic hierarchy process (AHP), and outranking methods, ELECTRE and PROMETHEE. A complete review of MCDM in forestry can be found in Diaz-Balteiro and Romero (2008).

During the 1990s, the core ideas of MAUT were developed for the purposes of natural resource planning. One of the most widespread methods was AHP (see, for example, Schmoldt et al. 2001) as it offers several advantages: first, it allows the integration of information on objective values with expert knowledge and subjective preferences; and secondly, qualitative criteria can be applied in the evaluation of alternative plans (Kangas et al. 2001). The disadvantage of most MAUT methods arises when dealing with data on the natural environment, since in many cases this cannot be expressed in quantitative terms or in intervals. Furthermore, MAUT techniques assume that there is a utility function (a value function) for integrating different criteria, which is not often the case.

Mendoza and Dalton (2005) developed an example of a web-based system to assess forest sustainability that was based on AHP (Mendoza and Dalton 2005). The program, called CIMCAT, used pairwise comparisons and weighting methods to include the stakeholders' preferences in the evaluation of forests in Ontario.

Moffet et al. (2005) developed a software (MultCSync) that combines a modified version of AHP and multi-attribute value theories (MAVT) for the selection of conservation area networks.

Lexer et al. (2005) developed "DSD v1.1" (Decision Support Dobrova), a decision support tool that uses AHP and an additive utility function to define the relative importance of different management objectives in Austrian forest stands.

Outranking methods can be considered a further development of voting theory. They seek to find a partially complete arrangement of preferred alternatives, so that any alternative that does not belong to the subset is exceeded by at least one of the subset. Therefore, the goal is to obtain as small a subset as possible, and a compromise alternative will then be chosen from that set (Figueira et al. 2004). The outranking methods enable the utilization of incomplete value information (indifference or fuzzy preferences) as well as judgments in an ordinal scale. The main advantage in comparison to MAUT techniques is that they do not require the assumption of a value function.

One example of a decision-making support system based on outranking methods is the HIPRE program developed by the Systems Analysis Laboratory at the Helsinki University of Technology (Hämäläinen and Lauri 1995).

Pauwels et al. (2007) resorted to ELECTRE to compare several silvicultural alternatives for larch (*Larix sp.*) stands in Belgium, taking into account biodiversity and resistance to windstorms.

There are also Internet-based software applications that use MCDM other than MAUT and outranking. For instance, MESTA (Pasanen et al. 2005) is a program developed by the Finnish Forest Research Institute that can be applied to numerous decision situations. The system allows the use of up to 10 criteria and 30 alternatives and helps to reach a decision by means of accepting border definitions from users and group negotiation (Hiltunen et al. 2009). Likewise, Hjortsø (2004) developed an application for tactical forest planning in Denmark, using strategic option development and analysis (SODA) and group decision-making techniques (GDM).

10.2   DESIGN GOALS

Computer applications designed to assess individual preferences basically consist of a display to input the preferences of the individual and their personal data, an algorithm (precalculated or not) that classifies the individuals into homogeneous groups, a value function (either dynamic or fixed) to aggregate the individual preferences of all voters, and a display to show the output results. Figure 10.1 shows the basic scheme of an Internet-based model to assess sustainability preferences.

When designing a program to assess preferences, the technical limitations of the implementation must be taken into account to avoid communication and processing failures.

It is highly desirable to allow for rapid implementation of the experiment, so that the system can be recalculated in the case that alternative methods for either preference representation or production optimization are proposed. However, many calculations concerning spatial analysis such as landscape planning, regression models like outreaching statistical values, and matrix algebra such as the aggregation of preferences require a high computation capacity, and therefore, the designer very often has to include precalculated modules in which some of the operations are already done. This circumstance forfeits some possibilities of the program but significantly improves the calculation time.

In new computer applications, the tendency is to develop programs with a strong modular orientation. Functional elements in the application can easily be replaced to allow different methods to be tested or to allow collaborative users to improve or modify certain aspects of the program. Interfaces between modules must be data oriented, so modules can be either of the static or dynamic type; that is, queries against fixed data, or "on the fly" calculations, are to be determined according to their computing requirements.

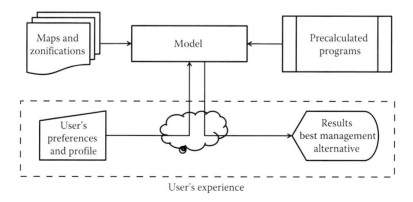

FIGURE 10.1   Basic scheme of an Internet-based program to assess forest sustainability.

10.2.1   DYNAMIC DESIGN

Preferences are not static: changes occur among individuals and over time. Predicting the future evolution of people's judgments is difficult and imprecise, and it is, therefore, much more convenient to develop an application in which the value function for the aggregation of preferences changes as new judgments from new users are received. The problem of dynamic changes is the burden of constant voting, and therefore, Internet-based systems—either fully programmed "in the cloud" or hosted on a server to be downloaded—provide a good solution for dynamic applications. Users can express their judgments and automatically send the results to a database hosted elsewhere. The computer application can be programmed to calculate the value function based on opinions or to give different weights to old and new opinions.

10.3   THE CURRENT COMPUTER APPLICATION

The theoretical issues of a decision support method were established in Chapters 7 through 9. User input—consisting of the user's answers to a set of comparisons—is processed to obtain a system of preferences for that user, represented as described in Chapter 7. Users also provide some data about occupation, age, gender, etc. that serve to assign them to a group. The value function is defined corresponding to this set of preferences, and its effect on the area of interest is determined by simulated annealing as described in Chapter 8.

Simultaneously, the user is assigned to one of the different groups previously generated using the polythetic divisive classification algorithm (Martínez-Falero et al. 1995).

The user is then given an output consisting of two maps: one providing the management actions according to his or her personal preferences, and the other representing the actions derived from his group's profile.

If the reader would like to download the computer application before continuing, the link and instructions can be found in Section 10.5. Figure 10.9 also contains the QR code for direct access to the download page.

10.3.1   OPERATION OF THE COMPUTER APPLICATION

From the user's standpoint, SiLVANET (Martínez-Falero et al. 2010) works according to the scheme shown in Figure 10.2.

The arrows represent the sequence of operations made by the application; the dashed lines represent optional processes that provide the evaluator with additional information.

The user of the application must take into account that certain algorithms require a high computation time to run and that in order to facilitate the operation, programs that involve significant delays have been previously calculated in the area of application. However, this version of SiLVANET includes the executable files of these programs (Table 10.1). They are separated from the rest of the application in a different folder (*Direct-Aux*), so that if the user wants, they can be activated from the directory mentioned (independent from the central core of the application).

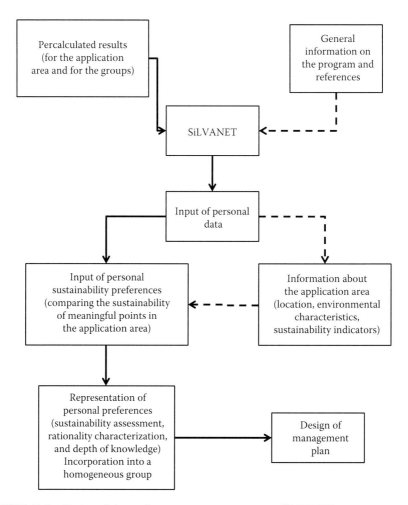

FIGURE 10.2   Design of the preference assessment program SiLVANET.

The precalculated programs are the following:

The resulting output files from previous programs are also stored in the directory (*Direct-Aux*). Bear in mind that SiLVANET comes with these programs already executed and that the resulting files have been transferred to the root directory (the directory where the application is activated).

Users should also be aware that some of the precalculated programs incorporate statistical simulations based on random number generation (whose sequence depends on your PC's internal clock); consequently, the results produced when they are executed will vary slightly depending on the time at which the programs are run. For this reason, the names of the files resulting from the implementation of precalculated programs have been slightly modified from those used in the main computer application. It is not advisable to change their names and copy them to the root directory for use by SiLVANET. In any case, if you do so, make sure before transferring them that you

TABLE 10.1
Description of the Precalculated Programs of SiLVANET

Project	Description	Runtime (Average)
Para-monte-id	Calculates the height distribution model for the trees in the reference location: before and after cutting Also determines the parameters for the dynamic model	Less than 1 h
IdenArboles	Allocates spatially the trees from the LiDAR image in the application area	4 h
Indices	Calculates the sustainability indicators for all points in the application area and for the points used in the comparison	24 h[a]
AgrupaPref	Generates 5000 random preference matrices, calculates for each one the 25 descriptors used in the grouping. Groups the matrices in analogous preference systems, determines the matrix of "average behavior," and characterizes the preferences of each group	24 h
PlanGestion	Designs and tests the forest management plan for the application area that best adapts to the sustainability preferences of the "average user" in each group defined previously	4 h

[a] This runtime corresponds to the calculation of the sustainability indicators in all 1-m-sided pixels. In order to make the display easier, the attached project (indices) is executed in every 20th pixel, both horizontally and vertically. Therefore, the runtime is reduced to a few minutes.

have fully completed the execution of the programs that generate the data, otherwise the execution will fail.

The application is started by clicking on the executable "*silvanet.exe*" which is stored in the root directory. The application searches the data files in the directory where the executable is located.

Since the zip folder in which the application comes does not allow you to work directly with the files, you need to unzip the entire content of the folder to a working directory on a rewritable device from which you can run the application.

The online part of the application is developed through a series of input, output, and display procedures (grouped in forms) and information-processing algorithms (adapted to the programming modules). Table 10.2 describes the forms used in SiLVANET, and Table 10.3 describes the modules.

The execution of the programming elements (precalculated programs, forms, and programming modules) responds to the operational diagram shown at the beginning of this section, and its activation and sequencing are shown in the figures later.

The information flow between elements of the program is done through "information transfer vectors," consisting of public variables (given in Module 1 of the main application), data files (containing start-up information or information which has been generated and modified by programming), and images. Figures 10.3 through 10.8 describe the algorithms included in the computer application.

TABLE 10.2
Description of the Forms of SiLVANET

Form	Description	
PalInfoEntra	SiLVANET's home screen	E
PalInfoRefer	General information on SiLVANET	i
PantaDatosPer1	References cited in SiLVANET	i
PantaInfoMont1	Entering personal data	E
ppppp0	General process of sustainability assessment	i
ppppp1	Real Forest: description of the application form	i
ppppp2	Real Forest: information coding	i
ppppp3	Parameter calculation for the Real Forest	i
ppppp31	Ideal Forest: distribution of the variable "tree height" for regular stands from yield tables	i
ppppp32	Ideal Forest: stems/ha for each age class	i
ppppp33	Ideal Forest: distribution of the variable "tree height"	i
PantaInfoMont2	Sustainability indicators	i
ppppppi1	Structural diversity index (1 of 2)	i
ppppppi12	Structural diversity index (2 of 2)	i
ppppppi2a	Timber revenue index (1 of 3)	i
ppppppi2b	Timber revenue index (2 of 3)	i
ppppppi2c	Timber revenue index (3 of 3)	i
ppppppi3	Biomass index	i
PantaInfoMont21	Fitting the LiDAR data to the calculation of sustainability indicators	i
PantaInfoMont3	Point selection for the pairwise comparison	i
PantaInfoMont4	Concept of sustainability assessment	i
PantaSelecc1	Establishment of preferences for comparing alternatives	E
PantaPresen1	Preference representation	i-S
ZInfovalor1	Preference representation: information about the calculation of sustainability value (1 of 2)	i
ZInfovalor2	Preference representation: information about the calculation of sustainability value (2 of 2)	
ZInfoPropPref1	Preference representation: information about the calculation of the type of rationality and the depth of knowledge of each group (1 of 2)	i
ZInfoPropPref2	Preference representation: information about the calculation of the type of rationality and the depth of knowledge of each group (2 of 2)	i
ZInfoRegres1	Preference representation: analytic expression of the value function (1 of 2)	i-S
ZInfoRegres2	Preference representation: analytic expression of the value function (2 of 2)	i
ZInfoCluster1	Preference representation: grouping users with analogous preference systems (1 of 2)	i
ZInfoCluster2	Preference representation: grouping users with analogous preference systems (2 of 2)	i-S
PntaPlanGest	Presentation of the management plan	i-S
ZZInfoAlgorOpti	Management plan: information on the optimization algorithm	i
ZZInfoRepreSolu	Management plan: spatial representation	i-S
SALIDA	Presentation and storage of the information obtained from the preference analysis carried out	S

TABLE 10.3
Description of the Modules of SiLVANET

Module	Description
Module 1	Draws the maps of the sustainability indicators; calculates the objective distance from each point to the point of maximal sustainability; draws this map and selects and characterizes the points for the comparison of sustainability
Module 2	Selects the images corresponding to each pair of points in which the evaluator compares the sustainability; displays the images of both points on the screen
Module 3	Presents a screen with information about the sustainability indicators for the two points compared. Facilitates the decision of the evaluator
Module 4	Displays information about past answers of pairwise comparison given by other evaluators that have used SiLVANET
Module 5	Calculates the parameters for the evaluator's classification
Module 6	Determines the group to which the user is assigned
Module 7	Identifies the management solution and characterizes it in terms of economic balance, sustainability value, and spatial distribution of activities
Module 8	Calculates the sustainability value in the points compared. Determines the type of rationality of each evaluator and their depth of knowledge. Outreaches the sustainability value to the entire application area and the analytic expression of the value function and its possible applications. Displays the previous information

10.3.2 RESULTS OF THE APPLICATION

As a final result of the preference analysis of each individual, the information is recorded and displayed on the last screen of SiLVANET.

This information is stored after the last record in the file *datos-acumulados.txt*, along with a first field that identifies the date and time of execution of the program. Here, two records of this file are shown as an example (Table 10.4).

This is the file that can be used by the person in charge of the final decision making in order to test the convergence in the preference of the different agents involved in sustainability. In any case (whether convergence is observed or not), this is the file that should be used to describe the final management alternative. *datos-acumulados.* txt is also checked to show each evaluator the history of answers that other agents have given, including the agents that have similar and different preferences to those of the evaluator.

10.3.3 TECHNICAL INFORMATION ABOUT PROGRAMMING

- Programming language: Visual Basic 6.0, programming using procedures.
- Operating environment: Windows XP, Windows Vista 6, and Windows 7.
- Minimum screen resolution: 1280×800 pixels. (With a lower resolution, not all the information provided by SiLVANET will be seen). Furthermore, to avoid the presentation of distorted information, the ratio of the number of pixels in each axis must be equal to the length ratio of the widths to the heights of the screen to which the application is submitted.

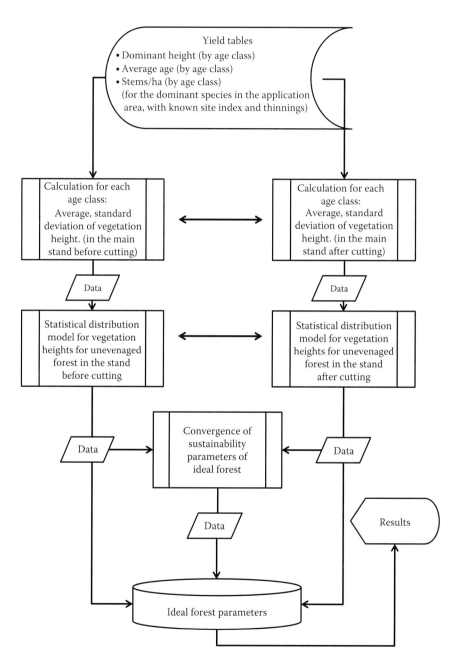

FIGURE 10.3   General parameters of the "Ideal Forest."

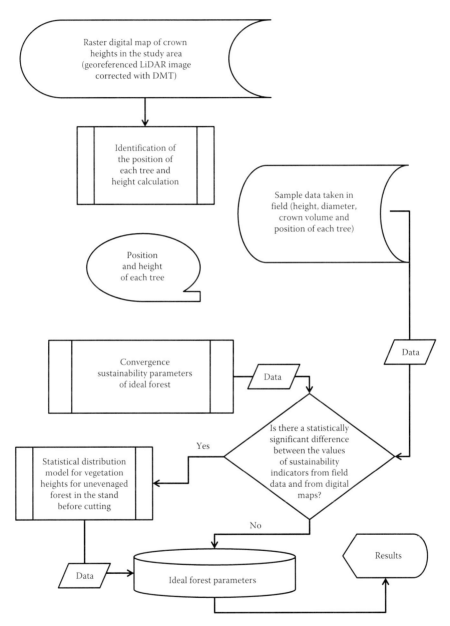

FIGURE 10.4 Parameters of the "Real Forest."

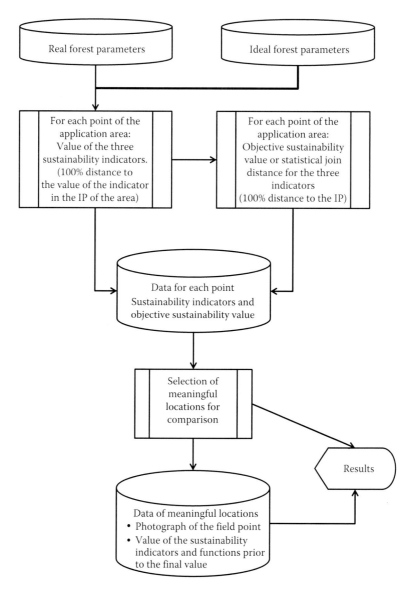

FIGURE 10.5 Sustainability indicators and the locations where sustainability is assessed.

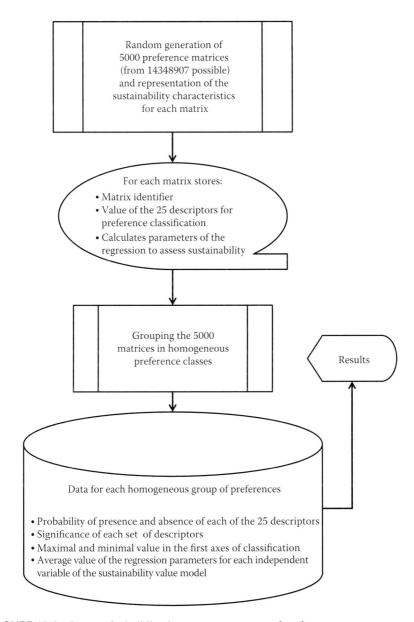

FIGURE 10.6 Process for building homogeneous groups of preferences.

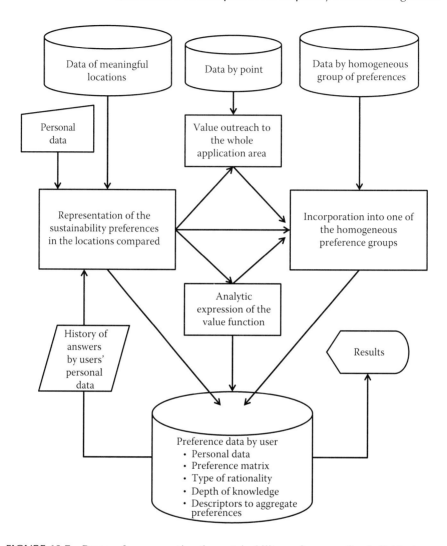

FIGURE 10.7   Process for representing the sustainability preferences of an individual.

10.3.4   Flowcharts

From a functional standpoint, programming elements are divided into the following calculation processes:

1. Calculation of the ideal stand parameters
2. Calculation of the parameters of each point of the real stand
3. Sustainability indicators and locations for comparison of sustainability judgments
4. Formation of homogeneous preference groups
5. Representation of sustainability preferences for any user
6. Management plan for each homogeneous group of preferences

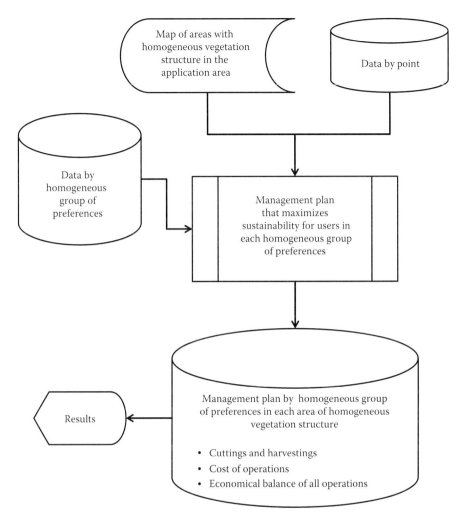

FIGURE 10.8 Process for obtaining the management plan for every homogeneous group of preferences.

TABLE 10.4

Individuals' Records of the Results of Their Preference Analysis

1001182012 1 1 2 54 2 2 3 1 1 1 1 2 3 2 2 2 1 2 1 3 1 1 1 2 1 2 3 2 1 2 1 2 1 3 1 2 2 2 2 2 3 0 1 0 0 0 1 0
 0 0 0 0 0 0 0 1 0 0 0 0 1 0 1 0 0 30

1001182015 1 3 3 58 2 2 3 1 1 1 1 1 2 3 1 2 2 2 2 2 3 1 1 1 1 2 3 2 1 2 1 2 1 3 1 2 1 2 2 2 3 0 1 0 0 0 0 0
 0 0 0 1 0 0 0 0 1 0 0 0 1 0 0 1 0 0 30

TABLE 10.5

Programming Elements in the Calculation Process

	Programming Elements		
Calculation Process	**Projects (Precalculated)**	**Forms**	**Modules**
I	Para-monte-id	Form1; PalInfoEntra; PalInfoRefer; PantaInfoMont1; PantaInfoMont4; SALIDA; ppppp3; ppppp31; ppppp32; ppppp33	
II	IdenArboles	Form1; PalInfoEntra; PalInfoRefer; PantaInfoMont1; PantaInfoMont4; SALIDA; ppppp0; ppppp1; ppppp2	
III	Indices	Form1; PalInfoEntra; PalInfoRefer; PantaInfoMont1; PantaInfoMont4; SALIDA; PantaInfoMont2; pppppi1; pppppi12; pppppi2a; pppppi2b; pppppi2c; pppppi3; PantaInfoMont21; PantaInfoMont3	Module1
IV	AgrupaPref	Form1; PalInfoEntra; PalInfoRefer; PantaInfoMont1; PantaInfoMont4; SALIDA; PantaDatosPer1; ZInfoCluster1; ZInfoCluster2	Module5; Module6
V		Form1; PalInfoEntra; PalInfoRefer; PantaInfoMont1; PantaInfoMont4; SALIDA; PantaDatosPer1; PantaSelecc1; PantaPresen1; ZInfovalor1; ZInfovalor2; ZInfoPropPref1; ZInfoPropPref2; ZInfoRegress1; ZInfoRegress2	Module2; Module3; Module8
VI	PlanGestion	Form1; PalInfoEntra; PalInfoRefer; PantaInfoMont1; PantaInfoMont4; SALIDA; PntaPlanGest; ZZInfoAlgorOpti; ZZInfoRepreSolu;	Module7

Calculation processes are related to programming elements as shown in Table 10.5 later:

The different calculation processes are described later with flowcharts. Figure 10.3 describes the general parameters of the "Ideal Forest," whereas Figure 10.4 describes the parameters of the "Real Forest." The sustainability indicators and the locations where the sustainability is assessed by judgments are shown in Figure 10.5. Figure 10.6 shows how to build homogeneous groups of preferences. The representation of the sustainability preferences of a generic user is explained in Figure 10.7. Figure 10.8 shows the management plan for every homogeneous group of preferences. Finally, to facilitate the download of SiLVANET, Figure 10.9 contains the QR code for direct access to the page where the computer application can be downloaded.

10.4   ROAD MAP FOR FURTHER DEVELOPMENTS

The computer application described provides a sustainability value and a management plan based on the preferences of each group of individuals; however, it does

FIGURE 10.9 QR code to directly download the SiLVANET application.

not take into account the opinions of people that do not vote. The outreaching of the preferences to the entire population can be estimated by statistics on the population (educational level, occupation, age intervals, etc.).

The development of cloud computing will significantly improve public participation, as it tends to provide a limitless scalable platform able to record and incorporate into its calculations the preferences of visiting voters and landscape managers. Cooperative, voluntary or commercial processing will allow for frequent recalculation of tables and feedback to users.

10.5 DOWNLOAD SILVANET

You can download the SiLVANET computer application totally free at the following link:

www.montes.upm.es/ETSIMontes/Silvanet

Or you can scan the QR code in Figure 10.9.

REFERENCES

Díaz-Balteiro, L. and C. Romero. 2008. Making forestry decisions with multiple criteria: A review and an assessment. *Forest Ecology and Management* 255:3222–3241.

Figueira, J., Greco, S., and M. Ehrgott. 2004. *Multiple Criteria Decision Analysis: State of the Art Surveys*. New York: Springer.

Hämäläinen, R. and H. Lauri. 1995. *Hipre3+ Users guide*. Helsinki University of Technology, Systems Analysis Laboratory, Helsinki, Finland.

Hiltunen, V., Kurttila, M., Leskinen, P., Pasanen, K., and J. Pykäläinen. 2009. Mesta: An internet-based decision-support application for participatory strategic-level natural resources planning. *Forest Policy and Economics* 11:1–9.

Hjortsø, C.N. 2004. Enhancing public participation in natural resource management using Soft OR- an application of strategic option development and analysis in tactical forest planning. *European Journal of Operational Research* 152:667–683.

Kangas, A., Kangas, J., and J. Pykäläinen. 2001. Outranking methods as tools in Strategic Natural Resources Planning. *Silva Fennica* 35(2):215–227.

Lexer, M.J., Vacik, H., Palmetzhofer, D., and G. Oitzinger. 2005. A decision support tool to improve forestry extension services for small private landowners in Southern Austria. *Computer and Electronics in Agriculture* 49:81–102.

Martínez-Falero, E. et al. 2010. *Silvanet, Participación pública para la gestión forestal sostenible.* Madrid, Spain: Fundación Conde del Valle de Salazar.

Martínez-Falero, E., González-Alonso, S., and A. Cazorla. 1995. *Quantitative Techniques in Landscape Planning.* Boca Ratón, FL: CRC Press.

Mendoza, G.A. and W.J. Dalton. 2005. Multi-stakeholder assessment of forest sustainability: Multi-criteria analysis and the case of the Ontario forest assessment system. *Forestry Chronicle* 81:222–228.

Moffett, A., Garson, J., and S. Sakar. 2005. MultCSync: A software package for incorporating multiple criteria in conservation planning. *Environmental Modelling Software* 20:1315–1322.

Myers, D.G. *Intuition: Its Powers and Perils.* New Haven, CT: Yale University Press, 2002.

Pasanen, K., Kurttila, M., Pykäläinen, J., and J. Kangas. 2005. MESTA-non industrial private forest owners' decision-support environment for the evaluation of alternative forest plans over the Internet. *International Journal of Information Technology & Decision Making* 4(4):601–620.

Pauwels, D., Lejeune, P., and J. Rondeux. 2007. A decision support system to simulate and compare silvicultural scenarios for pure even-aged larch stands. *Annals of Forest Science* 64:345–353.

Pykäläinen, J., Kangas, J., and T. Loikkanen. 1999. Interactive decision analysis in participatory strategic forest planning: Experiences from state owned boreal forests. *Journal of forest Economics* 5(3):341–364.

Schmoldt, D.L., Kangas, J., Mendoza, G.A., and M. Pesonen. 2001. *The Analytic Hierarchy Process in Natural Resource and Environmental Decision Making.* Dordrecht, the Netherlands: Kluwer Academic Publishers.

Sheppard, S.R.J. 2005. Participatory decision support for sustainable forest management: A framework for planning with local communities at the landscape level in Canada. *Canadian Journal of Forest Research* 35:1515–1526.

Index

Printed and bound by CPI Group (UK) Ltd, Croydon, CR0 4YY

18/10/2024

01776271-0012